California's Amazing Geology

California has some of the most distinctive and unique geology in the United States. It is the only state with all three types of plate boundaries, an extraordinary history of earthquakes and volcanoes, and many rocks and minerals found nowhere else. The Golden State includes both the highest and lowest points in the continental US and practically every conceivable geological feature known. This book discusses not only the important geologic features of each region in California but also the complex geologic four-dimensional puzzle of how California was assembled, beginning over two billion years ago. The author provides an up-to-date and authoritative review of the geology and geomorphology of each geologic province, as well as recent revelations of the tectonic history of California's past. There are separate chapters on some of California's distinctive geologic resources, including gold, oil, water, coastlines, and fossils. An introductory section describes basic rock and mineral types and fundamental aspects of plate tectonics, so that students and other readers can make sense of the bizarre, wild, and crazy jigsaw puzzle that is California's geological history.

In this second edition, the book has an entirely new final section, "California's Environmental Hazards and Challenges," with new chapters on California's landslides, air and water pollution, renewable energy, and the future of climate change in California.

KEY FEATURES

- Thoroughly updates the market-leading textbook on California's geology
- Is written by an author with 30 years of teaching geology and leading field trips in California
- Introduces California's unique geological history
- Covers fundamentals of geology
- Characterizes specific geographical regions of California
- Describes major geological resources of California
- Summarizes the paleontology of California
- Reviews the likely impact of climate change on California's environment

California's Amazing Geology

Second Edition

Donald R. Prothero

CRC Press
Taylor & Francis Group
Boca Raton London New York

CRC Press is an imprint of the
Taylor & Francis Group, an **informa** business

Cover image: Deeply eroded spheroidal concretions from Bowling Ball Beach, Mendocino County, California. The cemented concretionary layer is visible in the outcrops of the Miocene Galloway Formation in the cliffs in the background. They are then weathered and sculpted by the wave action along the beach into these huge spheroidal concretions. (Photo courtesy Wikimedia Commons).

Second edition published 2024
by CRC Press
2385 NW Executive Center Drive, Suite 320, Boca Raton FL 33431

and by CRC Press
4 Park Square, Milton Park, Abingdon, Oxon, OX14 4RN

CRC Press is an imprint of Taylor & Francis Group, LLC

© 2024 Taylor & Francis Group, LLC

First edition published by CRC Press 2016

Library of Congress Cataloging-in-Publication Data
Names: Prothero, Donald R., author.
Title: California's amazing geology / Donald R. Prothero.
Description: Second edition. | Boca Raton : CRC Press, 2024. | Includes bibliographical references and index.
Identifiers: LCCN 2023038125 (print) | LCCN 2023038126 (ebook) | ISBN 9781032294957 (hbk) |
    ISBN 9781032294902 (pbk) | ISBN 9781003301837 (ebk)
Subjects: LCSH: Geology—California. | Geophysics—California. | Plate tectonics—California. |
    Plate tectonics—West (U.S.)
Classification: LCC QE89 .P76 2024 (print) | LCC QE89 (ebook) | DDC 557.94—dc23/eng/20231025
LC record available at https://lccn.loc.gov/2023038125
LC ebook record available at https://lccn.loc.gov/2023038126

ISBN: 978-1-032-29495-7 (hbk)
ISBN: 978-1-032-29490-2 (pbk)
ISBN: 978-1-003-30183-7 (ebk)

DOI: 10.1201/9781003301837

Typeset in Times LT Std
by Apex CoVantage, LLC

Contents

Preface

This book was inspired when I began to teach California geology at the college level and found there were no books suitable for such a course. All the books on the market either were grossly out of date or didn't discuss a modern understanding of California tectonics. California has one of the most amazing and complex histories of any state in the United States, and so the simplistic approaches of the books currently on the market do not do it justice. In this book, I try to give a taste of this incredible jigsaw puzzle of California geology without going into too much depth or requiring too much specialized geologic training. In particular, I am familiar with the backgrounds of most college students taking a course like this at a community college or many of the four-year universities in the state, and so I do not assume that they have taken a previous course in introductory physical geology. For that reason, the first five chapters of the book are a quick review of the basic concepts needed to understand the fundamentals of geology assumed in the book. Other concepts, such as glacial geomorphology, coastal geology, and volcanology, are introduced in later chapters as appropriate.

An updated book on California's fascinating geology should also interest the general reader who has no previous background on the state. Again, the introductory chapters are written to give any reader a grasp of fundamental geologic concepts so that the rest of the story makes sense.

Fate has placed me in an unusually good position to write this book. I have taught college-level geology for almost 45 years, for 27 years at Occidental College in Los Angeles, but also at the California Institute of Technology in Pasadena, the California State Polytechnic University in Pomona, and community colleges such as Pierce College in Woodland Hills, Glendale College in Glendale, and Mt. San Antonio College in Walnut. Thus, I have a wealth of experience in teaching this material at the college level and leading hundreds of field trips to localities all over California. These include not only trips I led as a college instructor but also field trips I have led for the Pacific Section SEPM (Society for Sedimentary Geology), for which I have served as president and vice president, and for groups such as the Skeptic Society. In addition, I'm a second-generation Californian, hooked on dinosaurs at age four, and I never outgrew my love of paleontology. My first field trip experiences were as a Cub Scout collecting fossils in Old Topanga Canyon, and I've had many such experiences since then. I received my undergraduate education from the University of California, Riverside, where a very active department ran field trips out into the Mojave Desert and Transverse Ranges dozens of times each year. Finally, for many years, I have collected fossils and published research and done fieldwork in the Cenozoic strata of the West Coast (Prothero 2001), so I have seen more outcrops in California than most geologists alive today. My roots in the state go very deep: my great-great-grandfather William E. Prothero was in the California Gold Rush but returned to his family farm in Illinois after he did not strike it rich (as did most of the Forty-Niners) and later served during the Civil War in Sherman's Illinois regiments as they marched to the sea.

I hope this book helps the student and interested amateur better understand the incredible geological jigsaw puzzle that is the Golden State.

RESOURCE

Prothero, D.R. (ed.). 2001. *Magnetic Stratigraphy of the Pacific Coast Cenozoic*. Pacific Section SEPM Special Publication 91: 394. Pacific Section SEPM, Fullerton, CA.

Acknowledgments or Credits List

I thank my editor and fellow paleontologist, Dr. Charles R. Crumly, for supporting this project and helping in many ways. I thank project editor Iris Fahrer at CRC Press/Taylor & Francis for her help in the production process. I thank Mario Caputo, longtime secretary/treasurer of the Pacific Section SEPM, for his many helpful suggestions on this project, even if he could not participate as a coauthor. I thank John Wakabayashi, Steven Newton, and Korey Champe for their thoughtful reviews of the complete draft of this book; Gary Girty, Ray Ingersoll, and Mike Woodburne for helpful comments on many chapters; and Don Clarke for his feedback on the oil geology chapter. I thank Pat Linse and my son, Erik Prothero, for their help with illustrations for this book, and individual scientists and photographers are acknowledged for the outstanding images that they so generously provided. I thank my former professors at University of California, Riverside, especially Mike Woodburne, Mike Murphy, Lew Cohen, Paul Robinson, Peter Sadler, and Harry Cook, and at Columbia, especially Rich Schweickert, for exposing me to the wonders of California geology. I thank my friends and mentors among California geologists who have taught me so much, including Earl Brabb, Steve Graham, Jim Ingle, Ray Ingersoll, Bruce Luyendyk, Gary Girty, Pat Abbott, Dave Bottjer, and the late Bill Dickinson, Gene Fritsche, and John Cooper.

Finally, I thank my wonderful sons, Erik, Zachary, and Gabriel, for their love and support on this project, even as I dragged them all over the state on field trips. I especially thank my wonderful wife, Dr. Teresa LeVelle, for expertly driving me to field localities while I navigated, and for putting up with so much during the completion of this project.

About the Author

Donald R. Prothero has taught college geology and paleontology for 45 years, at Caltech, Columbia, Cal Poly Pomona, and Occidental, Knox, Vassar, Glendale, Mt. San Antonio, and Pierce Colleges. He earned his BA in geology and biology (highest honors, Phi Beta Kappa, College Award) from University of California, Riverside, in 1976, and his MA (1978), MPhil (1979), and PhD (1982) in geological sciences from Columbia University. He is the author of over 50 books (including eight leading geology textbooks and several trade books) and over 400 scientific papers, mostly on the evolution of fossil mammals (especially rhinos, camels, and pronghorns) and on using the Earth's magnetic field changes to date fossil-bearing strata. He has been on the editorial boards of journals such as *Geology*, *Paleobiology*, *Journal of Paleontology*, and *Skeptic* magazine. He is a fellow of the Linnean Society of London, the Paleontological Society, and the Geological Society of America and also received fellowships from the Guggenheim Foundation and National Science Foundation. He served as the president of Pacific Section SEPM (Society for Sedimentary Geology) in 2012 and served for five years as the program chair of the Society of Vertebrate Paleontology. In 1991, he received the Charles Schuchert Award for outstanding paleontologist under the age of 40. In 2013, he received the James Shea Award of the National Association of Geology Teachers for outstanding writing and editing in the geosciences. In 2015, he received the Joseph T. Gregory award for service to vertebrate paleontology. In 2016, he was named "Friend of Darwin" by the National Center for Science Education. In 2023, he was named "Distinguished Speaker" by the Palaeontological Association. He has been featured on numerous TV documentaries, including *Paleoworld*, *Walking with Prehistoric Beasts*, *Prehistoric Monsters Revealed*, *Monsterquest*, *Prehistoric Predators: Entelodon and Hyaenodon*, *Conspiracy Road Trip: Creationism*, as well as *Jeopardy!* and *Win Ben Stein's Money*.

1 The Golden State

Growing up in northern California has had a big influence on my love and respect for the outdoors. When I lived in Oakland, we would think nothing of driving to Half Moon Bay and Santa Cruz one day and then driving to the foothills of the Sierras the next day.

—Tom Hanks

Of all the 50 states of the United States, California has some of the most amazing geology and natural history. It has rocks dating back 1.8 billion years ago (Ga) as well as volcanoes that erupted just a century ago. It has a wide spectrum of minerals and rocks, from ancient volcanoes to young ones, granitic rocks that formed the mighty Sierra Nevada range, and peculiar rocks in the Coast Ranges found nowhere else in the United States. It has the highest mountains in the lower 48 states (Mt. Whitney, 4,421 m, or 14,495 ft, above sea level) and the lowest point in the Americas (Badwater in Death Valley, 86 m, or 283 ft, below sea level). Both are within a 135 mi running distance for ultramarathoners in the Badwater Marathon. Death Valley is one of the hottest and driest places on Earth, reaching 60°C (134°F) and receiving less than 2.5 cm (1 in) of average annual rainfall. The High Sierras are one of the snowiest places in the United States, with recorded snowfalls on Tamarack Mountain reaching 23 m (76 ft) in one year, 10 m (33 ft) in a single month, and 11 m (37 ft) on the ground at one time. California has the highest waterfall in North America (Yosemite Falls, 869 m, or 2,850 ft, of sheer drop). After Mt. Rainier, Mt. Shasta is the second tallest active volcano in the United States, reaching 4,317 m (14,162 ft). Until the eruption of Mt. Saint Helens in 1980, Mt. Lassen was the most recent active volcano in the lower 48 states, last erupting in 1914–1921.

Many of California's other geological features are extraordinary as well. It experienced the largest earthquakes in the United States outside of Alaska, including the 1906 San Francisco quake (magnitude 8.2), the 1872 Lone Pine quake (magnitude 8.25), and the 1857 Ft. Tejon quake (magnitude 8.2), as well as many smaller quakes, such as the 1994 Northridge quake (magnitude 6.7). One of the biggest explosive volcanic eruptions in the history of the planet occurred about 750,000 years ago, when Long Valley Caldera (near Mammoth Mountain ski resort and Bishop in the Owens Valley) blew its top and scattered 521 km³ (125 mi³) of ash across the entire western United States, reaching eastern Nebraska and Kansas. California is also one of the biggest producers of oil and gas in the United States, as well as yielding major deposits of gold, mercury, chromite, rare earth elements, and many other important natural resources.

Its natural history is also remarkable. It has the largest living things on Earth (the giant sequoias), the tallest on Earth (the coast redwoods), and some of the oldest living things on Earth (the bristlecone pines in the White Mountains are up to 5,000 years old). It has ecological regions ranging from coastal rain forests to dry deserts to snow-capped mountains to some of the most picturesque coastlines on Earth. It is home to some of the largest mammals, including large elk, cougar, and deer. There are immense elephant seals on its beaches, as well as tiny rodents and endemic species of foxes on its offshore islands. The grizzly bear once roamed widely over the whole state, which is why the state has the "Bear Flag" and the bear is the mascot of some of the University of California campuses. Due to the wide range of habitats and huge area, California is host to one of the widest diversities of birds, reptiles, amphibians, and fish in North America, as well as extremely diverse plant habitats, from misty redwood forests to cactus-filled desert.

California can claim many other distinctions as well, which make it one of the most important states. By population, California is the largest state in the United States (38 million people in the 2010 census), and it is the third largest state in area. In fact, if California were an independent nation, it would be the 35th largest nation in the world by population. This status as the largest state by population means California has major political clout: the largest delegation in Congress (53 representatives) and the most electoral votes in presidential elections (55). By itself, California provides about 20% of the electoral votes needed to win the presidency.

Even individual counties in California are larger than many states. San Bernardino County is the largest in the United States by area (52,070 km², or 20,105 mi²). It is bigger than nine states (Maryland, Hawaii, Massachusetts, Vermont, New Hampshire, New Jersey, Connecticut, Delaware, and Rhode Island), so if it were a state, it would be the 43rd largest in area. Los Angeles County has the largest population of any US county (almost 10 million people), more people than 42 of the 50 states. If Los Angeles County were to become a state, it would be the ninth largest in the United States by population.

California is home to 3 of the 10 largest cities in the United States, more than any other state except Texas, which also has 3. These include Los Angeles (number 2) and, surprisingly, San Diego (number 8) and San Jose (number 10). Although people think of San Francisco as huge, it is only number 13 on the list. Fresno is the 34th largest US city, Sacramento the 35th, Long Beach the 36th, and Oakland the 47th. That makes 8 of the top 50 largest cities, more than any other state. The Los Angeles Metropolitan Area is the second largest in the United States after New

DOI: 10.1201/9781003301837-1

York City, with 18 million people, larger than all but a few states. The San Francisco Bay Metropolitan Area ranks the 5th largest with 8 million, and the San Diego area is the 18th largest with 3 million.

This huge population and size are matched by incredible economic power, from the tech wizards of Silicon Valley to the center of the entertainment industry in Hollywood and environs (movies, TV, and music especially) to the incredible wealth in oil and agricultural products. California is the largest agricultural state in the union, producing 99% of all the almonds, walnuts, and pistachios; 95% of the broccoli and strawberries; 90% of the grapes and tomatoes; and 74% of all the lettuce consumed in the United States. California's gross domestic product (more than $14 trillion) is larger than that of all but eight countries in the world, so if California were a nation, it would be the world's ninth largest economy. When the governor of California speaks at international meetings of world leaders, he has more clout than any American official except the president or vice president. That influence has been translated into California leading the United States in clean energy policies, laws against air pollution, and many other progressive stances, such as political reform. California was the first state in the nation (and still one of the few) to institute measures such as initiatives (where voters can put measures on the ballot without their politicians), referendums, and recalls.

1.1 GEOGRAPHIC AND GEOLOGIC PROVINCES

California is particularly unusual and complicated in its geography and geology, especially compared to some of the flat states in the Midwest. Many of these states (such as Michigan or Illinois or Kansas) have almost no mountains or even tall hills, and the geology underlying them is fairly simple, with broad shallow dish-shaped basins and nearly flat-lying beds across most of the state. By contrast, California is the only state that has all three types of plate tectonic boundaries (subduction zone in the north, transform boundary along the San Andreas fault zone, and spreading ridge in the Salton Trough and Gulf of California) within it. It is also the only US state that straddles two tectonic plates.

Consequently, California has remarkably distinct geographic regions (Figure 1.1a), each of which has an almost entirely different geology from the other regions (Figure 1.1b). Each region has its own complicated geologic history, which we will look at one region at a time to better understand their local stories before we try to assemble the complete jigsaw puzzle of California in Chapter 15. We start with the regions with the simplest geology, such as the Cascade Range and Modoc Plateau, and then move to more complex regions, such as the Basin and Range and Sierras, before finishing with the wild and crazy geology of the Coast Ranges and Transverse Ranges.

The major geographic provinces of California are shown in Figure 1.1. We discuss each of them in the following chapters.

(a)

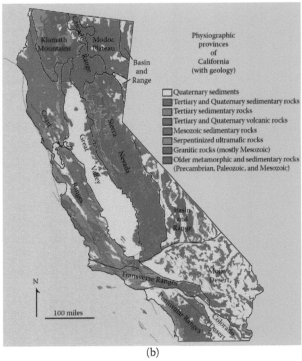

(b)

FIGURE 1.1   (a) Digital elevation map of California; (b) Geologic map and geographic provinces of California showing the geologic complexity of the state.

Source: Courtesy of California Division of Mines and Geology.

2 Building Blocks
Minerals and Rocks

The rock I'd seen in my life looked dull because in all ignorance I'd never thought to knock it open. People have cracked ordinary pegmatite—big, coarse granite—and laid bare clusters of red garnets, or topaz crystals, chrysoberyl, spodumene, emerald. They held in their hands crystals that had hung in a hole in the dark for a billion years unseen. I was all for it. I would lay about me right and left with a hammer, and bash the landscape to bits. I would crack the earth's crust like a piñata and spread to the light the vivid prizes in chunks within. Rock collecting was opening the mountains. It was like diving through my own interior blank blackness to remember the startling pieces of a dream: there was a blue lake, a witch, a lighthouse, a yellow path. It was like poking about in a grimy alley and finding an old, old coin. Nothing was as it seemed. The earth was like a shut eye. Mother's not dead, dear—she's only sleeping. Pry open the thin lid and find a crystalline intelligence inside, a rayed and sidereal beauty. Crystals grew inside rock like arithmetical flowers. They lengthened and spread, adding plane to plane in awed and perfect obedience to an absolute geometry that even the stones—maybe only the stones—understood.

—Annie Dillard
An American Childhood

Before we can discuss the details of California geology, we must briefly cover the fundamentals. For those who have already taken an introductory course on geology, this chapter may mostly be review or can be skipped altogether. However, this book is often used for college courses that assume no previous exposure to geology. In addition, many nongeologist readers of this book will be baffled if they encounter the names of important rocks and minerals without the proper explanation, so these are introduced here.

However, the material in this chapter is important to understand even if you have some geology background. California has some really unusual minerals and rocks, such as its state rock, serpentine, or some of its other peculiar rocks, such as ribbon cherts, blueschist, ophiolites, weird evaporite minerals, diatomites, and many others that are rare or unknown in most other states.

In the next few chapters, we also cover the basics of tectonics and structural geology, seismology, and other principles of geology that are usually covered in an introductory college course. When we discuss the geologic provinces, everyone should have at least some background to follow the discussion.

2.1 ATOMS AND ELEMENTS

First, let's quickly review the most basic principles of physics and chemistry. All matter is made of **atoms**, which are the smaller particle of matter that has the properties of a given **element**. For example, atoms of the element gold have its characteristic properties (such as its high density), but if you break a gold atom into its subatomic particles, they no longer have those same properties. The three main subatomic particles are the **proton** (which has a +1 charge and a mass of 1) and **neutron** (which has no charge and a mass of 1), both of which are found in the nucleus of the atom. The nucleus is surrounded by charged clouds of energy known as **electrons**, which have no mass but a −1 charge.

The proton is especially important because the number of protons in the nucleus is the **atomic number**, which determines which element you have. Any nucleus with one proton will be a hydrogen atom, a nucleus with two protons is part of a helium atom, and so on. If you change the number of protons in the nucleus, it becomes a different element. Because each proton carries a +1 charge, there should be an equal number of −1 electrons to balance the charge and make the atom electrically neutral. If the number of electrons does not match the number of protons, then the atom is a charged **ion**. If it has lost one or more electrons, then there is a net positive charge and it is a **cation**. If it has gained one or more electrons, then it has a net negative charge and it becomes an **anion**.

The number of neutrons in the nucleus has no effect on the charge of the atom (since neutrons are uncharged), but each neutron adds a mass of +1 to the nucleus, so they affect the **atomic weight**. For a given element (and thus for a fixed number of protons), there can be different atomic weights, depending on how many additional neutrons are in the nucleus. These different atoms with the same number of protons but different atomic weights are known as **isotopes**. For example, all isotopes of hydrogen have only one proton (they are no longer hydrogen if they don't), but hydrogen has several isotopes. Hydrogen with a mass of 1 (shown as ^1H, with the atomic weight in the left superscript) has only a single proton and no neutrons. This form of hydrogen (sometimes called hydrogen 1 or protium) is by far the most common in the universe and makes up 99.98% of the known hydrogen. However, about 0.01% of the hydrogen in the universe has not only a proton but also a neutron, giving an atomic weight of 2. It is known as hydrogen-2, or **deuterium**, and shown by the symbol ^2H. A third, very rare form of hydrogen is hydrogen-3, or **tritium**, with one proton and two neutrons. It is shown by the symbol ^3H. It is radioactive and produced only by nuclear reactions.

DOI: 10.1201/9781003301837-2

Nearly all the elements in nature have several different isotopes, and they are often very useful in geochemistry and in many other fields of science. For example, carbon has 15 known isotopes, only 2 of which are not radioactive but stable in geological settings. Normal carbon is carbon-12, or ^{12}C (six protons and six neutrons), and it makes up 99% of the carbon in your body and in almost anything carrying carbon. But carbon-13 (^{13}C, with six protons and seven neutrons) is also stable. Even though it makes up less than 1% of the carbon on the planet, its occurrence is a powerful tool in geology, geochemistry, and oceanography. Carbon-14 (^{14}C, with six protons and eight neutrons) is produced by bombardment of ^{14}N in the atmosphere that changes it into carbon-14. It is radioactive and unstable, decaying back to nitrogen at a known rate. This useful property allows it to be the basis of **carbon-14 dating** (or radiocarbon dating), the best tool geologists and archeologists have for measuring the age of human artifacts and anything less than about 60,000 years old (see Chapter 3).

There are more than a hundred elements in the periodic table, but most of them are extremely rare in geologic settings. Many of them exist only for milliseconds in a high-powered physics lab. In geology, the chemistry is even simpler. It turns out that there are only eight common elements (Table 2.1) in the Earth's crust that are worth remembering, because they make up more than 99% of all rocks on Earth.

The first surprise is that so many of most rocks in the Earth's crust are made up of oxygen (Table 2.1). An average rock is 46% oxygen by weight percent but 94% oxygen by volume. This is because not only is oxygen a light element, but also its ion has a large radius. You pick up a heavy rock, but in reality, it's made mostly of oxygen. Why is oxygen by far the most abundant? If you glance at Table 2.1,

you'll note that it is the only common anion on the table, and *something* has to balance all the positive charges of the remaining seven cations. Oxygen is very abundant in air and water, so it combines with almost any cation. Next in abundance is silicon, which (like carbon) is an element that readily bonds into long chains and forms complex three-dimensional structures. This is important in making the common rock-forming minerals, which are mostly combinations of silicon and oxygen, or **silicates**. In distant third place is aluminum, which is also an element that readily bonds into complex three-dimensional arrays with silicon, so many silicate minerals are rich in aluminum. These three elements just happened to be the common ones when the solar system and Earth formed, and they are found in most of the common rock-forming minerals as a result.

The next five elements in Table 2.1 are much less abundant than the "big three," with most of them making up only 5% or less in percentage of crustal rocks. All five are metallic cations, which bond with oxygen or complex silicate structures to make the huge variety of minerals. Note that they have different charges: sodium and potassium are both cations with a +1 charge. They can sometimes replace each other in a mineral, since they have the same charge. Likewise, calcium, magnesium, and ferrous iron (Fe^{+2}) are all common +2 cations and also can switch with each other in many minerals.

Even more surprising is what elements are *not* in the "big eight." Hydrogen and helium are the most common elements in much of the solar system, especially in the sun and outer planets (Jupiter, Saturn, Uranus, and Neptune). But they are rare on Earth, except where hydrogen is bonded with oxygen to make water. Why? When the Earth formed, it was not massive enough to have enough gravity to hold on to these elements and they floated off into space. Large planets like Jupiter and Saturn have much more gravity, and they held in their hydrogen and helium. Phosphorus and sulfur are also relatively rare and only concentrated in special settings. Perhaps most surprising of all is how rare carbon is in crustal rocks. After all, carbon is the building block of all life. There are a few minerals and rocks that have carbon in them, but they are rare compared to silicates.

2.2 MINERALS AND ROCKS

Now that we have the foundation of the elements available in the Earth's crust, let's see how they are combined into more complex molecules (groupings of atoms bonded together) known as **minerals**. The word *mineral* has all sorts of casual and inconsistent meanings in popular culture, but to geologists and chemists, the word *mineral* has a very strict and clear definition. A mineral is a

- Naturally occurring
- Inorganic
- Crystalline solid
- With a definite chemical composition and
- Characteristic physical properties

TABLE 2.1

Average Chemical Composition of Earth's Crust, Hydrosphere, and Troposphere

Element (Symbol)	Crust		Hydrosphere	Troposphere
Percent by Mass	Percent by Volume	Percent by Volume	Percent by Volume	
Oxygen (O₂)	46.40	94.04	33.0	21.0
Silicon (Si)	28.15	0.88	—	—
Aluminum (Al)	8.23	0.48	—	—
Iron (Fe)	5.63	0.49	—	—
Calcium (Ca)	4.15	1.18	—	—
Sodium (Na)	2.36	1.11	—	—
Magnesium (Mg)	2.33	0.33	—	—
Potassium (K)	2.09	1.42	—	—
Nitrogen (N₂)	—	—	—	78.0
Hydrogen (H₂)	—	—	66.0	—
Other	0.66	0.07	1.0	1.0

Let's discuss each of these components.

Naturally occurring. There are lots of complex compounds in the world, but if they are not produced naturally, they are not minerals. Thus, a synthetic diamond produced in a lab has all the properties of a diamond mined from the Earth, but it's not a mineral. Ice formed as ice crystals or snowflakes is a mineral, but not the ice in your ice cubes. Most of the stuff sold in a health food store that is called "mineral" was produced synthetically and therefore is not a mineral as a scientist uses the word.

Inorganic. Organic chemicals are built of the element carbon, so minerals are not made mostly of carbon. However, there are a handful of important minerals with carbon; we'll use *organic* in this context to mean compounds of carbon, oxygen, and hydrogen. Thus, sugar forms beautiful crystals, but it is organic and therefore not a mineral. Many of the "minerals" sold in a health food store are organic as well and thus not minerals.

Crystalline solid. Like the word *mineral*, the word *crystal* has a different meaning to a scientist than it does in popular culture. Typically, people use the word *crystal* to describe anything that sparkles. In a scientific definition, a *crystal* must have a *regular three-dimensional arrangement of atoms* in its internal structure, which repeats over and over. This three-dimensional array is called a **lattice**. It is analogous to the regular repeated pattern in wallpaper. For example, the atoms of the salt (sodium chloride, or $NaCl$) crystal (Figure 2.1a) are arranged in a cubic pattern, with each atom of sodium or chlorine forming 90° angles with the others. The same lattice is found in the mineral galena (Figure 2.1b), which is made of lead and sulfur in equal amounts (lead sulfide, or PbS). All minerals have a regular three-dimensional lattice of some kind, often very complex and with many other angles between atoms besides the 90° one seen in the simple cubic lattice.

Some things in nature may have a three-dimensional arrangement of atoms but they are not regular and repeating. Take, for example, volcanic glass or obsidian (Figure 2.2). At the molecular level, the atoms are not in any kind of repeated pattern but in a random tangle of long chains, like a bowl of spaghetti. Technically speaking, a glass isn't really a solid at all but a supercooled liquid. Over very long spans of time, the glass will slowly flow and change shape. This is clear if you ever see a piece of window glass in a very old house. If it has been in its window frame for about a century, the glass will be thicker at the bottom because it has slowly flowed downhill over the course of decades. Thus, a *glass is by definition not a crystal*. A popular item at many gift shops is "cut glass crystal" drink-

(a)

(b)

FIGURE 2.1 (a) Giant cubic crystal of salt (the mineral halite) next to a ball-and-stick model of the atomic structure of the lattice of a cubic mineral like halite. The blue balls could represent sodium atoms, while the green balls are chlorine atoms. (b) Crystals of the cubic mineral galena, or lead sulfide (PbS), showing its 90° cleavages and cubic habit.

Source: (a) Photo by the author. (b) Courtesy of Wikimedia Commons.

ing goblets and chandeliers, but this is not the definition of *crystal* that scientists use.

Definite chemical composition. Most minerals have a simple chemical formula, like most other compounds. There is a bit of substitution allowed if you replace one ion with another of a similar charge. For example, the mineral calcite (calcium carbonate, or $CaCO_3$) can have a certain percentage of magnesium replacing calcium sites in its lattice and still be calcite. However, if it gets to be 50:50 Ca/Mg, then it's no longer calcite but a different mineral, dolomite.

Characteristic physical properties. Most of the features of the minerals we have discussed occur at the atomic level. But to identify the mineral, you need to know what physical properties are typical of a hand sample of the mineral. These include its

FIGURE 2.2 Chunk of obsidian or volcanic glass, with the characteristic scallop-shaped fracture pattern ("conchoidal fracture") found in glass, and also minerals without cleavage, such as quartz. This natural fracture pattern makes sharp edges, which is why obsidian and quartz are often used for stone arrowheads and spearheads.

Source: Courtesy of Wikimedia Commons.

TABLE 2.2
Mineral Classes Classified Based on Their Dominant Anion

Mineral Classes
Minerals are classified by their dominant anion.

➤ Silicates	SiO_2^{4-}	*Most rock-forming minerals*
Oxides	O^{2-}	Magnetite, hematite
Sulfides	S^-	Pyrite, galena
Sulfates	SiO_2^{4-}	Gypsum
Halides	Cl^- or F^-	Fluorite, halite
Carbonates	CO_3^{2-}	Calcite, dolomite
Native elements	Cu, Au, C	Copper, gold, graphite

pg12_1.jpg

Note: Word endings are very important in chemistry! *Silicon* is the element on the periodic table. A *silicate* is a mineral made of silicon plus any number of atoms of oxygen. *Silica* is the compound silicon dioxide, or silicon plus two atoms of oxygen (SiO_2). But *silicone* is a synthetic compound made of silicon in a lab, used for lubricants, breast implants, and other purposes. Don't mix them up!

color, its hardness (from soft minerals like talc and gypsum to the hardest mineral, diamond), whether it fractures with an irregular surface or cleaves into many parallel planes, and less commonly used properties, like density (lead sulfide or galena, for example, is unusually dense because it contains lead), reaction to acid (the mineral calcite fizzes in dilute hydrochloric acid), and magnetism (the mineral magnetite is naturally magnetic).

Many of these properties of minerals can be understood by knowing the crystal lattice. For example, the cubic lattice of minerals like salt (NaCl) or galena (PbS) is demonstrated at the hand sample level, since any time you hit and break a piece of these minerals, they will naturally cleave to form cubic faces with 90° angles (Figure 2.1).

The atomic-level properties and crystal lattice can make a huge difference in the behavior of a mineral at the macroscopic level. Let's take as an example the two common minerals formed of pure carbon: diamond and graphite. One is the hardest substance in nature, and the other is one of the softest, yet they are chemically identical. Why are they so different? Diamond has a crystal lattice with all the atoms of carbon tightly bonded together and very short, strong chemical bonds. This structure will survive huge amounts of pressure, and an expert cutting a diamond has to know exactly how to cleave a large stone into several smaller ones. If he misses, the diamond is ruined. Graphite (the "lead" in a pencil), on the other hand, has all its carbon atoms arranged in sheets, with very long, very weak molecular bonds between the sheets. Pushing the graphite tip of a pencil across paper is enough to break those weak bonds, leaving tiny flakes of graphite behind on the paper as pencil markings.

Another example is calcium carbonate, or $CaCO_3$, in two different mineral lattices: calcite (the common mineral in limestones and marbles) and aragonite (also known as mother-of-pearl). They are the same chemistry, but their lattices are very different and lead to very different properties. The most obvious of these is that aragonite is much more soluble than calcite, so in weakly acidic conditions, aragonite will dissolve, but calcite won't. This is why if you own pearls, it is important to wash your acidic sweat off them once you put them away after wearing them. For some minerals, knowing the chemical composition is not enough; the crystal lattice makes a huge difference in the properties of the mineral as well.

There are literally thousands of different kinds of minerals, but most can be organized into just a few classes based on chemical composition (Table 2.2). There is no room to discuss them all in a brief introduction like this, but their major characteristics are detailed in Table 2.2. As we saw in Table 2.1, however, most of the Earth's crust is made of silicon and oxygen with minor aluminum, so it's no surprise that the most important rock-forming minerals are the **silicates**, made of silicon plus oxygen.

The major classes of silicate minerals are given in Table 2.2. Each has the same basic building block: the silicon–oxygen tetrahedron (Figure 2.3), or SiO_4 unit. Each class of silicate minerals uses these fundamental building blocks over and over and links them together in more and more complex structures.

The simplest silicate mineral structures are built of **isolated tetrahedra**, where the silicon–oxygen (SiO_4) building

SILICATE STRUCTURE	MINERAL/FORMULA	EXAMPLE
Single tetrahedral 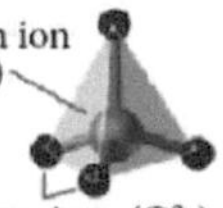 Silicon ion (Si^{4+}) Oxygen ions (O^{2-})	Olivine: Mg_2SiO_4 Garnet: $Fe_3Al_2(SiO_4)_3$	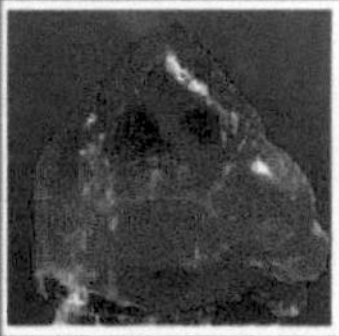 
Single chain 	Pyroxenes e.g., augite: $(Mg,Fe)SiO_3$	
Double chain 	Amphiboles e.g., hornblende: $Ca_2(Mg,Fe,Al)_3Si_8O_{22}(OH)_2$	
Sheet 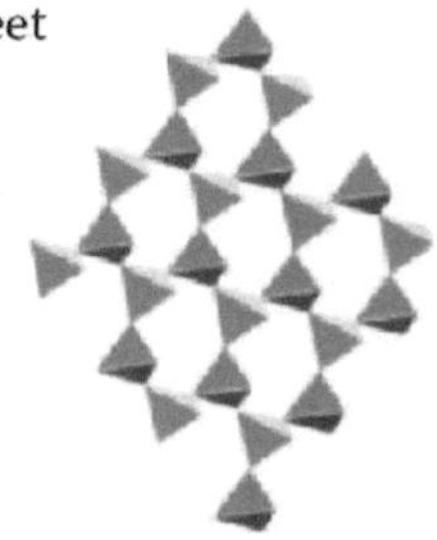	Micas and clays e.g., kaolin: $Al_2Si_2O_5(OH)_4$	
Framework 	Quartz: SiO_2	

	Feldspars
	    Potassium feldspar:
	    $KAlSi_3O_8$

	Plagioclase feldspar: $(Ca,Na)AlSi_2O_8$	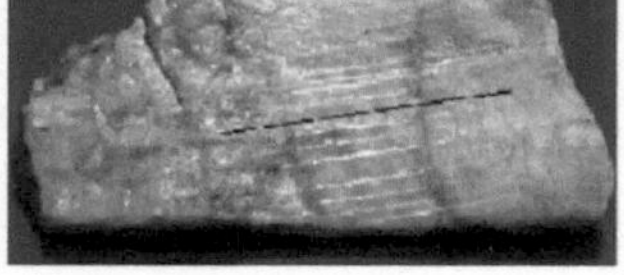

FIGURE 2.3 Classes of silicate minerals and their structures.

Source: Redrawn from several sources by E. T. Prothero; courtesy of Wikimedia Commons.

blocks do not bond directly to each other but are held in place by the electrostatic charges of the cations (especially Mg and Fe) between them in the lattice. Common single tetrahedral minerals are the green mineral **olivine**, which is (Mg, Fe) SiO_4, and **garnet**, which is built of SiO_4 tetrahedra with a variety of cations to make the six major types of garnets (from red-brown almandine to green grossular, and many others).

The next most complex arrangement is found in the class of minerals known as **pyroxenes** (PEER-ox-eens), which have linked the SiO_4 tetrahedra into long **single chains**. Among the common pyroxenes are the dull greenish-black mineral **augite** ($MgSiO_3$) as well as **jadeite**, one of the two minerals that can make the beautiful gemstone known as jade. Because of their crystal lattice made of single chains stacked together, pyroxenes have a 90° cleavage in hand samples, the most reliable physical property to recognize them.

The next step in silicate complexity is to link two single chains side by side to make a **double-chain** silicate structure, similar to the way the rails of a train track are linked together by wooden ties. Double-chain silicates are known as **amphiboles**, and their lattice structure gives the hand samples cleavages that are roughly either 60° or 120° (technically, 57° and 124°), their most diagnostic property. The most common amphibole is the shiny jet-black prismatic mineral known as **hornblende**, which is common in many igneous and metamorphic rocks discussed later. There are others, such as the greenish amphiboles tremolite and actinolite and the blue amphibole glaucophane, which we will discuss later.

Once you have a double-chain structure, the next, more complex arrangement is to bond the double chains together side by side to form **sheet silicates**. Sheet silicates are made of two layers of silicon–oxygen tetrahedral (*t*) sandwiching a layer of aluminum-oxygen octahedra (*o*). This *t-o-t* structure is like an Oreo cookie, with the *t* layers represented by the chocolate cookie layers, and the *o* layer representing the aluminum–oxygen creamy filling. Most sheet silicates are built of stacks of *t-o-t* structures with other materials (different kinds of cations and water molecules) trapped between the *t-o-t* layers. The most familiar sheet silicates are a class of minerals known as **micas**, which are distinctive in that they cleave into large flat thin sheets. The silver-white mica is known as **muscovite**, and before it was possible to make glass windows, large sheets of muscovite were used as windows and curtains (such as the "isinglass curtains" in the song "The Surrey with the Fringe on Top" from the Rogers and Hammerstein musical *Oklahoma*). There is also a common black mica known as **biotite**, a green mica called **chlorite**, and a lithium-rich lavender mica known as **lepidolite** that we will see in many different types of rocks. In addition to the micas, all the clay minerals that make up the muds of the world are sheet silicates; they are the most common minerals on Earth for this reason.

We have progressed from isolated tetrahedral to single chains to double chains to sheets, each structure being more and more complex and linked together (in chemical terms, **polymerized**). The final step is to link the silicon–oxygen tetrahedral into a complex three-dimensional **framework** structure that is almost impossible to render in a two-dimensional illustration. However, there are ball-and-stick models that capture their shape well, and now there are animations online that give some idea of their geometrically complex structure. Some of the most common and important minerals of all are framework silicates. The most important of these is **quartz**, made of pure silica (SiO_2), one of the most abundant sedimentary minerals on Earth. The other important classes of framework silicates are the aluminum-rich **feldspars**, the most common minerals in igneous rocks. Two types of feldspar are particularly important:

the pink-colored **potassium feldspars** ($KAlSi_3O_8$), which exist in three mineral forms (orthoclase, microcline, and sanidine), and the **plagioclase feldspars**, which transform continuously from pure calcium-rich plagioclase known as anorthite ($CaAl_2Si_2O_8$, typically bluish-gray in color) to intermediate plagioclases with a mixture of calcium and sodium, to pure sodium-rich plagioclase known as albite ($NaAlSi_3O_8$, white in color). Plagioclase crystals typically show lots of fine parallel lines known as **striations** on their cleavage surfaces. Plagioclases are typically the most common minerals in most igneous rocks and are found in all but a few of them.

These are a lot of mineral names to master, especially for beginning students, yet this is the minimum number of minerals needed to understand the rocks that occur on the Earth's surface and its interior. If you have the opportunity to study hand samples of each of these minerals and test their properties and compare them to each other, the names make more sense and become easy to remember with lots of experience and practice. Nearly every geology student masters these minerals after their first few classes in geology, so it's not that hard to do. It just takes practice and study. Once you know your minerals, you can understand the rocks that are made from them.

2.3 IGNEOUS ROCKS

Igneous rocks are formed by **magma** (molten rock) from deep in the Earth that rises up from deep **plutons** (magma chambers) and crystallizes as it cools. The size of the crystals in the rock depends on how fast it cools and solidifies. If the molten rock is spewed out of a volcano, the magma cools quickly and the crystals have little time to form, so they are **microcrystalline** ("aphanitic"), too small to see with the naked eye. They are only visible in thin polished sections of rocks when viewed under a special microscope. If the crystals cool slowly over years to hundreds of years deep in an underground pluton, then they have time to grow larger. Sometimes they are still just barely visible to the naked eye, but still they are **macrocrystalline** ("phaneritic") nonetheless. A few magmas cool extremely slowly over the course of decades or centuries or longer, producing **pegmatites** full of giant crystals.

Some igneous rocks have a composite texture, with macrocrystalline crystals (**phenocrysts**) floating in a rock that is mostly microcrystalline (called the **groundmass**). This hybrid texture is called **porphyritic** (the noun form of the word is **porphyry**) and results from a magma that had a two-stage cooling history: the phenocrysts cooled slowly in a large pluton, and then the semi-crystallized magma was blown out of a volcano, where the rest of the melt cooled quickly to form the groundmass surrounding the phenocrysts. The important thing to remember is that the *crystal size is a result of the time and mode of cooling*, so microcrystalline rocks are volcanic, macrocrystalline rocks cool slowly in plutons (plutonic rocks), and porphyritic rocks have both stages in their history.

The crystal size is one element of a classification of igneous rocks (Figure 2.4). The other axis is based on their chemical composition. Rocks coming from the mantle are relatively rich in magnesium, iron, and calcium and are known as **mafic** rocks for short ("ma" for magnesium, plus the chemical abbreviation "Fe" for iron). This magnesium–iron–calcium chemistry produces minerals rich in these elements, such as olivine and pyroxenes (magnesium- and iron-rich silicates) and calcium plagioclase. If a mafic magma cools quickly, it is the familiar black lava called **basalt** that erupts out of Kilauea on the Big Island of Hawaii and other mantle-sourced volcanoes. If it cools slowly, the same magma that makes black basalt instead produces a macrocrystalline rock with visible pyroxene and calcium plagioclase known as a **gabbro**.

Some magmas are so rich in magnesium and iron that they are known as **ultramafics**. These contain almost nothing but olivine derived directly from the mantle or lower crust. A macrocrystalline rock made of pure olivine is called a **peridotite** and represents a direct sample of the upper mantle. (The gem name for olivine crystals is *peridot*, so a rock made of olivine is a peridotite.) Currently, there are no ultramafic volcanic lavas erupting anywhere on Earth, but back about three billion years ago (Ga), they were common and made an olivine-rich lava called **komatiite**.

The other extreme of magma chemistry are melts that are rich in silicon, aluminum, potassium, and sodium. These rocks are often called **silicic** or **felsic** (an abbreviation based on combining *feldspar* and *silica*). With this kind of magma chemistry, you produce minerals such as quartz (pure silica), sodium feldspars (sodium, aluminum, and silica), potassium feldspars, and micas such as biotite and muscovite (rich in potassium, aluminum, and silica). If the rock contains quartz and two-thirds of its total feldspar is potassium feldspar, then it is a true **granite**. True granites have so much pink or red potassium feldspar in them that they tend to be red as well. There are volcanic eruptions of magmas with the composition of granites. These are known as **rhyolite**, and they are extremely fine-grained and generally pink or red in color due to the rusting of the iron in them.

The rocks in the Sierra Nevada Mountains and other California mountains that laypeople call "granites" do not have enough potassium feldspar in them to be true granites, as a geologist defines the term. Many of them have less than 33% of their total feldspar as potassium feldspar, as well as about 20% quartz. Most geologists call these rocks **granodiorite** rather than granite. Most of the plutonic rocks in California are granodiorites or diorites. However, in this book, we will often use the term *granitics* as a category name to refer to the entire classes of felsic plutonic rocks, including granodiorites, diorites, and many other felsic rocks we will not discuss here. The volcanic equivalent of a granodiorite is called a **dacite**. There are more complicated schemes of subdividing felsic igneous rocks, but we will not discuss them here because that requires much more background in geochemistry and mineralogy than is appropriate for this book.

Rocks that are intermediate in composition between basalt–gabbro and granodiorite–dacite are usually made of hornblende plus an intermediate mixed sodium–calcium plagioclase, but no quartz or potassium feldspar. These rocks are known as **diorite** if they are macrocrystalline and usually have a speckled black-and-white, "salt-and-pepper" appearance. If they cooled quickly, a diorite composition becomes a microcrystalline volcanic rock known as **andesite**. They were named after the Andes Mountains of South America, although it turns out there are few actual andesites in the Andes by the modern definition of that term. Most andesites are light to dark gray in color and porphyritic, with tiny black phenocrysts of hornblende floating in

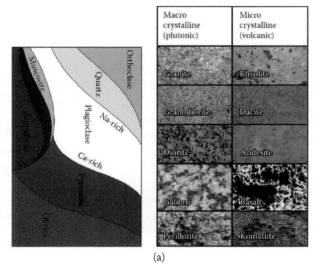

(a)

Bowen's reaction series

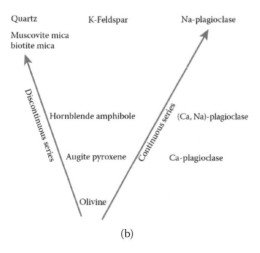

(b)

FIGURE 2.4 Classification chart of (a) igneous rocks and (b) Bowen's reaction series.

Source: Redrawn from several sources by E. T. Prothero; photos by the author.

a microcrystalline groundmass of intermediate plagioclase plus hornblende.

This is the simplest possible list of names for ten basic types of igneous rocks (Figure 2.4a). Igneous petrologists recognize hundreds of different types, but that level of detail isn't required here. Now that we have some names and understand some of the basics of magma chemistry, the next question is, *Why are there so many different types of igneous rocks?* If their magmas were all coming straight from the mantle, they would all be peridotite or komatiite in composition. Something happens to that ultramafic magma as it rises from the mantle and through the crust so that it changes chemistry as it rises, losing magnesium, iron, and calcium and gaining silicon, aluminum, potassium, and sodium, in order to make all the different kinds of magmas that produce the hundreds of different kinds of igneous rocks (**magma differentiation**).

This is the central question of all of igneous petrology. It perplexed and puzzled geologists for most of the nineteenth century, when many types of igneous rocks were named and described but no one could explain how they all formed. The breakthrough came in the early twentieth century, when Norman L. Bowen tried to simulate the process of magma differentiation in the laboratory. He would take a chip of peridotite, melt it in a high-temperature furnace, and then drop the molten blob into liquid mercury to instantly quench and chill it. Once the blob had cooled, he would separate out all the material that had crystallized (mostly olivine) and then melt the rest and go through the process again and again. (Although he didn't know it, working with mercury is very dangerous, and eventually Bowen lost his mind due to mercury poisoning, just as the "mad hatters" who used mercury to make felt for hats did in the days of *Alice in Wonderland*.) As Bowen did his experiments, the remaining material became more and more depleted in magnesium, iron, and calcium as he removed the crystals of olivine, pyroxene, and calcium plagioclase that had formed first. As the residue became depleted in those elements, the magma that remained became richer in silicon, aluminum, potassium, and sodium.

Bowen's experiments simulate what happens in many magma chambers. In a process called **fractional crystallization**, the first mineral to cool and crystallize will settle out to the bottom or sides of the magma chamber and remove their magnesium, iron, and calcium from the remaining magma. Over time, this changes the composition of the magma left behind, so eventually it can no longer crystallize olivine but must make pyroxene instead, and its plagioclase goes from calcium-rich to more sodium-rich. After enough olivine and pyroxene have been removed from the melt, it would no longer have enough magnesium or iron left, and then hornblende would be the next mineral formed. In other words, as each fraction crystallizes out and changes the melt left behind, the magma changes chemistry (hence "fractional crystallization").

As an analogy, imagine a room full of students standing up from their chairs. Let's say that the room is 60% men and 40% women. For this example, the men will represent iron and magnesium atoms, and the women silica atoms. Initially, all the students are standing, as if they were atoms floating around in a melt. Now we'll have two men sit down for every woman who sits down, and do this over and over. Sitting down represents a crystal that sinks out of the melt, removing itself from the chemistry of the magma. After enough time, all the men will be sitting, but quite a few women will still be standing. So the ratio of men to women at the start was about two men for every woman, but at the end, there are zero men standing and only women left (as if they were still floating in the melt). In the same manner, fractional crystallization pulls out the elements from the minerals that crystallize out first (such as iron, magnesium, and calcium) and deposits them on the walls and floor of the magma chamber. This leaves the remaining melt depleted in those elements and enriched in elements that were rare initially.

Bowen plotted this crystallization sequence of minerals, and today it is known as **Bowen's reaction series** (Figure 2.4b). This simple arrangement of minerals exactly parallels the mineral sequence of the igneous rock classification scheme (Figure 2.4a). It is considered one of the greatest discoveries in the science of geology and showed the power of experimental approaches and lab simulation when purely descriptive methods failed.

Bowen's reaction series has two pathways. The **discontinuous series** is the sequence of different classes of dark-colored minerals, from olivine to pyroxene to amphibole to biotite. At the same time, the magma is continuously crystallizing out plagioclase as well, forming a **continuous series**. The first plagioclase to form is calcium-rich, and each later plagioclase is poorer in calcium and richer in sodium. Eventually, the two series converge in felsic magmas, which crystallize out quartz, potassium feldspar, and sodium plagioclase, plus biotite or muscovite, or both—the minerals found in granites and granodiorites.

Bowen's reaction series was based on a lot of brilliant lab experiments derived from understanding the well-documented geochemical patterns of igneous rocks. It has since been confirmed by many field observations. There are many ancient layered magma chambers (now exposed to the surface and eroded) that cooled slowly and built up layers of crystals on their floors as they cooled. The bottom layers are always rich in ultramafic minerals and rocks (mostly peridotites and related rocks) and then gradually become enriched in gabbros as you go up through the layers into magmas cooled later in the process.

There are other proofs as well. When studying a thin section of a gabbro or basalt under the microscope, it is typical to see an early formed olivine crystal that is surrounded by a rim of pyroxene which crystallized around the olivine core. As the magma chemistry changed, it could no longer make olivine but made pyroxene instead. Also, microscopic examination of plagioclase crystals often shows that they have layers around the outside that are sodium-rich, while the inner core is calcium-rich, also demonstrating that the surrounding magma chemistry changed as the crystal grew.

Fractional crystallization is an important way to change an ultramafic magma direct from the mantle into a mafic or even intermediate magma. Another mechanism is called **partial melting**. Imagine that deep in the Earth's crust, there was a mass of already-cooled gabbro or diorite that was reheated. The first materials to separate away from the original rock would be minerals that melt at the lowest temperatures, such as quartz, potassium feldspar, and sodium plagioclase. If this low-temperature melt were then cooled, it would become a granodiorite or a granite.

Another mechanism for getting different magmas is to melt a different country rock. Thus, if you start with a rock of felsic or intermediate composition that was already in the crust, melting it will only produce a rock that is just as felsic, if not even more so.

Another important consideration is what causes rocks to melt. It turns out that for rocks that are in subduction zones (Chapter 4) plunging deep in the mantle, there is a lot of water entrapped in the ocean floor basalts that once formed on a mid-ocean ridge. Water and other volatiles (gases) dramatically lower the melting temperature of a rock. Many of the andesite–dacite–rhyolite magmas we will see in later chapters are formed because of different amounts of volatiles in the source material that melted.

Yet another likely mechanism for changing a gabbro to a diorite to a granodiorite magma is to have it melt its way up through existing crustal rocks, which are typically much richer in silicon, potassium, aluminum, and sodium. As these wall rocks of the magma chamber were melted and digested into the magma, they could change its chemistry from mafic to felsic. If pieces of wall rock are ripped away and melted into the new magma, it is known as **assimilation**. If the wall rock around the hot magma chamber is partially melted into felsic magma and then mixed with the more mafic magma, it is known as **contamination**. Either way, magmas formed deep in the Earth's crust that must melt their way through more than 100 km (60 mi) of crustal rock above them will end up being very different in chemistry. At one time, assimilation and contamination were thought to be crucial in making magmas more silicic, but now partial melting and fractional crystallization are considered more important.

Finally, there are a few rare instances where a mafic magma chamber from one source melted its way into a felsic magma chamber from a different source (or vice versa), forming a **mixed magma chamber** that is intermediate in composition (Figure 2.5).

This is the most basic (and highly oversimplified) summary of how the chemistry of magmas changes to form the hundreds of different types of igneous rocks known on the planet. Most of these processes happen deep in the crustal magma chambers to form plutonic rocks. Plutons intrude into older rocks ("country rock" or "wall rock") and melt their way to higher levels. They can take many shapes (Figure 2.6). Most plutons intrude to form a body of magma

(a)

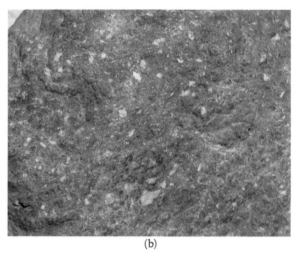

(b)

FIGURE 2.5 Evidence of magma mixing from Lassen Volcanic National Monument: (a) pink rhyolite with inclusions of dark andesite; (b) light andesite inclusions in dacite.

Source: Photos by the author.

cutting through older rock called a **dike** (spelled *dyke* in Britain). If a dike intrudes parallel to the bedding of a sedimentary rock, it is a special kind, known as a **sill**. A large chain of plutons that underlie a volcanic mountain range are known as a **batholith**. We will see all these features, as well as the features typical of volcanoes, in later chapters.

2.4 SEDIMENTARY ROCKS

The second major class of rocks is much more familiar to us, since they form at the Earth's surface, not in deep magma chambers or dangerous volcanoes. They are known as **sedimentary rocks**. They are of enormous importance, since nearly all the economic products we obtain from the Earth come from loose sediments or sedimentary rocks. These range from energy sources like oil, gas, coal, and uranium, to the groundwater we rely on, to materials for construction (building stone like sandstone or limestone, concrete made of crushed limestone plus sand and gravel,

FIGURE 2.6 Terminology of dikes, sills, and other types of intrusions: (3), huge magma chamber known as a batholith; (2), (4), and (6), dikes that either feed an (4) intrusion or a (6) volcano or cool after they reach a (2) dead end; (5), dike intruded parallel to bedding known as a sill; (1), blister- or dome-shaped intrusion known as a laccolith; (7), flat-topped intrusion called a lopolith.

Source: Courtesy of Wikimedia Commons.

gypsum for drywall, quartz sand for glass, etc.), to many of our metallic mineral resources (especially uranium, iron, and steel).

Even without these economic incentives, understanding sedimentary rocks is still supremely important. Sedimentary rocks are the source of nearly all our information about Earth history and ancient environments and the only source of fossils that demonstrate the history of life and help us tell geologic time. Finally, carbon-rich rocks such as coal and limestone are the thermostats that make our planet livable. These crustal reservoirs lock up or release carbon to the atmosphere so Earth is neither a hellish super-greenhouse like Venus nor a frozen ice ball like Mars.

All sedimentary rocks undergo some version of a basic pathway. They always start as *weathered material* of a pre-existing rock that can be igneous, metamorphic, or sedimentary. Once that weathered material is picked up by wind or water and *transported*, it becomes loose sediment. Eventually, the sediment in motion comes to a stop and is *deposited*. But it's still loose sediment until, sometime later in its history (usually after some deep burial), the loose sand grains are cemented together, or the mud particles are compressed, and the loose sediment is *lithified* into a sedimentary rock. Every step of this history can be detected in clues in the rock, and expert sedimentary geologists are like detectives, gleaning clues about the past from a sandstone or limestone that no one else even notices.

There are two versions of this pathway. Most sedimentary rocks are made of broken particles of pre-existing rocks or minerals, known as **clasts** (Greek *klastos*, meaning "broken fragments"). Therefore, **clastic** (or "detrital") sedimentary rocks are made of different-sized pieces of other rocks, from huge boulders down to fine clay. Eventually, these loose grains of gravel, sand, or mud must be lithified into sedimentary rock. The second pathway is much simpler. Instead of fragments of rocks or minerals, the ions from the pre-existing rock weather out and are dissolved in water, where they stay until something causes them to precipitate from the water and crystallize into minerals like halite (forming rock salt), gypsum (hydrous calcium sulfate), and calcite or aragonite (calcium carbonate). Thus, when they crystallize, they are already lithified into solid rock. Since this is a purely chemical process, we call these **chemical** sedimentary rocks.

2.4.1 CLASTIC SEDIMENTARY ROCKS

The most important property of rock fragments is their size, so classifications of clastic sediments are based on size. It's also important because the size of grains is a good indicator of the agents of deposition (wind, water, and glacial ice), and the grain size decreases downstream from the source, so it helps tell us about transport. Rather than the exotic and obscure names of igneous rocks (try to remember what minerals are in a *lherzolite* or a *jacupirangite*!), most sedimentary rock classifications are based on the practical Anglo-Saxon words that English speakers have used all our lives—*sand, gravel*, and *mud*—only those words have a strict definition in classification schemes (Figure 2.7). Any particle larger than 2 mm in diameter is gravel, which can be further broken down into granules (2–4 mm in diameter), pebbles (4–64 mm), cobbles (64–256 mm), and boulders (larger than 256 mm). A rock made of lithified gravel and finer sand is called a **conglomerate** if it has rounded gravel and a **breccia** if its clasts are angular. Conglomerates are the product of flood energies or debris flows and are diagnostic of sedimentary settings near the mountains where such floods occur. Breccias are even more specific, since it's hard for large pebbles not to become rounded, even with a few kilometers of washing downhill. Thus, breccias occur in unusual settings (collapsed cave ceilings, impact debris, and landslides) and are powerful indicators of these odd conditions.

We all know what sand is, and what it feels like (think sandpaper or beach sand), but to a sedimentologist, *sand* is strictly defined as particles between 2 and 1/16 mm in diameter. Many sands become cemented to form the rock known as **sandstone**. Most sandstones are rich in quartz, the most stable mineral on the Earth's surface, so quartz sandstone is the most common type. However, rare sandstones can be rich in feldspars, which are common in igneous rocks but rapidly break down into clays when weathered on the Earth's surface. Thus, a sandstone rich in feldspars (called an **arkose**) is very unusual and typically indicates

Millimeters (mm)		Wentworth size class		Rock type	
	4096	Boulder			
	256	Cobble			
	64	Pebble	Gravel	Conglomerate/ Breccia	
	4	Granule			
	2.00	Very coarse sand			
	1.00	Coarse sand			
1/2	0.50	Medium sand			
1/4	0.25	Fine sand	Sand	Sandstone/ Arkose	
1/8	0.125	Very fine sand			
1/16	0.0625	Coarse silt			
1/32	0.031	Medium silt			
1/64	0.0156	Fine silt	Silt	Mudstone Siltstone	Shale
1/128	0.0078	Very fine silt			
1/256	0.0039	Clay	Mud	Claystone	
	0.00006				

FIGURE 2.7 Classification of clastic sedimentary rocks.

Source: Redrawn based on several sources by E. T. Prothero; photos by the author.

that the sand grains have not traveled very far or have not been deeply weathered in wet, humid conditions.

Sediment finer than 1/16 mm can be called by the familiar name *mud*, and if it is compressed into a rock, it is called a **mudstone**. Mud can be further subdivided into the coarser material known as silt (1/16–1/256 mm) and clay (finer than 1/256 mm). A rock made entirely of silt is a **siltstone**, and clay makes a **claystone**. In the field, you can tell silt from clay because silt is still close enough to the sand size range that it is gritty to chew, whereas clay is "creamy" to chew. Almost all mud rocks, however, undergo some burial and pressure. This squeezes down the clay minerals (layered sheet silicates), forces out the water between the mineral grains (often almost 70% of the volume of mud is water), and turns it into a slightly different rock known as **shale**, which breaks into flat sheets along bedding (so it is "fissile"). Shales are by far the most common sedimentary rock on the Earth's surface.

2.4.2 CHEMICAL SEDIMENTARY ROCKS

As mentioned earlier, chemical sedimentary rocks form when ions from pre-existing rocks dissolve by weathering and go into solution in the waters of the Earth (groundwater, rivers, lakes, or the ocean) and then precipitate back out to form new sedimentary minerals. The most common way to do this is for organisms (plants or animals) to pull ions out of seawater (such as calcium and carbonate) and precipitate their shells with calcium carbonate minerals, such as calcite or aragonite. When these carbonate shells and coral skeletons of sea creatures accumulate, they build up carbonate sediment that eventually can crystallize into a rock known as **limestone**. Unlike sand and gravel and mud, *limestones are born, not made*. They are always built

from fossils (Figure 2.8a), even if recrystallization might make the fossils invisible. Most limestones forming today are restricted to tropical or subtropical settings, with warm, shallow, clear water and no clastic sand or mud. Places such as Florida, the Bahamas, parts of the Caribbean, the Persian Gulf, and the South Pacific are the main locations forming carbonate sediments today. During the geologic past, huge shallow tropical seas drowned the continents for millions of years, accumulating huge thicknesses of limestone in much of the world.

Another chemical found in water is silica. It can precipitate to form a rock known as **chert**, which is made of submicroscopic crystals of quartz. Chert comes in many colors based on impurities, so if it is black, we call it flint; if it is red, it is jasper; white chert is novaculite; and so on. Chert in the form of flint or jasper was once important for arrowheads and spearheads, for flint was used to start fires and fire flintlock muskets, among many other purposes.

Chert forms in two main ways. In places where plankton that use silica in their skeletons are extremely abundant, they accumulate to form a silica-rich shale known as **bedded chert** or "ribbon chert" (Figure 2.8d). This kind of chert is precipitated by organisms, just as limestones are. The other kinds of chert form when silica-rich groundwater percolates through other rocks (usually limestone) and replaces calcite with silica. These are known as **nodular cherts** (Figure 2.8e).

Most of the chemical sedimentary rocks just discussed were formed by organisms precipitating ions out of water, so they are often called **biochemical** (or organic) **sedimentary rocks**. But there are also rocks that form by straight chemical precipitation, without the help of organisms to drive precipitation. These are **inorganic chemical sedimentary rocks**. The main mechanism required to make

FIGURE 2.8 Selection of biochemical sedimentary rocks (limestones and cherts), including (a) a shell-rich coquina; (b) fossiliferous limestone full of brachiopods; (c) fossil coral, which was already a solid limestone while the corals were still alive; (d) bedded or "ribbon" cherts, which are made of layers rich in silica from plankton interbedded with shales formed from the deep-sea muds; and (e) nodular chert, in which the silica replaces fossiliferous limestone as it seeps through in the groundwater. Here, the chert nodule has replaced the limestone surrounding a calcite crinoid stem, which has dissolved away, leaving a crinoid stem-shaped void.

Source: Photos by the author.

minerals precipitate in this way is to evaporate all the water away and leave the minerals behind to crystallize out of a salty **brine**. Thus, these minerals are known as **evaporites**. Such evaporation is common in dry lake beds, but also in tropical lagoons and hot desert seas like the Persian Gulf, where the rate of evaporation is greater than the amount of seawater flowing in to replace the water evaporated. The most common evaporite minerals are salt (sodium chloride), gypsum (hydrous calcium sulfate), and sometimes calcite or aragonite, but there are hundreds of additional evaporite minerals, many of which are unique to some of California's dry lakes (see Chapter 7).

2.5 METAMORPHIC ROCKS

The third major class of rocks is metamorphic rocks (*meta* meaning "after" or "changed," and *morphos* meaning "form" in Greek). We are familiar with the word *metamorphosis* to describe many kinds of changes in form, such as the caterpillar changing to a pupa and then a butterfly. Metamorphic rocks are transformed or changed from some original "parent rock" or **protolith** (usually an igneous rock or sedimentary rock) into an entirely new and different kind of rock with new minerals and new textures or fabrics. Some metamorphic rocks are so completely transformed that they are unrecognizable, and we may never know what kind of protolith they started from.

2.5.1 PRESSURE AND TEMPERATURE

Metamorphism occurs when the protolith is buried deep in the crust and experiences extremely *high temperatures* and *directed pressure*. Thanks to the enormous heat flow coming up from the Earth's interior, the crustal rock below your feet gets hotter and hotter the deeper you go. In fact, the **geothermal gradient** is about a 30°C increase per kilometer of depth, so at 30 km, the crustal rocks would be about 900°C (above the melting temperature of many minerals). Most people can't even imagine this, but if you go down an old abandoned mine shaft with no air-conditioning, you can feel how it gets hot as you descend. South African diamond and gold miners work at depths close to 3.9 km (12,800 ft) and require a continuous supply of refrigerated fresh air to survive shifts of only a few hours, because the temperature of the rocks and air down there is 60°C (140°F).

The increase in pressure is also intense. The pressure gradient is 3 kbar per 10 km depth you descend. One kilobar (which is 1,000 bars in the metric system) is about 14,500 lb/in², so at 30 km down (very shallow crustal levels), pressures are about 9 kbar, or almost 45,000 lb/in². These pressures are far too great for any device to overcome by drilling (no drilling hole has gone much farther than 12 km, or 40,000 ft). You can laugh at any of those science fiction movies that imagine drilling to the Earth's interior or a "journey to the center of the Earth"! They are complete fantasy, because the pressures and heat are so intense even down 4 km that no human or machine or drill could survive.

The pressure is also **directed pressure**, or pressure applied in a specific direction, so that one axis of a rock is being crushed in the up–down direction, while there is less pressure in the perpendicular direction, so the rock can be squeezed sideways as it is flattened. This is very different from the uniform pressure that you experience on all sides of your body from the air around you, or from water around you when you dive. Directed pressure tends to squash or flatten most rocks, stretching and squeezing them outward. Old minerals that were present in the rock as it began metamorphism will be flattened out and rotate until they are perpendicular to the direction of maximum pressure. Any new minerals that grow during this metamorphic process (especially platy minerals like the micas muscovite, biotite, and chlorite) will tend to grow perpendicular to the direction of greatest stress as well. As a result, the rock will acquire a strongly planar fabric in the direction of least stress, which is called a **foliation** (after the Latin word *folium*, for "leaf"). Many metamorphic rocks have a planar fabric or layering and will split along this plane of foliation as a result.

2.5.2 TYPES OF METAMORPHIC ROCKS

Let's take a variety of different protoliths and see what happens during metamorphism (Figure 2.9). If we start with shale, for example, we have lots of chemistry (silicon, aluminum, potassium, sodium, and other elements) to work with. The first metamorphic product of a shale would be a rock known as a **slate**, which is platy and highly foliated, so it readily breaks into things like roofing tiles and (in the old days) slate for blackboards. As the temperature and directed pressure increase, the clay minerals in the shale are transformed into tiny flakes of micas like muscovite or biotite (not yet visible to the naked eye), and the rock acquires a distinctive sheen; this is called a **phyllite**. Further pressure and temperature allow the new metamorphic minerals (muscovite and biotite plus garnet, hornblende, and others) to grow large enough to be visible to the eye; this kind of rock is called a **schist**. Finally, at extremely high pressures and temperatures, some of the minerals begin to melt and segregate into bands of light-colored minerals (typically quartz and plagioclase) and dark-colored minerals (typically biotite and hornblende). This *compositional banding* is the diagnostic feature of a rock known as **gneiss**. Any further increase in pressure and temperature and the gneiss melts completely and can become a magma that could cool to form an igneous rock.

Let's try a different protolith (Figure 2.9). If we start with an olivine peridotite or pyroxene-rich gabbro, we have only magnesium, iron, and silica to work with. Under high pressures and temperatures, you transform olivine or pyroxene into a new mineral known as **serpentine**, which gets its name from its snake-like green color and waxy, smooth, scaly "snakeskin" feel to the touch. A rock made of the mineral serpentine is called a **serpentinite**. Serpentinite is the official state rock of California because it is common in the Coast Ranges, the Klamaths, and the Sierra Nevada Mountains, where ultramafic olivine-rich slices of oceanic crust have been metamorphosed to form serpentinite.

What if you start with a different protolith, such as a quartz sandstone? Quartz sandstone contains only one chemical, silicon dioxide, and you can't make anything but quartz from that chemistry, no matter how high the pressure and temperature. So under metamorphism, the quartz sandstone becomes a different quartz-rich rock, a **quartzite** (Figure 2.9). The mineralogy won't change (always forming quartz), but the original fabric of spherical quartz sand grains packed together will vanish, and the grain boundaries will fuse together and become interlocked, like pieces of a jigsaw puzzle. If you hit

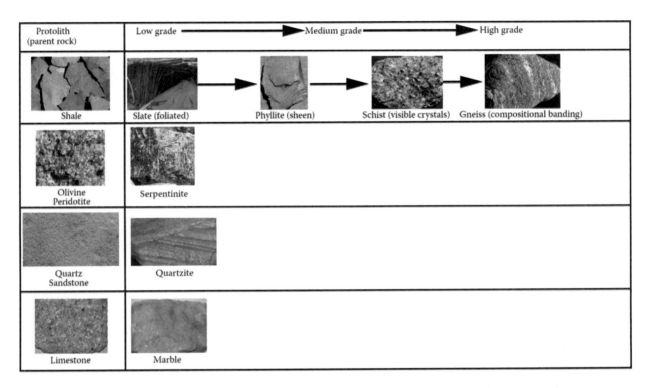

FIGURE 2.9 Different protoliths produce different kinds of metamorphic products as pressure and temperature increase.

Source: Redrawn from several sources by E. T. Prothero; photos by the author.

a sandstone, the rock will break between the grains, but if you hit a quartzite, it will fracture right through the original sand grains, since they are fused into one mass of quartz.

Let's consider one more protolith: limestone. Limestone has only one mineral available, calcite (calcium carbonate), so it can only form calcite no matter what the pressure and temperature. Thus, a fossil-rich limestone protolith transforms into a **marble** under metamorphism (Figure 2.9). A true marble is still made entirely of calcite, but the minerals and fossil fragments have completely recrystallized so that no fossils are visible anymore and only large shiny crystals of calcite are left.

So far, most of the metamorphic rocks we have discussed (shale, phyllite, schist, gneiss, and serpentinite) are highly foliated. But quartzite and marble never undergo foliation, no matter how high the pressures and temperatures. This is because they are not made of platy minerals like micas or elongated prismatic minerals like hornblende. Minerals like quartz and calcite have no long axis or preferred orientation, so no matter how much you cook or squeeze them, they will never foliate.

2.5.3 METAMORPHIC GRADE AND FACIES

We can plot all the different possible regimes of pressure and temperature on the graph shown in Figure 2.10. Temperature increases to the right on the *x*-axis, and pressure increases downward (as it does in the real world) on the *y*-axis. In the upper left part of the diagram is the region of lowest pressure and temperature (Earth surface

conditions or very shallow burial), and no metamorphism takes place at this level. If we descend the geothermal gradient line until temperatures are around 330–500°C and pressures are about 2–8 kbar, we are in a region of relatively low pressures and temperatures, producing **low-grade metamorphism**. Geologists use a shorthand term for this region, calling it the **greenschist facies** (*facies* is the Latin word for "appearance"). Rocks that have undergone low pressures and temperatures *appear* as green schists because they usually grow the green mica chlorite, plus sometimes other green minerals, such as amphiboles known as actinolite–tremolite, and maybe even the pistachio-green mineral epidote. Moving farther down the geothermal gradient, we reach a region of intermediate pressures and temperatures (4–12 kbar and 500—700°C). The geologists' term for **intermediate-grade metamorphism** is **amphibolite facies**, because these rocks tend to be rich in the black amphibole hornblende. Finally, we go to the very deepest part of the continental crust, where there are extreme pressures (6–14 kbar) and temperatures (at least 700°C), making the highest-grade metamorphics. These rocks have a gneissic texture and are sometimes called **granulite facies**. All three of these facies are produced by **regional metamorphism**, such as when a collision between continents creates a huge uplifted mountain belt (such as the Himalayas today). Rocks that undergo shallow burial will only reach greenschist facies, while those with even deeper burial will become amphibolites or, finally, granulite gneisses, if they are many kilometers down into the crust at the root of the mountains.

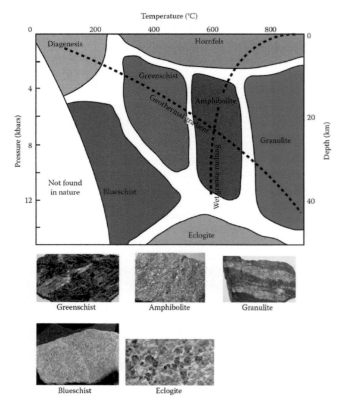

FIGURE 2.10 Metamorphic facies diagram showing how rocks change their characteristic appearance (*facies* in Latin) as they change minerals and texture. Rocks subjected to very low temperatures and pressure are only slightly changed ("diagenesis") but not truly metamorphic. Low temperatures and pressures produce greenschist-grade metamorphics. If those rocks are buried deeper in the roots of mountains ("regional metamorphism"), they experience higher pressures and temperatures as they proceed down the geothermal gradient (dashed line going from upper left to lower right), turning into amphibolite facies or granulite gneiss facies. Rocks that are heated by an igneous intrusion without much pressure of burial are known as the hornfels facies. Rocks formed in a subduction zone, which is relatively cool despite extreme pressure of burial, form blueschist metamorphics. The boundary for melting metamorphic rocks into magma is also shown.

Source: Redrawn based on several sources by E. T. Prothero; photos by the author.

Look at Figure 2.10 again. We can see how the three main facies are found in increasing depths and pressures as they descend deeper into the crust (and further down the geothermal gradient). But what about the peculiar region across the top of the diagram labeled **hornfels**? A glance at the axes of the plot suggests that these would be formed under high temperatures but low pressures. This could not happen by simple burial at greater depths, because both temperature and pressure increase together. Where do such peculiar conditions exist? The only possibility is to heat a rock to high temperatures without burying it deeply. This can only happen when a magma body or dike intrudes through the country rock, cooking it without much depth of burial. Such metamorphism caused by the contact of an intruded magma is called **contact metamorphism** and

produces hornfels. Such rocks often don't show much in the way of different minerals, but their fabric is welded and baked compared to the unheated rock at some distance from the magma intrusion.

Finally, there is one more peculiar region in Figure 2.10 to explain. On the lower left is a region marked **blueschist**. These rocks get this name because they are often a bluish-gray or even deep-blue color (Figure 2.10), due to the blue amphibole glaucophane, plus another blue or white mineral called lawsonite. The axes in Figure 2.10 indicate that it is a region of very high pressure but relatively low (less than 400°C) temperatures. How could this happen? Most rocks descending into the deep crust get hot when they are under such high pressures. The answer to the mystery of blueschists was discovered in California, because they are particularly common in the Coast Ranges (see Chapter 12) and rare elsewhere in the United States. The only place blueschists are found is in the remnants of ancient subduction zones (Chapter 4), where the cold down-going plate plunges into the hot mantle and reaches depths of 20 km or greater. At this point, the rocks are surrounded by the high pressures of such deep burial, but they are still relatively cold because the old oceanic plate retains a lot of water and heats up slowly. Under these unusual conditions, the pressures can get very high, but the temperatures low enough to form glaucophane, lawsonite, and some other distinctive minerals. Then when some of this plate gets scraped off in an accretionary wedge (see Chapter 4), the blueschist can rise to the surface after having been more than 20 km underground in a subduction zone.

2.6 ROCK CYCLE

As we have suggested already, minerals and rocks can transition from one category to another quite easily. Take the example we used earlier of the sedimentary rock known as shale. It can transition from sedimentary rock to metamorphic rock as it experiences high pressures and temperatures. It then goes from shale to slate, phyllite, schist, and gneiss. Eventually, it gets hotter and hotter until it melts. Then it has become a magma and can cool into an igneous rock (Figure 2.9). Thus, we have a sequence of sedimentary to metamorphic to igneous rock (Figure 2.11). Then if that igneous rock reaches the Earth's surface, it will weather and break down into loose sand and mud and return to the sedimentary beginning of this loop all over again.

This is the **rock cycle**, and it is a demonstration of the fact that rocks can transform from one category to the next and eventually return to their starting point over millions of years. One of the great lessons you learn from geology is not only that time is immense (millions and billions of years), but given enough time, ocean bottoms can turn into mountains and then weather down into sediment and return to the ocean—or be forced down into the lower crust and transform into metamorphic rocks or even melt into a magma. With enough time, any of these extremely slow processes are inevitable.

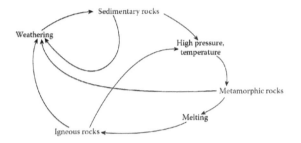

FIGURE 2.11 The rock cycle. Over millions of years, no rock is permanent, but it is part of a continuous slow cycle, changing from one class of rock (e.g., an igneous rock) to another (e.g., weathered sediments from the igneous rocks) to another (e.g., metamorphism of sedimentary rocks into metasedimentary rocks). If the metamorphic rock is heated enough, it melts and returns to the igneous part of the loop.

Source: Redrawn based on several sources by E. T. Prothero.

RESOURCES

Bowen, N.L. 1956. *The Evolution of Igneous Rocks.* Dover Publications, New York.

Deer, W., and Howie, J. 1996. *An Introduction to the Rock-Forming Minerals.* Prentice Hall, Englewood Cliffs, NJ.

Dillard, A. 2008. *An American Childhood.* Harper & Row, New York.

Garlick, S. 2014. *National Geographic Pocket Guide to Rocks and Minerals.* National Geographic Society, Washington, DC.

Klein, C., and Philpotts, A. 2012. *Earth Materials: Introduction to Mineralogy and Petrology.* Cambridge University Press, Cambridge, UK.

Pellant, C. 2002. *Smithsonian Handbook of Rocks and Minerals.* Smithsonian Institution Press, Washington, DC.

Philpotts, A., and Ague, J. 2009. *Principles of Igneous and Metamorphic Petrology.* Cambridge University Press, Cambridge, UK.

Prothero, D.R., and Schwab, F. 2013. *Sedimentary Geology.* W.H. Freeman, New York.

Winter, J.D. 2009. *Principles of Igneous and Metamorphic Petrology.* Prentice Hall, Englewood Cliffs, NJ.

VIDEOS

Three kinds of rocks: www.youtube.com/watch?v=sN7AficX9e0
Series on minerals and rocks:
www.youtube.com/watch?v=8a7p1NFn64s
www.youtube.com/watch?v=32NG9aeZ7_c
www.youtube.com/watch?v=ZkHp_nnU9DY
www.youtube.com/watch?v=aCnAF1Opt8M
www.youtube.com/watch?v=Etu9BWbuDlY
www.youtube.com/watch?v=1oQ1J0w3x0o

Animations

Clever site that allows you to build silicate structures with interactive animation: https://ees.as.uky.edu/sites/default/files/elearning/module09swf.swf

Excellent 3D animation of complex silicate structures: http://web.visionlearning.com/silica_molecules.shtml

Learning module on silicate structures: www.visionlearning.com/en/library/Earth-Science/6/The-Silicate-Minerals/140

3 Dating California
Stratigraphy and Geochronology

[The concept of geologic time] makes you schizophrenic. The two time scales—the one human and emotional, the other geologic—are so disparate. But a sense of geologic time is the important thing to get across to the non-geologist: the slow rate of geologic processes—centimeters per year—with huge effects if continued for enough years. A million years is a small number on the geologic time scale, while human experience is truly fleeting—all human experience, from its beginning, not just one lifetime. Only occasionally do the two time scales coincide.

—Eldridge Moores
in Assembling California, *by John McPhee*

3.1 STRATIGRAPHY

One of the oldest branches of geology is known as **stratigraphy**. It literally means "the study and description of layered rocks," but over the past 250 years, it has come to mean any study of rocks that helps us determine their age, how they were deposited, and what they tell us about geologic history. Today, we can think of stratigraphy as the toolbox we use to understand the geologic past.

3.2 DATING ROCKS

There are two basic ways of determining the age of a geologic event or rock body. They are **relative age** and **numerical age**.

Relative age is the age of one object or event in relation to another, that is, this rock is younger than that one. There are several common ways to determine this. Most of these concepts (Figure 3.1) were first formulated by the Danish physician Niels Steensen (Nicholas Steno in Latin) in 1669:

Principle of superposition. In any layered sequence of rocks, it makes sense that the lowest rocks in the stack have to be older than the upper ones; you can't put one thing on top of another if it isn't already there. Thus, in the layered sequences we will see repeatedly in this book, the rocks get younger as you move up the section, and this is the primary method by which we determine relative age. A good analogy for this is a stack of papers on a messy desk that have piled up for months undisturbed. The oldest papers would be at the base of the stack, and the most recently used papers would be at the top.

Crosscutting relationships. Whenever one rock unit or fault cuts across another, the rock that does the cutting must be younger than whatever it cuts through. This applies primarily to dikes of molten rock (magma) from the Earth's interior, which can cut through (intrude) the surrounding rock as they melt their way through the Earth's crust. Likewise, a fault can only cut through rocks that are already there.

Principle of original continuity. Whenever you see two rock units that are now separated by erosion, it is safe to assume that they were originally connected and continuous, and that the erosion has since cut through them. Thus, the erosion cuts through the layers, and it must be younger. For example, just below the rim of the Grand Canyon, you can see a distinctive white layer known as the Coconino Sandstone on both the North Rim and South Rim side. Even though they are separated by a huge canyon, we can visualize how they were once connected and conclude that the cutting of the canyon must be younger than the (now discontinuous) layers it cut through.

Principle of original horizontality. When we see sedimentary beds of sand or mud laid down in rivers or oceans today, they are nearly always in horizontal layers due to Earth's gravity. Thus, if we find these beds tilted or folded in any way, then the deformation must have happened after the sand or mud was laid down and turned into sandstone or shale.

In addition to determining the sequence of rock units and faults through these principles, we must also be aware of erosional gaps in the rock sequence, known as **unconformities**. These occur when one rock unit is uplifted and eroded away and then down-dropped again and the erosional surface covered by a new rock layer. Most of these unconformities take millions of years to develop and form, and each erosional surface represents a time gap that may be millions of years long, often longer than the time represented by actual rocks in the sequence. There are three common kinds of unconformities (Figure 3.2):

Angular unconformity. The rocks below the erosional surface are tilted and eroded off and then covered by the upper sequence (Figure 3.2b). This is the easiest unconformity to spot in the field.

DOI: 10.1201/9781003301837-3

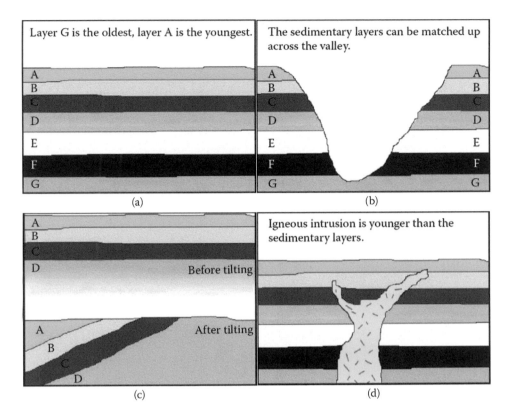

Layer G is the oldest, layer A is the youngest.	The sedimentary layers can be matched up across the valley.

(a)   (b)

Before tilting | Igneous intrusion is younger than the sedimentary layers.
After tilting |

(c)   (d)

FIGURE 3.1 Steno's principles. (a) *Superposition*: the layers at the bottom of the stack are older than the layers placed above them. (b) *Original continuity*: if you find a canyon or valley cut through layers that match on both sides, then the canyon is younger than the layers it cuts. (c) *Original horizontality*: rock layers are normally deposited as horizontal sheets of sediments or horizontal lava flows and volcanic ash layers, so if you find them tilted or folded, their deformation occurred after they were deposited or erupted. (d) *Crosscutting relationships*: any event (igneous intrusion, fault, etc.) that cuts through pre-existing rocks must be younger than what it cuts.

Source: Drawing by M. P. Williams.

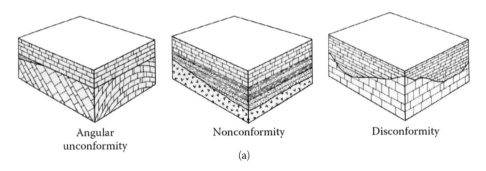

Angular   Nonconformity   Disconformity
unconformity
(a)

FIGURE 3.2 The major types of unconformities: (a) Diagram representing the three main types. (b) Angular unconformity between the 12-million-year-old Mint Canyon Formation (tilted gray sandstones) and 1-million-year-old Saugus Formation (horizontal gravels incised into the lower unit), so there is an 11-million-year time gap between the lower and upper units. This outcrop is on California Highway 14 northeast of the Soledad Canyon Road exit. (c) Nonconformity between Precambrian granitic rocks (deeply weathered and crumbling in lower left) and hard, resistant, layered Lower Cambrian Wood Canyon Formation, Marble Mountains. (d) Disconformity, showing one sequence of tan sandstones full of gravels eroded down into channels cut into a lower sequence of gray-pink sandstones and siltstones, lower Miocene Vasquez Formation, Vasquez Rocks.

Source: (a) redrawn from several sources by P. Linse. Photos (b) and (d) by the author; photo (c) by J. Foster.

(b)

(c)

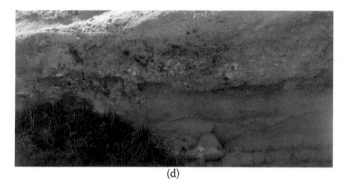

(d)

FIGURE 3.2 (Continued)

An angular unconformity represents at least five separate events: the slow deposition of the lower sequence of sediments, the hardening of those loose sediments into sedimentary rock, tilting the lower sequence, eroding it off, and then depositing the upper sequence. Each angular unconformity by itself represents millions of years of time.
Nonconformity. Here the rocks below the erosional surface (Figure 3.2c) are a nonsedimentary rock (in other words, igneous or metamorphic) formed by crystallization or high temperatures and pressures

deep in the Earth's crust. These must be uplifted, exposed, and eroded and then covered by sediments to form the erosional surface.
Disconformity. This is the subtlest type of unconformity, since the beds above and below the erosional surface are parallel. However, between them lies a slight erosional surface that usually must be detected by scouring or channeling into the lower unit (Figure 3.2d) that is then filled by sediments of the upper unit. In many cases, there is no obvious field evidence of erosion, and the gap can only be determined by the age of the fossils above and below.

In the field, geologists must reconstruct a long sequence of events using superposition of the bedded rocks (sedimentary rocks and lava flows) plus the crosscutting relationships of the igneous intrusions, faults, and unconformities. From this method, we can determine the relative age in any complex series of events.

3.3 GEOCHRONOLOGY

Numerical dating (formerly but incorrectly called by the obsolete term *absolute dating*) or **geochronology** is the method of determining the age of geologic event given in some sort of numerical age, for example, "this rock is X number of years old." Although the principles of relative dating and the geologic time scale were developed more than 200 years ago, there was no reliable way of determining the numerical age of events until about 1913, when radioactive dating was developed. All radioactive dating works on similar principles. Certain elements in nature (such as uranium-238 or potassium-40) are naturally radioactive and decay to a daughter element (e.g., lead-207 or argon-40) while giving off heat and radioactive particles (alpha, beta, and gamma radiation). This decay occurs in a predictable fashion, with an exponential or logarithmic decay curve, so that if you know the amount of parent material and daughter material and the decay rate for this system, you can determine the age.

There are many different radioactive elements, but only a few are useful in geologic settings. The most commonly used method is potassium–argon dating, where potassium-40 decays to argon-40. Because potassium is one of the eight most common elements in the Earth's crust (Table 2.1), it can be found in many different minerals (especially micas and potassium feldspars) and can date rocks as old as 4.6 billion years old and as young as a million years old. The main weakness of this system is that argon is a gas, which tends to escape from the system and give erroneous ages if the material has been weathered.

There are also systems of uranium–lead and rubidium–strontium dating, but these are used mostly for rocks that are more than 600 million years old, since their decay rates are very slow. In addition, uranium and rubidium are relatively rare elements, so a lot of time is needed to accumulate a measurable amount of daughter atoms.

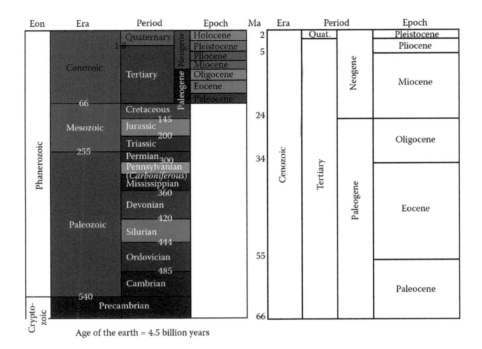

FIGURE 3.3   Standard geologic time scale. On the left is the entire Phanerozoic eon. On the right is a detail of the Cenozoic epochs.

Source: Drawn from several sources by E. T. Prothero.

Finally, most people have heard of carbon-14 dating, where you can measure the decay of radioactive carbon-14 back to nitrogen-14 in any carbon-bearing substance, including bones, wood, charcoal, pottery, baskets, and shells. The main limitation of this method is that the decay rate is very fast, so that nothing older than about 60,000–80,000 years old can be dated by carbon-14. Thus, it is a common method in archeology and in studying fossils and rocks from the last ice age, but useless to any other kind of geologic setting.

In all these systems (except carbon-14), you are measuring the ratio of parent and daughter elements that accumulate in a closed, sealed crystal. This means that you need the crystal to cool in a melt and lock in the radioactive materials in its crystal structure so that none can escape. For this reason, the only kinds of rocks that can be numerically dated by most methods (e.g., potassium–argon) are igneous rocks formed from a molted magma, such as volcanic rocks erupted out of the Earth, or magma that has cooled and crystallized in a deep underground magma chamber (plutonic rocks). Sedimentary rocks are not formed this way. Their grains of sand or mud are weathered and eroded out of pre-existing rock, so there is no way to radioactively date sedimentary rocks directly. When a geologist gives a numerical age estimate to a sedimentary rock, it is based on knowing the relative age (which is based on its distinctive fossils) and then finding some place on Earth where a volcanic lava flow or ash layer is interbedded with rocks bearing the same fossils. That is how the international geologic time scale was produced (Figure 3.3).

RESOURCES

McPhee, J. 1993. *Assembling California*. Farrar, Straus, and Giroux, New York.

Prothero, D.R., and Schwab, F. 2013. *Sedimentary Geology: Principles of Sedimentology and Stratigraphy*. W.H. Freeman, New York.

Videos

Numerical dating

Bill Nye on radiometric dating: www.youtube.com/watch?v=Q6O LBVDbJoI

NOVA on radiocarbon dating: www.pbs.org/wgbh/nova/tech/radio carbon-dating.html

Radiocarbon dating

www.youtube.com/watch?v=54wR-zwuDGo

www.youtube.com/watch?v=udkQwW6aLik

www.youtube.com/watch?v=phZeE7Att_s

Relative dating (note that these use the obsolete term *absolute dating*):

www.youtube.com/watch?v=fYSeM63Fv0s

Animations

http://science.jburroughs.org/mbahe/BioA/starranimations/ chapter16/videos_animations/carbon_14_dating_v2.swf

4 The Big Picture
Tectonics and Structural Geology

Scientists still do not appear to understand sufficiently that all earth sciences must contribute evidence toward unveiling the state of our planet in earlier times, and that the truth of the matter can only be reached by combining all this evidence. . . . It is only by combining the information furnished by all the earth sciences that we can hope to determine 'truth' here, that is to say, to find the picture that sets out all the known facts in the best arrangement and that therefore has the highest degree of probability. Further, we have to be prepared always for the possibility that each new discovery, no matter what science furnishes it, may modify the conclusions we draw.

—**Alfred Wegener**
The Origin of Continents and Oceans, 1915

4.1 HOW ARE ROCKS DEFORMED?

One of the first things a geologist looks for when studying and mapping rocks are signs of deformation or tilting or other changes that occurred to a rock since it was formed. Small, subtle structures like folding, faulting, stretching, crinkling, and crumpling of rock are valuable clues as to the large-scale forces that affected those rocks. This information, in turn, can be useful in reconstructing the history of ancient mountain belts, plate collisions, plate separation, and other tectonic events. **Structural geology** is extremely important not just in understanding how mountain belts formed and how plates moved but also in understanding where to find mineral resources, especially oil and gas, coal, and many other valuable minerals. Oil and gas, for example, are often trapped and accumulated under the crests of folds, or against fault lines, so mapping these features helps predict the next discovery. The field of structural geology is highly specialized, so I will introduce only the basic terms here.

The first important concepts to understand are **stress** and **strain**. Although most people use these words interchangeably, they have a very specific meaning in geology. A stress is *a force applied per unit area*. This is familiar with units such as pounds per square inch in the English units or the kilobar (kbar) and Pascal (Pa) in the metric system. These stresses (Figure 4.1) can be **compressional** (forces coming together and squeezing the material between them), **tensional** (forces pulling something apart), or **shear** (forces sliding past one another). *Strain is the deformation resulting from stress.* Thus, *stress* is the force and *strain* is the

response. Even though people use both words interchangeably, they are not synonyms in geology. An analogy might be to say that when a person is under a lot of stress, that person shows the strain by loss of sleep, erratic behavior, or nervousness.

Objects can react to stress by straining in three different ways: **brittle** (breaking or shattering), **ductile** or **plastic** (slowly flowing and deforming), and **elastic** (bouncing back rapidly after deforming). The way an object deforms is often dependent on its temperature and how quickly the stress is applied. For example, if you apply a quick force to bend a glass tube, it will break in a brittle fashion. But if you bend it slowly as you heat it in a Bunsen burner (as skilled glassblowers and chemists do), it will slowly deform and bend without breaking. Most objects only exhibit one or two types of strain responses. There is one substance (silicone putty, sold as "Silly Putty" in toy stores) that can do all three: if you hit it with a hammer, it will shatter into pieces in a brittle fashion; if you slowly knead it in your fingers, it will stretch in a ductile fashion; and if you roll it into a ball and slam it against the ground, it will bounce and rebound in an elastic fashion.

If the stress is very shallow where the pressure is low, rocks behave in a brittle fashion and break along faults. If the rocks are deeply buried, however, the huge pressures and high temperatures allow the rocks to deform in a plastic fashion, and the rocks can become folded. Faults and folds are very valuable clues to the stresses and strains that the crustal rocks have experienced.

Faults are fractures in the Earth's crust that have moved or slipped. If the fracture does not move, it is called a **joint**. Faults are classified by the sense of motion that they show.

Normal faults (Figure 4.2) are faults that have slipped down the incline of the fault plane so that the overhanging block ("hanging wall") is down relative to the other block ("foot wall"). They are particularly typical in areas where the crust has been stretched out or sagged downward. If a block drops between two normal faults and forms a long down-dropped fault valley, this is known as a **graben** (German for "trench" or "ditch"). The raised block between two grabens is known as a **horst** (German for "heap" or "pile"). Normal faults are the most common type of fault (hence their name) and always indicate that the crust has been stretched and undergone tension.

If the overhanging block is pushed *up* the fault plane due to compression, then it is a **reverse fault** (Figure 4.2). A reverse fault that pushes one block over another at a very low angle or even horizontal is known as a **thrust fault**. Reverse faults cause crustal rocks to overlap and stack on

DOI: 10.1201/9781003301837-4

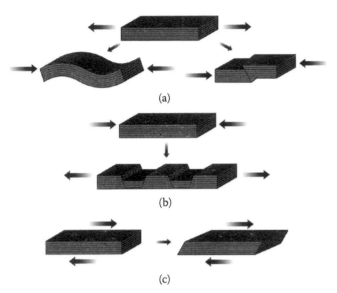

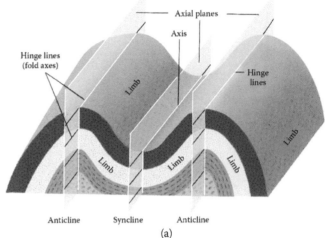

FIGURE 4.1 Three types of stresses acting on rocks: (a) compressional, (b) tensional, and (c) shear.

Source: Courtesy of Wikimedia Commons.

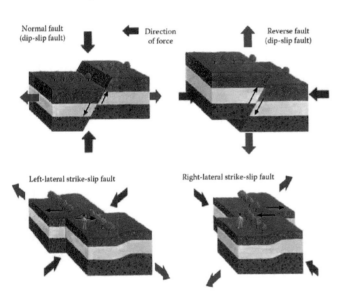

FIGURE 4.2 Fault terminology showing the stress directions and resulting fault slip.

Source: Courtesy of Wikimedia Commons.

FIGURE 4.3 (a) Terminology of folds showing synclines and anticlines. (b) Complex of tight synclinal and anticlinal folds (with minor faulting) in the Barstow Formation, west side of the parking lot at Calico Ghost Town.

Source: (a) Redrawn from several sources by P. Linse. (b) Photo by the author.

top of one another, usually to accommodate shortening and compression. Thus, they are diagnostic of regions in the crust where there has been collision and intense crushing and deformation.

There are also faults that do not move in the vertical sense at all but only slide in a horizontal sense. These are known as **strike-slip faults** (Figure 4.2). The San Andreas fault is a famous example of a strike-slip fault. There is a standard way to describe the motion on a strike-slip fault. If you stand on one side and the rocks and landmarks on the opposite side appear to have moved to the left, then it is a left-lateral strike-slip fault. If they appear to have moved to

the right relative to the block you are standing on, then it is a right-lateral strike-slip fault. Most real faults in nature have a bit of both dip-slip (either normal or reverse faulting) and strike-slip in their motions, so they are known as **oblique slip faults**.

If rocks are buried deeply enough, they will flow and fold and bend rather than break. **Folding** is almost always the result of compressional forces in the Earth's crust, so folds are often associated with compressional faulting, such as reverse and thrust faults. The simplest way to view a fold is to imagine beds folded up into a broad arch, or **anticline**, or downward into a trough-shaped bend, or **syncline** (Figure 4.3). (In Greek, *anticline* means "bent away from the earth" and *syncline* means "bent toward the earth.") Each of these folds has an axis down the center and an axial plane that divides the fold in half. On each side of the axis are the flanks or limbs of the fold. Keep in mind: folds like anticlines and synclines are structures *within* the crust of the Earth. They do not always correspond to ridges or valleys on the surface. In fact, it is common for anticlines to form valleys if the rock in the center is more easily eroded, and for synclines to form ridges if the rocks in the flank are easily eroded.

Simple folds with a horizontal fold axis are easy to visualize, but most folds are more complex than this. Typically, the fold axis plunges into the ground, so the fold itself plunges into the ground as well. When this happens, erosion often bevels away the uplifted part of the fold and exposes the core of the fold at the surface. Beveled and eroded plunging folds typically produce a parabolic outcrop pattern where they meet the ground (Figure 4.4). Plunging anticlines produce a parabolic outcrop pattern, and the oldest rocks are exposed at the core, while the younger rocks are on the flanks. The point of the parabola in a plunging anticline points in the direction of plunge. Plunging synclines, on the other hand, produce a reverse of this pattern: the youngest beds are in the center, and the oldest beds on the flanks, and the point of the parabola is in the opposite direction from the plunge of the fold.

A good example of this pattern can be seen at Devil's Punchbowl County Park on the north side of the San Gabriel Mountains southeast of Littlerock (Figure 4.4b–f). The reddish-tan sandstones of the Miocene Punchbowl Formation are folded into a broad syncline (Figure 4.4b), with the axis trending east–west and plunging to the west. In map view (Figure 4.4c), the geologic map pattern of the Punchbowl Formation shows up as a tight parabola, with the crest of the parabola pointing east and the opening to the west. This is also apparent from the satellite image (Figure 4.4d). If you stand at the overlook just east of the Punchbowl Visitors' Center, you can see the south limb of the fold, with the sandstones all dipping vertically to the north. A short hike down the trail to the northeast before it plunges steeply down into the canyon of Punchbowl Creek places you right on the fold axis, so you can stand in one spot and look north to see the south-dipping beds of the north limb, east to see the west-dipping beds of the axis, and south to see the north-dipping beds of the south limb back at the visitors' center (Figure 4.4e and f).

This is a quick summary of the major concepts and terminology of structural geology, which is a huge research field unto itself. Now we will look at the large-scale tectonic forces that build mountains and move continents and tear them apart. These forces and processes can be deciphered using the structures and analyses we just reviewed.

4.2 THE WAY THE EARTH WORKS

4.2.1 A Scientific Revolution

Philosopher of science Thomas Kuhn pointed out in his influential 1962 book *The Structure of Scientific Revolutions* that science operates very differently from the way that most people think it does. It is not a slow, steady, uninterrupted march toward final truth. Instead, it goes through periods of time where everyone practices "normal science" and accepts certain basic premises and assumptions (a **paradigm** in Kuhn's sense). Eventually, however, anomalies and problems and inconsistencies with the prevailing paradigm start to accumulate, and then someone "thinks outside the box" and comes up with a totally new paradigm for their science, rejecting the assumptions of the old paradigm. This paradigm shift results in a **scientific revolution**.

Kuhn's example was the Copernican revolution in astronomy, where just one simple change in the basic model (putting the sun, rather than the Earth, at the center of the solar system) solved lots of problems with the old Ptolemaic geocentric system and led to a whole new worldview. Likewise, Newtonian physics transformed the fields of mechanics that were still stuck in the false notions of Aristotle. Einsteinian relativity revolutionized physics once again, since Newtonian mechanics does not apply in the realm of things moving near the speed of light.

Likewise, Darwinian evolution overthrew the old creationist notions of life, and biology has never been the same since. It's not clear whether there has been a true scientific revolution in chemistry, although some key ideas have been proposed, like Mendeleev's invention of the periodic table.

Unlike other sciences, geology underwent its scientific revolution very recently. The old paradigm long assumed that continents were fixed and stable, and the first real challenge to this idea came in 1915 when German meteorologist Alfred Wegener published *The Origin of Continents and Oceans*. But the idea was rejected for decades, until new data from marine geology and from studying the ancient magnetic fields of rocks on land accumulated in the 1950s to show that continents had indeed moved. In 1962 and 1963, a series of key discoveries launched the new paradigm of geology called **plate tectonics**, and geology hasn't been the same ever since. Most of the key discoveries were made by young scientists in the 1960s, and many of them are still alive and active. (I was fortunate to be trained in geology at a key moment in this transition. My undergraduate professors accepted the new ideas in the early 1970s, although all our textbooks were out of date. Then I was a graduate student at Columbia University and Lamont-Doherty Geological Observatory, where the American pioneers of the plate tectonic revolution were my professors.)

4.2.2 Plate Tectonics

The key discovery was the idea that the earth's surface is covered by **crustal plates** that move around on the fluid mantle beneath. They slide around on the curved surface of the Earth like pieces of eggshell on a hard-boiled egg. These plates are driven by large convection currents in the mantle which originate from heat rising from the Earth's interior, bringing up hot mantle plumes, and then other areas in the mantle where it cools and sinks down to the core–mantle boundary. The crustal plates have two basic types (Figure 4.5). **Oceanic crust** is relatively thin (only 10 km thick), dense, and made of basalt. **Continental crust**

FIGURE 4.4 (a) Plunging anticlines and synclines. (b) Geologic cross section of the Punchbowl Syncline looking west (south to the left, north to the right), showing the folded middle Miocene Punchbowl Formation (Tp) and the upper Paleocene San Francisquito Formation (Tss and Tsf) and its bounding faults. (c) Geologic map pattern of the Punchbowl Syncline in Devil's Punchbowl. The Punchbowl Formation (colored in orange, with the "Tps" label) forms a broad parabola, with the youngest beds in the middle and the oldest on the flanks. The axial plane is almost east–west, and the fold plunges to the west. (d) Satellite image of the Devil's Punchbowl, showing how the outcrop of the ridges of Punchbowl Formation also demonstrates a broad parabolic curvature pointing east and opening west. (e) A 180° panorama of the fold photographed standing on its axial plane. The beds in the left of the photo (to the north of the photographer) dip south toward the camera. The beds in the center (to the east of the photographer) dip west and lie along the axis; the beds to the right (to the south of the photographer) dip north toward the camera and are visible from the Punchbowl Visitors' Center. (f) Close-up of the fold axis.

Source: (a) Redrawn from several sources by P. Linse. (b) Courtesy of Pacific Section SEPM. (c–d) Courtesy of US Geological Survey. (e–f) Photo by the author.

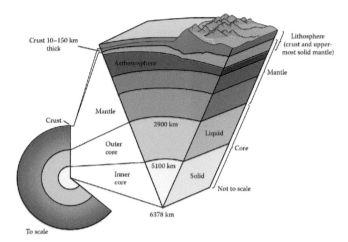

FIGURE 4.5 Diagram showing the Earth's layers and the distinction between *crust*, *mantle*, *lithosphere*, and *asthenosphere*.

Source: Courtesy of Wikimedia Commons.

is about 5–10 times thicker than oceanic crust (50–150 km thick on average) and is made of much less dense rocks, such as granitic rocks and gneisses. The sharp boundary at the base of the crust is called the **Mohorovicic discontinuity** ("Moho" for short). The uppermost semirigid part of the mantle moves with the overlying crustal plates, and together these two layers are called the **lithosphere**. Beneath the lithosphere is the next layer of the mantle (from about 150–200 km to about 400 km down), called the **asthenosphere**, where the mantle is semifluid; this is where the main convection currents that push and drag the plates along occur.

A good analogy for this process would be a pot of hot cocoa with marshmallows on a stove. The stove burners provide the heat (analogous to the heat coming from the Earth's interior), and the pot has warm currents of cocoa rising in some areas and cool cocoa sinking in others, like the convection currents in the mantle. At the top surface of the cocoa (especially as it cools) is a thin scum, which is a good analogy for the oceanic crust. The cocoa scum can easily be destroyed, melted, or sink back into the cocoa. The floating marshmallows are like continental crust. They can collide with each other and move around, but they never sink down into the cocoa. If the cocoa scum collides with the marshmallows, the scum sinks down into the cocoa.

The Earth's tectonic plates (made of both oceanic and continental crust) behave in a similar fashion. Three types of plate boundaries are possible (Figure 4.6). In a **divergent boundary**, plates pull apart and produce new oceanic crust by the process known as seafloor spreading (discussed later). There are only small earthquakes and volcanoes on these boundaries, so divergent boundaries are also called **passive margins**. The second type of plate boundary occurs when two plates collide, so it is called a **convergent boundary**. On most convergent boundaries, one of the plates is made of thin, dense oceanic crust, which sinks down beneath the other plate in a **subduction zone**. As it does so, the down-going plate generates most of the world's

earthquakes and also melts as it goes down to produce volcanoes, so this type of plate boundary is called an **active margin**. The third possibility is that two plates are neither converging nor diverging but sliding past one another. This kind of plate boundary is known as a **transform boundary**. It is typified by huge strike-slip faults, such as the San Andreas fault in California.

4.2.3 PASSIVE MARGINS

New crustal rock is produced on **mid-ocean ridges**, which are the main part of passive margins. The mid-ocean ridges were one of the key discoveries of the early days of marine geology in the mid-1950s and were the first key piece of the puzzle that led to plate tectonics. They are the longest range of mountains on Earth, running for many thousands of kilometers down the middle of the Atlantic, the eastern Pacific, across the Indian Ocean, and several other places, like the seams on a giant baseball. They are also some of the highest ranges of mountains on Earth (more than 4,500 km, or 15,000 ft), with a huge fault graben valley down their entire length deeper than the Grand Canyon. Yet no one knew they were there, or just how big they were, until the late 1950s, when Marie Tharp and Bruce Heezen first mapped the ocean floor and discovered their true size.

In the early 1960s, marine geologists discovered that the ocean floor pulls apart at the mid-ocean ridges, which is why there is a giant fault graben valley (Figure 4.7) down their entire length. As the plates pull apart, new magma wells up from gabbroic magma chambers below them and chills into new oceanic crust made of basalt. Thanks to many geophysical measurements, as well as observation in small submarines, scientists have witnessed this process over and over.

The magma flowing up through the extensional cracks in the graben valley hits seawater and then instantly chills into blobs of congealed magma called **pillow lavas** (Figure 4.8). Pillow lavas are produced as hot magma chills in contact with seawater. Hot new magma from below forces its way through cracks in the older chilled lava and then extrudes from the crack like a blob of toothpaste and quickly chills into a rounded, pillow-like shape. This amazing sight has been observed and filmed many times, and you can watch footage of it online with just a search for "pillow lava video."

Beneath the ocean floor layer of pillow lavas (Figures 4.7 through 4.9a), the magma that flows up through the vertical cracks and fissures to feed the pillows eventually congeals in those cracks to form hundreds of vertical slabs of basalt (Figure 4.9b) called **sheeted dikes**. Beneath the dikes is a gabbroic magma chamber that fed the dikes and pillows. When it finally cools, it forms a layered gabbroic magma chamber (Figure 4.9c). All three of these rocks together (pillow lavas, sheeted dikes, and layered gabbro) are the standard components of oceanic crust all over the world (Figure 4.7). Since oceanic crust makes up about 70% of the Earth's surface, they are the most abundant rock types on the planet.

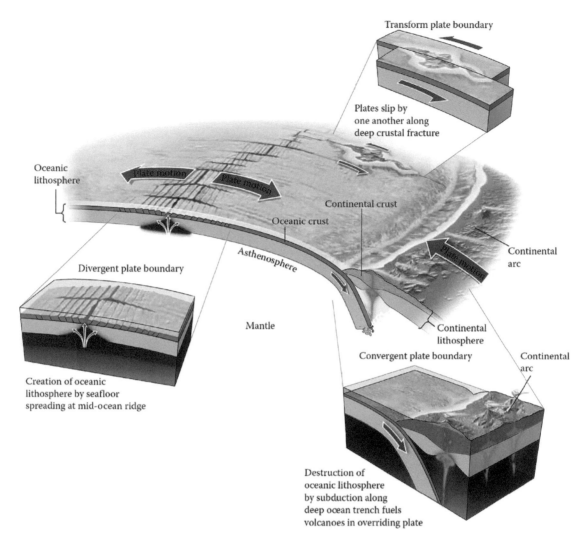

FIGURE 4.6   Diagram of the different types of plate boundaries.

Source: Courtesy of Oxford University Press.

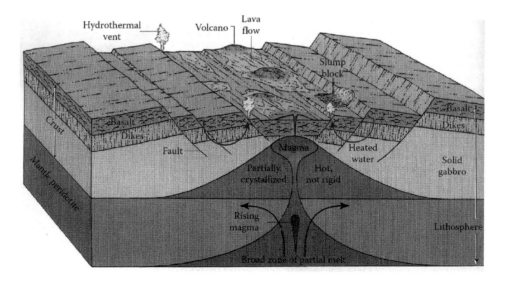

FIGURE 4.7   Diagram of the mid-ocean ridge, showing how it produces the three characteristic rocks of the ophiolite suite.

Source: Redrawn from several sources.

FIGURE 4.8 Pillow lavas. (a) Cracks open in old lava flows as hot magma from inside the flow oozes out into pillow-shaped blobs and then is chilled by seawater. (b) Ancient pillow lavas exposed west of Port San Luis Pier near Avila Beach.

Source: (a) Photo from NOAA. (b) Courtesy by D. Patton.

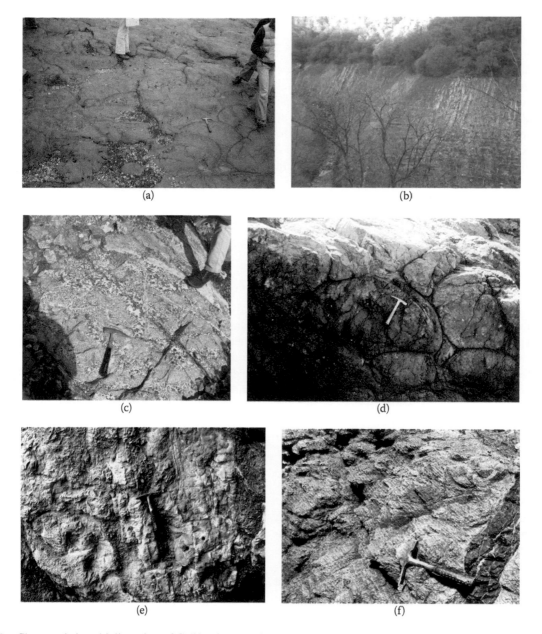

FIGURE 4.9 Characteristic ophiolite suites of California. (a–c) Smartville ophiolite complex on eastern foothills of the Sierra Nevada mountains: (a) pillow lavas, (b) sheeted dikes near Oroville Dam, and (c) coarsely crystalline layered gabbro from the magma chamber that fed the volcanics. (d—f) Point Sal ophiolite complex, Santa Barbara County: (d) pillow lavas, (e) sheeted dikes, and (f) layered gabbros.

Source: (a–c) Photos by E. Moores. (d–f) Photos by R. Stanley.

The association of these three distinctive rocks was named the **ophiolite suite** (Greek *ophis*, "snake," and *lithos*, "rock," since these gabbros and basalts were usually metamorphosed into serpentine). This peculiar threefold association was first recognized in the Alps by Alexandre Brongniart in 1813. In 1927, the association between pillow lavas, serpentized peridotites or gabbros, and the overlying marine sediments, such as ribbon cherts, was formally described by Gustav Steinmann, so it is often called the "Steinmann trinity." In 1892, Steinmann visited the Marin Headlands on the north end of the Golden Gate Bridge and realized that it matched the sequence of rocks he knew in Europe. Thus, California was a key place in discovering this association.

The association of pillow lavas, sheeted dikes, and layered gabbros, often capped by oceanic sediments, was found in many other mountain ranges in Europe (especially in the Alps and in the Macedonian region of Greece) and elsewhere (such as the island of Cyprus and in Oman in the Persian Gulf). No one could explain what it meant until plate tectonics came along in the 1960s and studies of active mid-ocean ridges showed they were producing ophiolites all the time. Today, we recognize ophiolites as slivers of oceanic crust that have been sliced off a down-going subducting oceanic plate and stuck onto the land. They are very common in the Coast Ranges, the Klamath Mountains, and the Sierra Nevada mountains of California, and we will look at them in greater detail in later chapters.

In most places, the mid-ocean ridge is completely beneath the ocean (often 3–4 km deep) and can only be studied by submersibles, as they were when the RV *Alvin* first dived on the mid-ocean ridges in 1977. In one place, however, the mid-Atlantic ridge rises above sea level: Iceland. That is because Iceland also sits on a mantle "hot spot," which raises the whole mid-Atlantic ridge off the deep seafloor. With the seawater drained away, you can see a giant rift valley that runs right down the middle of Iceland (Figure 4.10). It forms a narrow chasm with faulted basaltic cliffs on the edges, and the central valley dropped down on the extensional faults as Iceland steadily pulls apart. Every once in a while, one of the rift volcanoes erupt, like the eruption of Eldfell volcano on the island of Heimaey in 1973. This is what most mid-ocean ridges might look like if all the water were drained off them.

Over millions of years, the cooled oceanic crust pulls away from the center of the mid-ocean ridge and new magma wells up the cracks to replace it. From this point onward, all the oceanic crust gradually spreads and pulls away from the mid-ocean ridge like a gigantic pair of conveyer belts. As the old oceanic crust spreads away, it also sinks because it is cooling and shrinking.

Eventually, the oceanic crust reaches the original edge of the passive margin, which is a thick **passive margin wedge** (Figure 4.11). The passive margin wedge is a thick prism of sediments forming at the edge of the continental crust where it meets the oceanic crust. The wedge tends to thin toward the onshore direction and thicken rapidly

FIGURE 4.10 The mid-Atlantic rift valley, which runs down the center of Iceland, shows two plates pulling apart. In Thingvellier National Park, this is Flossja Canyon, with tall cliffs of basalt on each side and a down-dropped fault valley in the middle. Here it has a lake at the bottom, but in many places in Iceland, you can walk right down the center of the rift valley. The North American plate is on the left, and the Eurasian plate is on the right.

Source: Courtesy Wikimedia Commons.

in the offshore direction (Figure 4.10). It is composed of thousands of meters of thickness of shallow marine shelf sandstones and shales on the landward side and deep-water shales and cherts on the deep-water edge. It continually sinks and subsides because it is tied to the old oceanic crust far from the mid-ocean ridge, which continues to shrink and sink due to thermal cooling.

4.2.4 ACTIVE MARGINS

Active margins have completely different behavior and characteristics from passive margins of tectonic plates. As stated earlier, they are caused when two plates collide, and the usual result is that oceanic plate slides down beneath the other plate to make a subduction zone (Figure 4.12). As the down-going slab scrapes beneath the overlying slab, it causes many huge earthquakes. The subduction zones of the world are the most seismically active regions in the world, with all the world's biggest quakes.

Then, as the subducting plate plunges deeper into the mantle, the mantle rocks above the slab begin to melt due to the abundance of water and volatiles in the weathered ophiolitic basalts, which lowers their melting temperature. This molten rock rises up through the overlying plate and creates a chain of volcanoes, making the "Ring of Fire" around the Pacific. Such a volcanic chain is often called an "island arc," or just called "arc volcanoes," because many of them form chains of islands and have an arcuate shape because the globe is spherical.

The boundary between the plates, where one plate plunges beneath the other, is a deep valley on the ocean floor known as an **oceanic trench**. Trenches form above nearly every subduction zone on the planet, and they are

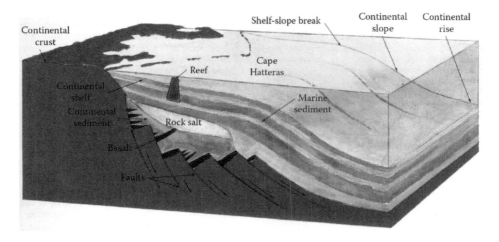

FIGURE 4.11 Passive margin wedge, based on the modern margin of the Atlantic Coast of North America. Shallow marine continental shelf sediments form the top of the wedge, and deep-water slope and rise turbidites and shales form offshore. Typically, the earliest stage of the passive margin, the rift valley graben, can be found deeply buried under the younger sediments.

Source: From Prothero, D. R., and R. H. Dott Jr., *Evolution of the Earth* (8th ed.). McGraw-Hill: New York, 2010.

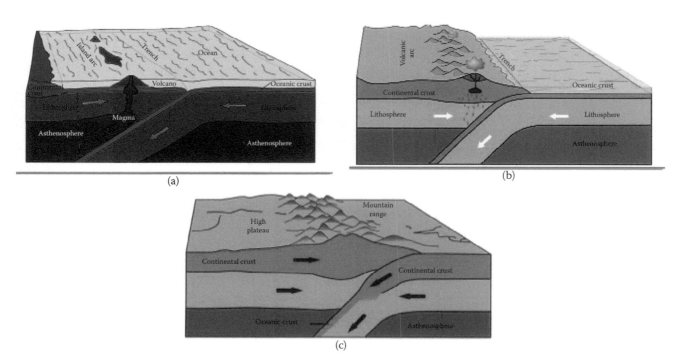

FIGURE 4.12 Convergent boundaries. (a) Japan-style arc formed when oceanic plate plunges beneath another oceanic plate. (b) Andean-style arc, when oceanic plate plunges beneath continental crust. (c) Himalayan-style boundary, where two continental blocks collide with one another and form uplifted mountain ranges.

Source: Redrawn from several sources.

by far the deepest spots in the ocean. The Marianas Trench is the world record holder at 11,034 m (36,201 ft) below sea level, deep enough to hold Mt. Everest and several other mountains as well. But there are many other trenches that are 8,000–10,000 m (26,000–33,000 ft) deep around the world. Others, like the Cascadia Trench off the coast of British Columbia, Oregon, and Washington, are nearly completely filled with the sediment eroding down from the big rivers in the region, like the Columbia River.

Between the trench and the volcanic arc are two other important parts of an active margin complex that we see often in California. The boundary zone between the two plates is marked by an **accretionary wedge**, or accretionary prism. Accretionary wedges are amazing geological phenomena. They are formed when slices of the subducting slab are scraped off the plate as it goes down and plastered up against the bottom of the overlying slab (Figure 4.13). Consequently, the rocks are continuously added to the

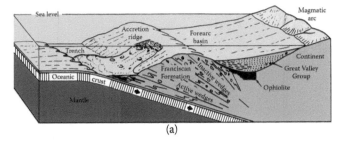

(a)

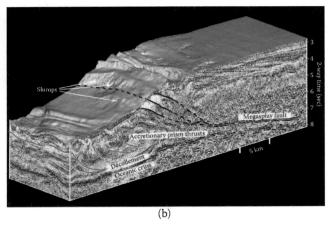

(b)

FIGURE 4.13 Accretionary wedge and forearc basin. (a) Diagram of the geometry of the accretionary wedge (Franciscan) and forearc basin (Great Valley) in California. (a) Modified from Prothero and Dott, 2010. (b) Three-dimensional seismic cross section of a modern accretionary wedge, showing the intense shearing and deformation of the tectonic slices as they are chopped off the down-going slab. The oldest slices tend to be pushed to the top as younger slabs are underplated beneath them.

Source: From Moore, G. F., et al., *Science* 138 (5853): 1128–1131, 2007.

bottom of the stack, and they get *older* as you go to the top. This is the reverse of normal superposition, where old rocks are at the bottom and young rocks on the top (see Chapter 3).

These rocks undergo a tremendous amount of shearing and slicing and dicing and being run through the blender, so they no longer have any continuity or bedding or remnants of their original order. They are so mixed up that we use the French word *mélange* for them ("mixture"). Several different and unique rock types are found almost exclusively in mélange from accretionary wedges:

1. *Old oceanic sediments.* The most commonly expected rock to get scraped off are the sediments that used to lie on the old plate before it sank down the subduction zone. These are deep-ocean shales with chert layers in them (**ribbon cherts**) (Figure 2.8d), along with shreds of sandstones that flowed down in submarine gravity slides (**turbidites**). Rocks like these are everywhere in the Franciscan mélange (Chapter 12).
2. *Slices of oceanic crust (ophiolite).* Sometimes, not only the sedimentary cover of the down-going

slab is scraped off but also big chunks of the oceanic crust itself. These are the ophiolites we have already discussed earlier (Figure 4.8).
3. *Blueschist.* Subduction zones are the only places in the world where you find a peculiar metamorphic rock known as blueschist (Figure 4.14), discussed in Chapter 2. The subducting plate may be 50–100 km down into the crust, so it experiences extremely high pressures. Yet the cold, old oceanic crust in the slab resists heating up and melting for a very long time as it sinks into the mantle, so the region around it is high pressure without being the normal high temperature. Thus, blueschists are made deep in subduction zones and then sliced off and pushed up into the accretionary wedge as newer slices are added beneath them.

Finally, between the accretionary wedge and the volcanic arc is a trough-like basin that warps downward. It is usually drowned by seawater, so it fills with marine sandstones and shales, many of which are derived from the volcanic arc above it. This is known as the **forearc basin** (Figures 4.13a and 4.15). We discuss the Great Valley forearc basin in greater detail in Chapter 10. Many of the volcanic arcs have another basin behind them, called a **back-arc basin**.

There are three possible configurations for convergent margins:

1. If one oceanic plate slides beneath another oceanic plate, it is called a **Japan-style volcanic arc**. Not only is this setting well studied and exemplified, but also many of the other volcanic chains around the Pacific Ring of Fire are Japan-style arcs, including the Aleutians, the Philippines, the Indonesian arc, and the Tonga–Kermadec arc. Japan-style arcs tend to be mostly chains of islands, with only a small forearc basin. The back-arc basin, if present, is completely under the ocean and fills with marine sediments.
2. If an oceanic plate slides beneath a continental plate, then it is called an **Andean-style volcanic arc**. The Andes (Figure 4.15) are the most famous example, but the Cascade Range in California, Oregon, Washington, and southern British Columbia is another. Because the overlying plate is continental crust, the volcanic chain erupts on land, and the forearc basin may be filled with either seawater and marine sediments or sometimes a nonmarine basin. The accretionary prism also tends to be uplifted, forming a coastal chain of islands (such as the islands off the Pacific Coast of South America, or offshore from the Cascades). The back-arc basin is nearly always filled with nonmarine sediments (rivers, deltas, and lakes), and (in the case of the Andes) there is usually a big zone of thrust faults just behind the arc that push out over the back-arc basin.

(a) (b)

FIGURE 4.14 Blueschists are high-pressure, low-temperature metamorphic rocks only produced in subduction zones. (a) Outcrop of a classic blueschist, with its distinctive deep blue color, near Jenner. (b) Tightly folded and sheared Catalina blueschist from the Catalina mélange, north end of Descanso Beach, Catalina Island.

Source: (a) Photo by C. A. Lee. (b) Photo by the author.

3. Beyond oceanic–oceanic and oceanic–continental collisions, the third possible scenario is the collision between two continental plates. Since continental crust is light and buoyant ("marshmallows in cocoa"), it cannot subduct, so instead there is a huge collision zone between the plates. This produces enormous folded and thrust-faulted mountains, which are often the highest mountains in the world. Of course, the best modern example is the Himalayas. For this reason, a continental–continental collision is often called a **Himalayan-style margin**, although the Alps and most of the mountain ranges across Europe between them (such as the mountains in the Balkans, Turkey, Iran, and Afghanistan) are part of the same crush zone. The Alps and Himalayas started forming about 50 Ma, when Indian collided with the belly of Asia, and Africa began to push up against southern Europe. These mountains are still rising as the plates continue to push against one another after 50 Ma, as evidenced by the frequent deadly earthquakes in western China, Nepal, Pakistan, Afghanistan, Iran, Armenia, Turkey, and Greece.

4.2.5 TRANSFORM MARGINS

We have seen how plates can converge or diverge, but there is a third possibility: a plate boundary where the plates are neither spreading nor colliding. Instead, the plates slide past one another in a gigantic strike-slip fault. The most familiar and best-studied example of a transform is the San Andreas fault, which we will discuss at length in Chapter 11. But there are others, including the Queen Charlotte transform on the northern coast of British Columbia and the Alpine transform that runs down the spine of the two main islands of New Zealand.

4.3 SUMMARY

Plate tectonics revolutionized geology only 50 years ago, and nothing has been the same since. Sadly, many geologic writings about California and books on California geology do only the minimal effort to provide a plate tectonic explanation for the geological features that make California distinctive. In this book, we use the latest plate tectonic models wherever possible to give an up-to-date interpretation and explanations for many of the features in the Golden State that were long unexplained.

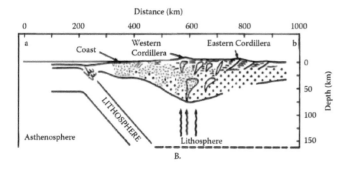

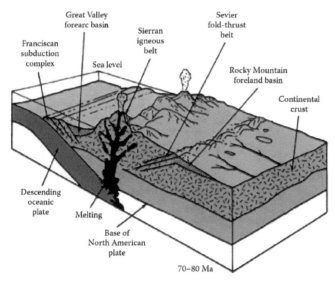

FIGURE 4.15   The accretionary wedge and forearc basin in front of the volcanic arc. At the top is a true-scale cross section of the modern Andean forearc basin and accretionary wedge in Chile and Argentina. On the bottom is the equivalent tectonic geometry for California during the Mesozoic, when the Great Valley was the forearc basin for the Sierran volcanic arc, and an accretionary wedge developed that is part of the modern Coast Ranges.

Source: From Prothero, D. R., and R. H. Dott Jr. *Evolution of the Earth* (8th ed.). McGraw-Hill: New York, 2010.

RESOURCES

Cox, A., and Hart, R.P. 1986. *Plate Tectonics: How It Works*. Wiley-Blackwell, New York.

Davis, G.H., Reynolds, S.J., and Kluth, C.L. 2011. *Structural Geology of Rocks and Regions*. John Wiley, New York.

Kearney, P., Klepeis, K.A., and Vine, F.J. 2009. *Global Tectonics*. Wiley-Blackwell, New York.

Molnar, P. 2015. *Plate Tectonics: A Very Short Introduction*. Oxford University Press, Oxford.

Moore, G.F., Bangs, N.L., Taira, A., Kuramoto, S., Pangborn, E., and Tobin, H.J. 2007. Three-dimensional splay fault geometry and implications for tsunami generation. *Science, 138*(5853), 1128–1131.

Prothero, D.R., and Dott, R.H., Jr. 2010. *Evolution of the Earth* (8th ed.). McGraw-Hill, New York.

Van der Pluijm, B.A., and Marshak, S. 2003. *Earth Structure: An Introduction to Structural Geology and Tectonics*. W.W. Norton, New York.

Wegener, A. 1966. *New English translation of 4th revised edition of Die Entstehung der Kontinente und Ozeane*, published in 1929 by Friedrich Vieweg and Sohn. Dover Publications, Inc., New York.

VIDEOS

Planet Earth, Episode 1: The Living Machine: www.youtube.com/watch?v=kvz2hkWRyRQ

Plate Tectonics

www.youtube.com/watch?v=1-HwPR_4mP4
www.youtube.com/watch?v=KCSJNBMOjJs

Pillow Lavas

www.youtube.com/watch?v=xsJn8izcKtg
www.youtube.com/watch?v=o1Y2mu0qrus
www.youtube.com/watch?v=dpz4fXH7nuo
www.youtube.com/watch?v=n3aAi-9YWT4

5 Earthquakes and Seismology

I was awakened by a tremendous earthquake, and though I hadn't ever before enjoyed a storm of this sort, the strange thrilling motion could not be mistaken, and I ran out of my cabin, both glad and frightened, shouting, "A noble earthquake! A noble earthquake" feeling sure I was going to learn something.

—John Muir
The Wild Muir: Twenty-Two of John Muir's Greatest Adventures

God will break California from the surface of the continent like someone breaking off a piece of chocolate. It will become its own floating paradise of underweight movie stars and dot-commers, like a fat-free Atlantis with superfast Wi-Fi.

—Laura Ruby
Bad Apple

5.1 EARTHQUAKE MYTHS

Perhaps no other phenomenon in geology is so burdened with myths and misconceptions than earthquakes. Thanks to generations of stupid, misleading, and badly done movies and TV shows, most of what people think they know about earthquakes is wrong. Earthquake fault lines are not giant chasms in the ground and don't have magma at the bottom, as you see in movies like *San Andreas* or the first *Superman* movie with Christopher Reeve. They tend to form long, straight valleys that are hard to detect until there is motion along the fault. The chasms that actually form during earthquakes happen due to landsliding, usually far from the fault line itself.

California will not fall into the sea, as some believe. The portion of California west of the San Andreas fault is sliding north toward Alaska each time the fault moves, but it is not sinking (or rising) more than a few centimeters in an earthquake. Another example of an urban myth: there is no such thing as "earthquake weather." Earthquakes happen at any time of year and in any weather and any time of day. The slip on the fault line cannot be affected by temperature at the surface or any other weather effect, because the fault plane is usually many kilometers down in the crust, while the daily fluctuations of temperature due to weather only penetrate down a few meters at most from the surface. (This is how desert animals survive by burrowing down into the cool earth below the surface.)

Earthquakes also cause more fear and panic in people than any other natural disaster. Yet your chances of being killed in an earthquake in the United States are less than being struck by lightning or bitten by a snake. Over the past century, fewer than six Americans per year (on average) have died in quakes, while many times that number die from disasters like floods, hurricanes, tornadoes, and landslides. Most people are far more afraid of quakes than the truly deadliest natural disasters of all: winter and summer. If you just look at the numbers, hundreds of Americans die each year from killer heat waves and dangerous blizzards and cold, the deadliest natural disasters of all (even though most people don't realize it).

If earthquakes are not really deadly to Americans, why do people fear them way out of proportion to their real risk? In part, it's because we hear about lots of deaths from earthquakes in the underdeveloped world, where they build their structures out of unreinforced brick and stone, the deadliest possible materials in earthquake country. Such construction is illegal in California since the state legislature passed the Field Act right after the 1933 Long Beach quake. In less-developed countries, where poor people have no access to wood-frame housing (the safest type of construction) or steel structures, they rebuild their houses out of the same unreinforced bricks and stone and mortar that came down in the last quake—resetting their death traps to kill more people in the next quake.

Another reason for this irrational fear is the deep psychological effect of having the ground beneath your feet, *terra firma*, suddenly stop being so firm and steady and reliable. But probably the biggest reason is that (unlike most other natural disasters) earthquakes are completely unpredictable—and probably always will be. Thanks to weather satellites and meteorological monitoring, we can carefully track hurricanes, blizzards, and heat waves and issue tornado warnings. Earthquakes happen without warning, catch us by surprise, and then literally shake our world down to its foundations. This is the kind of thing that stirs deep-seated psychological fears that no amount of rational argument about the low risk to human life can ever dispel.

5.2 WHAT ARE EARTHQUAKES?

Earthquakes were long attributed to supernatural beings, or the wrath of the gods—so immense and powerful and destructive did they seem. The great Lisbon quake of 1755 not only caused immense death and destruction but also profoundly altered the religious and philosophical landscape of Europe. The practical, enlightened prime minister Pombal, who helped Portugal recover from the quake, also commissioned the first efforts to collect data about how widespread the damage was, and what people felt during the quake.

But the event that marked the birth of modern seismology was the San Francisco earthquake of 1906. Even as the

DOI: 10.1201/9781003301837-5

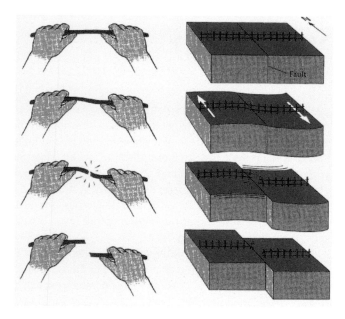

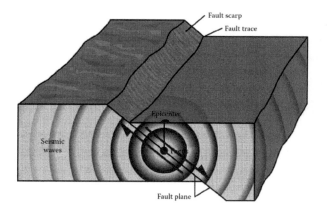

FIGURE 5.2   Epicenter, focus, and hypocenter, showing how the epicenter might not be on the fault line if it dips steeply.

Source: Courtesy of Wikimedia Commons.

FIGURE 5.1   Elastic rebound. When a stick is put under increasing stress, it begins to bend and crack but mostly builds up the stress energy, until it breaks and snaps rapidly apart. Similarly, the Earth's crust builds up stress until it snaps and moves rapidly to release the stress.

Source: Courtesy of Wikimedia Commons.

city was in ruins and burning to the ground, early seismographs at Stanford and Berkeley had recorded it. Pioneering geologist Grove Karl Gilbert rode around the Bay Area, documenting the surface ruptures. The big multiauthor scientific report that came out afterward laid the foundations for modern seismology.

One of the first major conclusions of that report was that earthquakes are caused by **elastic rebound** on fault lines (Figure 5.1). The crust around a fault builds up stress, often for decades, and the strain even causes the ground to bend and warp in some cases. Eventually, the friction on the fault line is not enough to resist all that stress, and the fault slips and the ground snaps back to a less-strained configuration. It's analogous to breaking a stick. The wood will bend and strain for some time as you put more stress on it, until finally you overcome its mechanical strength and it fails and breaks. Then it springs back to an unstrained configuration after it breaks. Meanwhile, the "crack" of the stick you hear is the energy being released as the stick breaks, and it's analogous to the energy release (some of it in sound waves) that happens after a quake.

The actual place on the fault plane where the slippage occurs is called the **focus** or **hypocenter**. Most of the time, however, earthquakes are plotted on two-dimensional maps, so they are plotted as an **epicenter**, the spot on the map immediately above the hypocenter. If the fault plane is vertical, then the epicenter will lie above the hypocenter, but if the fault plane is dipping (Figure 5.2), then the epicenter might plot nowhere near the spot where the fault

meets the Earth's surface. This is important when interpreting plots of many epicenters on a map, because they may not actually mark the fault line.

Energy propagates away from the quake in huge ripples in all directions through the ground. The energy released emerges as a series of different waves, which all pass through the Earth's crust, and some even go through the mantle and core (Figure 5.3). In really big quakes, the energy causes the entire Earth to vibrate, like the ringing of a bell. The fastest of the waves to emerge are called **P-waves** (primary waves), because they are the first to arrive and be detected on seismographs. P-waves produce pulses of compression and extension in the direction they travel, just like sound waves passing through air. They travel extremely fast (about 5.5 km/s, or about 15,000 mph, in crustal rock, and much faster in the mantle and core), although they travel slower through less-dense media, like water (only 1.5 km/s).

Originating at the same time but spreading out much more slowly are **S-waves** (secondary waves), so called because they are the second wave to arrive on the seismograph after the P-waves. S-waves have a vertical shearing motion, like snapping a rope up and down. They move about 3.0 km/s through rock (about half as fast as P-waves) but will not pass through fluids. This is important, because the fact that S-waves passing through the Earth's interior are blocked by the core tells us that the outer core must be fluid.

Even slower are **surface waves**, which travel through only the top layer of the crust. They are much slower than P-waves and S-waves, which travel through the body of the Earth's interior. Surface waves arrive at a seismic station later than do the P-waves and S-waves and only travel relatively short distances across the Earth's surface. One kind of surface wave, the **love wave**, travels in a side-to-side wiggle like the motion of a snake. The other, the **Rayleigh wave**, causes the surface rocks and soils to go through a shallow vertical looping motion, like the orbital motion of water particles in surf (Figure 5.3).

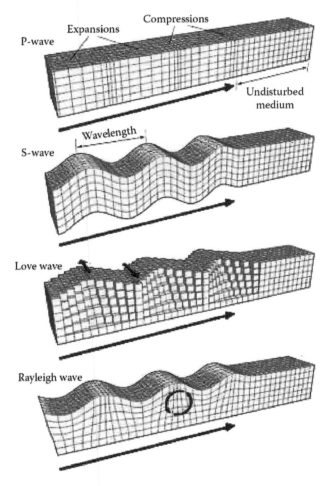

P-wave — Expansions — Compressions — Undisturbed medium

S-wave — Wavelength

Love wave

Rayleigh wave

FIGURE 5.3 The four types of seismic waves. P-waves are body waves that move by a pulse of compression and extension traveling in the direction of motion of the wave, just like sound waves in air. S-waves are body waves that have a vertical shear motion. The two types of surface waves travel only through the shallow crust. Love waves have a sinuous snake-like side-to-side motion, while Rayleigh waves move in small shallow orbits like the waves in the ocean (see Chapter 19).

Source: Courtesy of Wikimedia Commons.

5.2.1 MEASURING EARTHQUAKES

Ground motions caused by earthquakes are detected by an instrument called a **seismograph**. It is basically composed of a suspended weight, usually hanging from a spring to measure motion up and down in the vertical plane, or a spring-controlled pendulum arm to measure motion in the horizontal planes. When the ground (and the frame of the seismograph) vibrates with the quake, the inertia of the suspended weight keeps it from moving at all. Attached to the weight is typically a long needle with a pen at the tip, which will produce the wiggly line on the rotating drum covered by paper that is so familiar from news broadcasts about quakes (Figure 5.4). Even though it appears that the needle is kicking back and forth on a piece of paper, in reality the needle (attached to the suspended weight) is holding still, and the paper on the seismograph drum (and the rest of

the world attached to it) is actually wiggling back and forth beneath the stationary pen.

The old drum-and-paper seismographs are now antiques used for display only. Modern seismographs contain a suspended magnetic weight that bounces up and down around a coil of wire with a charge in it, so their motions can be converted to an electronic digital signal that is much easier to analyze and plot on a computer and store as digital files. No more storing shelves and shelves of old paper traces off the drum!

Before the invention of the seismograph, there was no objective or quantitative way to measure the strength of an earthquake. For that reason, the only measurement scale available for a long time was a more qualitative measure known as the **Mercalli scale of intensity** (Table 5.1). It ranks quakes by the amount of damage to structures, the amount of ground motion, and the way people perceive the quake, which is obviously not very precise. The Mercalli intensity is given by Roman numerals from I to XII, but the criteria are quite vague. For example, a "weak" quake of intensity II is "felt only by a few persons at rest, especially on upper floors of buildings." A "very strong" quake of Mercalli intensity VII is "damage negligible in buildings of good design and construction; slight to moderate in well-built ordinary structures; considerable damage in poorly built or badly designed structures; some chimneys broken." And an "extreme" quake of Mercalli intensity XII is "damage total. Waves seen on ground surfaces. Lines of sight and level distorted. Objects thrown upward into the air."

By contrast, the famous **Richter magnitude scale** (invented by Caltech seismologist Charles Richter in the 1930s) gets its readings from the measurement of the highest needle movement on a seismograph, then scaling it for the distance from the source. It is a logarithmic scale, so the difference between an M7 and M8 is a peak 10 times higher than the lower value, or about 31 times as much energy released between an M7 and an M8. Most M3 quakes are barely felt by people at all unless they are right below you, but M5–M6 can cause extensive damage. Anything about an M7 is a "major earthquake," and anything above M8 is a "great earthquake." There have only been a few quakes this big in recent history, such as San Francisco 1906 (M = 8.3), Alaska 1964 (M = 8.6), and Chile 1960 (M = 8.8, the largest quake ever recorded). Richter magnitudes don't go above 9.0, the elastic limit of rocks, so if you hear a movie or TV show say something about "10 on the Richter scale," that's nonsense.

Since the 1970s, however, the Richter scale has been replaced by the **moment magnitude scale** (M_w), which uses the measurement of the largest peak on the seismogram but also takes into account the strength of the rocks and the length and depth of the rupture. Bigger, stronger quakes rupture more rock and stronger rock, something that is not apparent by just the record on the seismograph alone. Thus, certain really large quakes get different results between the two scales. The 1964 Alaska quake was a Richter M = 8.4, but its M_w = 9.2 (much stronger rocks in

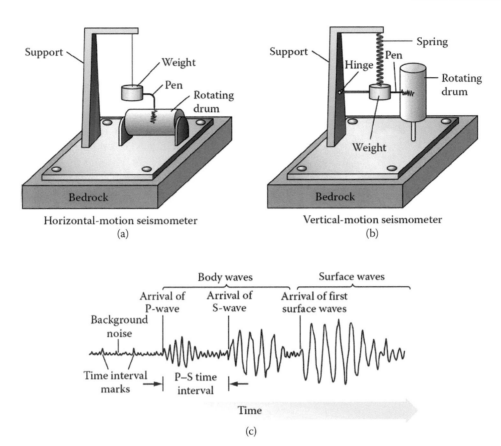

FIGURE 5.4 How a seismograph works. They are built around a suspended weight that stays still during a quake due to its inertia. The earth (and the rest of the seismograph) move with respect to the stationary weight, making the wiggly lines on the seismic tracing.

Source: Courtesy of Oxford University Press.

TABLE 5.1
Mercalli Scale of Intensity

Intensity	Definition
I	Detected only by sensitive instruments.
II	Felt by a few persons at rest, especially on upper floors. Delicate suspended objects may swing.
III	Felt noticeably indoors; not always recognized as an earthquake. Standing automobiles rock slightly. Vibration similar to that caused by a passing truck.
IV	Felt indoors by many, outdoors by few; at night some awaken. Windows, dishes, doors rattle. Standing automobiles rock noticeably.
V	Felt by nearly everyone. Some breakage of plaster, windows, and dishes. Tall objects disturbed.
VI	Felt by all; many frightened and run outdoors. Falling plaster and damaged chimneys.
VII	Everyone runs outdoors. Damage of buildings negligible to slight, depending on quality of construction. Noticeable to drivers of automobiles.
VIII	Damage slight to considerable in substantial buildings, great in poorly constructed structures. Walls thrown out of frames; walls, chimneys, monuments fall; sand and mud ejected.
IX	Considerable damage to well-designed structures; structures shifted off foundations; buildings thrown out of plumb; underground pipes damaged. Ground cracked conspicuously.
X	Many masonry and frame structures destroyed; rails bent; water splashed over banks; landslides; ground cracked.
XI	Bridges destroyed; rails bent greatly; most masonry structures destroyed; underground service pipes out of commission; landslides; broad fissures in ground.
XII	Total damage. Waves seen in surface level; lines of sight and level distorted; objects thrown into air.

the Alaska subduction zone), while the 1906 San Francisco quake was a Richter M = 8.3, but M_w = 7.9 (much weaker rocks in the San Andreas fault zone). The difference in the scales is only apparent in the largest quakes (above M or M_w = 7); for smaller quakes, the Richter and moment magnitudes are virtually the same.

One of the big problems with comparing the Mercalli scale with the two seismograph-based scales is that *they are not measuring the same thing*. The Mercalli scale is dependent on what types of structures are on the ground and what the soil and bedrock are made of, but the seismograph-based scales are not influenced by these factors. Buildings sitting on soft valley fill will vibrate much longer and more strongly (like a bowl of gelatin) than buildings sitting on hard bedrock; these experience one quick shock and they're done. This difference in response to the same energy of seismic waves can produce a lot more damage and change the Mercalli scale reading, even though the Richter and moment magnitudes are the same for every building. During the 1985 Mexico City quake, for example, the actual fault line was out on the Pacific Coast, and the shaking there was strong but ended quickly. However, 400 km away in Mexico City, the shaking was much stronger and lasted longer, and the damage much more severe, since the city is built on ancient lake beds that kept vibrating like a bowl full of gelatin for minutes after the quake first arrived. This same quake had several different Mercalli intensities, but only one Richter magnitude, so any diagram or table that attempts to convert Mercalli intensity to Richter or moment magnitude is false and misleading.

5.3 EARTHQUAKE PREDICTION

One of the longstanding goals of seismology has long been to predict earthquakes and thus give us some warning that would make them less dangerous. However, the entire effort is fraught with problems. This has been true ever since the beginning, and there are still many con artists and quacks and frauds out there who claim to have successfully predicted a quake. Once you look more closely, however, you will find that either their prediction was so vague as to be useless or they made so many predictions (most of them wrong) that sooner or later one was bound to be right by sheer numbers and chance.

In the 1970s, seismologists thought they had the problem solved. A new idea about the behavior of rocks called **dilatancy theory** offered great promise that we finally understood how rocks behaved before a quake. Scientists placed all sorts of instruments (seismographs to measure foreshocks, plus strain gauges, electrical conductivity meters, and many others) in hopes of finding reliable precursors to earthquakes. In 1975, Chinese seismologists used these precursors and successfully predicted a quake in Haicheng, in northern Manchuria, and everyone thought reliable quake prediction was about to become a reality. However, just a few months later, a huge quake struck Tangshan, in Sichuan Province, China, killing 650,000 people—and it

had no precursors. Since then, repeated failed attempts to find reliable precursors have discouraged seismologists from ever making bold statements that quake prediction is "just around the corner." The lesson in China is that no two quakes are alike or give the same precursors—and many have no precursors, so prediction is truly impossible in any real, useful sense of hours to days to weeks before a quake.

Of course, the fact that the P-wave arrives a few seconds to tens of seconds earlier than the S-wave is used to put quake sensors and alarms into modern buildings so that you would get a few seconds' warning if you're far enough from the epicenter. This happened with the Sendai quake in Japan in 2011. The P-wave arrivals set off alarms in Tokyo, and people were able to brace and prepare a few seconds ahead of the shaking. However, it did no good for the people close to the epicenter, since both the P- and S-waves arrive at about the same time.

Many people have noted the strange behavior of animals before quakes. Although lots of research has been done on this, there is no reliable way to use it to predict earthquakes. It has no basis in science, except possibly that animals are more sensitive to P-waves than humans are and might give a few seconds' warning. The problem lies in the fact that animals exhibit strange behavior for many reasons all the time, so it's not a practical way to predict quakes. People often use hindsight bias to remember animal behavior just before a quake and forget all the other instances of strange behavior that didn't precede a big event.

More realistic hopes are being tied to long-term prediction: probabilities for a quake in a given region over years. They aren't helpful in deciding when to take cover, but they are much more scientifically sound and successful than the long, sad record of failed attempts at short-term prediction. Seismologists plot the quake activity in major seismic zones and look for areas that are quiet—"too quiet," as they say in the movies. These regions are called **seismic gaps** and are the ones considered to be overdue for a quake. The 1989 Loma Prieta quake occurred in a seismic gap that was predicted and published a year earlier. The long quiet area of the 1857 Fort Tejon quake from Parkfield to San Bernardino is at least 159 years since its last quake and widely considered to be overdue. As discussed further in Chapter 11, this stretch has a recurrence interval of about 137 ± 8 years, and a later study suggested a recurrent interval of 145 ± 8 years. Consider that in 2017 it will have been 160 years since the 1857 Fort Tejon quake, and you can see why seismologists are very worried about this seismic gap. When it does slip, it will probably be offset about 10 m (33 ft) in seconds, just as it did in 1857, and be the "big one" that all Southern Californians have awaited for so long.

Another problem with earthquake prediction in the United States, however, is not scientific but legal. Any seismologist who makes too specific a prediction is almost certain to get sued, whether they, he, or she is right or wrong. If you make a prediction and are wrong, people will sue you for all the trouble and money that your prediction cost them. Even if you are right, someone will sue you for not predicting it closely enough to allow them to prepare, or some other failed aspect of your prediction.

This is no laughing matter. In Italy, an unqualified lab technician issued a "prediction" using the lab's name, based on no evidence, and people believed him. His superiors disavowed his prediction, but then a quake did occur (not when or where the lab technician predicted), as they often do in Italy. The seismologists were then taken to court and convicted for not predicting a quake that did happen, something that horrified the entire global scientific community.

5.4   FLUID INJECTION AND EARTHQUAKES

Another controversial and politicized issue about earthquakes is whether they can be triggered by human activity, especially by pumping fluids underground (a topic known as **induced seismicity**). It is thought that injecting fluids into a fault zone increases the pore pressure between the rocks and even inside the pores between individual mineral grains and lubricates them, allowing the fault to move more easily and more often. This was first observed as early as 1932, when it was noticed that when a reservoir level in Algeria was at its highest, there were frequent small earthquakes, but when the reservoir level was low, seismicity stopped. It suggested that high water levels in the reservoir increased the pore pressure in the groundwater and helped lubricate the faults. This phenomenon has since been observed in lakes and reservoirs in Russia, China, India, Italy, Zambia, Lesotho, Tajikstan, and possibly even behind Oroville Dam in California.

Another demonstration of the effects of changing fluid pressure and affecting earthquakes was accidentally demonstrated at the Rocky Mountain Arsenal northwest of Denver, built during World War II. In 1961, seismographs noticed that every time the Army pumped waste fluids down the wells on the base, small earthquakes increased, and when they pumped fluids out of the water table, the quakes stopped. In 1992, the site was closed and designated a Superfund site, because the US Army had pumped so many nasty chemicals into the ground that it was an environmental disaster. It is now a wildlife refuge, since it's unsafe for humans to live there, even though the Denver suburbs have grown and completely surrounded it.

In recent years, the issue of induced seismicity has come up in the context of fracking, or by fluid injection of waste fluids from drilling. Texas, southern Kansas, Colorado, Ohio, but especially Oklahoma have now become some of the most seismically active states in the union because of fluid injection triggering many strong earthquakes. For a long time, the oil companies tried to deny it, but the science is indisputable (Ellsworth, 2013), and even the US Geological Survey has officially established a link between fluid injection in oil fields and small quakes.[1] Most of the quakes are small (magnitude < 3.0), but some as big as 5.8 have occurred in Oklahoma (near Pawnee, Oklahoma, in 2016). Thanks to the power that the oil lobby has over Oklahoma politics, there has been little response or regulation of the issue, but if enough quakes of M = 5.8 or bigger happen, Oklahoma citizens may finally rise up and vote against the powerful oil lobby and the politicians they control.

How about California? For a long time, there was no big impetus to study the problem, and no one could point to a quake directly triggered by fluid injection by oil companies. Part of the problem, however, is that almost all parts of California have small quakes all the time, so they would be hard to detect. In a previously seismically quiet state like Oklahoma, the oil-related quakes are obvious, but not so in California. However, more recent efforts have found ways to separate and analyze natural small quakes and fluid-injection-triggered quakes.[2] Several studies have shown that fracking done in the Salton Sea area has produced earthquakes,[3] as well as the Salinas Basin.[4] And research published in 2016 has shown that oil field activities may have triggered the huge M6.4 1933 Long Beach quake, as well as some other major quakes.[5] If this is confirmed (and especially if oil-related fluid injection triggers a huge quake in California soon), then the political pressure to end oil field fluid injection in California would be immense. Already, California banned fracking after 2024, but this would further curb any oil field activities that pump fluids into the ground. The oil lobby may control politics in states like Texas and Oklahoma, but they do not have that power in California.

5.5   WHAT SHOULD YOU DO IN A QUAKE?

Earthquakes are so common in California that everyone who spends time in this state should know what to do. Sadly, most people are terrible procrastinators or never take the possibility of a quake seriously until after it has happened and it's too late to undo the damage that their lack of preparation caused. As they say, it's like closing the barn door after the horses are gone. Invariably, after nearly every earthquake in California, the TV news reporters interview hundreds of people who admit they never got around to preparing an earthquake kit or promise the reporter they will do it now that they just experienced a smaller quake. If you went back and interviewed most of those same people right now, you'd probably find that none of them actually did what they knew they should do.

Most California schools and businesses take part in annual earthquake drills, especially during the "Great California ShakeOut" in October, but more preparation would be better. In countries like Japan, they drill many times in a year, so there is no panic or confusion when one of their frequent quakes strikes.

The essentials things to remember are as follows:

- "Duck and cover." If you are outdoors, stay there. If you are indoors, duck under a desk or table and protect your head and body. The biggest danger indoors is the collapse of the ceiling, especially all the suspended light fixtures and ceiling panels.

Also, stay away from windows, which can shatter and shower you with glass.

- If you are in a building, *don't* run out the door! The most dangerous place outdoors are the outsides of buildings, where the exterior features often shake loose during a quake. The sole victim of the 1987 Whittier quake was a student at CA State Los Angeles who ran out of a building and then was killed by falling debris from its façade.

- At home, make sure all your heavy furniture is secured to the walls, and that valuable objects are stuck to the shelves with earthquake putty.

- By California law, your water heater must be strapped to a stud in your house, and you should have a special wrench by your gas heater to shut off the gas after a quake. Your water heater is also a valuable source of fresh water, which will be scarce after a big quake.

- If you have the option when buying a home, avoid buying houses on soft valley fill that will shake much worse than those built on bedrock—unless it's an unstable hillside prone to landslides. It's a big investment—hire a qualified geological engineer to make sure.

- When a big quake happens, expect days without water, power, gas, and most other utilities. All electronic banking will be down, so have some cash stowed away. You should have several days' supply of food and water in your earthquake kit, along with flashlights and batteries, candles, a first aid kit, and other essential materials. Over and over, people experience a small quake, say they're going to get their earthquake kit together, and then put it off until it's too late.

- Expect the roads to be closed due to damage, so you won't be able to drive far—and most gas stations will be closed without their power.

- All cell phone service will be down for days, as will all computers and Internet connections, so have an old plug-in landline phone near the wall jack. If a quake happens, use a landline phone to call a contact out of state who can be your central communication hub with your family. You will not be able to reach anyone directly by phone or computer in the earthquake zone.

Remember, it is not a question of *if* a quake will happen but *when*. Make your plans, and get that earthquake kit together and stowed in a place where you could reach it if your building is damaged. It *will* happen, so if you procrastinate again, you have only yourself to blame when you're caught unprepared.

Earthquakes are a part of life in California. They kill fewer people than do lightning strikes or snake bites, so be grateful you don't experience the deadlier disasters of blizzards, tornadoes, hurricanes, or other such disasters found in most parts of the country. If you are prepared, you will get through the experience and recover much more quickly than your neighbors.

NOTES

1 www.usgs.gov/faqs/does-fracking-cause-earthquakes.
2 www.ecowatch.com/fracking-earthquakes-california-2648860865.html.
3 www.kpcc.org/show/take-two/2013-07-15/study-fracking-in-salton-sea-area-triggers-earthquakes.
4 www.ecowatch.com/fracking-earthquakes-california-2648860865.html.
5 www.ocregister.com/2016/10/31/study-drilling-may-have-caused-deadly-1933-long-beach-quake/amp/?fbclid=IwAR019ALeCnsLzfhHRF-qbejnQg7_KmyKIL7ajbdwUdftzD5mtfviOmmSOjo.

RESOURCES

Collier, M. 1999. *A Land in Motion: California's San Andreas Fault.* University of California Press, Berkeley.

Dvorak, J. 2014. *Earthquake Storms: The Fascinating History and Volatile Future of the San Andreas Fault.* Pegasus, New York.

Ellsworth, W.L. 2013. Injection-induced earthquakes. *Science, 341,* 225942. https://doi.org/10.1126/science.1225942

Hough, S.E. 2004. *Finding Fault in California: An Earthquake Tourist's Guide.* Mountain Press, Missoula, MT.

Iacopi, R. 1971. *Earthquake Country: How, Why, and Where Earthquakes Strike in California.* Sunset Publishing, Oakland, CA.

Lynch, D.K. 2015. *The Field Guide to the San Andreas Fault.* Sunbelt Publications, El Cajon, CA.

Muir, J. (edited by L. Stetson). 2013. *The Wild Muir: Twenty-Two of John Muir's Greatest Adventures.* Yosemite Conservancy, San Francisco, CA.

Powell, R.E., Weldon, R.J., II, and Matti, J.C. 1993. *San Andreas Fault System: Displacement, Palinspastic Reconstruction, and Geologic Evolution.* Geological Society of America Memoir 178. Geological Society of America, Boulder, CO.

Ruby, L. 2009. *Bad Apple.* Harper & Row, New York.

U.S. Geological Survey. N.d. Does fracking cause earthquakes. www.usgs.gov/faqs/does-fracking-cause-earthquakes

Winchester, S. 2006. *A Crack in the Edge of the World: America and the Great California Earthquake of 1906.* Harper Perennial, New York.

Yeats, R.S., Sieh, K.E., and Allen, C.R. 1997. *Geology of Earthquakes.* Oxford University Press, Oxford.

VIDEOS

http://video.nationalgeographic.com/video/101-videos/earthquake-101

http://video.nationalgeographic.com/video/inside-earthquake

www.youtube.com/watch?v=vbLjzZg79ZU

www.youtube.com/watch?v=3owcJB0x6m0

www.youtube.com/watch?v=K2FSI2cnJGw

6 Young Volcanoes
The Cascades and Modoc Plateau

Each volcano is an independent machine—nay, each vent and monticule is for the time being engaged in its own peculiar business, cooking as it were its special dish, which in due time is to be separately served. We have instances of vents within hailing distance of each other pouring out totally different kinds of lava, neither sympathizing with the other in any discernible manner nor influencing other in any appreciable degree.

—**Clarence Edward Dutton**
1880

6.1 DATA

Highest points: Mt. Shasta: 4,319 m (14,162 ft)

Lassen peak: 3,188 m (10,457 ft)

Modoc plateau: Average elevation 1,370 m (4,500 ft)

Predominant rocks: Late Cenozoic volcanics, especially basalts in the Modoc Plateau, andesites and dacites in the Cascades

Plate tectonic setting: Andean-style volcanic arc in the Cascades, Basin and Range volcanoes in the Modoc Plateau

Most recent eruptions: Lassen, 1914–1921

Geologic resources: Pumice, building materials, landscaping materials, mining

National parks: Lassen Volcanic National Park, Lava Beds National Monument

State parks: Shasta, Castle Crags, Burney Falls, among many others

6.2 GEOGRAPHY OF THE CASCADES AND MODOC PLATEAU

The northeast corner of California (Figure 6.1) is one of the most sparsely populated regions in the state, with the lowest density of humans per square mile—only the peaks of the High Sierras and the blazing hot wastes of the Mojave Desert have fewer people. Only 9,100 people live in the huge area (10,886 km², or 4,203 mi²) of Modoc County, or 2.25 people per square mile. Its largest city is the county seat, Alturas, with only 2,700 people. Siskiyou County has only 43,000 people in 16,400 km² (6,347 mi²), for only 7.1 people per square mile. The "Empty Northeast" is also one of the least-traveled areas in all of California, with only a handful of small cities and small towns. Outside the corridor around Interstate 5, which connects the larger cities of Red Bluff and Redding on the rim of the Central Valley and passes through small towns like Weed and Yreka to the Oregon border, there is very little additional traffic in the region. Yreka is the county seat of Siskiyou County, but it has only 7,700 people. The economy of the northeastern part of California is based primarily on ranching and logging, tourism to the region's major parks, and a significant number of public employees serving the national and state parks and forests, plus various government agencies.

Most of the region is high in elevation (at least 1,370 m, or 4,500 ft or higher). It is covered by piñon-juniper forests at the lower elevations and dense firs and Ponderosa pine and Jeffrey pine forests at higher elevations. Much of northeastern California is national forest, national grassland, state forest, and wildlife refuge, and there are several national monuments and two major national parks and monuments: Lassen and Lava Beds. The Cascades (especially Mt. Shasta) get about 260 cm (102 in) of snow and 2.5 m (83 ft) of total precipitation each year. During the winter, the Modoc Plateau often averages about 79 cm (31 in) of snow as well. However, the Modoc Plateau lies behind the rain shadow of the Cascades, so it averages only 38 cm (15 in) of rainfall a year, making it a semidesert. Most of the region stays near freezing much of the winter but averages in the 30–36°C (86–96°F) range in the summer.

6.3 VOLCANISM

Nearly the entire northeastern corner of California is covered by late Cenozoic volcanic rocks (Figure 6.2). These include both huge lava flows on the Modoc Plateau and young volcanic ash from the Cascades. The Cascades are an example of a volcanic chain produced by the subduction of oceanic plate beneath a continental plate, known as an Andean-style arc (see Figure 4.11). As discussed in Chapter 4, melting of the mantle above the subducting oceanic crust basalts produces magma that erupts to form Andean-style volcanic arcs. As these magmas rise through the silica-rich continental crust above the subduction zone, they become differentiated into more felsic magmas, especially by wall rock contamination, and assimilation from felsic melts of the country rock produced by partial melting. Thus, they erupt at the surface as andesites, dacites, and even some rhyolites.

These magmas are much stickier and more viscous (like peanut butter or molasses) than the more fluid oceanic basalt lavas, which flow like water. Instead of erupting with big lava flows (as is typical of basaltic volcanoes like Hawaii),

DOI: 10.1201/9781003301837-6

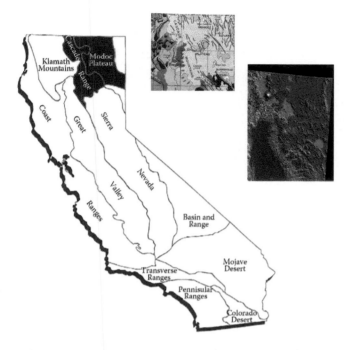

FIGURE 6.1 Index map of California, highlighting the Cascade Range and Modoc Plateau. In the upper right is the geologic map of the region and the digital shaded relief map of northeastern California, showing the elevation of the Modoc Plateau and Mt. Shasta.

Source: Courtesy of US Geological Survey.

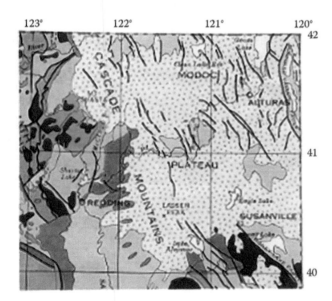

FIGURE 6.2 Geologic map of the northeastern corner of California, showing the huge area of late Cenozoic volcanic rocks (pink color).

Source: Courtesy of California Division of Mines and Geology.

rhyolite–dacite magmas choke and clog the throat of the volcano until enough pressure builds up, resulting in explosive eruptions that send huge clouds of **pyroclastics** (volcanic material broken into pieces, from the Greek *pyro*, "fire,"

and *clastos*, "broken fragment"). Pyroclastics come in lots of different sizes. They are usually separated into volcanic ash or dust, lapilli, and volcanic bombs (Figure 6.3).

Volcanic ash or dust (Figure 6.3a) is the tiniest material, made of shards of volcanic glass less than 2 mm in diameter. The bulk of most dacite–rhyolite volcanic eruptions is in the form of huge clouds of volcanic ash, which can spread hundreds of kilometers from the vent. If the ash is injected into the stratosphere, it can stay up there for years and travel around the globe and even change the climate briefly. The clouds of stratospheric dust can block sunlight and cause "volcanic winter," as well as brilliant orange sunsets. This was demonstrated when Mt. Tambora in Indonesia erupted in 1815. As a result, 1816 was the "year without a summer," when it froze and snowed in June in the northeastern United States and much of Eurasia, causing crop failure, widespread starvation and famine, and even epidemics of disease like typhus. When a big eruption spills enormous volumes of ash across the landscape, it can kill huge amounts of vegetation and wildlife. After big eruptions, the weight of wet ash on rooftops often destroys more buildings than the volcano itself.

Sand- and pebble-sized (between 2 and 64 mm) pieces of volcanic material (Figure 6.3b) are called **lapilli**. The biggest volcanic pyroclastics (bigger than 64 mm, or cobble- to boulder-sized or larger) are called volcanic **bombs** (Figure 6.3c—e). They are aptly named because huge blobs of volcanic rock flying through the air can have the same effect as a bomb landing on the ground. Volcanic bombs often have a streamlined aerodynamic shape (such as a spindle or football or teardrop shape) because they fly out of the volcano as a blob of hot fluid magma but then cool and harden as the air flows over them and they spin in midflight (Figure 6.3d). A mixture of any of these size classes of pyroclastics, such as ash plus lapilli, is given the broader term *tephra*.

When a volcano erupts in a huge pyroclastic eruption, it can form lots of different volcanic deposits. One of the most common and by far the deadliest of these is **pyroclastic flows**, or *nuées ardentes* (French for "glowing clouds") (Figure 6.4). These are superheated gas clouds of gas and ash traveling down the slope of the volcano at speeds up to 700 kph (450 mph) and typically at 1,000°C (1,850°F). As these hot ash clouds sweep down and across the landscape at speeds faster than any human or animal can escape them, they can incinerate and vaporize their victims in seconds. *Nuées ardentes* got their French name after the disastrous 1902 eruption of Mt. Pelée in the Caribbean island of Martinique, which incinerated the town of St. Pierre and its 28,000 residents in a matter of minutes. The eruptions from Mt. Vesuvius in AD 79 were also pyroclastic flows, and they wiped out Pompeii and Herculaneum and thousands of people who lived there. The deposits of these eruptions are called **ignimbrites**, and they are frequently welded together by their extreme heat into **welded tuff**. If the hot ash settles into a lake, it will make a stratified volcanic deposit called a **water-laid tuff**.

(a)

(b)

(c)

(d)

(e)

FIGURE 6.3 Pyroclastics come in many sizes. (a) Volcanic ash from Mt. Lassen. (b) Coarse sand- and pebble-sized pyroclastics are called lapilli, from Panum Crater. (c) Large (more than 1 m across) lava bomb, Lassen Volcanic National Park. (d) Two examples of smaller lava bombs that have streamlined shapes from the air currents that flowed over them as they spun and cooled in flight.

Source: (a–d) Photos by the author. (e) An enormous volcanic bomb in the cinder quarries near Clear Lake. Courtesy of US Geological Survey.

If hot pyroclastics mix with water (often from the melting of glaciers on the top of the volcano), they can make a hot volcanic mudflow known by the Javanese word **lahar** (Figure 6.5). Lahars have the consistency of wet concrete and can move at speeds greater than 48 kph (30 mph), so they are impossible to outrun. Since they are so dense, they can pick up huge boulders, cars, trucks, and even houses on their destructive path down the side of the volcano. One of the deadliest of recent lahar eruptions occurred when the Nevado del Ruiz volcano erupted in Colombia in 1985, forming lahars that killed 25,000 people in the town of Armero as it buried them in flows 5 m (15 ft) thick. The 1991 eruption of Mt. Pinatubo in the Philippines occurred

FIGURE 6.4 Ash clouds and pyroclastic flows erupting from Mayon volcano in the Philippines.

Source: Courtesy of US Geological Survey.

at the same time as Typhoon Yunya, making huge numbers of lahars and killing more than 1,500 people. The 1991 eruption of Mt. Unzen in Japan killed a number of volcanologists with its pyroclastic flows and produced big lahars (Figure 6.5b).

Volcanoes of andesite–dacite composition are formed by a mixture of both small lava flows (especially if they are of more fluid andesitic composition) and pyroclastics of various types (especially pyroclastic flows). Thus, they are known as **stratovolcanoes** (because they are stratified with different types of volcanic products) or **composite volcanoes** (because they are a composite of lava flows and pyroclastics). Stratovolcanoes form the classic conical shape (Figure 6.6) most people think of when they hear the word *volcano* and can have quite steep slopes if the lavas and pyroclastics are sticky and viscous. Stratovolcanoes are very different from the low, domed shield volcanoes produced by the fluid eruptions of basaltic lava flows, such as on the Big Island of Hawaii. For one thing, shield volcanoes are tens to hundreds of times bigger than any stratovolcano, even though they have very gentle slopes. Shield volcanoes are made entirely of hundreds of fluid basaltic lava flows, which stack up as they flow down the gentle slopes of the volcano.

As we discussed in Chapter 4, stratovolcanoes are formed above subduction zones, such as the Ring of Fire of volcanoes and earthquakes that surrounds the Pacific Ocean (Figure 4.11b). They are products of the melting of the subducting slab of oceanic basalt (Figure 6.7). One of the interesting aspects of many volcanic arcs like the Cascades is the **calc-alkaline trend**. *Calc* refers to calcium, and *alkaline* sodium and potassium. Volcanoes that are near the trench and the subduction zone boundary tend to be richer in calcium, and their magmas are mostly andesites (made of calcium-rich plagioclase plus hornblende). But as you move farther back from the trench and the plate behind the main arc volcanoes, you get magmas that are richer in sodium and potassium (the "alkali elements"), producing more and more sodium and potassium feldspar, and thus magmas that are more dacitic or even rhyolitic. This makes

(a)

(b)

FIGURE 6.5   (a) Muddy lahar deposits moving like wet concrete and carrying a bridge ripped away from its moorings, Mt. Saint Helens, 1980. (b) Lahars flowing down the flanks after the 1991 eruption of Mt. Unzen, Japan.

Source: Courtesy of US Geological Survey.

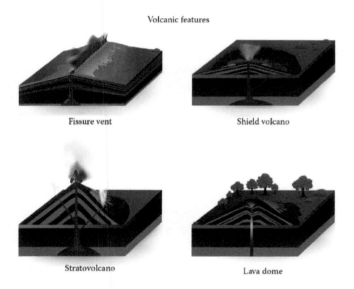

FIGURE 6.6   Diagrammatic cross section of the different types of volcanoes. Fluid basaltic eruptions typically emerge through a fissure vent to form flood basalts, or a shield volcano, which is broad and relatively low in slope and relief. By contrast, sticky andesite or rhyolite lavas form a steep-sided conical composite volcano, or stratovolcano, which is built of interbedded andesitic flows and pyroclastics. Small rhyolitic eruptions that produce nothing but cinders and pyroclastics often make a lava dome.

Source: Courtesy of Wikimedia Commons.

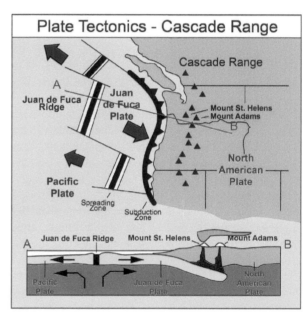

FIGURE 6.7   Plate tectonic geometry of the Cascades. The subducting Juan de Fuca plate plunges beneath the Cascadia subduction zone to generate the andesites and dacites of the Cascade volcanic chain.

Source: Courtesy of US Geological Survey.

sense if you think about how the down-going plate dips beneath the overlying plate (Figure 6.7). The slab closest to the trench and plate boundary melts at a relatively shallow depth and encounters relatively little felsic crustal rock above it as it melts its way upward, so it is less differentiated and becomes andesite. But as the plate plunges deeper and deeper, it releases magmas that must melt their way through more and more felsic crust overlying it. This increases the chances for assimilation and contamination, as well as opportunities for the magma chamber to remain buried at depths where fractional crystallization can occur. For all

these reasons, the further from the trench the magma, the more likely it is more felsic and produces dacite or even rhyolite lavas. This trend is well demonstrated in the Cascades, where the major volcanoes in the western part of the chain tend to be andesites or dacites, but there are numerous rhyolitic volcanics to the east of and behind the main volcanic chain, such as the rhyolites of Lassen Volcanic Park or the rhyolites east of the Cascades near Bend, Oregon.

6.4   THE CASCADE RANGE

Only the southern end of the Cascade Range is found in California, but it is a single unified geographic and geologic

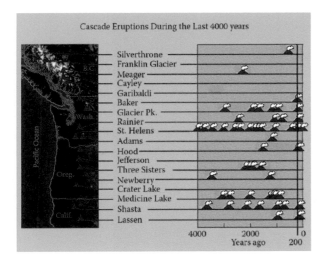

FIGURE 6.8 Location and eruption history of the Cascade volcanoes. Some volcanoes, like Mt. Saint Helens, erupt nearly every few centuries for thousands of years, while others, like Mt. Mazama (Crater Lake), have not erupted in 7,700 years.

Source: Courtesy of US Geological Survey.

province, so we will first look at the range in its entirety (Figures 6.7 and 6.8). It is the longest chain of volcanoes in North America, running from British Columbia in the north down to Mt. Lassen at its southern end, a distance of 800 km (500 mi). Even though most of it lies outside California, Mt. Shasta and Lassen Peak are the largest and youngest volcanoes in the state. Twelve of the Cascade volcanoes are more than 3,000 m (10,000 ft) in elevation. Two of them, Mt. Shasta and Mt. Rainier, are more than 14,000 ft high (a "fourteener" in the lingo of hikers and climbers). These peaks are among the few mountains this tall in the United States outside the Rockies, the Sierras, and Alaska.

The majority of the Cascade Range volcanoes are less than 2 million years old, and many are very recent; these are often called the "High Cascades." However, between 37 and 17 Ma (late Eocene to middle Miocene), there was another chain of arc volcanoes, known as the "West Cascades," "Old Cascades," or "Ancestral Cascades" (Figure 7.22). As the name implies, in Oregon they lie just west of the High Cascades and east of Oregon's Willamette Valley, although their deposits extend far to the east of the entire Cascade Range. Because they are so old, most of them are deeply eroded and largely covered by later volcanic deposits. Some of the eruptive events of these older predecessors of the High Cascades include the Mt. Aix volcanic complex, about 24 mi east of Mt. Rainier in Washington, and numerous lava flows found in North Cascades National Park in Washington. In addition, the Chilliwack batholith, the ancient cooled magma chamber of the Ancestral Cascades, is also exposed in North Cascades National Park. Another product of these eruptions are the extensive sheets of upper Eocene through upper Oligocene volcanics and volcaniclastics of the John Day region in north-central Oregon, which are also famous for their fossils of mammals and leaves. The trend continued through Nevada, which formed a high

volcanic plateau called the Nevadaplano, over 3,000 m (10,000 ft) in elevation and stood higher than the eroded Sierra Nevada Mountains (see Chapter 8). The volcanic chain continued through central Utah and then joined with the immense Oligocene volcanic field in the southwestern corner of Colorado known as the San Juan volcanic field, which covered over 25,000 km^2 (Figure 7.22). Further volcanics of Eocene and Oligocene age can be traced down through the New Mexico–Arizona border region and then down into Mexico.

Then, about 17 Ma, arc volcanism ceased abruptly across the region, and most of the Ancestral Cascades volcanoes became extinct. From about 17 Ma until 6 Ma, there were no more arc volcano eruptions in the area. Instead, volcanism shifted to the huge Columbia River basalts in eastern Oregon and Washington. During the middle Miocene (about 15–16 Ma), huge fissures opened up in eastern Oregon and Washington, and an immense volume of mantle-derived basalts poured across the landscape like flood. These are known as the **Columbia River flood basalts**, because today the Columbia River cuts across them. But back in the middle Miocene, these enormous eruptions poured across some 40,000 km^2 in a matter of days, covering the landscape with thick basaltic lava flows over and over. The flows moved about 5 kph, and each was about 30 m (100 ft) thick and about 100 km wide and reached temperatures of over 1,100°C. Over about 3.5 million years, flow after flow incinerated the landscape again and again, until they covered 300,000 km^2 of eastern Oregon and Washington with a sequence of lavas over 4,000 m (13,000 ft) thick. Between eruptions, the lava flows cooled and became forested landscapes, such as was preserved in Ginkgo Petrified Forest near Vantage, Washington, before the landscape was again incinerated by the next volcanic flood. In many places in Oregon and Washington, you can find cliffs with many stacked lava flows hundreds of feet thick. In most of the region, however, the Columbia River lavas form the hard basement rock that covers all the more ancient rocks (except the Blue Mountains in Oregon and the Wallowa Mountains in Washington, which have Paleozoic and Mesozoic remnants still exposed).

It is probably not a coincidence that the Ancestral Cascade volcanoes shut off at the same time that the Columbia River lavas erupted—clearly, there is a connection. Various tectonic models have been proposed as to why this happened, but there is no consensus explanation as to how they are linked. Whatever the reason, by 6 Ma, the modern High Cascades began to erupt their earliest phase of volcanism, especially with the spectacular eruption of the Deschutes volcanoes in Oregon, between 6.25 and 5.45 Ma. During this 800,000-year span, approximately 400 km^3 to 675 km^3 of pyroclastic material was expelled in 78 distinct eruptions. The basement rock of most of the active High Cascades volcanoes was also erupting at that time.

Most of the Cascade volcanoes (Figure 6.8) have been active in the past 2 million years, some of them very recently (such as Lassen, which erupted in 1914–1921). Some have erupted only rarely, like Mt. Mazama, which blew its top to form Crater Lake about 7,700 years ago, while others erupt frequently, like Mt. Saint Helens, which erupts every few centuries, most recently in 1980 (Figure 6.8). Often, the shape of the volcano helps indicate how recent its activity was. Volcanoes that have not had significant activity since the last glacial maximum 20,000 years ago are often heavily glaciated, with numerous glacial valleys descending from their summit. But if they have erupted often in the past few thousand years, they destroy their glacial valleys and have a smooth conical summit.

There are at least 20 major volcanoes in the Cascades, with about 4,000 smaller volcanic vents. They each have their own distinctive histories.

6.4.1 MT. RAINIER

The highest peak of all is Mt. Rainier (4,392 m, or 14,411 ft in elevation), which looms above Seattle and the cities of Puget Sound in western Washington. Its earliest dated eruptions are 2.9 million to about 840,000 years old, which produced a "proto-Rainier." Most of the present volcano is more than 500,000 years old, although it is heavily carved by glaciers from the late ice age, so it has not completely blown its top since 20,000 years ago. Nevertheless, there have been small eruptions in parts of the volcano, the most recent in the time interval 1820–1854. The most dramatic geologic event at Rainier was an eruption about 5,600 years ago that produced the Osceola mudflows. These lahars removed almost 2–3 km³ (0.5–0.7 mi³) of rock from the summit of Rainier and spread across the river valleys. It covered about 550 km² (212 mi²) of the modern location of nearly all the towns to the northeast of the mountain, especially Tacoma, Auburn, Kent, Enumclaw, Sumner, and Puyallap, and towns located in the White River Valley and adjacent valleys closer to the mountain. Almost all the towns built in these river valleys are right on top of the old lahar deposits. During eruption of the Osceola lahars, the summit of Rainier collapsed, making it about 7 km (2,100 ft) lower. This avalanche of hot volcanic mud, rocks, and trees traveled down the river valleys at more than 20 fps, destroying everything in its path. It even entered the legends of the Native Americans who lived there.

About 500 years ago, a smaller lahar known as the electron mudflow swept down the Puyallap Valley, knocking down trees 3 m (10 ft) across; the town of Orting is built on these ancient deposits. There has not been a significant large eruption since, but many geologists worry that one of the greatest hazards presented by Rainier is not a gigantic eruption (which hasn't happened in a long time) but a smaller eruption that would unleash similar lahars on the giant populations of the Seattle–Tacoma suburbs that lie in the path of disaster.

6.4.2 MT. MAZAMA—CRATER LAKE

A very different volcanic eruption is the catastrophic explosion of Mt. Mazama, which became Crater Lake (Figure 6.9). Mazama began to grow about 400,000 years ago as one of the tallest stratovolcanoes in the Cascade chain. Andesite lavas flowed down the north and southwest slopes about 50,000 years ago, making the volcano about 3,400 m (11,000 ft) high. As the magmas evolved, however, they became more and more felsic and silica-rich, so they also got more viscous and sticky. About 40,000 years ago, there were numerous dacitic eruptions that formed a series of domes on the mountain (much like the dacite dome currently forming in Mt. Saint Helens). These were all destroyed in a series of eruptions and lahars, which left large landslide deposits on all sides of Mt. Mazama. The next eruptions were 25,000–30,000 years ago, with the eruption of magma with a composition between a dacite and a rhyolite (called rhyodacite). These thick pasty flows erupted from the vents on the northwest flank of Mt. Mazama, forming Redcloud Cliff and the dome above Steel Bay. Then the volcano was dormant for 20,000 years. During this time, it was carved by large glacial valleys, which are still visible on the lower slopes of what remains of Mt. Mazama.

The final phase began 7,677 years ago with a series of huge eruptions of rhyodacite, generating giant pyroclastic flows all across the region. After centuries of these intermittent eruptions, the final cataclysmic explosion blew the top off Mt. Mazama about 6,800 years ago, sending a column of volcanic ash a mile (1.6 km) wide and hot tephra up to 16 km (10 mi) into the stratosphere at twice the speed of sound. The huge column then collapsed upon itself, just as the mushroom cloud of an atomic bomb does, and sent monstrous pyroclastic flows in all directions. Most of them were so hot they formed welded tuffs. The Mazama ash covered much of the western United States as far as Saskatchewan, Wyoming, and Utah. The eruptions were huge (46–58 km³, or 11–14 mi³ of rock) and so fast that the magma chamber beneath the summit didn't have time to refill, and the entire mountain collapsed into its own empty interior. This large ring-shaped depression formed by the original collapsed crater is known as a **caldera** to volcanologists. Technically speaking, Crater Lake is not a true crater at all but a caldera. Unfortunately, it's impossible to get people to start calling it by its correct name, Caldera Lake. There are a number of other calderas on the east side of the Cascades, especially the huge Newberry Caldera near Bend, Oregon.

6.4.3 MT. SAINT HELENS

The most famous of recent Cascade eruptions was Mt. Saint Helens in 1980. As is shown in Figure 6.8, it is by far the most active volcano in the chain, with several eruptions over the course of a few centuries. The oldest eruptions were 37,600 and 35,000 years ago (called the Ape Canyon stage), followed by glaciation during a quiet phase. The second phase occurred from 20,000 to 18,000 years ago (Cougar stage), and then another period of relative quiescence. The third major eruptions occurred 4,500 years ago (Smith Creek stage), which erupted so violently the ash can be found as far as Alberta. The fourth phase occurred 3,200 years ago (Pine Creek

(a)   (b)   (c)

(d)

FIGURE 6.9 Crater Lake. (a–c) Diagrammatic history of the collapse of Mt. Mazama to form the caldera of Crater Lake. (d) Photograph of the caldera, now filled with a lake. Wizard Island in the middle is a later eruption, called a parasitic cone.

Source: Courtesy of Oxford University Press.

eruption), which lasted until about 2,800 years ago, and a fifth phase (Castle Creek eruption) started to build the modern cone that was blown off in 1980. There were additional eruptions in AD 1480 (Kalama period), which were recorded in the legends of Native Americans, and another phase around AD 1790 that lasted 57 years. It was witnessed in 1792 by British captain George Vancouver sailing up the coast, one of the first European explorers to pass by the region.

Earthquakes preceded the most recent eruption in March 1980, and then venting of steam, and the north flank of the mountain began to bulge outward. Most residents and tourists were evacuated from the "red zone" close to the volcano. There were geologists monitoring it from what they thought was a safe distance, and some campers and photographers as well. The owner of Spirit Lake Lodge, Harry Truman (no relation to the former president), refused to leave his home and amused reporters for weeks with his colorful chatter and defiance of the mountain. Then early on the morning of May 18, 1980, all hell broke loose (Figure 6.10). First, an earthquake of magnitude 5.1 shook the region and then the entire north flank of the mountain broke loose and began to form a gigantic debris avalanche, the largest in recorded history, which covered the area to the north of the mountain for about 600 km² (230 mi²). The blast was directed not only upward but also sideways to the north, which endangered and even killed people to the north of the volcano who thought they were safe. The force of the blast and the moving debris

was so powerful, and it flattened all the trees in the area as if they had been knocked over by a giant comb. Geologists modeling the force of the blast used the dynamics of jet engines to understand the flow patterns. In some cases, the gases swirled backward in turbulent eddies, as evidenced by trees that were knocked down pointing toward the mountain. US Geological Survey (USGS) geologist David Johnston, who was monitoring the activity, radioed his last words to his headquarters in Vancouver, Washington, "Vancouver, Vancouver, this is it!" and then died instantly. Most of the landscape vanished under the debris flows, including Spirit Lake and Harry Truman.

Gigantic lahars roared down the Toutle and Cowlitz River Valleys, destroying bridges with huge logjams and carrying cars and houses and logging trucks (Figure 6.5a). These lahars transported about 3 million m³ (3.9 million yd³) of material over 27 km (17 mi) down to the Columbia River. The eruption kept sending plumes of ash into the stratosphere for over nine hours, sprinkling a dense blanket of ash across all of eastern Washington and parts of Idaho and Montana and even reaching as far as Edmonton, Alberta. When it was all over, 57 people had died, and 250 homes, 47 bridges, 24 km (15 mi) of railways, and 298 km (185 mi) of roads were destroyed.

Today, the horseshoe-shaped crater has a gap to the north side (Figure 6.10c), where you can see the dacite dome that has been slowly growing (with small ash and steam eruptions) in the 36 years since this eruption. You can visit the

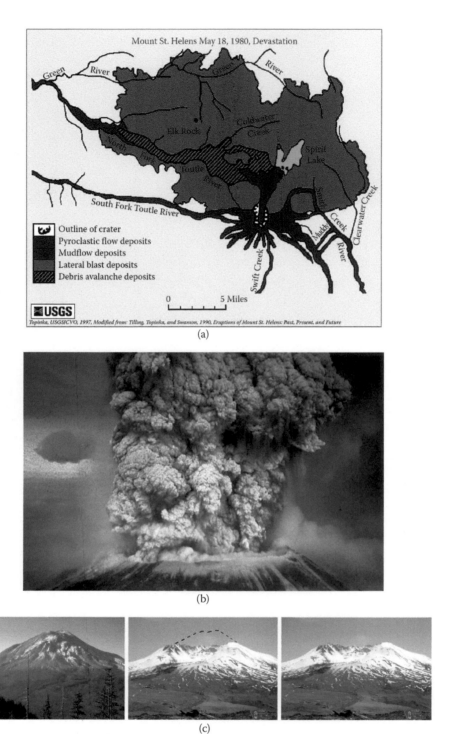

FIGURE 6.10 Mt. Saint Helens eruption, May 18, 1980. (a) Map showing the areas of major volcanic damage. Most of the volcanic blast blew horizontally to the north (yellow area of "lateral blast deposits"), so the devastated area fans out from the north flank of the volcano. (b) Ash plume during the peak eruption. (c) Mountain before and after the eruption. The new dacite dome can be seen in the center of the crater.

Source: Courtesy of US Geological Survey.

Mt. Saint Helens National Volcanic Monument, which has a modern visitors' center on Johnston Ridge. There, the entire eruption is explained, and you drive through miles of moonscape where the broken trees lie where they were blasted over, and only a small amount of vegetation and wildlife has come back.

6.4.4 Mt. Shasta

California's biggest volcano is Mt. Shasta (Figure 6.11), which is 4,322 m (14,179 ft) in elevation, second only to Mt. Rainier among Cascade peaks, and the fifth highest mountain in California after four Sierra peaks. In many ways, it

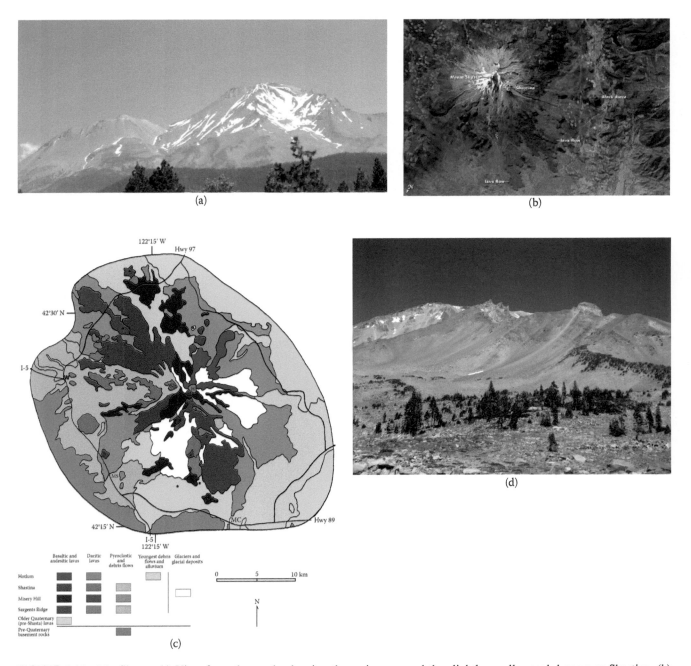

FIGURE 6.11 Mt. Shasta. (a) View from the north, showing the main cone and the slightly smaller peak known as Shastina. (b) Satellite image of the major features of Mt. Shasta. (c) Geologic map of the sequence of volcanic eruptions around Mt. Shasta. (d) Misery Hill, one of the smaller volcanic cones on the north flank of the volcano.

Source: (a) Photo by the author. (b–c) Courtesy of US Geological Survey. (d) Courtesy of G. Hayes.

is more impressive than any other peak, since it is built of 354 km³ (85 mi³) of rock, bigger in volume than Mt. Rainier or any other Cascade volcano. It towers more than 3,000 m (10,000 ft) above the surrounding landscape, with no other volcanoes or other Sierra-like peaks nearby. On a clear day, its snowcapped summit can be seen for 230 km (140 mi) to the south down in the Sacramento Valley.

Mt. Shasta is not a simple conical volcano but built of four overlapping cones that combine to form a complex shape, including the main summit and a secondary peak

called Shastina, which is 3,760 m (12,330 ft) in elevation (Figure 6.11a). Shastina by itself would be the fourth largest peak in the Cascades, since it is lower only than two summits on Rainier and the main summit of Shasta. Together, the two summits give Shasta a distinctive saddle-shaped profile, making it look very different from the simple cones of most Cascade peaks. Most of the surface of Mt. Shasta is smooth and unglaciated, except for the seven glaciers on the upper slopes of the north and east sides; the largest valley (Avalanche Gulch) no longer has a glacier in it. Its glaciers are larger than those that remain in the Sierras, with the longest glacier in California (Whitney Glacier) and the most voluminous (Hotlum Glacier).

The oldest-dated deposits near the volcano tell us that Shasta began to form about 593,000 years ago (Figure 6.11c). About 300,000 years ago, the entire north side of the mountain collapsed, sending a debris avalanche with about 27 km² (6.5 mi³) of material about 45 km (28 mi) down into the Shasta Valley; it has since been deeply eroded by the Shasta River. The oldest-dated cone is Sargents Ridge on the south side, which was an andesite–dacite eruption dating about 100,000 years ago (Figure 6.11c). These ancient flows were then glaciated as recently as 20,000 years ago. About 20,000 years ago, another eruption occurred, forming Misery Hill, a subsummit peak just south of the main summit (Figure 6.11d). The next eruption was Shastina itself about 9,800 years ago, sending flows that reached as far as Black Butte, 11 km (7 mi) away. Since Shastina formed after the last ice age, it is unglaciated and has no glacier valleys. Then came the eruption of Hotlum Cone, about 8,000 years ago, producing the 150 m (500 ft) thick Military Pass flow that roared as far as 9 km (5.5 mi) down the northeast face. A dacite dome then grew in Hotlum Cone, which has erupted nine times since the initial eruption 8,000 years ago, producing at least four big lahars that rumbled 12.1 km (7.5 mi) down from the summit. Its most recent eruption was witnessed by the French explorer La Perouse sailing off the coast of California in 1786. Shasta's eruptions are spaced roughly 600–800 years apart, and the volcano has been dormant since 1786. So it is apparently not overdue for an event but is certainly capable of a catastrophic eruption at any time.

In addition to the direct volcanic hazards, in 2014 Mt. Shasta produced giant cold wet mudflows without any eruptions. The global warming of the past century has rapidly melted its ice back so that there is a considerable reservoir of water trapped beneath some of the glaciers. All it takes is some event to dislodge the rocks and ice at the snout of the glacier and this huge volume of water is released, creating rapidly flowing masses of mud, rocks, and water called **debris flows**.

6.4.5 Lassen Volcanic National Park

Southeast of Mt. Shasta is the other young volcano in California, Lassen Peak and its surrounding volcanoes (Figure 6.12). At 3,817 m (10,457 ft), Lassen is among the highest Cascade peaks, and it stands about 600 m (2,000 ft) above the surrounding landscape. It also is the southernmost active volcano in the Cascades. Lassen Peak is actually the shattered remnant bigger volcano, Mt. Tehama, which once stood more than 300 m (1,000 ft) taller. Lassen Peak is the largest peak on the old Tehama caldera rim, with Brokeoff Mountain forming a peak at the other side of the caldera ring (Figure 6.12b). Lassen Peak is also one of the snowiest places in California, with an average annual snowfall of 1,676 cm (660 in), and some years, snowfalls of 2,500 cm (1,000 in, or more than 12.3 ft) are recorded.

The oldest rocks in the area are Pliocene lahar deposits known as the Tuscan Formation, which is not exposed within the national park but underlies much of the region to the south, erupting from the Yana volcanic center. There are also Pliocene basalt flows, which helped build the lava plateau that is part of the Modoc volcanic field. About 3.27 million years ago (Ma), an early volcano erupted and sent huge ash flows and lahars down the Sacramento Valley for 130 km (80 mi), reaching as far south as Willows. This eruption, called the Nomiaki tuff, is more than 100 m (300 ft) thick near Redding, and its ash clouds blew as far south as New Mexico.

During the Pleistocene, a series of andesite flows called the Juniper lavas and the Twin Lakes lavas erupted, along with flows called the Flatiron, which covered the southwestern part of the park. In the late Pleistocene, the first eruptions to form Mt. Tehama occurred. Like many stratovolcanoes, Tehama was built of alternating layers of andesite flows and pyroclastics. About 350,000 years ago, Mt. Tehama collapsed on itself (much like Mt. Mazama did) and created a caldera 3.2 km (2 mi) wide after the volcano emptied its throat. The remnants of this old caldera can be seen by looking at the profile from Brokeoff Mountain to Lassen Peak (Figure 6.12b).

Some of this eruption formed a fluid black glassy dacite flow more than 460 m (1,500 ft) thick, which can be seen at the base of Mt. Lassen. During the rest of the ice ages, glaciers sculpted what was left of Mt. Tehama, although most of these glacial features have since been buried by younger eruptions.

About 27,000 years ago, the main part of Lassen Peak formed as a big dacite dome, which built high above the old rim of the Tehama caldera. It shattered the overlying remnants of the Tehama eruptions as it rose, and those rocks have accumulated as huge talus piles on the slopes of Lassen Peak. During the last ice age (peak glaciation 20,000 years ago), more glaciers grew on the Lassen dacite dome, some of which extended as much as 11 km (7 mi) from the volcano. Additional dacite domes, such as Chaos Crags, occur just north of Lassen Peak (Figure 6.12c). To the north of Chaos Crags is Chaos Jumbles, a pile of andesitic boulders that broke loose 300 years ago in a huge rock avalanche. Buoyed by a carpet of air trapped beneath the mass of boulders, they slid down the north slope at speeds of 160 kph (100 mph), flattening the forest in front of them, before coming to rest where they are found today.

The entire area continues to show signs of magma and geothermal heat not far underground, since there are many hot springs, geysers, and mud pots in the Lassen area (Figure 6.12d and e). They are very active most of the time, stinking of hydrogen sulfide (rotten egg smell) and coated with yellow sulfur deposits. They bear a number of colorful names, such as the Sulfur Works, Devil's Kitchen, Terminal Geyser, Cold Boiling Lake, Growler Hot Spring, and Boiling Springs Lake. One of the most famous of these is called Bumpass Hell (Figure 6.12d). It was named for a miner named Kendall Vanhook Bumpass, who was the

FIGURE 6.12 Lassen Volcanic National Park. (a) Map of the major features of the Lassen region. (b) Panorama of Lassen Peak (left) and Brokeoff Mountain (right), both remnants of the rim of Mt. Tehama. Dashed line shows the original profile of Mt. Tehama before it exploded. (c) Chaos Crags (in the background) with the landslide deposit called Chaos Jumbles in the foreground. (d) Bumpass Hell hot springs and boiling mud pots. (e) Boiling mud pot at Hot Sulphur Springs.

Source: (a) Courtesy of National Park Service. (b) Courtesy of Wikimedia Commons. All other photos (c–e) by the author.

FIGURE 6.13 Historic images of the 1914–1915 eruptions of Mt. Lassen, taken by pioneering Lassen geologist and photographer B. F. Loomis. (a) Small plume from the October 1915 eruption. (b) Larger plume erupting on June 14, 1914. (c) Another photo of the big plumes from October 1915. (d) Glowing lava flow on May 19, 1915, melting its way through the snowpack that was 40 ft deep, making a lahar more than 18 mi long. (e) "Hot Rock," an enormous boulder carried down from the peak in a lahar in May 1915. Even after it was immersed in mud, it was still hot to the touch for visitors.

Source: Courtesy of US Geological Survey.

first to visit in the 1860s and scalded his leg. When he brought the news of his "hell on earth" to the newspaper editor in Red Bluff, both of them went back to confirm what Bumpass had claimed. During this second trip, Bumpass broke through the thin crust over a boiling mud pot, scalding his leg so badly it had to be amputated.

Lassen holds the distinction as being the most recent volcano to erupt in the lower 48 states other than the 1980 eruption of Mt. Saint Helens (Figure 6.13). The eruption occurred with rocks of both andesite and dacite composition, suggesting the imperfect mixing of two different magma chambers (Figure 2.5b). After 27,000 years of dormancy, the first rumblings of this activity occurred on May 30, 1914, when area residents heard noises and witnessed the beginning of a year with more than 180 steam explosions, which opened a new crater in the dacite dome. About a year later (May 14, 1915), incandescent blobs of lava could be seen more than 20 mi away. Five days afterward, the main eruption occurred (Figure 6.13a—c) and lasted from May 19 through May 22. In its first phase, the crater blew its top, and lavas spilled 300 m (1,000 ft) down the side. Then melting snows formed lahars that flowed 11 km (7 mi) down the river valleys, wiping out six houses, but no one died. The final climactic eruption occurred on May 22, 1915, when a column of ash blasted from the mountain and rose 12 km (40,000 ft) up into the stratosphere. Pyroclastic flows spread across the landscape, especially down the river valleys, where the lahars had just occurred, and covered 7.7 km² (3 mi²), and many new lahars were produced. Ash and pumice blew eastward from the stratospheric dust cloud and reached as far as Winnemucca, Nevada, more than 320 km (200 mi) from the volcano. After the paroxysmal blast, the volcano settled down, and only steam eruptions were reported over the next two years, ending around June 1917. The volcano now appears to be dormant, although every once in a while there are steam eruptions on the flanks. However, there are still areas of geologic hazard in the region, from lahars formed by rain and snow mobilizing old pyroclastics, to the possibility of future pyroclastic flows and explosive eruptions.

6.5   THE MODOC PLATEAU

To the east of the Cascades is a huge, almost-empty region of California known as the Modoc Plateau, after the Modoc tribes that once lived there. Unlike the Cascades, which are stratovolcanoes made largely of andesite–dacite–rhyolite magma, the Modoc Plateau is covered by huge flows of basalt that erupted from shield volcanoes. The largest of these is **Medicine Lake volcano** (Figure 6.14), which is even larger in total volume than Mt. Shasta or any other mountain in the state. However, it is not obvious how large it is because it is low and domed (Figure 6.14b), not steep and conical in shape. Medicine Lake shield volcano is about 48 km (30 mi) long from north to south, and 35 km (22 mi) from east to west. It rises from the average plateau elevation of 6,440 m (4,000 ft) to more than 12,734 m (7,913 ft)

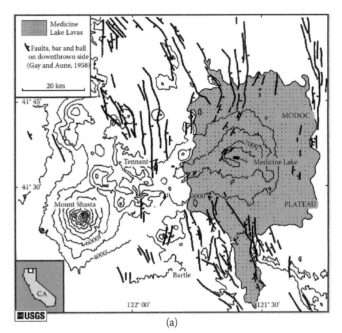

(a)

(b)

FIGURE 6.14 Medicine Lake shield volcano. (a) Map of Medicine Lake volcano showing its location with respect to Mt. Shasta. (b) View of Medicine Lake volcano in the distance, showing its low-domed shield shape. Photo taken from Captain Jack's Stronghold in Lava Beds National Monument, looking south.

Source: (a) Courtesy of US Geological Survey. (b) Photo by the author.

above sea level at its summit. The summit contains a crater of 11 km (7 mi) long by 6.4 km (4 mi) wide, with Medicine Lake in the center.

Most of the flows in the Modoc Plateau are basalts, although there are parts of the Medicine Lake volcano that are more andesitic in composition higher on the slopes. The first eruptions date back to only 1.9 Ma, followed by eruptions dated at 700,000 years ago, producing basalt flows that are seen east of Highlands. Medicine Lake volcano has erupted at least six times from four different vents in the past 2,000 years. Its volcanic history is peculiar in that it produces some eruptions that are pure basalt and others that are rhyolite (presumably from two different magma chambers in different parts of the crust), but no intermediate rocks like andesite. In a number of places, there are rhyolitic eruptions that produced the obsidian flows of Glass Mountain and Little Glass Mountain, along with abundant pumice and rhyolite. Some of these flows are only 200–300 years old. Modoc peoples mined this abundant obsidian for spearheads and arrowheads and used it to trade with other tribes. On the north side of Medicine Lake volcano is the Callahan andesite flow, which is only 1,100 years old. Another flow, called Mammoth Crater, is less than 100 years old.

To the north of Medicine Lake volcano is Lava Beds National Monument (Figure 6.15). This is a park that encompasses one of the younger lava flows in the region, famous for its caves called **lava tubes**. These are formed when a basaltic lava flows down a mountain and the top of the flow cools and forms a ceiling over the hot magma flowing beneath it, resulting in a tubular conduit for the lava (Figure 6.15a and b). If the eruption stops, the lava drains out the bottom of the tube, forming a long cave system with walls made entirely of basalt (rather than limestone, like most subterranean caves). Sometimes the ceiling of the tunnel collapses into the molten magma flowing beneath, forming a skylight (Figure 6.15c). Volcanologists on Kilauea on

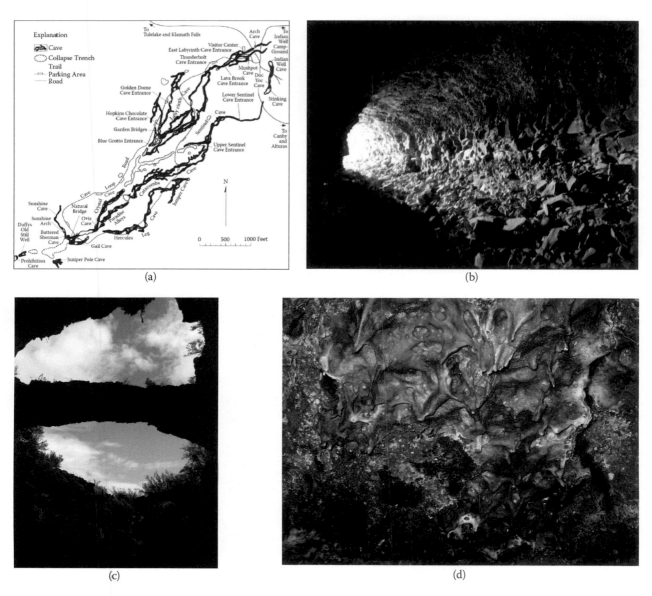

FIGURE 6.15 Lava Beds National Monument. (a) Map of the major lava tubes and caves. They all tend to trend from southwest to northeast because they are the products of flows from Mammoth Crater in the southwest corner of the park. (b) Photograph of the inside of Skull Cave, a lava tube with frozen icicles at the bottom. (c) The double skylight formed by the collapse of the ceiling of Tichner Cave formed this lava bridge. (d) Lavacicles, formed by molten lava that was dripping from the ceiling before it cooled.

Source: (a) Courtesy of US Geological Survey. (b) Photo by the author. (c) Photo by G. Hayes. (d) Courtesy of Wikimedia Commons.

the Big Island of Hawaii have studied many active lava tubes as magma was flowing beneath and use the skylights to lower instruments down into the magma for measurements and samples.

There are even lava stalactites ("lavacicles") that once dripped from the ceiling before cooling (Figure 6.15d), and many other peculiar features. Lava Beds National Monument is famous for its many miles of lava tubes, many of which can be explored during a visit to the park.

The entire Modoc Plateau demonstrates the full range of volcanic features typical of basaltic eruptions. In many places, you can see wrinkled ropy-looking pahoehoe flows, formed when the soft cooled "skin" on a hot lava flow got crumpled up by the moving flow beneath (Figure 6.16a). You can find many instances of jagged a'a lava, formed into blocky lavas full of gas bubbles (Figure 6.16b). And in some places, it is possible to see columnar jointing, formed when lavas cool and contract (Figure 6.16c).

The Modocs were among the last Native Americans to resist the American conquest of California, fighting a bitter war in 1871–1872 that ended when their chief, "Captain Jack," was holed up in what is now Lava Beds National Monument (Figure 6.16b). He and his people hid in the natural fortifications and defenses of the lava tubes and in the crevices of the lava fields, where they had an advantage defending their positions against attack and where no search parties could find them. After three failed assaults by the US Army spanning almost a year, they were all captured, thanks to the guidance of Native American scouts who knew the terrain that white soldiers couldn't navigate. The Modocs were executed for attacking white settlers or sent off to the Indian reservations in eastern Oklahoma.

Beneath the very young lavas that cover most of the Modoc Plateau are much older lavas that are exposed in just a few areas not covered by recent eruptions. The oldest ones appear to be similar to the Steens Mountain basalt just across the border in Oregon. It covered about 50,000 km² (19,000 mi²) of southwestern Oregon, and the southern edge poured into Modoc County, California. In some places, the numerous stacked flows are more than 1,000 m (3,300 ft) thick. Most of the volcanic material is basalt, although there were some rhyolitic eruptions in Oregon as well that show up as white sparkling volcanic ash in Modoc County, often interlaid with dark basalt lava flows.

The Steens Mountain eruption is dated at 16.7 Ma, and it is the oldest of a series of eruptions known as the **Columbia River flood basalts** that covered most of eastern Washington and Oregon (and parts of California, northwestern Nevada, and Idaho) with 174,300 km³ (41,800 mi³) of lava flows during the middle Miocene, from 16.7 Ma to 14 Ma. Flood basalts don't erupt from a single crater like most volcanoes but instead from big fissures (Figure 6.6) in the ground that bring up mantle material in enormous volumes. Since basalts are as fluid as water, these eruptions spread from the fissure in the ground like enormous water floods and can cover huge areas just as fast as floodwaters spread. The enormous thickness (1,800 m, or 5,900 ft) of

(a)

(b)

(c)

FIGURE 6.16 Typical basaltic lava structures of the Modoc Plateau. (a) Ropy, wrinkled pahoehoe flows, south flank of Medicine Lake volcano. (b) Blocky, bubble-filled a'a flows, in Captain Jack's Stronghold, one of the largest lava flows in Lava Beds National Monument, looking north toward Oregon. (c) Hexagonal tops of columnar joints, south flank of Medicine Lake volcano.

Source: Photos by the author.

the Steens Mountain eruption and the dozens that covered the same region over the next 3 m.y. completely covered the older bedrock in eastern Oregon and Washington, so just about the only geology you can find in these areas are Columbia River basalts. The source of the lavas that flowed up from fissures is thought to be a huge hot spot or mantle plume. Since that eruption, the North American plate has shifted westward, and the hot spot burned a path across southern Idaho in the Pliocene (Snake River floodplain); some geologists think it is the same mantle plume that lies beneath Yellowstone today (although it currently erupts as rhyolite, not basalt).

At the edges of the young basalts of the Modoc Plateau are large Pleistocene lake beds which filled the valley with fine-grained lake sediments. Some of the large lakes near the Oregon border (Tule Lake and Goose Lake) sit in those basins and are still contributing lake sediments today.

The eastern edge of the lava plateaus run up against the Warner Range, just west of the Nevada border. The Warner Range is more typical of the Basin and Range Province, with a thick sequence of Eocene and Oligocene volcanic rocks in most places, famous for its fossil plant localities (see Chapter 20). Its geology and faulting style are really part of the story of the Basin and Range, so the province boundary between the Modoc Plateau and the Basin and Range can be considered the western edge of the Warner Range. The Basin and Range Province is the topic of Chapter 7.

6.6 SUMMARY

The northeastern corner of California is almost entirely built of young (Neogene, mostly Quaternary) volcanic rocks. They include the Andean-style volcanic arc of the Cascades, especially its youngest large volcanoes, Mt. Shasta and Mt. Lassen. The Modoc Plateau is covered by young lava flows erupted from the Medicine Lake volcano.

RESOURCES

Alt, D., and Hyndman, D.W. 2000. *Roadside Geology of Northern and Central California*. Mountain Press, Missoula, MT.

Clynne, M.A., Robinson, J.E., Nathenson, M., and Muffler, L.J.P. 2012. *Volcanic Hazards Assessment of the Lassen Region, Northern California*. U.S. Geological Survey Scientific Investigations Report 2012–5176-A. U.S. Geological Survey, Reston, VA.

Dutton, C.E. 1880. *Report on the Geology of the High Plateaus of Utah*. U.S. Geog. and Geol. Survey of the Rocky Mountain Region, vol. 32.

Francis, P., and Oppenheimer, C. 2003. *Volcanoes* (2nd ed.). Oxford University Press, Oxford.

Harris, S.L. 2005. *Fire Mountains of the West: The Cascades and Mono Lake Volcanoes*. Mountain Press, Missoula, MT.

Hill, R. 2004. *Volcanoes of the Cascades: Their Rise and Risk*. Falcon Guides, New York.

Lamb, S. 1991. *Lava Beds National Monument*. Lava Beds Natural History Association, Tulelake, CA.

Lockwood, J.P., and Hazlett, R.W. 2010. *Volcanoes: A Global Perspective*. Wiley-Blackwell, New York.

Miller, C.D. 1980. *Potential Hazards from Future Eruptions in the Vicinity of Mt. Shasta Volcano, Northern California*. U.S. Geological Survey Bulletin 1503. U.S. Geological Survey, Reston, VA.

Sigurdsson, H., Houghton, B., Rymer, H., Stix, J., and McNutt, S. 1999. *Encyclopedia of Volcanoes*. Academic Press, New York.

VIDEOS

Lassen Volcanic National Park:
www.youtube.com/watch?v=SDkpRbrNk8M
www.youtube.com/watch?v=951GD_GnPKI
www.youtube.com/watch?v=0-85WYYi8PU
Lava Beds National Monument:
www.youtube.com/watch?v=U-nt__ZxSKs
www.youtube.com/watch?v=1HlVqnMF3uA
www.youtube.com/watch?v=YGziBQyphKw
www.youtube.com/watch?v=o1TL0a40J4c
Modoc War and Captain Jack's Stronghold
www.youtube.com/watch?v=pCXIoNvNabE
Mt. Shasta:
www.youtube.com/watch?v=0-85WYYi8PU
www.youtube.com/watch?v=qwwjl6ke9Qs
www.youtube.com/watch?v=JqtQ6BczcI4
Volcanoes:
http://video.nationalgeographic.com/video/101-videos/volcanoes-101
http://video.nationalgeographic.com/video/volcano-eruptions
www.youtube.com/watch?v=LB8DLqQXREM
www.youtube.com/watch?v=EHWSTkrWKSg

7 The Broken Land
The Basin and Range Province

Basin. Fault. Range. Basin. Fault. Range. A mile of relief between basin and range. Stillwater Range. Pleasant Valley. Tobin Range. Jersey Valley. Sonoma Range. Pumpernickel Valley. Shoshone Range. Reese River Valley. Pequop Mountains. Steptoe Valley. Ondographic rhythms of the Basin and Range. . . . Each range is like a warship standing on its own, and the Great Basin is an ocean of loose sediment with these mountain ranges standing in it as if they were members of fleet without precedent, assembled at Guam to assault Japan. Some of the ranges are forty miles long, others a hundred, a hundred and fifty. They point generally north. The basins that separate them—ten and fifteen miles wide—will run on for fifty, a hundred, two hundred and fifty miles with lone, daisy-petalled windmills standing over sage and wild rye.

—**John Mcphee**
Basin and Range

7.1 DATA

Highest point: Boundary Peak, White Mountains: 4,341 m (14,242 ft)

Lowest point: Badwater, Death Valley: −86 m (−282 ft)

Predominant rocks: Precambrian, Paleozoic, and Mesozoic rocks in the ranges; Quaternary alluvium in the basins; abundant Cenozoic volcanics

Plate tectonic setting: Passive margin in the Paleozoic, back-arc of a subduction zone in the Mesozoic, extension since the early Miocene

Most recent eruptions: Long Valley Caldera, Mono-Inyo craters, Coso volcanic field

Geologic resources: Gold, silver, copper, rare earth elements, and other valuable metals; evaporite minerals; building stone, cement, sand, and gravel

National parks: Death Valley National Park, Mojave National Recreation Area, Inyo National Forest

State parks: Red Rock Canyon

7.2 GEOGRAPHY OF THE BASIN AND RANGE

The Basin and Range Province (Figure 7.1) actually covers a large portion of western North America: southern Oregon and Idaho, all of Nevada and western Utah, the eastern part of California, southern Arizona and New Mexico, and much of northern Mexico. For the purposes of this book,

we will focus on the California portion and some parts of adjacent Nevada, although we will look at the tectonics of the entire province. Even though most of the Basin and Range is outside California, a huge area of California (especially the Mojave Desert and Owens Valley) is part of this region. The entire province spans 800 km (500 mi) from the Wasatch fault in central Utah to the eastern edge of the Sierra Nevada in California, with the Colorado Plateau on its southeastern boundary and the Snake River Plain on its northern boundary.

It's important to distinguish the **Basin and Range Province** (a geological entity defined by distinctive crustal rocks and their structure) from the **Great Basin**, which is a topographic description of the surface features of a region and its drainage pattern. A large portion of the Great Basin has no rivers that reach the ocean (Figure 7.2). Instead, the intermittent rivers are dry much of the year. When they have water after seasonal thunderstorms and flash floods, they flow across the landscape until they sink into the ground, a pattern known as **internal drainage**. For example, the Humboldt River arises from snowmelt in the mountains of northeastern Nevada and crosses 530 km (330 mi) west along the northern half of the state, making it one of the largest rivers in the Great Basin. It creates gaps in the mountains that made it the natural pathway for the pioneer trails across the state; today it is the route of Interstate 80. It then disappears into the ground at Humboldt Sink near Carson City. The Mojave River arises from snowmelt in the San Gabriel and San Bernardino Mountains, drains north past Victorville and Barstow, and then vanishes underground through most of its course except in the wettest years.

The entire Great Basin is a huge desert, with most of it receiving less than 25 cm (10 in) of rain in a year. What little rain does arrive comes in the form of sudden thundershowers and downpours that cause an entire year's worth of erosion of the landscape in minutes. Enormous flash floods and debris flows carry enormous volumes of rock and sediment. For the rest of the time, the region gets no rainfall and often reaches extreme temperatures. The hottest spot in the Americas (and probably the world) is Death Valley, which is routinely above 40°C (104°F) in the summer and holds the record temperature of 60°C (134°F) while receiving less than 2.5 cm (1 in) of average rainfall.

Much of the extreme dryness of the Great Basin is due to the fact that it is in the **rain shadow** (Figure 7.3) of some of the highest mountains in North America, especially the Sierra Nevada and the Cascades. In this latitude, the prevailing winds come out of the west ("westerlies") and bring Pacific moisture to the west side of these mountain ranges.

DOI: 10.1201/9781003301837-7

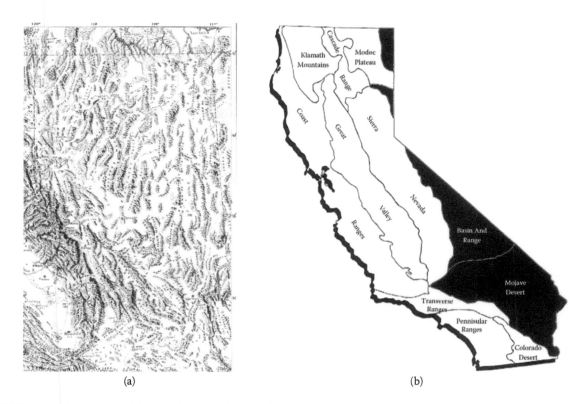

(a)   (b)

FIGURE 7.1 (a) Hand-drawn physiographic map showing the extent of the Basin and Range topography across the western United States. (b) Index map showing the location of the Basin and Range and Mojave Desert Province in California.

Source: Courtesy of US Geological Survey.

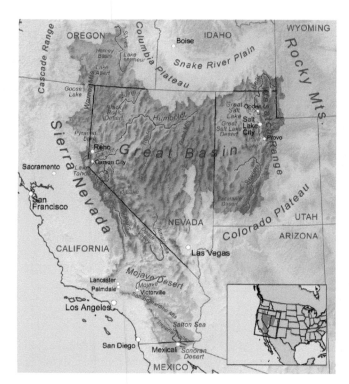

FIGURE 7.2 Simplified hydrologic map of the Great Basin (area shaded in light brown), showing the internal drainages that never reach the ocean but end up sinking into the gravels of the river bottoms, or into "sinks" like Humboldt Sink.

Source: Courtesy of Wikimedia Commons.

FIGURE 7.3 Rain shadow effect. Rising moist clouds from the ocean blow up the slope of the mountains (in this case, the Sierras and Cascades). As the clouds rise, they cool and condense and release most of their moisture as rain or snow on the ocean side of the range. Coming down the other side, the dry air warms and expands, sucking further moisture from the ground and creating a desert behind the range.

Source: Courtesy of Wikimedia Commons.

In the process of rising above the mountains, the clouds cool and they rain or snow most of their moisture on the western slope of the Sierras. As the air descends the eastern slope, it has been stripped of much of its moisture. Descending high-pressure air also warms up from compression and absorbs heat from the ground below, so it becomes warm and dry and sucks all the moisture out of the regions behind the mountains.

The region is also very dry because of its geographic location. The parts of the globe between 10° and 40° north and south of the equator are known as the **subtropical high-pressure belts** of the atmospheric circulation system. Low-pressure air that rises from the tropical rain forests (and rains back most its moisture) descends as part of a global pattern of air circulation cells. In the subtropical latitudes, this creates permanent high pressure and thus warm, dry conditions. Thus, the desert conditions of the Great Basin are similar to those of some of the other large deserts of the world, almost all of which are located between 10° and 40° north and south of the equator.

7.3 BASIN AND RANGE GEOLOGY

7.3.1 Structure

Geologically and topographically, the distinctive feature of the province are the parallel basins and ranges that cover the entire landscape. They trend roughly north–south and are evenly spaced across the entire region (Figures 7.1 and 7.4). The great geologist C. P. Dutton looked at a hand-drawn physiographic map of the region (Figure 7.1a) and said it reminded him of an "army of caterpillars marching to Mexico." The striking parallelism in the north–sound trends, and even spacing of the ranges, shows up not only on traditional hand-drawn maps (Figure 7.1) but also on satellite imagery and digital maps (Figure 7.4a) and aerial photos (Figure 7.4b).

There is a good reason for this remarkable pattern of parallel ranges and basins. The crust of Nevada has undergone tremendous east–west extension in the past 23 m.y., stretching to about twice its original dimensions (Figure 7.5). The brittle crust at the surface fractures into thousands of normal faults to accommodate this extension, with hundreds of horsts and grabens (see Chapter 4 and Figure 4.2).

As the crust pulls apart, the grabens drop down between the horsts and sediment eroded from the ranges fills the grabens, often with thousands of feet of sands and gravels and lake deposits (Figure 7.6). The basin fill varies depending on where you encounter it. In the foothills of the ranges and against the fault scarps, it tends to be coarse conglomerates, arkoses, and other deposits of flash floods and debris flows. These roar out of confinement of the mountain canyons and then spread out once they reach the valley floor, where they lose energy and drop most of their coarse boulders, gravel, and sand to form alluvial fan conglomerates (also called **fanglomerates**). In the middle of the basin, however, the fine-grained muds accumulate to form a **playa lake**. In many basins, the water floods during the rains and then evaporates completely, so many playa lakes are covered with extensive evaporite deposits of salt, gypsum, and other chemical sediments (discussed further later).

Not all the ranges in the province are simple horsts and grabens. There is a lot of strike-slip faulting in certain areas

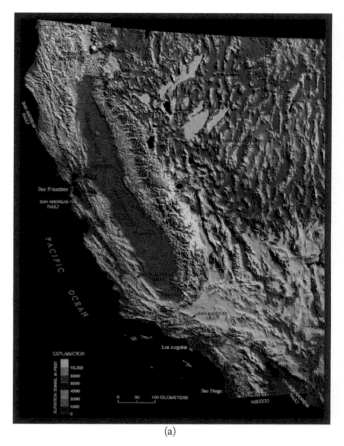

(a)

(b)

FIGURE 7.4 (a) Digital relief map showing the details of the parallel Basin and Range topography. (b) Low-angle aerial photograph, looking across the tops of the sequence of parallel ranges.

Source: (a) Courtesy of US Geological Survey. (b) Photos by M. B. Miller.

that may have contributed to the tension and shearing that produced the normal faulting. Recent research shows that most of the ranges are tilted blocks sliding down a ramp known as a **detachment fault** (Figure 7.7). These detachment surfaces tend to form steep fault scarps near the surface but then flatten out to horizontal at depth so the crustal blocks are sliding down a ramp like roller coaster cars, one after another. The tilted blocks are bounded by a normal

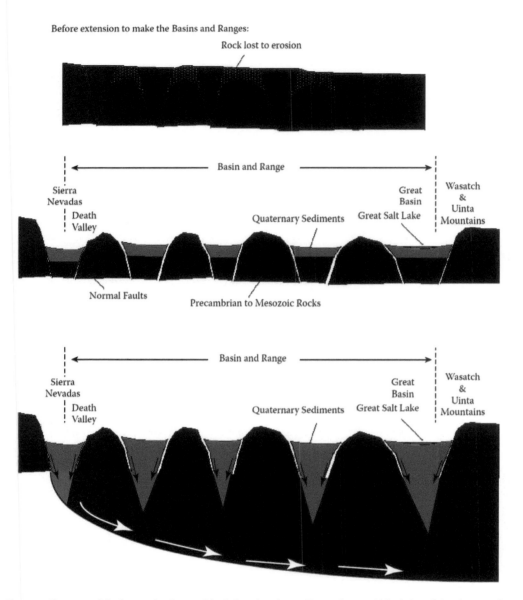

FIGURE 7.5 Cartoon diagram of the horsts (upthrown blocks) and grabens (down-dropped blocks) and the degree of crustal extension represented by the Basin and Range structure. At the top is the crust before extension. In the middle is the conventional model of horsts and grabens. At the bottom is the more modern understanding of the structure, with the fault blocks sliding down the detachment fault to form half-grabens.

Source: Redrawn from several sources by E. T. Prothero.

fault on one side of the valley, but the other side is formed by the dip slope of the top of the tilted block. Since they are formed by one normal fault, not two, they are called **half-grabens**.

So when you travel through the Basin and Range Province, many mountain fronts are probably large normal faults, often with displacements of many kilometers. The other half of the rocks you see high on the range are located deep beneath your feet on your side of the fault.

We will look at the deep crustal structure of the Basin and Range later in this chapter. First, let's look at a lot of other important geological and geophysical features of the region.

7.3.2 Heat Flow

The Basin and Range Province has the highest heat flow in the United States, a measure of how much heat is coming up from the mantle below (Figure 7.8). In most places, the heat flow is well over 100 mW/m², which is extraordinarily high and results in ground temperatures close to the boiling point of water (100°C) in many places. Within this large region of high heat flow, the northern part of Nevada has the highest heat flow in all the United States. It is known as the Battle Mountain heat flow high. Hot springs and geothermal operations in the United States have their highest concentrations in the Basin and Range, where boiling

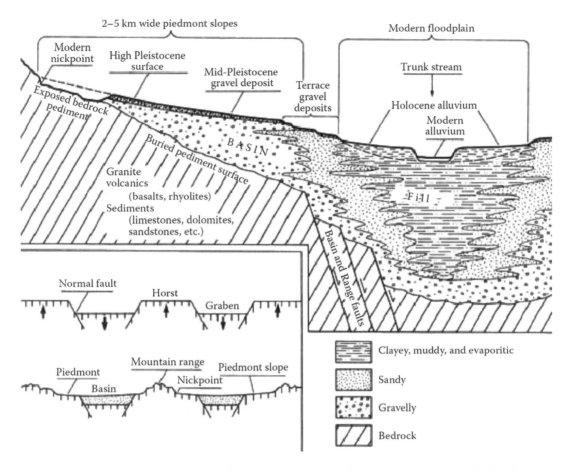

FIGURE 7.6 Schematic diagram of the geometry of the graben basins and their typical sedimentary valley FILL.

Source: By permission of the Geological Society of America.

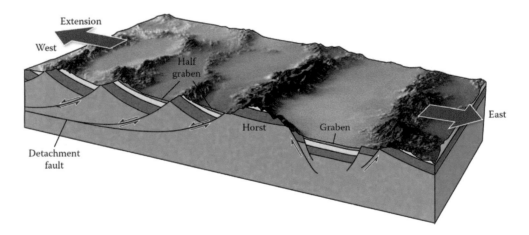

FIGURE 7.7 Diagram showing the geometry of the detachment faults and half-grabens, forming many of the Basin and Range features.

Source: Courtesy of Oxford University Press.

water from heat by shallow underground magma chambers rises to the surface. Such high heat flow is almost always regarded as an indication of a shallow heat source in the area, especially magma chambers, or even very shallow crust and mantle near the surface. Notice that the heat flow is very low (Figure 7.8) in the major mountain chains of the United States (Sierras, Cascades, and Appalachians) because they have deep crustal roots and float like a cork on the mantle. Thus, the colors of the map are a rough indication of the depth of the mantle.

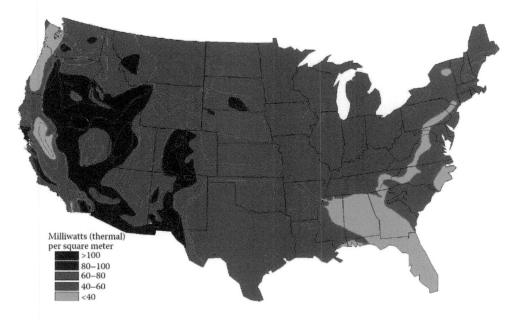

FIGURE 7.8 Heat flow map of the United States showing the high heat flow in the Basin and Range.

Source: Courtesy of US Geological Survey.

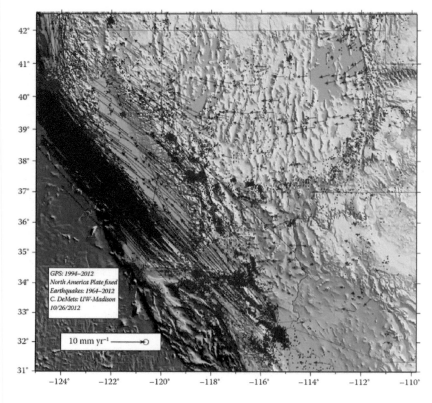

FIGURE 7.9 Map showing the seismicity in the western United States. Blue dots are earthquake epicenters; red arrows indicate direction and amount of fault movement. The most active and rapidly shifting faults occur in California in association with the San Andreas fault, but there are many east–west extensional normal faults in the Basin and Range (short red arrows pulling apart in the east–west axis).

Source: Courtesy of US Geological Survey.

7.3.3 SEISMICITY

After the strike-fault zones associated with the San Andreas fault, the next most seismically active region in the United States is the Basin and Range (Figure 7.9). Almost all the faults in the region are extensional normal faults, all pulling apart in an east–west direction, as predicted by the structure. Although these faults have not produced as many great earthquakes as the San Andreas and its related faults (Chapter 11), there are many moderate quakes along these faults. This is especially true of the faults in the Owens Valley along the east flank of the Sierras and the Wasatch fault that runs down the center of Utah. Both of these faults mark the extreme limits of the province, where the Basin and Range extension changes abruptly into a different type of geologic province on the edges (Colorado Plateau on the east and Sierra Nevada on the west).

7.3.4 Volcanism

One of the most characteristic features of the Basin and Range is the prevalence of volcanic eruptions. Such eruptions have gone on through most of the Cenozoic and are still active (Figure 7.10). During the middle Eocene through Miocene, huge rhyolitic and dacitic eruptions were common across the entire region, covering it with ash blankets and ignimbrites that were later chopped up as the Basin and Range began expanding and stretching in the early to middle Miocene. There are also many places with basaltic eruptions, suggesting that the magmas came from both deep milliwatts (thermal) per square meter mantle sources (the basalts) and the shallow granitic magma chambers (rhyolites and dacites). This combination of deep mantle basalts erupting alongside shallow magma chambers producing rhyolites is known a **bimodal volcanism** and is very typical of the Basin and Range Province.

The volcanism is not finished, by any means. The big late Cenozoic (some only a few thousand years old) eruptions on the Modoc Plateau (Chapter 6) are probably related to Basin and Range extension. There are young volcanic basaltic eruptions in many places across the region, such as the extensive lava flows in southern Utah and Nevada (especially near St. George and Hurricane, Utah), the huge lava flows and cinder cones in the southern Mojave Desert (such as Amboy and Pisgah craters), and the lava flows at Devil's Postpile near Mammoth Mountain Ski Resort.

The biggest and most recent eruption of all, however, was the explosion of Long Valley Caldera, in the Owens Valley, near Bishop and the Mammoth Mountain Ski Resort (Figure 7.11a and b). Dated 758,900 years ago, this was one of the largest eruptions in North American prehistory. It released 600 km³ (140 mi³) of ash, which spread across almost all of western North America as far east as Nebraska and Kansas (Figure 7.11c). Huge ignimbrites at 820°C (1,500°F) covered the local region, forming the thick Bishop Tuff near the vent (Figure 7.11a and d), which filled the 3 km (2 mi) deep caldera nearly to the rim after it collapsed into its own magma chamber (Figure 7.11b). This volcano has been dormant since that time, but there are many hot springs in the area, showing that the magma is still not far beneath the surface. In the 1980s, there was much seismicity in the region (including several M6 quakes), causing geologists to worry that a new eruption was on the way. However, that seismicity has died down for a few decades, so the alarms have been scaled back.

The presence of so much volcanism, especially from mantle-derived basalts, suggests that the mantle is not far

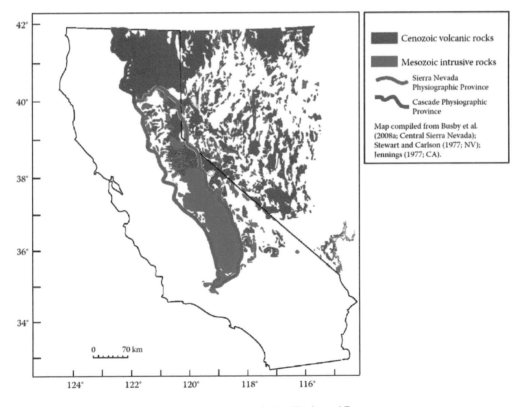

FIGURE 7.10   Map showing location of Cenozoic volcanic rocks in the Basin and Range.

Source: With permission from the Geological Society of America.

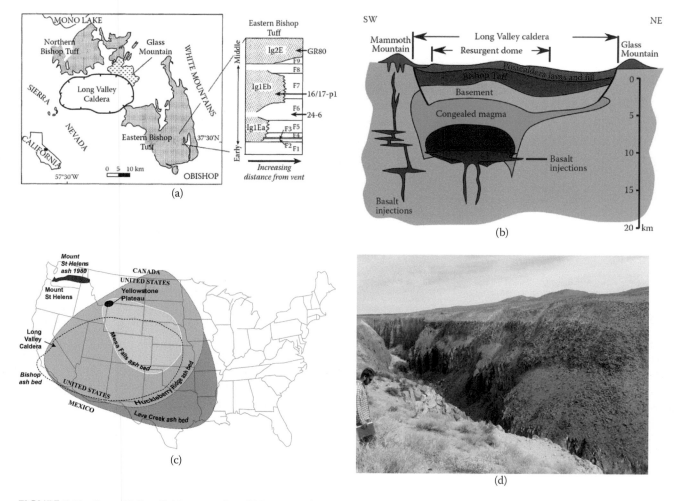

FIGURE 7.11 Long Valley Caldera eruption. (a) Map showing the location of Long Valley Caldera and the major remaining sheets of the Bishop Tuff. On the right is the stratigraphic column of the various eruption phases, interfingering with nonvolcanic rocks beyond the eruption. (b) Cross section of Long Valley Caldera. (c) Map showing how much of North America was covered by the Bishop Tuff in comparison to the similar eruptions from Yellowstone (Huckleberry Ridge, Lava Creek, and Mesa Falls eruptions) and the "tiny" eruption of Mt. Saint Helens. (d) Thick deposits (more than 200 m thick) of several different flow units of Bishop Tuff with distinctive columnar jointing, exposed by the erosion of Owens River in Owens Gorge.

Source: (a) With permission from the Geological Society of America. (b) Courtesy of US Geological Survey. (c) Courtesy of Wikimedia Commons. (d) Photo by the author.

beneath the thin crust of the Basin and Range Province. The many faults provide conduits for volcanic eruptions derived from both shallow rhyolitic magma chambers and deep basaltic magma chambers all erupting in close proximity.

7.3.5 MINERALIZATION

The Basin and Range is legendary for its mineral wealth. Nevada received its nickname, "The Silver State," based on the huge silver mining rush in the 1860s–1880s. Many of the miners had been in the California gold fields in 1849 and later and then switched to silver when they failed to get rich from California gold. The main silver rush started in 1859 by the discovery of rich crusts of silver just lying on the ground, followed by rich silver lode deposits, which came to be known as the Comstock Lode. From this huge silver boom, big cities like Gold Hill and Virginia City sprang

up overnight, and by 1865, so many people had come to Nevada that it was admitted to the Union as a state. At the peak of mining in 1877, the Comstock Lode was producing $310 million in gold and $465 million in silver a year (in today's dollars). By the late 1880s, most of the ore had been mined out, and the boomtowns became ghost towns by the 1890s. Gold and silver are still being mined in other places in the state (Tonopah, Eureka, Pioche, Rochester, and Reese River), although nowhere near the production as during the peak of the Comstock Lode. Silver and gold were also found in many places in California's Basin and Range mountains, including the Mojave Desert (Calico and Randsburg) and the Owens Valley (Cerro Gordo and Panamint).

After silver and gold, copper is the most abundant mineral in the Basin and Range Province, and the region is still one of the biggest copper producers in the world. Arizona's Basin and Range mountains have produced more

FIGURE 7.12 Bingham Canyon Copper Mine west of Salt Lake City, the biggest mine in the world.

Source: Courtesy of Wikimedia Commons.

than $5 billion in 2006 alone, and there are big copper mines in other states, like Utah, as well. Most of the Basin and Range copper comes from strip-mining away whole mountains of igneous intrusions known as **porphyry copper deposits**, where tiny veins of copper and silver have penetrated magma as they cooled. These enormous open-pit mines are larger than meteorite craters and can be seen from space. Bingham Canyon Mine near Salt Lake City, Utah (Figure 7.12), is 4 km (2.5 mi) wide, covers 770 ha (1,900 ac), and is 1 km (0.6 mi) deep. This single mine has produced 17 million tons of copper, 715 tons of gold, 5900 tons of silver, and 386,000 tons of molybdenum, worth more than the California Gold Rush, the Nevada Comstock Lode, and the Klondike Gold Rush in Alaska and Canada combined. Morenci Copper Mine in southern Arizona is the largest producer of copper in the world, with more than 3.2 billion tons produced so far.

Gold, silver, copper, molybdenum, tungsten, lead, zinc, and many other metals have been produced in the Basin and Range, making it one of the richest mining areas in the world. All these deposits are due to magma intrusions along normal faults, bringing up fluids that concentrate these rare metals. Many are due to the fact that the magmas intrude Paleozoic sedimentary host rocks, which interact with their chemistry and contribute to the unique mineralization. Thus, the Basin and Range owes its entire wealth to the unique pattern of faulting and igneous intrusions.

7.3.6 Paleomagnetism

Geologists knew about the general idea of the Basin and Range extension for more than a century, but in the 1980s, another key piece of evidence came from paleomagnetism. Stanford geophysicist Allan Cox and his students and colleagues looked at ancient magnetic directions recorded in rocks from different ages in the Sierras and Cascades.

Instead of pointing due north in their current position, these rocks indicated that the Sierra–Cascade ranges had rotated and swung like a huge door hinging on the Olympic Peninsula of Washington. In particular, the Sierras had swung southwestward from their position before the Basin and Range opened in the Miocene and today are about 210–340 km (130–211 mi) west of where they were in the early Miocene (Figure 7.13).

Geologists have long known from geophysical data that large mountain ranges float on the mantle like corks, or like icebergs in the ocean. Most of their mass floats below the fluid level as a deep "root" for the mountain range, and only the top fraction sticks up above sea level. The High Sierras and Cascades must also float like icebergs on the mantle, with a very deep root displacing the mantle material as it floats. If the range has swung this far to the west, then something must be pushing the root of the ranges to the west, and then the crust gets stretched behind the "swinging door of the ranges." This suggests that some bulge of mantle material has come up behind the Sierras and beneath the Basin and Range, and as it did so, it pushed the root of the Sierras to the west.

It is analogous to a paper fan (Figure 7.13). The Sierras would be one frame of the fan, and the edge of the continent in central Utah the other. As the Sierras swing the fan open and southwestward away from North America, the pleats of the paper fan (the basins and ranges) are stretched apart and begin to open from the southern tip of the fan to the northern area near the hinge, which opens last.

This discovery solved another great mystery: why the Garlock fault (labeled *G* in Figure 7.13) was the only significant left-lateral fault in California. Nearly all California faults are right lateral, dominated by the right-lateral shear on the San Andreas transform. The Garlock fault, however, takes up the slack of the southern tip of the Sierras as it swings to the southwest, and if you look at the relative motion of the fault, it is left lateral.

7.3.7 Thin Crust

Putting all these distinctive features of the Basin and Range together, we can conclude that it has had its crust stretched and fractured to twice its original width. This comes from many different geological indicators: the pattern of east–west extension on thousands of normal faults; the evidence of high heat flow, suggesting thin crust with mantle not far below; the high seismicity, suggesting that the normal faults are still moving and stretching and taking up stress; the abundance of volcanic rocks that have flowed up through this thin crust, including mantle-derived basalts; the mineralization that followed these intrusions; and the displacement of the root of the Sierras to the west by some sort of mantle upwelling. All these suggest that the mantle is very shallow beneath the region.

Just how shallow? By bouncing seismic waves off the deeper structure of the crust, seismologists have found that the brittle surface rocks with normal faulting soften into

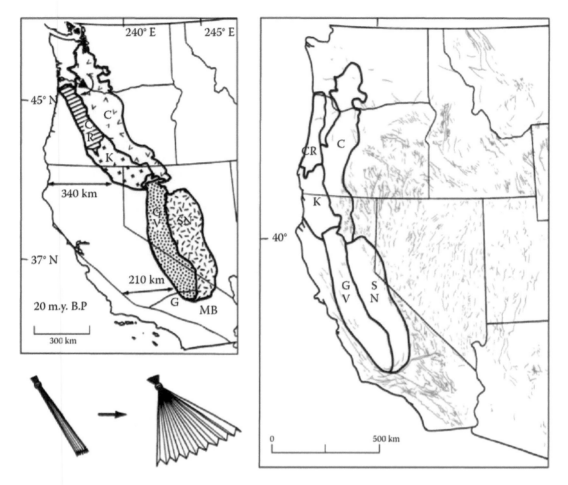

FIGURE 7.13 Clockwise westward tectonic rotation of the Sierra Nevada (SN) as indicated by paleomagnetic data. C = Cascades; CR = Coast Ranges; GV = Great Valley; K = Klamaths. The roughly 250 km of westward movement of the Sierra Nevada block and its deep roots has stretched and expanded the Basin and Range to the east, much like the opening of a paper fan.

Source: After Prothero, D. R., and R. H. Dott Jr. *Evolution of the Earth* (8th ed.). McGraw-Hill: New York, 2010.

rocks that undergo ductile flow at depths of about 10 km, especially along the detachment faults (Figure 7.14). Below that is some more stretched and thinned ductile continental crust. The Moho (crust–mantle boundary) is only 20–35 km (12–21 mi) below the surface (Figure 7.14). This is by far the largest area of continental crust this thin anywhere on the planet. It is stretched so thin that it is only a bit thicker than oceanic crust (10 km thick), and only 20% of the typical crustal thickness of 100–150 km. Some kinds of tectonic forces are bringing an enormous bulge of mantle material just beneath the thinned stretched skin of the Basin and Range. We will discuss some possible reasons for this in Chapter 15.

7.4 GEOLOGIC HISTORY OF THE BASIN AND RANGE PROVINCE

The oldest rocks in California occur in the Basin and Range Province, including Precambrian metamorphic rocks, and thick sequences of Paleozoic and Mesozoic rocks. Few Precambrian rocks are found in any other part of California (except for exotic blocks in the northwest Sierras and

Klamaths that didn't start in California), so we can assume that most of California did not exist until the early Paleozoic or later. This section will briefly review the geologic events that created the complex history of the Basin and Range Province in this chapter. Chapter 15 will combine this history with all the events in the rest of California to give a complete story of the assembly of the Golden State.

7.4.1 PRECAMBRIAN

The oldest rocks in California are metamorphic rocks that date between 1.6 and 1.8 Ga (Ga = giga-annum = billion years ago). They form the basement rocks in the many different uplifted ranges across the region, especially in places like Death Valley and in many ranges in the Mojave Desert and Arizona–Nevada. Such ancient rocks are part of a tectonic terrane that accreted to the core of North America 1.6–1.8 Ga, known as the **Mojavia terrane** (most of the basement of the Mojave Desert) and the **Yavapai terrane** (a wedge between the Mojavia and Mazatzal terranes) (Figure 7.15).

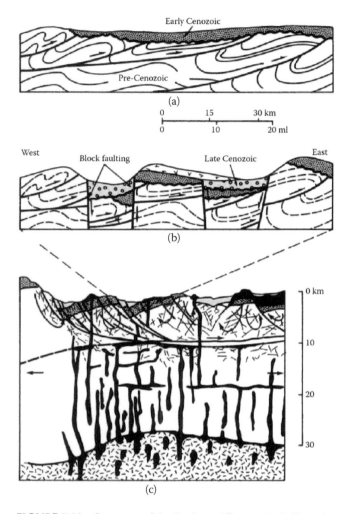

FIGURE 7.14 Structure of the Basin and Range. (a–b) Near the surface are ancient Paleozoic and Mesozoic thrust faults from the Antler and Sonoma and Sierra–Sevier Orogenies, cut by Miocene and younger shallow normal faults leading to detachments, plus many magma intrusions. (c) At depth, the entire continental crust is tremendously thin and stretched, and the mantle is very close to the surface.

Source: After Prothero, D. R., and R. H. Dott Jr. *Evolution of the Earth* (8th ed.). McGraw-Hill: New York, 2010.

All these units accreted in the late Proterozoic (1.6–1.8 Ga) and have formed the stable core of the central and western parts of North America ever since (see Chapter 15).

In much of the Mazatzal–Yavapai–Mojavia region, the only way to expose such ancient basement rocks to the surface is to have a deep valley like the Grand Canyon cutting all the way through miles of sedimentary rocks to reach them. Given that the Basin and Range is a region of extension, it seems surprising that rocks from deep in the basement could be uplifted at all. In most tectonic settings, compressional thrusting and folding are required to bring rocks up from so deep in the crust. This was long a puzzle, until geologists began to look at some of the ranges with exposed Precambrian metamorphic rocks (**metamorphic core complexes**) in a different way. They realized that what

exposed such deep basement rock were younger rocks sliding away on detachment faults (Figure 7.7), exposing what was beneath the detachment surface (Figure 7.16). As the Basin and Range extension causes the overlying blocks to slide down the detachment, they release the pressure holding the underlying basement rocks down, which then rise up in the gap by crustal rebound. Most of the exposed Precambrian rocks in the southern Basin and Range are metamorphic core complexes exposed by detachment faults in this manner. Particularly famous examples are the "turtleback" surfaces on the slopes of Death Valley, especially along the flank of the Black Mountains. These exposed Precambrian rocks are domed upward like the curve of the shell of a turtle (hence the name). Their top surface is also the detachment, above which all the rocks that used to lie above the basement complex have slid away and been removed, exposing deep crustal rocks at the surface.

7.4.2 LATE PROTEROZOIC–PALEOZOIC

In the latest Precambrian, the core of what would become North America was part of a supercontinent named **Rodinia**. The modern "western" coast of North America was actually on the north coast, since the North American block was rotated 90° clockwise from its present position. In addition, it straddled the equator, so it was in a tropical to subtropical climate. The East Antarctica plate adjoined southern North America, and the Australian plate nestled against western Canada. As these future continents ripped apart, huge rift valleys were formed and filled with many thousands of meters (thousands of feet) of sediment. Among those late Proterozoic valleys was a rift known as the **Amargosa aulacogen** (a failed rift), which today is found at the base of the sequence in Death Valley. It is a narrow fault-bounded valley (Figure 7.17) filled with a 5 km (16,000 ft) thick sequence known as the Pahrump Group. This unit is made of carbonate rocks (Crystal Springs Formation and Beck Springs Dolomite) at the bottom and capped with another carbonate unit, the Noonday Dolomite. Between these carbonate units that were deposited in shallow, warm tropical seas is a thick deposit of the Kingston Peak Formation (Figure 7.17). This unit is from ancient glaciers that once covered the entire planet right to the equator during one of several "Snowball Earth" events that nearly froze the planet solid.

Once the Rodinia continents had pulled apart and the rifting stopped, the final stage in this process was the slow sinking of the edge of the continent to produce a thick passive margin wedge (see Figure 4.10). The subsidence of this shallow marine shelf (Figure 7.18) off the coast of North America meant that nearshore sediments accumulated almost continuously through many millions of years, making an extraordinarily rich and complete sequence with few unconformities. These deposits were largely shallow marine and beach sandstones, offshore shales, and carbonates even farther offshore in shallow clear water. In places such as Death Valley (Figure 7.19), the Proterozoic and Paleozoic

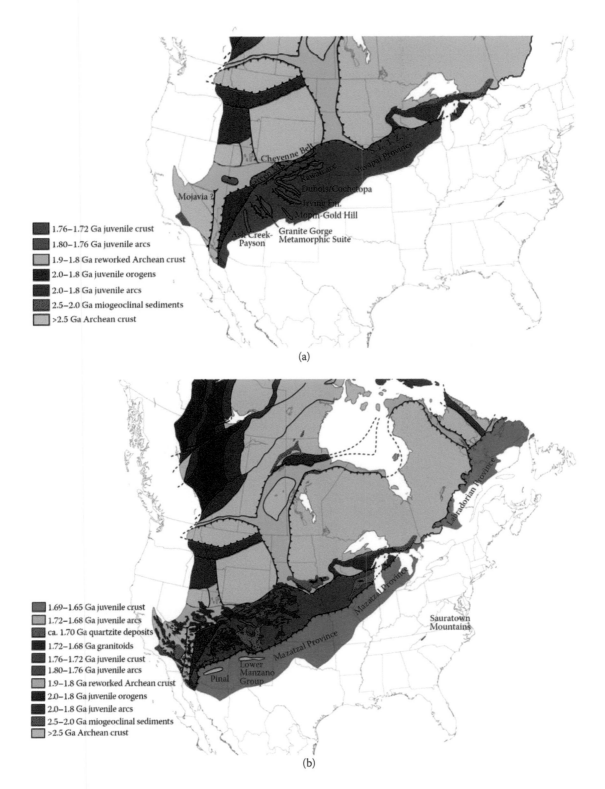

FIGURE 7.15 Early and middle Proterozoic terranes of the United States, focusing on the location of the Archean Mojavia terrane underlying most of the (a) Mojave Desert and (b) the later docking of the Yavapai terrane (light green) and Mazatzal terrane in adjacent Arizona.

Source: From Whitmeyer, S. J., and K. E. Karlstrom. *Geosphere* 3(4): 220–259, 2007.

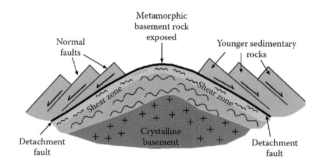

FIGURE 7.16 Metamorphic core complexes expose ancient Precambrian metamorphics due to the sliding of blocks along detachment faults.

Source: Redrawn from several sources.

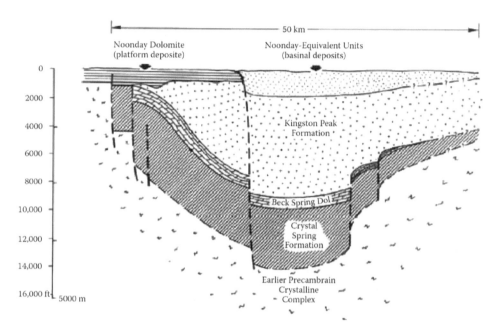

FIGURE 7.17 Cross section of the Amargosa aulacogen in Death Valley.

Source: Courtesy of Pacific Section SEPM.

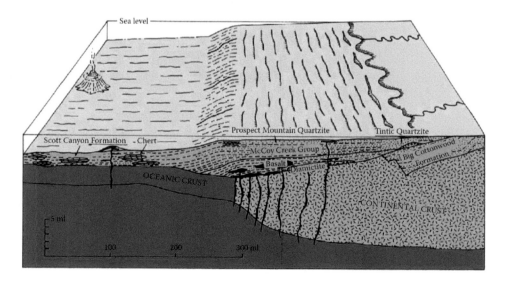

FIGURE 7.18 Late Proterozoic and Paleozoic passive margin wedge that formed on the western edge of Laurentia (future North America) as it rifted from other parts of Rodinia. Thick deposits of shallow marine shelf deposits covered most of the rifted margin, with deeper marine shales offshore.

Source: After Prothero, D. R., and R. H. Dott Jr. *Evolution of the Earth* (8th ed.). McGraw-Hill: New York, 2010.

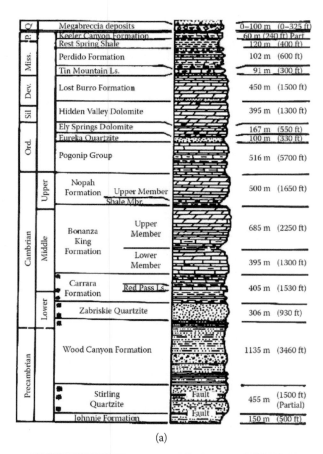

(a)

(b)

FIGURE 7.19 Paleozoic stratigraphic sequence at Death Valley. (a) Stratigraphic column showing the major units and unconformities. (b) View from Aguereberry Point on the top of the Panamint Range looking east into Death Valley. Dipping away from the viewer are the Upper Proterozoic Wood Canyon Formation (bottom of stack), the light band of the shallow nearshore and beach sandstones of the Lower Cambrian Zabriskie Sandstone (now a quartzite), and the shallow marine shales and limestones of the Carrara Formation. At the top of the nearest peaks is the middle Cambrian Bonanza King Formation (originally a shallow marine limestone). Further down the slope, out of view and plunging beneath the gravels of Death Valley are the rest of the Paleozoic formations.

Source: (a) Modified from Hunt, 1981, courtesy of US Geological Survey. (b) Photo by the author.

sequence spans almost 6 km (20,000 ft) in total thickness. In the Arrow Canyon Range northeast of Las Vegas, a similar sequence is more than 8 km (25,000 ft) thick.

While these deposits accumulated slowly over millions of years on the gradually subsiding shelf of the passive margin, there was much less thick and continuous deposition on the stable core of the continent (known as the **craton**) to the east and south of Death Valley. While there is 6–8 km of sediment spanning the Paleozoic in the passive margin wedge, on the craton the same amount of time is represented by just a few hundred meters of sediment. In Death Valley, for example, the Cambrian alone spans 2,300 m (7,000 ft) in total thickness (Figure 7.19), while in the Grand Canyon in Arizona, or the Marble Mountains in the central Mojave Desert, it is barely 300 m (1,000 ft) thick. This striking difference in thickness is caused by the very slow steady subsidence of the passive margin, while there is almost no subsidence on the craton, and it accumulates sediment only during the highest sea level episodes.

7.4.3   LATE PALEOZOIC–MESOZOIC EVENTS

The late Paleozoic–Mesozoic history of Nevada and Northern California (Chapter 15) is a very busy one, with several mountain-building events. In northwestern Nevada, there were two huge collisional events that brought in a piece of crust from across the ocean. The first of these was the **Antler orogeny** in the Late Devonian–Mississippian, which pushed a slice of central Nevada on top of the Roberts Mountain thrust fault (Figure 7.20). The second event collided during the Late Permian and Early Triassic (Figure 7.20). Known as the **Sonoma terrane** (or Sonomia), this huge collision between this exotic block from across the ocean adds a large piece of new crust to North America, an event called the **Sonoma orogeny**. The entire Sonoma terrane was pushed up and over North America many hundreds of kilometers to the west on a giant thrust fault known as the **Golconda thrust**. We will see much more of these events when we look at the Sierras and Klamaths in Chapters 8 and 9.

By contrast, the Basin and Range Province in California has almost no Mesozoic record. There are just few important Triassic exposures in the Owens Valley, such as Union Wash (Chapter 20), which has a thick sequence of Permian and Triassic limestones, famous for their Early Triassic ammonites. Most of the rest of western Nevada and adjacent California were drowned by another shallow sea that produced not only abundant Triassic ammonoids but also gigantic whale-sized ichthyosaurs known as *Shonisaurus* from Berlin-Ichthyosaur State Park near Gabbs, Nevada. Likewise, the Jurassic record in the Mojave Desert or Owens Valley is virtually nonexistent as well, although in adjacent Nevada and Utah and northern Arizona, the majority of the Basin and Range was under Jurassic seas. At the fringe

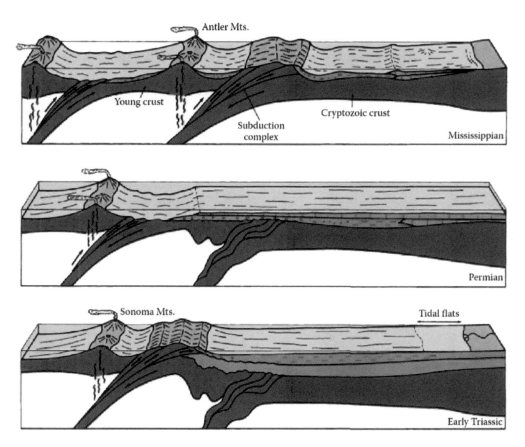

FIGURE 7.20 Paleozoic tectonics of the Basin and Range. Top: In the Late Devonian–Mississippian, the Antler terrane accreted along the Roberts Mountain thrust fault. The Sonomia terrane was not far offshore to the west (left). Middle and bottom: In the Late Permian–Triassic, the Sonomia terrane accreted along the Golconda thrust fault.

Source: From Prothero, D. R., and R. H. Dott Jr. *Evolution of the Earth* (8th ed.). McGraw-Hill: New York, 2010.

of these Jurassic seas, there were gigantic lower Jurassic coastal sand dunes that ran from the Aztec Sandstone at Red Rocks and Valley of Fire, near Las Vegas, to the amazing cross-bedded dune sands of the Navajo Sandstone exposed widely over Arizona, Utah, New Mexico, and Colorado, to the small fringe of that dune field, known as the Nugget Sandstone, way up in central Wyoming.

In the Cretaceous, however, the Basin and Range regions felt the effects of the eruption of the Andean-style arc that made the Sierra Nevada volcanoes (Chapter 8). Most Andean-style arcs have a large thrust belt behind them, facing toward the back-arc basin. This is true of the Argentinian foothills of the Andes Mountains today, and it was true of the Cretaceous Sierran arc as well (Figure 4.14). Huge thrust faults pushed blocks of Paleozoic limestones from California on top of Jurassic sandstones in the Keystone thrust fault, just west of Las Vegas in the Spring Mountains of Nevada. This thrust is part of a back-arc thrust belt that goes north from Las Vegas and parallels the track of Interstate 15 through Utah, continues into the Idaho–Wyoming thrust belt, and carries right on to the overthrusts seen in Glacier National Park in Montana and

on into Jasper and Banff National Parks in Alberta. The tectonic event that caused this thrust belt has earned the name *Sevier orogeny* for such enormous compressional thrusting and mountain-building. However, in modern plate tectonic theory, we now understand that it is just the back-arc thrust belt of a typical Andean-style arc, so most geologists lump them together as the **Sierra–Sevier orogeny** (also called the Nevadan orogeny).

7.4.4 LARAMIDE OROGENY (80–40 MA)

As we shall discuss in greater detail in Chapter 8, about 80 Ma (latest Cretaceous), the eruptions that formed the Sierra volcanic arc ended (Figure 7.21). After nearly continuous volcanism and deep intrusions since the Jurassic, the volcanic chain of the Sierras shut down and became an extinct volcanic chain. Some geologists have labeled this peculiar arc shutoff the "magmatic null." No volcanoes in the Sierras have erupted since then. Similarly, the back-arc thrusting of the Sevier belt ceased as well. Instead, a totally new form of tectonism appeared far to the east of the Basin and Range in California and Nevada. Huge folds and thrust

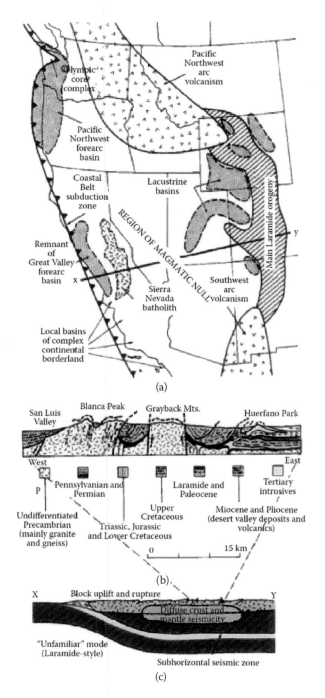

(a)

(b)

(c)

FIGURE 7.21 Tectonics of the Laramide orogeny. (a) Map showing the location of the Laramide uplifts ("main Laramide orogeny") in the future Rocky Mountains, and the arc shutoff in the Sierras (magmatic null). (b) Cross section of typical Laramide ranges in Colorado or Wyoming, with block-faulted and folded uplifted ranges separated from deep subsiding basins. (c) The Dickinson–Snyder model of horizontal subduction, which prevents the down-going slab from sinking deep enough in the mantle to melt and form magmas (magmatic null) while transferring the stresses of crustal compression to the Rocky Mountain region 1,600 km inland.

Source: After Prothero, D. R., and R. H. Dott Jr. *Evolution of the Earth* (8th ed.). McGraw-Hill: New York, 2010.)

faults brought Precambrian and Paleozoic basement rocks from deep in the crust up to the surface to form mountain ranges in central Colorado, New Mexico, Wyoming, and Montana. During the Cretaceous, all these areas had been immersed in giant epicontinental seaways that drowned most of the continent. Now, these mountain ranges rose high in the sky over the next 30 m.y. to form a predecessor of the modern Rocky Mountains. This spectacular event is known as the **Laramide orogeny** (Figure 7.21), and it lasted from the latest Cretaceous to the middle Eocene (70–40 Ma). Named for the Laramie Range in Wyoming, it is the second time a mountain range arose in the area of the Rocky Mountains. The first event, the Ancestral Rockies, had occurred back in the Pennsylvanian (Late Carboniferous) about 300–330 Ma. Those earliest Rockies had long since eroded by the time the seas drowned the region in the Late Jurassic.

The peculiar tectonic events of the Laramide orogeny were long a puzzle to geologists. Why had all volcanism ceased where the Sierran volcanic chain had long been active? And why did the region almost 1,600 km (1,000 mi) inland from the plate margin suddenly buckle upward and thrust basement rocks in the air to form the Laramide Rockies? In the 1970s, William Dickinson and Walter Snyder proposed an ingenious tectonic solution (Figure 7.21): the subducting plate that had been producing the Sierras stopped plunging down into the mantle. Instead, it started to slide horizontally beneath the overlying North American plate. Without sinking down into the mantle and melting, it could no longer supply the volcanoes to the Sierras. This explains the arc shutoff. If the plate slides horizontally and transfers its stresses far inland, this might explain the faulting and folding that produced the Laramide ranges far inland as well. Although there are disputes about the details, the idea that the Laramide orogeny was caused by flat or extremely shallow subduction is now widely accepted among geologists.

7.4.5 ARC VOLCANISM RESUMES (40–20 MA)

After 30 million years without volcanism in the California and Nevada (from the latest Cretaceous to the middle Eocene), arc volcanism came back with a vengeance in the late middle Eocene and persisted through the early Miocene (40–20 Ma). But this new volcanic arc was not in the location of the old, extinct Sierran volcanoes. Instead, it erupted far to the east, in a belt that ran from central Oregon and Washington (the "Ancestral Cascades") through Nevada and Utah, through the enormous San Juan volcanic field in southwestern Colorado, and down the border between New Mexico and Arizona and into Mexico (Figure 7.22). Huge Oligocene and Miocene ignimbrites and ash flows are common all across the Basin and Range, especially in the extreme northeast corner of California (Warner Range—see Chapter 20), in most of the Nevada mountains, and on down into Arizona.

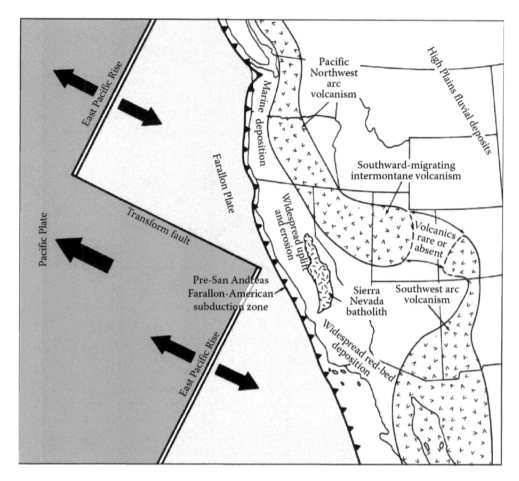

FIGURE 7.22 During the middle Cenozoic (40–20 Ma, or late middle Eocene to early Miocene), arc volcanism returned to western North America, but not in the old location of the Sierran arc. Instead, it ran considerably to the east, from the Ancestral Cascades in eastern Oregon and Washington, through Nevada and Utah, to the San Juan volcanic field in the southwestern corner of Colorado, through the volcanics of the Arizona–New Mexico border region. This suggests that the horizontal Laramide plate had peeled off the overlying plate and begun to sink down into the mantle again, but at a much shallower angle, so the melting occurred much farther inland than the Sierras.

Source: After Prothero, D. R., and R. H. Dott Jr. *Evolution of the Earth* (8th ed.). McGraw-Hill: New York, 2010.

One of these eruptions, the La Garita Caldera eruption in the San Juan Mountains of Colorado, was the biggest explosion in North American prehistory. It formed a huge hole in the ground, 35×75 km in size (22×47 mi), which produced a large irregular oblong valley in the San Juan Mountains. Some of the caldera and its ash deposits can be seen in the Wheeler Geologic Area northwest of South Fork, Colorado. In most places, the caldera has been covered up and partially filled by eruptions of the smaller but younger Creede Caldera and other younger calderas. When La Garita exploded 27.8 Ma, it blew more than 5,000 km³ (1200 mi³) of material all over the United States, enough material to entirely fill Lake Michigan. It deposited a widespread ash layer known as the Fish Canyon Tuff, which can still be seen in the Arkansas River Canyon, 100 km northeast of the caldera, and beneath the surface of the Alamosa area, 100 km east of the volcano. At one time, it was one of the largest volcanic deposits in the world and probably covered much of the Rockies and the western and central plains of the United States. It is calculated that the energy released by the La Garita explosion is 10,000 times more powerful than the largest nuclear device that humans have ever detonated.

So what brought this volcanic arc back after 30 million years of magmatic null, and why didn't it return to its old location in the Sierras? Dickinson and Snyder suggested that after the horizontal subduction of the Laramide orogeny, the down-going slab began to slowly peel away from the underside of the North American plate and sink back into the mantle, but at a very shallow angle. When the plate angle is so shallow, the portion that reaches the depths in the mantle to melt is much farther inland than if the plate subducts at a steep angle. This might explain why the new arc was at least 500 km (300 mi) east of the Sierran chain.

7.4.6 Late Cenozoic Extension and Faulting (20 Ma–Present)

After 20 m.y. (40–20 Ma) of explosive volcanism across most of the Southwest, the arc volcanoes were replaced by the beginning of the Basin and Range extension that we discussed earlier in the chapter. The Basin and Range began to stretch and pull apart, breaking up into normal faults, horsts, and grabens, and especially large detachment faults that brought up deeply buried metamorphic core complexes.

In many places in the Basin and Range (especially in the Mojave Desert), these Miocene basins filled with sediments deposited off the flanks of the rising range. In some cases, such as in the Barstow Basin and near Red Rock Canyon (Chapter 20), the basins sunk down for millions of years and produced a thick stack of deposits that span much of the middle or late Miocene, rich with mammal fossils of the time: mastodonts, horses, camels, rhinos, pronghorns, peccaries, bear dogs, and many other extinct groups that once roamed the savannas of the American Miocene (see Chapter 20).

During this period of rapid extension, explosive rhyolite and dacite volcanism was common over much of the region. In some cases, the huge explosive eruptions mantled the whole region with a blanket of volcanic ash. Perhaps the largest and most spectacular such eruption was the Peach Springs tuff ignimbrite, about 19.2 Ma. Erupting from a long-gone caldera near Laughlin, Nevada, enormous ash flow sheets and airfall tuffs were spread from western Arizona (where they are exposed near Peach Spring, the unit's type area) and especially near Kingman, Arizona, where they form a series of spectacular roadcuts just southwest of town on Interstate 40. Here the flows are more than 300 m (1000 ft) thick and include thick pumice beds, welded tuffs, and ignimbrites, many with faults in them caused by contraction and shrinkage as they cooled. But the tuff blanket spread across the entire eastern Mojave Desert as well and can be found in nearly any outcrop that dates to about 19–20 Ma, including the Barstow beds.

One of the interesting aspects of this stretching and extension is that it occurred from the south (Arizona–New Mexico) to the north (Nevada–Utah) over the course of the Miocene and Pliocene. The first region to break open was the broad desert landscape of southern Arizona, while the youngest faulting is still occurring in places like Death Valley and even into southern Oregon and Idaho. This is reflected in the type of desert topography you see in those regions (Figure 7.23). Phoenix and Tucson are typified by their broad flat landscape of eroded desert gravels (**pediment**), with isolated rocky crags and small mountains (**inselbergs**) here and there poking up through the pediment.

This is what the Basin and Range extension looks like after almost 20 million years of erosion. All the distinctive fault-block mountains (horsts and half-grabens) have been tectonically inactive for so long and so deeply eroded that only tiny remnants are left, sticking above the thick veneer of gravel eroded from them. By contrast, in areas

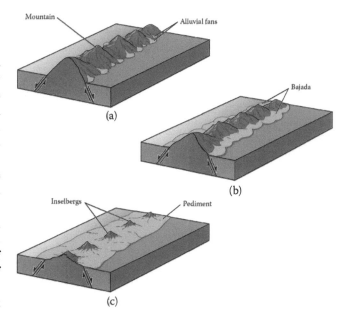

FIGURE 7.23 Evolution of desert topography. (a) Recently faulted ranges form "young" Basin and Range topography, with steep fault scarps and small alluvial fans in the canyons. (b) Over millions of years, the mountain ranges wear down and erode and the basins fill with sediment, so the relief is shallower and not as sharp or distinct. The alluvial fans from each canyon coalesce into one big fan or scree slope, called a bajada. (c) In the late state ("mature") of Basin and Range topography, the mountains are almost completely worn away, and only remnants (inselbergs) still poke above the thick blanket of sediment that covers the region (known as pediment).

Source: Courtesy of Oxford University Press.

like Death Valley, the Owens Valley, or northern Nevada–southern Idaho and Oregon, the faulting is very young and active, and the fault scarps along the foothills of the ranges are straight and distinct. They are still uplifting and down-dropping on active faults, and the alluvium rushing out of the canyons is just beginning to fill their basins. In between these extremes are Basin and Range valleys that are more mature, with more deeply eroded, less-distinct fault block mountains, and lots of alluvial fans coalescing into a continuous apron of gravel in the foothills of each range (known as a **bajada**). In addition, these valleys in the intermediate stage of erosion have much more sedimentary fill in their grabens and often trap large lakes in the playas that accumulate muds or evaporite minerals. The fingerprint of the south-to-north opening of the Basin and Range is written all over the topography.

We will discuss tectonic explanations for why the Basin and Range opened up in the early Miocene, and why it opens from south to north, in Chapter 15.

7.5 PLEISTOCENE OF THE GREAT BASIN

Superimposed on all the older geologic features of the region is the imprint of the last 2 m.y. of ice ages. The continental glaciers didn't flow far south enough to reach the Great Basin, nor did most of the glaciers in the high mountains, such as those in the Sierras, descend very far into the fault-graben basins. Instead, the prevailing effect of cooler global temperatures during Ice Age was a much wetter, milder climate. The dry desert mountains we have now were once densely forested, and the valleys were much greener and more densely vegetated, often with large fresh-water lakes in the middle.

These lakes covered nearly all the Great Basin (Figure 7.24). The largest was Glacial Lake Bonneville, which flooded most of northwestern Utah. The "Great" Salt Lake and the Bonneville Salt Flats are the remnants of this huge body of water. Northwestern Nevada hosted a huge lake known as Glacial Lake Lahontan. Today, all that remain are ancient shorelines along the much smaller Pyramid Lake, and the Black Rock Desert, site of the "Burning Man" festival. In fact, nearly every valley in the Great Basin typically has lots of soft tan or pink lake sediments somewhere in the basin, often forming terraces of different lake levels along the foothills of the ranges. If you travel the major roads through the Mojave, Nevada, or western Utah, you will see abundant examples of ancient lake deposits.

During the wetter peak glacial intervals, dense vegetation ran from the top of the mountains to the bottom of the valley, and animals could travel over a wide area of hospitable habitat. Then, during the Holocene, the entire region became hotter and drier, and the vegetation changed

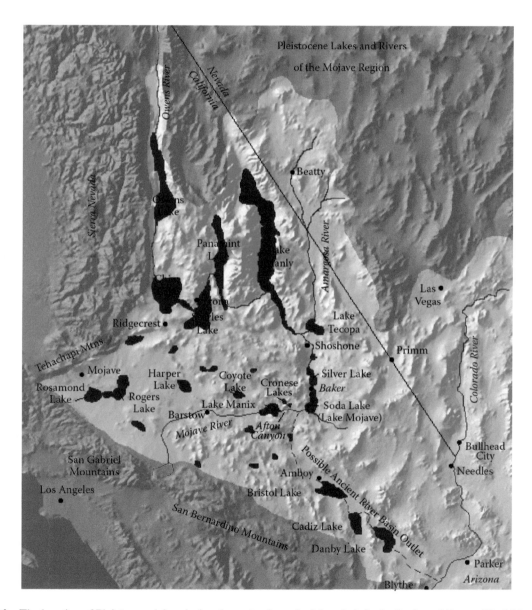

FIGURE 7.24 The location of Pleistocene lakes during the wetter interglacial periods in the Basin and Range. The highest was Mono Lake, which then drained down to Owens Lake, then China Lake, Searles Lake, Lake Panamint, and finally, Lake Manly at the bottom of Death Valley. Other lakes from the central Mojave also drained down to Lake Manly.

Source: Courtesy of Wikimedia Commons.

drastically. The tops of the ranges are often snowcapped and get much more rain than do the bottoms of the valleys. Many of the ranges are cloaked in pine forests, while the valleys between are covered in bare sand flats with dry sagebrush and mesquite. This means that the tops of the ranges are like ecological islands, with their own distinct plants and animals that cannot leave their habitat and spread to a different range. In many cases, the animals on each range have undergone evolutionary change and diverged genetically from their relatives on adjacent ranges, so they are in the process of turning into new species. Likewise, some of the animals adapted to the harsh conditions of the basins, such as the desert pupfish, which have become isolated, both geographically and genetically, so that each natural spring has its own species of pupfish.

The drainage system of Ice Age lakes in the Mojave and Death Valley is particularly interesting. Most of the major lakes in the region were interconnected during the wettest part of the ice ages, although nearly all these lakes and rivers have since dried up (Figure 7.24). The beginning of the chain of lakes is the highest ones, Mono Lake and Owens Lake up in the Owens Valley, which receive water from the Sierra snowmelts. Mono Lake is the highest at 1,946 m (6,383 ft) above sea level. When lake levels were high enough, the water from Mono Lake drained down to Owens Lake (1,084 m, or 3,556 ft) in elevation. From there, another outlet carried water down to China Lake (home to a large military bombing range now), and then over to Lake Panamint in the Panamint Valley, and finally down to Pleistocene Lake Manly on the floor of Death Valley, which is almost entirely below sea level. Meanwhile, the ancient drainage of the Mojave River carried snowmelt from the San Gabriel and San Bernardino Mountains to Ice Age Lake Manix just northwest of Barstow, an area famous for its fossils of Ice Age mammals, birds, and many other groups. Lake Manix then drained through a huge water gap called Afton Gorge before reaching Lake Mojave (now Soda Lake), Lake Dumont, and then Lake Manly on the floor of Death Valley. To the east, Lake Pahrump in Nevada and Lake Tecopa (famous for its hot springs and Ice Age mammals) drained west into Lake Manly (Figure 7.24).

7.6 EVAPORITE MINERALS AND DRY LAKES

In each of these valleys, you can find abundant lake bottom silts and clays, and in some places (like Shoreline Butte in southern Death Valley), you can see terraces carved by lakes when they stood at different levels. Today, however, most of the lakes are dry or nearly dry. There is no throughgoing flow to connect them, so nearly all the water leaves by evaporation. The result is extensive evaporite deposits.

In some of them, like Mono Lake, the water is highly alkaline. Its pH is about 10, so no fish can live in waters this salty and alkaline. It only supports a minimal food pyramid of planktonic algae, which support brine shrimp and alkali flies, which in turn support large populations of migratory birds during the spring and summer. The water has a lot of dissolved salt in it (more than 280 million tons), but it's not completely dry, so it mainly precipitates calcium carbonate in the form of **tufa**, aided by the calcareous algae growing in the water. During the wettest periods, Mono Lake was almost 300 m (900 ft) deep; now only few meters of water are left. On the shorelines of Mono Lake are many areas of tufa towers, where these deposits have built up and then been etched and dissolved by rainwater (Figure 7.25a). The same thing has happened south of Trona and Searles Lake, where enormous tufa towers up to 43 m (140 ft) high grow in several parts of the basin (Figure 7.25b). They, too, were precipitated by algae, aided by hot springs coming up through the valley floor below, which determine where the towers occurred. They were formed in water that was at least 70 m (220 ft) deep initially and then gradually dried up, leaving these pinnacles behind.

Death Valley (Figure 7.26), on the other hand, produced not only carbonate minerals that formed when evaporation began but also a **zonation of evaporates** that is typical of many drying bodies of water. These zones of evaporite minerals change from the edge of the basin to the center to form concentric rings, like a bull's-eye, as the water dries up (Figure 7.26). Numerous experiments show that carbonates (like aragonite) typically form when about 50% of the

(a)

FIGURE 7.25 (a-b) Mono Lake tufa towers, formed as crystals of carbonate nucleated due to bacteria and algae in the highly alkaline lake water. (c) Trona Pinnacles, towers of tufa formed when the lake basin was filled. Old lake terraces can be seen on the surrounding mountains.

Source: (a) Photos by the author. (b) Photos by the author. (c) Courtesy of Wikimedia Commons.

(b)

(c)

FIGURE 7.25   (Continued)

(a)

(b)

(c)

FIGURE 7.26   (a) Bull's-eye zonation of evaporates in Death Valley. The first minerals to precipitate when about 50% of the water has evaporated are carbonates (aragonite, natron, and trona) and borates, which form the outer ring of minerals in the bull's-eye. When about 80% of the original water has evaporated, sulfates like gypsum precipitate in a ring just inside the carbonate ring. When 90% of the water is gone and only an extremely salty brine remains, then the bitter salts (halite, sylvite, etc.) precipitate in the center of the basin. (b) Devil's Golf Course, covered by a thick layer of halite in the center of Death Valley as the last bit of water evaporated. Each time enough rain comes to fill the lake basin, the salts dissolve and then precipitate again when the lake dries up. (c) Harmony Borax Works, in the outer carbonate zone, where carbonates and borates (like cotton ball ulexite) crystallized first out of the lake water as it dried up and increased in salinity. Today only the ruins of the Borax Works and the 20-mule-team wagons remain, while the ground is coated with white cotton ball ulexite.

Source: (a) After Hunt, C.B., *US Geological Survey Professional Paper*, 494-A, 1966. (b–c) Photo by D. Patton.

original water has been evaporated away; they form a **carbonate zone** on the outer rim of the basin. If about 80% of the water is lost, then the next most soluble mineral, gypsum (hydrated calcium sulfate), forms. It precipitates in the **sulfate zone**, between the outer carbonate zone and the center of the bull's-eye. If the water is more than 90% evaporated, it becomes a supersaturated brine and loaded with the most soluble ions, especially sodium and chlorine. Once the water evaporates to this extreme, these "bitter salts" combine to form sodium chloride, or halite (NaCl, or rock salt). The **chloride zone** also may evaporate other minerals, like potassium chloride (KCl), the bitter-tasting mineral known as sylvite, or calcium chloride, $CaCl_2$. In the center of evaporite bull's-eye in Death Valley is a spectacular landscape known as the Devil's Golf Course, where large pinnacles of rock salt grow from the valley floor. These are formed after rare wet spring rains dissolve all the valley floor salts then grow as the lake evaporates again. However, after the first rains, they are etched and end up with many bizarre shapes and jagged edges.

Finally, California's dry lake basins often have a peculiar chemistry that produces a wide spectrum of minerals besides the common salt and gypsum. Many of these brines are rich in the element boron, a relatively rare substance in most places on Earth. The boron combines with the sulfate, carbonate, and chlorine in these waters to form a spectrum of **borate minerals**, some of which are unique to California (Figure 7.27). The most common of these is colemanite, but there are also minerals like borax, kernite, ulexite (the fiber-optic mineral used to make "TV rock" for collectors), hanksite (a rare potassium borate), and dozens of others.

Borate minerals are extremely valuable and have been commercially mined for soaps, detergents, reflux materials, and many other chemical and industrial uses. Miners have gone to extraordinary lengths to recover this rare substance. From 1883 to 1889, Death Valley had extensive borate mining (for fibrous ulexite known as "cotton ball" for its texture). Chinese laborers worked in the blazing heat shoveling the thin films of cotton ball ulexite into wheelbarrows and then into a boiling vat, where it was dissolved, re-evaporated, refined, and then loaded into the famous "20-mule teams" of immense wagons. The 18 mules and two horses (horses were better as the wheelers, the ones closest to the wagon) hauled two enormous wagons plus a water tank weighing 33 mt (73,000 lb) from Harmony Borax Works in northern Death Valley across 275 km (165 mi) of the worst desert conditions, up mountains and down across scorching valleys, to the nearest railhead in Mojave, California.

In the 1890s, the Death Valley operations shut down as less remote and richer sources of borate were found near Calico, Daggett, and Boron, just northeast of Barstow and much closer to the rail lines. These deposits began to run out in 1907.

Another remarkable lake deposit of minerals is the legendary Searles Lake Facility, on the south end of Glacial Lake Panamint and just east of Trona, California. Its

(a)

(b)

FIGURE 7.27 Common borate minerals of lake beds in the Mojave Desert. (a) Colemanite. (b) Kernite. (c) Borax. (d) Fiber-optic ulexite, which transmits images between one cut and polished face and the next, so it is nicknamed TV rock. (e) Hanksite.

Source: (a–b) Courtesy of Wikimedia Commons. (c) Courtesy of Wikimedia Commons. (d) Photo by the author. (e) Photo by the author.

unusual chemistry (rich in boron, potassium, and many other rare chemicals) means it produces more than 100 different kinds of minerals, including borates, chlorides, sulfates, and carbonates (especially natron and trona, which are sodium carbonate and sodium bicarbonate). Some of the minerals are only known from Searles Lake. This huge mining operation pumps salty brines from beneath the lake surface into huge evaporating pans, where (after drying out) the evaporated product is then screened, washed, dried, and further refined to recover all the rare chemicals that are

(c)

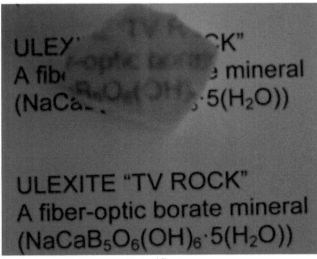

ULEXITE "TV ROCK"
A fiber-optic borate mineral
$(NaCaB_5O_6(OH)_6 \cdot 5(H_2O))$

(d)

(e)

FIGURE 7.27 (Continued)

valuable. The Searles Lake Facility mines about 1.7 million tons of minerals every year and supplies an enormous quantity of rare chemicals for the world's industrial uses. Normally, it is closed to the public, but one weekend in mid-October every year, it has an open house and allows thousands of geology students and mineral clubs to go in and

FIGURE 7.28 Low-angle aerial view of the immense operations near Boron, California.

Source: Courtesy Wikimedia Commons.

collect muddy crystals of rare minerals like hanksite to their heart's content.

The biggest borate mine in the Mojave today is the huge open-pit US Borax Mine (now owned by Rio Tinto Mining) near Boron, California, the largest open-pit borate mine in the world and the largest open-pit mine in California (Figure 7.28). It supplies nearly half the world's resources of refined borates, so it is an extremely valuable deposit. First discovered by John K. Suckow in 1913 while drilling for water, initially it was just thought to be another gypsum deposit. But testing showed that it was mostly colemanite instead. In 1925, Francis Marion "Borax" Smith bought it for his Pacific Coast Borax Company, and shaft mining was started in the late 1920s. Eventually, the company renamed itself as US Borax and started open-pit mining in 1957. Geological analysis showed that it was once a shallow lake fed by volcanic thermal springs rich in sodium and boron. The lake deposits were eventually buried to a depth of 457 m (1,500 ft), producing a temperature above 53°C, enough to create kernite. The sodium borates, together with claystone, are hosted as a core facies within the middle Miocene (16 Ma) Kramer beds and are completely enveloped by ulexite-bearing shales. Stratigraphic and structural studies indicate the Kramer borates were deposited in a small structural, nonmarine basin, elongated in an east–west direction and limited on the south by the Western Borax fault. The mine measures 2.8 km (1.74 mi) wide, 3.2 km (2 mi) long, and is up to 230 m (755 ft) deep. More than 80 minerals are found at this geologically unique site, including the four boron-based minerals in greatest demand by industry: tincalconite, kernite, ulexite, and colemanite. Recently, the mining operation has also begun to recover lithium from its spoil piles, an important component of lithium batteries and other forms of high technology.

7.7 RARE EARTH ELEMENTS

In addition to valuable minerals like gold, silver, copper, and borates, there is one extremely important mineral resource found in the Mojave Desert: rare earth elements (REEs). At the summit of the Clark Mountains (visible on the north of Interstate 15 just west of the Nevada border) is the Mountain Pass mining region (Figure 7.29), one of the largest REE deposits in the world, which once had a tonnage of rare earths greater than all the other REE mines in the world combined.

REEs are the elements near the bottom center row of the periodic table of the elements. As the name implies, they are normally in parts per billion in most rocks and rare by any measure in most natural settings. But when some process, such as peculiar magma chemistry, concentrates them, they can form an extremely valuable deposit. Some of the elements and uses for them include cerium, used in ultraviolet-absorbing glass and lighter flints; lanthanum, samarium, and gadolinium, used for infrared absorption in glass, improving the refractive index of glass and microwave oven temperature controls; neodymium, used to absorb ultraviolet light and, with cerium, to decolor glass; praseodymium, used as a coloring agent in glass when the index of refraction must not change; and europium, used extensively for the red phosphor in television tubes. These elements occur in a strange ore mineral known as bastnaesite (rare earth fluorocarbonate), which occurs in the carbonatites, magmas very high in carbon dioxide and poor in silica, and other typical elements. This ore is rich in the following elements: cerium 50.0%, lanthanum 34.0%, neodymium 11.0%, praseodymium 4.0%, samarium 0.5%, gadolinium 0.2%, and europium 0.1%.

About 1.4 Ga, magma from about eight different intrusions cut through the 2-billion-year old metamorphic rocks here and mineralized the region with a bizarre combination of chemistries. There are also intrusions of weird K-feldspar-rich magmas known as shonkinites, which look like red granites, except they are so low in silica they have no quartz normally found in granitic rocks. Igneous petrologists have studied these intrusions for many years, but

there is no consensus on what caused such unusual magmas to form and intrude in this area.

Originally, the area was prospected for uranium but turned out to have a much more profitable resource once geologists recognized its value: REEs. The mine produced most of the world's supply of REEs from the 1960s to the 1990s. Then enormous deposits were found in Mongolia

FIGURE 7.29 Aerial view of the enormous Mountain Pass REE mine showing the huge open-pit mining operation and the giant tailings piles. Interstate 15 is visible on the right.

Source: Courtesy of US Geological Survey.

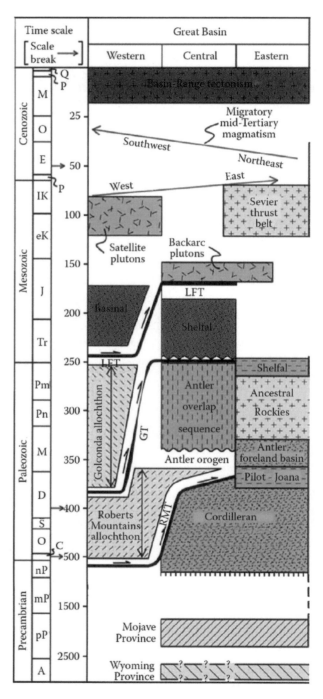

FIGURE 7.30 Summary of the geologic history of the Basin and Range Province. GT = Golconda thrust; RMT = Roberts Mountain thrust.

Source: Drawing modified by E. T. Prothero based on several sources.

and China, and the price of REEs dropped too far to justify mining. In addition, the cost of handling the toxic and frequently leaking waste products was too expensive. The mine shut down from 2002 to 2012, and only the tailings piles were processed for leftover REEs. But in recent years, China has come to own more than 95% of the world's REE supply and has used this power to squeeze Japan and its electronics industry by denying them their supplies of crucial REEs. This raised the price of REEs and also made American politicians worried that China could disrupt the American electronics industry as well if we don't continue to mine our own deposits. In 2012, Congress passed a bill (Critical Minerals Policy Act of 2012) to spur production of REEs and reduce our vulnerability. By 2015, the mine was back to full operations and now competes with China to provide REEs for the booming electronics industries around the world.

The entire complicated history of the Basin and Range Province is summarized in Figure 7.30.

7.8   BASIN AND RANGE

Spectacular mountains capped with snow. Searing deserts with the hottest temperatures in the world. Amazing landscapes with their geology well exposed and easy to interpret. A region complete with sand dunes, hot springs, active volcanoes, frequent earthquakes and flash floods, and some of the richest mineral deposits in the world. These are some of the reasons the Basin and Range Province has drawn people, especially miners and geologists, to study its complex geologic history and decipher its complicated structure. We look at some of the tectonic models for the Basin and Range in Chapter 15.

RESOURCES

Burchfiel, B.C., and Davis, G.A. 1981. Mojave Desert and environs, 217–252. In Ernst, W.G. (ed.), *The Geotectonic Development of California*. Prentice Hall, Englewood Cliffs, NJ.

DeCourten, F. 2003. *The Broken Land: Adventures in Great Basin Geology*. University of Utah Press, Salt Lake City.

Dickinson, W.R., and Snyder, W.S. 1978. Plate tectonics of the Laramide Orogeny. *Geological Society of America Memoirs*, *151*, 355–366.

Fiero, B. 2009. *Geology of the Great Basin*. University of Nevada Press, Reno.

Hunt, C.B. 1966. Stratigraphy and structure, Death Valley, California. *U.S. Geological Survey Professional Paper 494-A.*

Hunt, C.B. 1976. *Death Valley: Geology, Ecology, Archeology*. University of California Press, Berkeley, CA.

McPhee, J. 1982. *Basin and Range*. Farrar Straus Giroux, New York.

Meldahl, K.H. 2013. *The Rough-Hewn Land: A Geologic Journey from California to the Rocky Mountains*. University of California Press, Berkeley.

Nelson, C.A. 1981. Basin and range province, 203–216. In Ernst, W.G. (ed.), *The Geotectonic Development of California*. Prentice Hall, Englewood Cliffs, NJ.

Oldow, J.S., and Cashman, P.H. 2009. *Late Cenozoic Structure and Evolution of the Great Basin*. Geological Society of America Special Paper 000. Geological Society of America, Boulder, CO.

Prothero, D.R., and Dott, R.H., Jr. 2010. *Evolution of the Earth* (8th ed.). McGraw-Hill, New York.

Sharp, R.P. 1997. *Geology Underfoot in Death Valley and Owens Valley*. Mountain Press, Missoula, MT.

Whitmeyer, S.J., and Karlstrom, K.E. 2007. Tectonic model for the Proterozoic growth of North America. *Geosphere*, *3*(4), 220–259.

Videos

Basin and Range in Arizona: www.youtube.com/watch?v=_fnHBzSQJkg

Basin and Range structure: www.youtube.com/watch?v=TvvWqAdNV84

Basin and Range structure: a series of short video animations beginning with this link: www.youtube.com/watch?v=TvvWqAdNV84&list=PL8FDF28B8FD0C2E56

"Great Basin National Park," video about the Basin and Range at Great Basin: www.youtube.com/watch?v=Owjt_WJKrBo

Land of Sleeping Mountains, about the geology and tectonics of the Basin and Range: https://archive.org/details/themakingofacontinentpart2thelandofsleepingmountainsreel2

Death Valley:
www.youtube.com/watch?v=jhoYARaYvr8

Mono Lake:
www.youtube.com/watch?v=s_8iZn-skjM
www.youtube.com/watch?v=8rPpdemnO5Y

Owens Valley volcanics:
www.youtube.com/watch?v=Wjy5xE9od-8
www.youtube.com/watch?v=BRS6as2hOzQ
www.youtube.com/watch?v=tbOSsHQnPkc

8 Gold, Glaciers, and Granitics
The Sierra Nevada

Another glorious Sierra day in which one seems to be dissolved and absorbed and sent pulsing onward we know not where. Life seems neither long nor short, and we take no more heed to save time or make haste than do the trees and stars. This is true freedom, a good practical sort of immortality.

—**John Muir**
My First Summer in the Sierra

8.1 DATA

Highest point: Mt. Whitney: 4,421 m (14,505 ft); 10 other peaks above 14,000 ft

Predominant rocks: Mesozoic granitic plutons, intruded into late Paleozoic and Mesozoic metasedimentary rocks

Plate tectonic setting: Plutonic source of Andean-style arc volcanoes during the Jurassic and Cretaceous, plus accreted Paleozoic exotic terranes and Triassic–Jurassic accretionary wedges

Other events: Extensive Pleistocene mountain glaciation

Geologic resources: Gold, water (snowpack and rivers)

National parks: Yosemite, Kings Canyon, Sequoia

State parks: Marshall Gold Discovery Site State Historic Park, Emerald Bay on Lake Tahoe, Empire Mine, Malakoff Diggings, State Mining and Mineral Museum

8.2 "SNOWY RANGE"

The Sierra Nevada (Figure 8.1) is the highest mountain range in the lower 48 states, with the tallest peak (Mt. Whitney at 4,421 m, or 14,505 ft) and ten other peaks above 14,000 ft. There are 500 peaks that are more than 3,700 m (12,000 ft) high. Most of the foothills are at least 1,500 m (about a mile) in elevation, and the crest of the Sierras is 2,700–3,000 m (9,000–10,000 ft) high. The range runs almost 640 km (400 mi) from north to south and is 105 km (65 mi) wide. It covers an area of 63,100 km² (24,370 mi²).

This gigantic mountain range has been a big influence on California and adjacent areas. As discussed in Chapter 7 (Figure 7.3), it creates a rain shadow that turns all the lands east of it in the Great Basin into desert. Conversely, the huge amount of rain and snow dumped in the western

Sierras is the main source of water for much of California (Chapter 18). Not only do the state's biggest river drainages (Sacramento and San Joaquin River systems) get their supply from the Sierras, but also the two biggest aqueducts that bring water to parched Southern California are supplied by Sierra snowmelt. These are the California Aqueduct from the Feather River drainage in the northern Sierras and the Los Angeles Aqueduct from the eastern Sierras (Chapter 18). Without this rain and snow, California would not have the fertile, mild climate it has or could not support the 38 million people (and growing) that live here.

The Sierras have been profoundly influential in California history as well. In 1776, the Anza Expedition came up from Mexico across the deserts of southeastern California and then up the coast to San Francisco. On a hill near the mouth of the Sacramento River, Father Pedro Font described the snowy peaks he could see off in the distance as *un gran sierra nevada*—"a great snowy range" in Spanish. The impenetrable barrier of the Sierras discouraged trappers, explorers, and settlers from the east and diverted most of the trails to the Pacific Coast to the north (Oregon Trail) or down toward southern California (Old Spanish Trail). Mountain man Jedediah Smith was the first American to reach California by crossing the continent in 1837 and struggled back home over the snow-packed Sierras over what is now Ebbetts Pass. Many other settlers tried to cross the rugged Sierras, often getting trapped in the snow, with disastrous consequences. The most famous was the Donner Party in 1846–1847, which resorted to cannibalism after being snowbound for four months and losing half their party. The Sierras were the site of the key event that transformed California and made it into a state: the Gold Rush (discussed in Chapter 16). Without the Sierras, and the water, gold, and forests they provided, California would not have been among the first western states to be settled or admitted to the union.

8.3 SIERRA NEVADA GEOGRAPHY

If you've ever driven over the Sierras on Interstate 80, or via Tioga Pass from Yosemite Valley, or other routes, the first thing you'll notice is that the Sierras are not a symmetrical range. The eastern slope is much steeper than the western slope (Figure 8.2a). On the east, facing the Owens Valley, the Sierras drop off abruptly (Figure 8.2b). However, from the summit downward, the western flank of the Sierras is one gentle downward gradient, so the climb or descent is not nearly so steep or rapid but slow and gradual over many

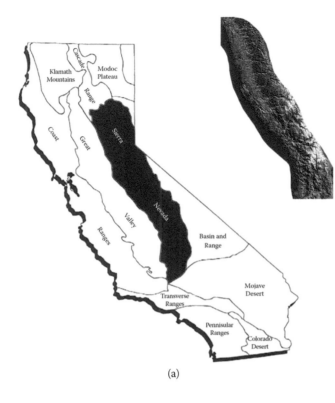

(a)

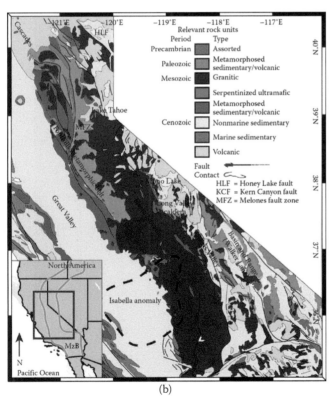

(b)

FIGURE 8.1   (a) Index map of California highlighting the Sierra Nevada. In the upper right is the digital shaded relief map of the Sierras. (b) Geologic map of the Sierra Nevada.

Source: (a) Courtesy of US Geological Survey. (b) Courtesy of the California Division of Mines and Geology.

miles (Figure 8.2c). For example, the trip east from the western edge of the range near Roseville up to Donner Pass is about 130 km (80 mi), but the steep descent down from Donner Pass to Reno is only about 75 km (47 mi).

This asymmetry is due to block faulting (Figure 8.2a), which tilts off the west. In *Assembling California*, author John McPhee compares the Sierras to a "trapdoor," with the opening edge on the east and the hinge as the western edge of the range. The eastern flank of the range is rising fastest, tilting the "door" (the western slope) to the west. Thus, the entire steep eastern face of the range is one huge set of normal fault scarps facing the Owens Valley (Figure 8.2b). It represents the westernmost fault of the Basin and Range Province to the east and the boundary between provinces. The entire Sierran block apparently tilted and rose in the east while sinking in the west so that it is shaped like a big playground slide: steep climb on one side, gradual shallow ramp on the other.

8.4   THE SIERRA BATHOLITH

As mentioned in Chapter 2, a **batholith** is a large complex of cooled magma chambers that once fed a chain of volcanoes above them. The Sierra Nevada batholith is made of a variety of granitic rocks (Figure 8.3), which are rich in silicon, aluminum, potassium, and sodium. They include rocks such as **diorite** (mixed calcium–sodium plagioclase plus hornblende), **quartz diorite** (mixed calcium–sodium plagioclase, hornblende, and quartz), **tonalite** (sodium plagioclase plus quartz, hornblende, or biotite, and less than 10% potassium feldspar) (Figure 8.3), **granodiorite** (sodium plagioclase plus quartz, hornblende, or biotite, and 10–33% potassium feldspar), and **quartz monzonite** (sodium plagioclase = 33–66%, potassium feldspar = 33–66%, plus quartz, hornblende, or biotite). As discussed in Chapter 2, most modern igneous rock classification schemes do not follow the old "wastebasket" usage of "granite" for any plutonic rock with quartz and plagioclase and potassium feldspar. Instead, they define a true granite to have at least 66% or higher potassium feldspar (in other classifications, two-thirds of the total feldspar is potassium feldspar). This tends to make the rocks pink or even red in color (see Figure 2.4).

By this criterion, *there are few or no true granites in the Sierras*, contrary to popular myth. The most common rock is tonalite (60% or more of the Sierran granitics, dominant on the western part of the batholith), followed by granodiorite (more common in the east), with quartz monzonite being third most common. Nevertheless, there are many geology books and websites that talk about Sierra Nevada "granite" with no awareness or regard for modern igneous petrology. We follow the proper petrological usage of the term *granite* in this book, but it is still acceptable to call the suite of rocks found in the Sierras (tonalities, granodiorites, diorites, quartz diorites, and quartz monzonites) "granitics."

The Sierra Nevada batholith is not a single huge magma chamber but a complex of hundreds of separate intrusions

FIGURE 8.2 (a) Block faulting of the Sierras. (b) Panoramic photograph of the steep eastern Sierra front looking west from the Owens Valley, showing the vertical scarps of the normal faults that chop off the Sierras on their eastern edge. Mono Lake is in the right at the bottom of the valley. (c) The summit of the Sierras is not a row of jagged peaks, as commonly imagined, but a relatively flat upland surface with small valleys carved by later glaciers. This is the summit at Donner Pass (notch on the left), with Interstate 80 on the right, facing west.

Source: (a) Courtesy of US Geological Survey. (b) Courtesy of US Geological Survey. (c) Photo by the author.

that came up at different times and places (Figure 8.4a). Each one melted its way upward through the rocks that were already there. Thus, the plutons form a complex pattern of one intrusion cut by another and yet another over a time span of about 80 million years. Individual plutons can be less than 0.6 km (1 mi) in diameter to a typical size of 1,300 km² (500 mi²). There are some that are many tens of kilometers (or miles) across. Most are roughly oval in shape. In one study area in Yosemite National Park, 53

different plutons were mapped, averaging 12 km (7.5 mi) in length, although some were 80 km (50 mi) long; they tend to be about half as wide as they are long. In *Assembling California*, John McPhee compares it to a big rigid airship or blimp. A modern airship is not a single big gas bag that can blow up (like happened during the *Hindenburg* disaster). Instead, the inside is filled with rows and rows of many smaller containers for helium gas, each of which provides lift. So too the Sierra Nevada batholith is not a single large

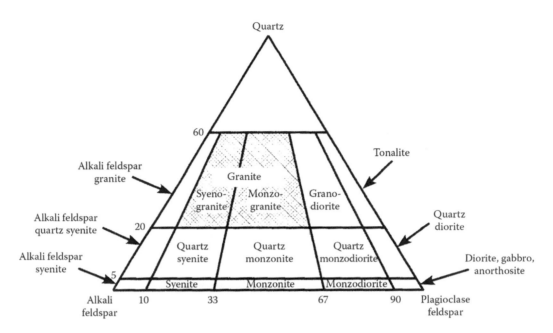

FIGURE 8.3   Standard classification of felsic igneous rocks showing the ratios of quartz (apex of triangle) and alkali feldspars (potassium feldspar) in the lower left corner, and plagioclases (lower right corner). The boundaries that define the different classes of granitic rocks (diorites, granodiorites, quartz diorites, quartz monzonites, tonalites, and true granites) are shown.

Source: Based on several sources.

intrusion but many smaller plutons side by side and cutting across one another through a long history of intrusion and cooling.

These magma chambers were then cut by thousands of smaller dikes (Figure 8.4b) that carried the last of the melt through cracks in the older cooled and solidified magma chambers. In many places, you can see where the walls of the magma chamber (the "wall rock" or "country rock") came in contact with the hot magma (known as contact metamorphism, as discussed in Chapter 2). There are many places where you can find chunks of wall rock that were ripped off the wall of the chamber and bathed in magma but never completely melted or digested before the magma cooled and solidified (Figure 8.4c and d). These form unusual and exotic inclusions in the granitic material known as **xenoliths** (Greek for "foreign stones").

The history of magma intrusion shows that not all parts of the Sierras formed at the same time. The oldest intrusions are on the far eastern edge of the Sierras, near Mono Lake, and they are Triassic in age (at least 210 Ma) (Figure 8.5a). The next youngest intrusions are Jurassic in age, and they, too, are to the east of the main range, especially in the Owens Valley area (Figure 8.5a). Then in the Early Cretaceous (about 120 Ma), the main episode of intrusion occurred, far away in the western Sierras. Over the rest of the Cretaceous, the locus of intrusion begins to migrate eastward again at a rate of roughly 3 mm/year so that most of the Late Cretaceous intrusions are in the central and eastern part of the main range (Figure 8.5a). The youngest are dated about 83 Ma. To many geologists, this suggests that the angle of dipping plate changed over

time (Figure 8.5b). In the Triassic and Jurassic, the plate must have slid into the mantle at a shallow angle, so it formed intrusions far to the east of the trench. Apparently, there was an abrupt steepening of the plate in the Early Cretaceous, because the line of intrusions shifted westward by more than 200 km. In the Late Cretaceous, the plate must have resumed its earlier plunge at a shallower angle, because the line of plutons migrated east as the position where the down-going slab reached the depth in the mantle where it could melt shifted east (Figure 8.5b). By 70 m.y., the intrusions were finished, and the volcanic arc shut off completely. There was no arc volcanism in the region for the next 30 m.y., and when it returned, it was far to the east of the Sierras (see Chapter 7).

So if the bulk of the Sierras represent extinct magma chambers that fed a chain of Andean-style arc volcanoes, where are the volcanoes that used to lie on top? Most erupted before 70 Ma, so they have been eroded away as the region uplifted and the entire volcanic edifice succumbed to at least 70 m.y. of erosion. However, in one place in the north-central Sierra, there is a tiny remnant of this original chain of volcanoes. It is called the Minarets Caldera, just up the mountain from and to the west of Mammoth Mountain Ski Resort, and it dates to 100 Ma (Figure 8.6). The Minarets themselves are part of the original caldera collapse deposit (Figure 8.6b), and the rocks around the Minarets are made of ash flow tuff, with lakes that reside in the center of the ancient caldera.

The only other traces of the Sierran volcanoes that erupted between 120 and 70 Ma are volcanic ash layers that blew to the east and settled at the bottom of the Western

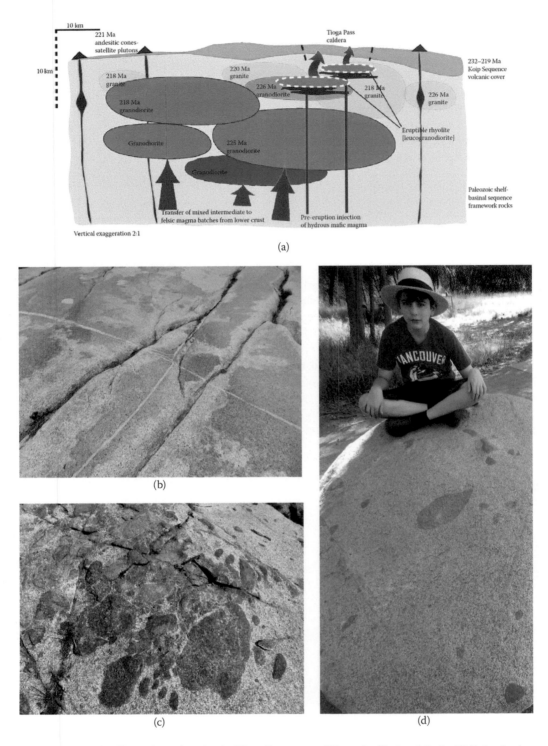

FIGURE 8.4 (a) Sequence of different intrusions in the Tioga Pass area of Yosemite National Park. (b) Pair of crisscrossing white aplite dikes, both of which cut through a xenolith. Taken at Olmstead Point, Yosemite National Park. (c) Angular xenoliths ripped from the wall of the magma chamber. (d) Rounded xenoliths in a glacial erratic boulder, Donner Pass.

Source: (a) With permission from the Geological Society of America. (b) Photo by the author. (c) Photo by C. T. Lee. (d) Photo by the author.

Interior Seaway during the Cretaceous in what is now Colorado, Utah, Wyoming, and the Dakotas. These abundant ash layers are found in nearly every Cretaceous shale deposit of that seaway. Once the ash landed on the ocean surface and settled on the seafloor, it altered to **bentonite**, a product of the weathering and alteration of volcanic ash on the sea bottom.

The Sierra Nevada batholith is itself a remnant of a larger chain of similar batholiths that run all the way from the Coast Range batholith in British Columbia to the

Idaho batholith down to isolated intrusions in the Klamath Mountains (see Chapter 9), and then to the Sierras, and finally to the Peninsular Range batholith (also called the Southern California batholith) (Figure 8.7). Today, they appear to be disjointed and discontinuous because they have been displaced since they intruded between 120 and 70 Ma. The expansion of the Basin and Range Province behind the Sierras has moved them west and away from the core of North America, while faults have also offset the Idaho batholith. If you restore them to their original positions (Figure 8.7b), they would have once formed a continuous chain of volcanoes, just as the Andes do today.

8.5 ROOF PENDANTS

Xenoliths (Figure 8.4) are small pieces of wall rock ripped off from the sides of the magma chamber being ripped off, but there used to be a much larger body of country rock surrounding the magma chambers. At one time, geologists thought that large chunks of wall rock still sitting on top of the old Sierra magma chambers were remnants of the old roof of the magma chamber. These were called **roof pendants** (literally, hanging down from the roof of the magma chamber like a pendant). In more recent years, detailed mapping has shown that these are typically not the roof of a single magma chamber but a large body of

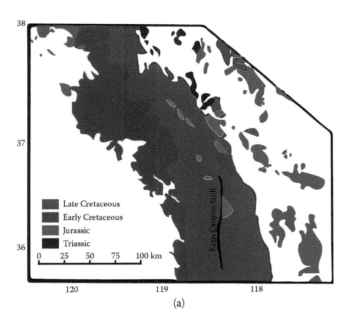

(a)

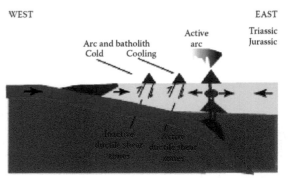

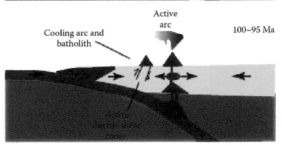

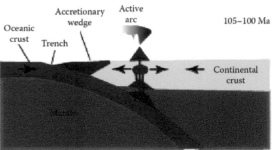

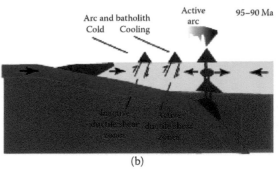

(b)

(c)

FIGURE 8.5 (a) Age trend of magmas in the Sierras. (b) Migration of subduction zone angle during the Mesozoic. (c) Tungsten Hills (foreground) in the Owens Valley northwest of Bishop are Triassic eastern intrusions of the early phase of Sierran volcanism. Behind them are the snowcapped peaks of the High Sierras, which are Late Cretaceous in age, as a consequence of the arc angle returning to its position in the Triassic and Jurassic (see also Figure 8.9a, "Triassic scheelite intrusive suite").

Source: (a–b) Redrawn from several sources. (c) Photo by the author.

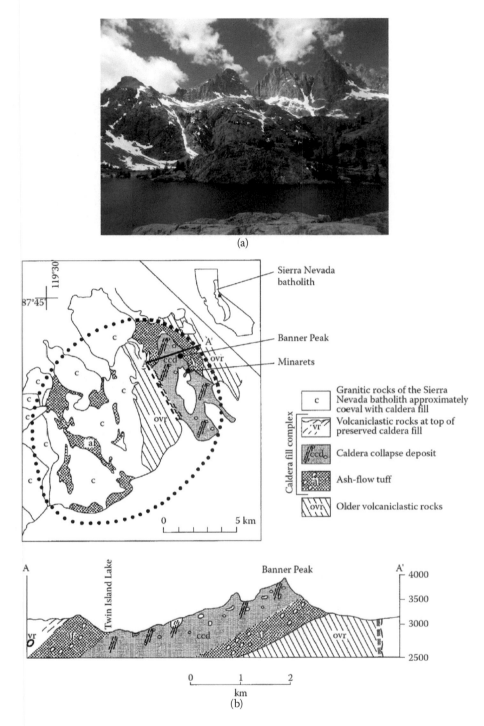

FIGURE 8.6 Minarets Caldera. (a) Photo of the Minarets towering above the remnant caldera (now filled with a lake). (b) Diagram showing a map and cross section of the Minarets Caldera Complex.

Source: (a) Courtesy of Wikimedia Commons. (b) From Fiske, R. S., and Tobisch, O. T. *Geological Society of America Bulletin*, 106, 582–593, 1994, with permission from the Geological Society of America.

wall rock intruded by multiple magma chambers coming up from below (Figure 8.3a). More than 100 of these features are known. Most are just tiny remnants of what was once a huge area of wall rock, but they are interesting clues to what surrounded the magma chambers before they melted their way up from the depths. They are scattered across the entire batholith (Figure 8.8), although they are most common in the central and southern Sierras.

Some of the largest are the Ritter Range pendant, the Mt. Tom pendant, the Mt. Morrison pendant, and the Saddlebag Lake pendant (Figure 8.9a). The Mt. Morrison pendant (Figures 8.8 and 8.9) is in the eastern Sierras and visible in

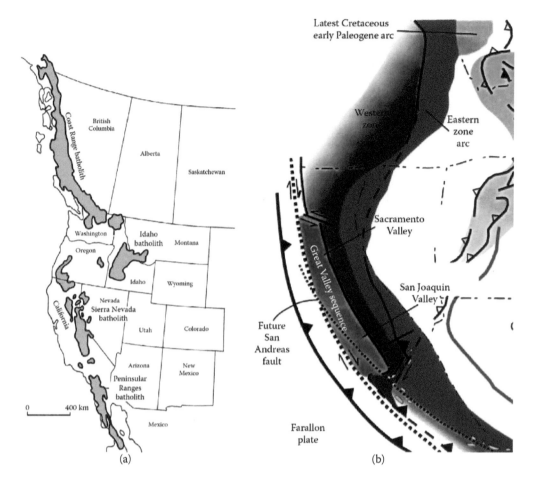

FIGURE 8.7 (a) Chain of Mesozoic arc volcanoes and their batholiths that ran up the western edge of North America during the Mesozoic. (b) Restored (or "palinspastic") position of the volcanic chain before later tectonism disrupted its alignment. Before Basin and Range extension, Nevada was half its present width, and the Sierras lined up better with the Idaho batholith and the Coast Range batholith. During the Miocene, the Basin and Range extension pushed the Sierras westward.

Source: (a) From Prothero, D. R., and R. H. Dott Jr. *Evolution of the Earth* (8th ed.). McGraw-Hill: New York, 2010. (b) After Stephenson, D. B., and Girty, G. H. "Constraining the Source of the Orocopia Schist, SE California: A Comparative Study of the Major and Trace Element Chemistry of Cretaceous Sandstones Deposited in Forearc Settings." *Pacific Section SEPM Book*, 101, 77–96, 2006, with permission from Pacific Section SEPM.

the Owens Valley between Mono Lake and Bishop. It covers an area of 414 km² (160 mi²). It consists of a package of lower Paleozoic metasedimentary and metavolcanic rocks that is more than 15,000 m (50,000 ft) thick, mostly of the Cambrian–Silurian and Pennsylvanian–Permian ages. These rocks are complexly faulted and folded, as would be expected for something that was once sitting adjacent to a hot, expanding magma chamber and lying beneath a volcanic chain.

The Saddlebag Lake pendant, due west of Mono Lake (Figures 8.8 and 8.9a–d), consists of three different rock sequences, all intensely deformed and slightly metamorphosed. The oldest packages of rocks are metamorphosed marine sediments of the Ordovician–Silurian age. The second is a sequence of Permian metavolcanic rocks that overlies the first package with an angular unconformity. Another metamorphosed sequence above this also consists of metavolcanic rocks, also of the Permo–Triassic age.

8.6 WESTERN FOOTHILLS

The roof pendants are not the only prebatholithic rocks that were in place before the big intrusions during the Jurassic and Cretaceous. The first thing that catches your eye when you glance at the geologic map of the Sierras (Figure 8.1b) is the huge area of Mesozoic granitics (shown in red in Figure 8.1b). But the western foothills of the Sierras (especially in the northwestern foothills) are shown on the map by bands of pale blue (Paleozoic) and green (Mesozoic) rocks that were the country rock when the Sierran magmas came up beneath them. As just discussed, some ended up as tiny remnant roof pendants in the central and southern part of the range. But the bulk of the western foothills is made of these old Paleozoic and early Mesozoic rocks, with only minor areas where the granitics (colored red in Figure 8.1b) managed to melt through them.

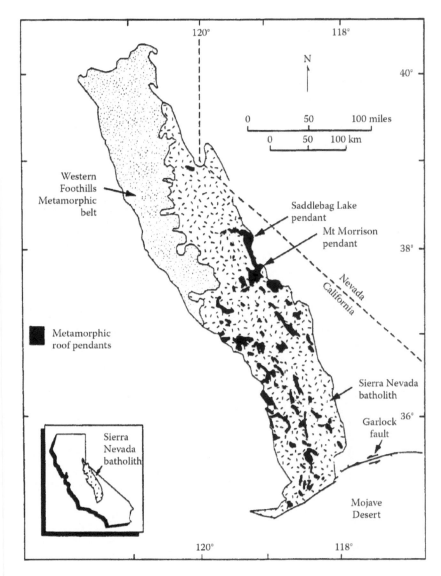

FIGURE 8.8 Map of Sierra roof pendants.

Source: From Schweickert, R. A., and Lahren, M. *Geological Society of America Bulletin*, 111, 1714–1722, 1999.

The western foothills are in fact ancient exotic terranes that were plastered onto North America during different events in the Paleozoic and Mesozoic. Some of them are connected to events we have already discussed in Chapter 7. They are also connected to similar rocks that make up the Klamath Mountains (Chapter 9). Many of these terranes have come from enormous distances across the early ancestor of the Pacific Ocean to be accreted to North America at various times in the past 400 million years. None were in California, when the oldest Precambrian and early Paleozoic rocks were formed, rocks now found in the Basin and Range.

Let us look at the western foothills in greater detail. The easternmost outcrops above the Sierran granitics are Paleozoic rocks called the **Shoo Fly Complex** (most of the pale blue color in Figure 8.1b). It is composed of Cambrian through Devonian metasedimentary rocks and

metavolcanics, almost all representing deep-water settings, probably from an ancient volcanic arc complex. The lower unit of the Shoo Fly is mostly deep-water shales, cherts, and turbidite sandstone layers (Figure 8.10a). Some of these units closely match the Upper Cambrian Harmony Formation in central Nevada and the quartzites of the Antelope Mountains in the Klamaths, near Yreka, California. The middle Shoo Fly is a highly sheared mixture of shales, sandstones, and cherts with minor volcanics and limestones and ophiolites; the limestones yield Ordovician fossils. The upper Shoo Fly consists of shales and cherts, with minor volcanic ash; its age is not well constrained but is between Ordovician and the Late Devonian. Deposited on top of the Shoo Fly metasedimentary sequence is a thick slice of pre–Late Devonian pyroclastic volcanics and sediments. Taken together, the Shoo Fly sediments and volcanics are thought to represent an early Paleozoic island arc complex (like

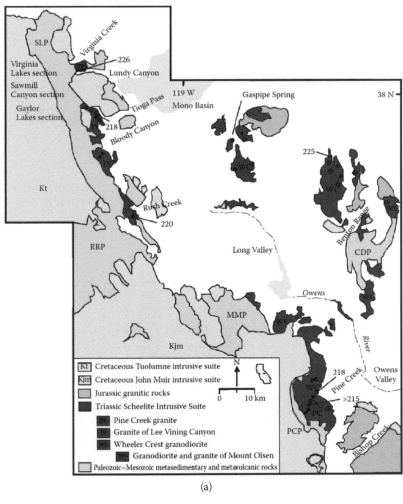

(a)

(b)

FIGURE 8.9 Roof pendants of the Sierra Nevadas near Long Valley and Mono Lake. (a) Map showing the location of the major roof pendants west of Mono Lake and Bishop. CDP = Casa Diablo pendant; MMP = Mt. Morrison pendant; PCP = Pine Creek pendant; RRP = Ritter Range pendant; SLP = Saddlebag Lake pendant. (b) Photo of the Saddlebag Lake pendant, the dark rock in the background of the lake. Lighter rocks are Sierran granitic intrusions. (c) Geologic cross sections of different parts of the Saddlebag Lake pendant, showing the incredibly complicated structural geology of the multiple fault slices all heavily metamorphosed. (d) Intensely sheared Jurassic metasandstones with crosscutting white feldspar dike, Saddlebag Lake pendant near Steelhead Lake. (e) Folded metasandstones of the Cooney Lake Conglomerate, Saddlebag Lake pendant. (f) Contact of the granitics (light rocks) with the dark-colored metamorphics of the Mt. Morrison roof pendant. (g) Color-banded metasedimentary rocks of the Laurel Mountain roof pendant. (h) The sharp contact between the dark metamorphic rocks and the light granitic intrusions, Pine Creek roof pendant.

Source: (a) From Bartley, J. M., et al., *Geosphere*, 8, 1086–1103, 2012, with permission from the Geological Society of America. (b) Courtesy of Wikimedia Commons. (c) After Schweickert, R. H., and Lahren, M. *Geologic Evolution of the Saddlebag Lake Pendant, Eastern Sierra Nevada, California: Tectonic Implications. Pacific Section SEPM Book*, 101, 27–56, 2006, with permission from the Pacific Section SEPM. (d–e) Photos by R. Hollister. (f–g) Photos by R. Hollister. (h) Photo by G. Hayes.

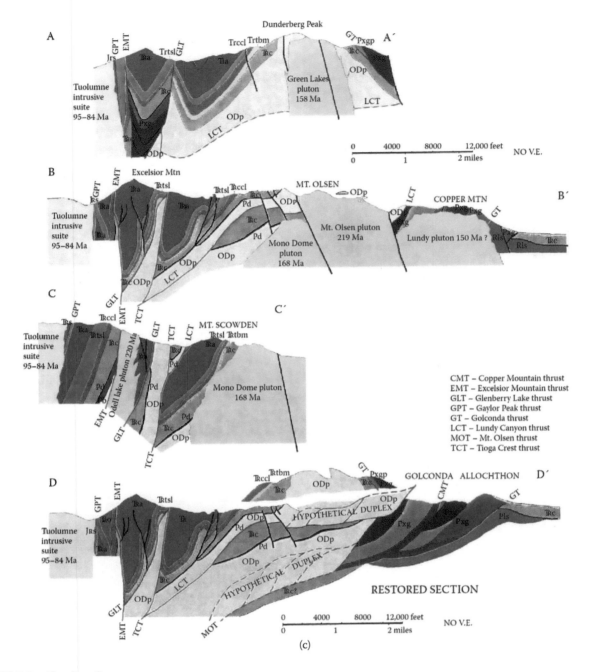

FIGURE 8.9 (Continued)

Japan—Figure 4.11b) that crashed into North America as it consumed all the oceanic crust between the two plates. All these rocks were accreted to the edge of North America during the Late Devonian—Early Mississippian **Antler orogeny**, which also brought the central block of Nevada as well (Figure 7.20).

Caught between the Shoo Fly and the belts of rock to the west is a sliver of Paleozoic oceanic crust, the **Feather River peridotite** (formerly called the Feather River ophiolite; colored purple in Figure 8.1b). It consists of serpentinized pillow lavas, sheeted dikes, and gabbros, typical of ophiolites the world over (Figures 4.8 and 8.10b).

Next to arrive was the **Northern Sierra terrane** (northern part of the light blue in Figure 8.1b). The Northern Sierra terrane are the remnants of another accreted volcanic arc, with deep marine metavolcanics and metasediments of the Mississippian through the Permian age (Figure 8.11), all accreted to North America during the Late Permian **Sonoma orogeny** (the same event that brought the Sonomia terrane to become the basement of northwestern Nevada—see Figure 7.20). In places such as Sierra Buttes, the Northern Sierra terrane rocks form an angular unconformity on top of Shoo Fly rocks.

The next unit west is a block of Triassic and Jurassic rocks called the **Calaveras Complex** (eastern portion of the

FIGURE 8.9   (Continued)

green areas in Figure 8.1b). It is composed mostly of an accretionary wedge that was beginning to form to the west of the Triassic–Jurassic predecessor of the Sierran volcanic arc (Figure 8.5). Like other such accretionary prism deposits, it consists mostly of highly sheared oceanic sediments (Figure 8.12a), plus slivers of oceanic crust (ophiolite), plus metamorphic and plutonic blocks brought up from deeper in the trench by the upward pushing of the pile during accretion.

Finally, the youngest unit is the **Foothills terrane**, also called the Western Jurassic terrane (western part of the green in Figure 8.1b), separated from the Calaveras

Complex by a large suture zone. It is yet another accretionary wedge that formed during the Middle-Late Jurassic eruptions of the early Sierran arc, including a famous gold-bearing unit known as the Mariposa Slate and numerous metavolcanic units (Figure 8.12b and c). It was plastered onto the western edge of the Calaveras Complex in the Late Jurassic–Early Cretaceous, forming a stack of three consecutive slivers: the Paleozoic Shoo Fly–Northern Sierra terranes, the early Mesozoic Calaveras terrane, and the Jurassic Foothills terrane (Figure 8.13). Just as we saw how the earliest eruptions to form the Sierras in the Triassic and Jurassic occurred in the Owens Valley area to the east of the

FIGURE 8.10 (a) Lower Paleozoic Shoo Fly phyllites and schists, west of Ladies Canyon, Highway 49. (b) Serpentinites of the Feather River ophiolite, near Goodyear Creek Road, Highway 49.

Source: Photos by the author.

FIGURE 8.11 Sierra Buttes, made of later Paleozoic metavolcanics of the Northern Sierra terrane, unconformably overlying Shoo Fly rocks to the west.

Source: Photo by the author.

FIGURE 8.12 Outcrop images of exotic blocks in the western Sierras. (a) Sheared Calaveras Complex metasediments, Highway 49, 2 mi south of bridge over North Fork of Yuba River. (b) Mariposa Slate, part of the Foothills terrane, Highway 49 near Gold Hill. Outcrop images of exotic blocks in the western Sierras. (c) Dark-green pillow lavas and metavolcanics, Foothills terrane, Highway 49 south of El Dorado.

Source: Photos by the author.

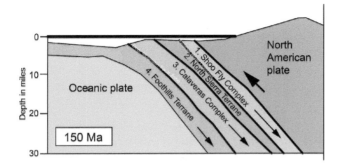

FIGURE 8.13 Diagrammatic cross section of the accretionary terranes in the western foothills of the Sierras.

Source: Redrawn from several sources by E. T. Prothero.

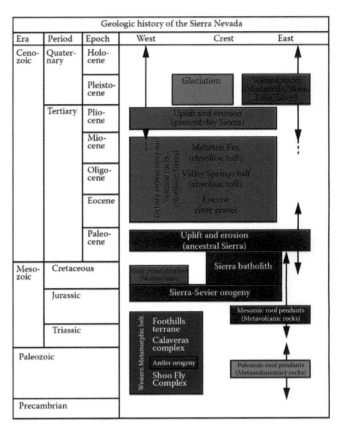

FIGURE 8.14 Geologic history of the Sierras.

Source: Redrawn from several sources by E. T. Prothero.

Cretaceous eruptions of the main Sierran batholith, so too we see the accretionary wedge for those Triassic–Jurassic arcs far to the east of the Cretaceous accretionary wedge, which occurs in the Coast Ranges (Chapter 12).

Following is a summary of the history of the western foothills (Figures 8.10, 8.13, and 8.14):

1. *Shoo Fly terrane.* Cambrian through Devonian deep-water sediments and volcanics accumulate somewhere west of the Sierras.
2. In the *Late Devonian*, the Shoo Fly Complex and the central block of Nevada (above the Roberts Mountain thrust) are plastered onto North America during the Antler orogeny.
3. *Northern Sierra terrane.* Mississippian through Permian deep marine and volcanic arc rocks are formed west of the Sierras; Feather River ophiolites are formed on the seafloor.
4. *Late Permian–Triassic.* The Northern Sierra terrane, Feather River ophiolite, and Sonomia terrane of northwest Nevada slam into the continent during the Sonoma orogeny.
5. *Late Triassic–Early Jurassic.* The early phase of Sierran volcanic arc is formed in Owens Valley, and the Calaveras Complex is the accretionary wedge for this event.
6. *Middle–Late Jurassic.* The next phase of early Sierran eruptions form in the Owens Valley, and the Foothills terrane is the accretionary wedge for this event.
7. *Early Cretaceous.* The main phase of Sierran eruptions and intrusions rises beneath the western foothills and the roof pendants and other pre-existing country rocks, melting its way through and forming an Andean-style volcanic chain. Meanwhile, the forearc basin forms in what is now the Central Valley of California, and the accretionary wedge shifts from the Sierra foothills to what is now the Coast Ranges.

8.7 CENOZOIC HISTORY OF THE SIERRAS

Although the Sierra Nevada mountains consist largely of Mesozoic and older rocks, there are younger deposits in many places that are crucially important, not only for understanding their tectonics, but also because they are important gold-bearing deposits. Chief among these are many deposits of unconsolidated Eocene gravels and sands that accumulated in paleovalleys carved across the crest of the Sierras, mostly near the course of modern-day rivers, such as the Yuba, American, Calaveras, Mokelumne, and Tuolumne Rivers (Figures 8.15a and 16.3b). Often called the "Ione Formation" or the "auriferous gravels," these loose sediments accumulated much of the gold that eroded from the Sierras in the Eocene. Once the Gold Rush miners discovered this, these deposits were extensively mined using giant fire hoses in a method called hydraulic mining (see Chapter 16). In some places, the Eocene sediments form a veneer that sits on top of an Eocene erosion surface that runs down the entire western slope of the Sierras from Donner Pass in the east to the Central Valley. This is the ancient surface that still can be seen on the western slope of the Sierras today (Figure 8.2c).

In the late Eocene and Oligocene, volcanic ashes from the huge eruptions in Nevada and Utah began to flow down the slope of the ancient Sierras. They were often trapped in the Eocene paleovalleys above the middle Eocene gravels

(Figure 8.15c). These volcanic ashes include the late Eocene (34.275 Ma) ash that preserved the famous LaPorte flora (Chapter 20), as well as the Oligocene Valley Springs tuffs that can be found at Donner Pass on the Sierra summit (Figure 8.15c). Nearly all these ashes show that there was an elevated area of volcanics to the east in what is now Nevada, which erupted ashes that flowed down the gentle slope of the Sierran block (Figure 8.15e).

In the Miocene, there were further volcanic eruptions of volcanic ash, which are preserved as the middle–upper Miocene Mehrten Formation (Figure 8.15e), found in many places along the western Sierran foothills. To many geologists, these ashes suggest that the Sierras were nearly buried in the older deposits of Oligocene and Miocene ash and formed a gently rolling or nearly flat surface, rather than the high rugged barrier they do now. These are capped by the Table Mountain lavas, dated at 9 Ma (see following).

8.8 WHEN DID THE SIERRAS RISE?

When did the uplift and tilting of the Sierras take place? As we have seen, the Sierras were a huge extinct Jurassic–Cretaceous magma chamber, so their first incarnation as a mountain range was a giant chain of volcanoes somewhat like the modern Andes. As these eroded down, they exposed the magma chambers that fed them. But how long has it been since the deep, long-cooled magma chambers rose to the surface and began to erode?

As early as 1911, legendary mining geologist Waldemar Lindgren argued that the Sierras were a relatively recent uplift. Most thought that the uplift occurred about 9–10 Ma. This is based on the geometry of the drainage system and the development of the ancient peneplane on the Sierras, which appears to be very young, with a very shallow gradient (Figure 8.15a). The ancient gold-bearing river deposits (especially Eocene gravels) are trapped in the valleys of the Sierras, and they seem to have derived much of their sediment from the Great Basin. According to this kind of thinking, they flowed across the region that became the Sierras without hindrance, so if there was a low mountain range there in the Eocene, it was no barrier.

Another feature that supports a rather recent uplift of the Sierras is numerous examples of **inverted topography** (Figure 8.16). The best example of this is Table Mountain, on the Stanislaus River, near Highway 108. The feature is well-named, because the mountain is flat-topped

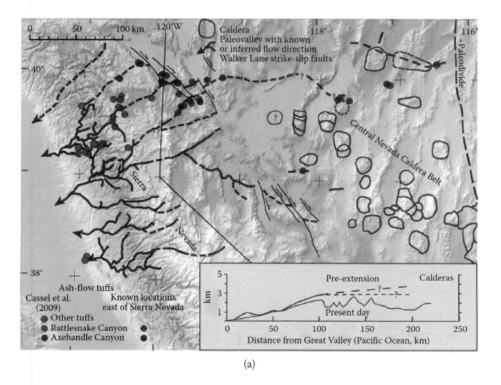

(a)

FIGURE 8.15 (a) Map showing the drainages from Nevada that once crossed the Sierras, filling the ancient river channels with Eocene gravels and sands, and then Oligocene volcanics. (b) Eocene gravels at the summit of the Sierras near Soda Springs. (c) Oligocene Valley Springs volcanic deposits, Sugar Bowl Ski Resort, near Donner Pass. (d) Reconstructed elevation profile of the Nevadaplano sloping down to the slightly lower summits of the Sierras. (e) Middle Miocene ashes and river deposits of the Mehrten Formation, near Turlock.

Source: (a) From Cassel, E. J., et al., *Geology*, 37, 547–550, 2009. (b–c) Photos by the author. (d) Redrawn from several sources. (e) Photo by J. Sankey.

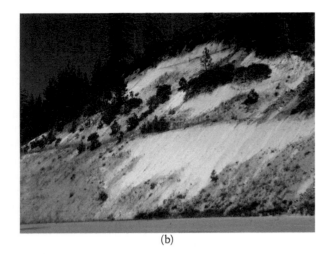

(b)

(c)

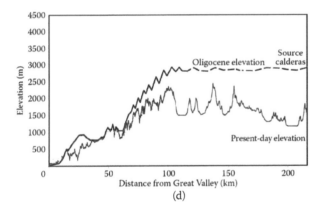

(d)

(e)

FIGURE 8.15   (Continued)

like a table. The reason for this shape is a caprock of a 9-million-year-old basaltic lava flow on top of the plateau, which sits on top of soft gravels that would normally erode quickly without the hard basalt cap. Yet when it was molten rock, basaltic lava is very fluid and would flow in the bottoms of the stream valleys as the water does today (Figure 8.16d). The only way for this topography to occur is if the lava flowed down at the bottom of the valley 9 Ma, and then the entire region was uplifted enough since the former valley bottom became the top of the Table Mountain, while the former valley walls of soft sediment eroded away. Thus, the highs became lows, and the low spots became the high spots, so the topography was turned inside out, or **inverted** (Figure 8.16d). For this region, at least, such geology is strong evidence that the area has undergone at least a kilometer or so of uplift since the lavas flowed 9 Ma. This and other basalt flows came from a source near Mono Lake, so the Sierras were not blocking the way yet. Analysis of the stream gradients in the western flank of the Sierras is consistent with a fairly recent uplift as well.

However, in recent years, this argument for recent uplift has been reconsidered. In addition, there is evidence from the vegetation and from geochemistry that the Basin and Range was already quite dry and semidesert by 15 Ma. If this is true, then the Sierras were high enough by 15 Ma to create the rain shadow effect.

Remember from Chapter 7 that Nevada then was not yet the downfaulted "Great Basin" but probably an uplifted plateau. Whatever elevation the Sierras had in the Eocene, Nevada was higher. DeCelles (2004) called it the "Nevadaplano" in analogy with the Altiplano highlands behind the Andes in South America (Figures 8.15a and d and 8.17). In other words, the elevation of the Sierras was not high enough to obstruct the water draining west from the much higher Nevadaplano to the east. In addition to the gold-bearing Eocene gravels, abundant Oligocene volcanic ash sheets flowed westward across the Sierras from high volcanic ranges and calderas in the Nevadaplano (Figure 7.22). Today, remnants of these ash flows are found in the paleovalleys that cross the Sierras (Figure 8.15a and c).

Other data suggest that the Sierras have been uplifted for a long time (Henry, 2009). Studies of the geochemistry of the clays in the paleovalleys (Mulch et al., 2006) suggested that the Sierra crest was about 2,200 m (7,217 ft) in the Eocene, much like the present elevation of Lake Tahoe. Cassel et al. (2009) analyzed the geochemistry of volcanic ashes in those Oligocene tuffs from Nevada and came up with an even higher estimate of 2,800 m (9,187 ft) for the crest of the Sierras near Lake Tahoe (even higher than today), and the calderas in Nevada to the east were at the same elevation (Figure 8.15d).

Thus, it appears that the Sierras have a complex multistage history of uplift (Figure 8.14). The recent geochemical data suggest that the Sierra crest was at 2,200–2,800 km elevation in the Eocene (50–35 Ma), but other data are

(a)

(b)

(c)

FIGURE 8.16 (a) Resistant ridge-capping basalt lava flows of Table Mountain near California Highway 108, dated at 9 Ma, sitting on slopes of the middle Miocene Mehrten Formation. (b) Close-up of the columnar jointing in the flow. (c) Aerial view of the ancient river valley now filled by a resistant sinuous ridge of caprock basalt. (d) Cartoon of the processes that form inverted topography. In the top panel, the valleys carved in soft sediments were filled by liquid lava, which then cools and hardens. Once it was cooled, it was harder and more resistant to erosion than the soft sediments around it, so the valley-bottom lava flow ends up as a caprock, while the soft walls of the valley erode more rapidly.

Source: (a–b) Photos by the author. (c) Photo from G. Hayes. (d) Figure drawn by E. T. Prothero.

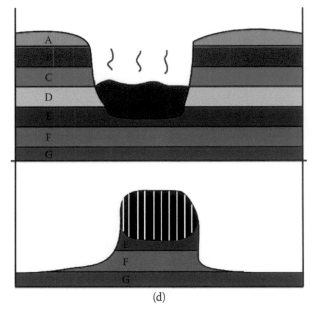

(d)

FIGURE 8.16 (Continued)

consistent with uplift only since 9 Ma. One solution might be a slight uplift of the range during the Eocene, as it formed a ramp down from the Nevadaplano (Figure 8.17), followed by a slow period of erosion in the Oligocene and early Miocene, resulting in almost complete burial of the range by volcanic ashes by the late Miocene. Finally, there was renewed uplift in the late Miocene, which incised the ancient river valleys and formed the inverted topography. This appears to be the consensus for the present (Jones et al., 2004). However, the topic remains controversial.

8.9 THE MODERN SIERRAS: EXFOLIATION AND GLACIATION

As the Sierra batholiths have risen into the sky over the past few million years, they have formed the mighty mountain range that we know today. Two major processes are responsible for the bulk of the modern geomorphology of the range: exfoliation and glaciation.

8.9.1 Exfoliation

The word *exfoliation* (literally, "peeling off leaves" in Latin) is commonly used to describe peeling off old layers of dead skin. In geology, however, it has another meaning: the peeling off of sheets of rock as they weather on the surface (Figure 8.18). Exfoliation is a very specific kind of weathering of rocks caused by pressure release. The granitic rocks

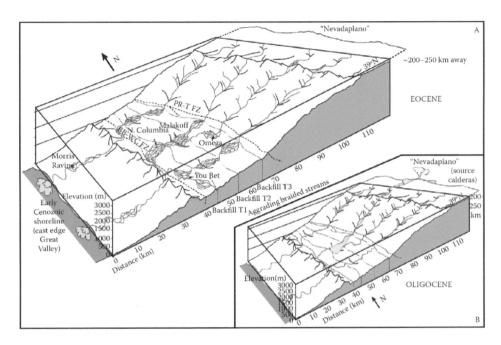

FIGURE 8.17 The Nevadaplano and the paleotopography of the Sierras in the Eocene and Oligocene.

Source: With permission from the Geological Society of America.

of the Sierras were once deep in the crust and under tremendous pressure. As they have risen up into the sky over the past few millions of years, that pressure is slowly released, and the rock expands (Figure 8.18a). However, such rapid expansion means that the brittle low-pressure rock cracks and shatters instead of slowly and smoothly expanding without breaking. As these sheets of rock peel away, they form curved fractures (joints) and the layers peel off in a manner that is similar to the peeling off of the layers of an onion (Figure 8.18b and c).

This tendency to peel off in onion skin layers is the predominant form of weathering and breakdown operating in the High Sierras today. It gives many of the mountain peaks in the Sierras their characteristic rounded and knob-like shape. Exfoliation is enhanced by a process known as frost wedging. When water seeps down into the cracks in the rock and then freezes, it expands by about 9.8%. (Think of how water in a tightly sealed container will bulge or crack the container if it freezes with no room for expansion.) This creates a huge amount of pressure (about 2,000 psi), which will shatter almost any kind of rock with even the tiniest crack in it. If the ice melts in the day, it seeps farther into the crack and then freezes again in the night, pushing the crack open. In a region like the Sierras, which will freeze and thaw many times in a winter as the snow melts and then new snow and ice form, this breaks down nearly all the rocks very rapidly. Freeze–thaw and frost wedging are especially important in forcing apart the originally tiny joints caused by pressure release exfoliation into wide cracks, which eventually

causes huge sheets of rocks to spall away and slide down the mountains.

8.9.2 GLACIATION

The High Sierras were the only part of California that was extensively glaciated during the past 2.5 m.y. of the ice ages. Most of California was too low in elevation (except for the highest peaks of the Klamaths and Mt. Shasta) to develop glaciers during the peak glacial episodes. The southernmost extent of the continental ice sheet, which once covered Canada and Alaska and New England and the northern Midwest, did not reach much farther south than northern Washington. Most of the high peaks of the Cascades, such as Mt. Rainier, Mt. Shasta, and Mt. Hood, were glaciated, as discussed in Chapter 6.

Where glaciers have appeared, they are an extremely powerful force in shaping the landscape. They form where there is more snow and ice accumulating than is melting away. Falling snow builds up winter after winter, and if there is not enough melting in the summertime, the ice accumulates. The light "powder" snow that skiers love is almost 90% air, but as this gets buried and compressed into packed, "granular" snow, the amount of trapped air is only 50%. As the years go by and the snowfall builds up, the pressure transforms the snowflakes into tiny granules known as **firn**. Over thousands of years, the weight of more and more snow and ice above further compresses the firn into fine-grained ice crystals, which have only 20% air bubbles by volume. If the glacier is thick enough and builds

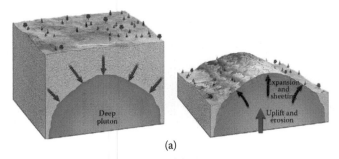

(a)

(b)

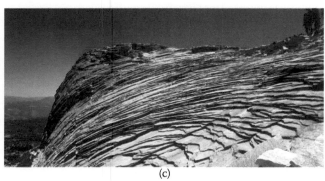

(c)

FIGURE 8.18 Exfoliation: (a) Diagram showing pressure release. Exfoliation: (b–c) Examples of Sierran exfoliation joints, resembling the peeling of the layers of an onion.

Source: (a) Courtesy of Wikimedia Commons. (b–c) Photos by the author.

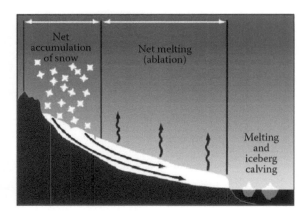

FIGURE 8.19 Dynamics of a mountain glacier flowing down a valley. At the head of the glacier in the mountains is the zone of accumulation, where more snow builds up than melts back. At the bottom is the zone of ablation, where more snow is lost to melting and evaporation than accumulates. The imaginary boundary where these two forces are balanced is known as the equilibrium line.

Source: Courtesy of Wikimedia Commons.

up over hundreds of thousands of years, the deepest ice is transformed from fine-grained crystals to coarse-grained crystals, again with about 20% air bubbles.

Glaciers can be thought of as "rivers of ice." Even though the top layers are brittle, the ice below about 50 m (160 ft) is under the weight and pressure of so much ice that it flows in a plastic fashion (see Chapter 3). It is easy to understand why a flowing mass of ice would move downhill in the mountains, but glaciers flow no matter where they are. Even continental glaciers that covered huge flat landmasses (such as Greenland and Antarctica today) still flow from the high spot of ice near their center out toward the edges. A good analogy is pouring a thick, viscous fluid, like honey, on the table. The honey makes a small pile but soon flows away from the high area in the center out to the edge as it spreads.

Glaciers are not only rivers of ice but also change dramatically in shape and size (Figure 8.19). The upper reaches of the glacier (where the snow and ice fall) are called the **zone of accumulation**. The lower part of the glacier (where the melting and evaporation dominate) is called the **zone of ablation**. In the middle of them is an imaginary line where the ablation exactly matches accumulation; it is called the equilibrium line. If the glacier is advancing, the equilibrium line shifts downhill. If it is melting back, the equilibrium shifts uphill.

It's important to remember that the *ice is always flowing downhill*, no matter whether the snout is advancing or retreating. This is easy to imagine when the glacier is advancing. When the glacier is stationary and the snout is neither advancing nor retreating, then ablation is exactly matching accumulation. Under these circumstances, the snout remains stalled in one place, and yet the ice flowing downhill piles up a huge mass of boulders, gravel, sand, and even silt, called a **glacial moraine** (Figure 8.20). But even when the snout is retreating, the ice *still* flows downhill. Its rate of downhill flow just cannot keep up with the rate at which the snout is melting away.

Glaciers are enormously powerful agents of erosion, capable of changing the land surface in ways that water, wind, and gravity cannot (Figure 8.21). In mountains, the glaciers begin by filling up normally V-shaped stream valleys with a thick flowing stream of ice. At the top of a glacier, the valley walls are plucked by the ice pulling away to form a crevasse at the top called a **bergschrund**. The continual plucking eventually rips enough rock away that the shallow stream valley is converted into a big bowl-shaped amphitheater known as a **cirque** (French for "circle"). If

FIGURE 8.20   Glacial moraines are deposits of poorly sorted sediments (from boulders down to fine clay) dumped by glaciers, with no bedding or stratification.

Source: Photo by the author.

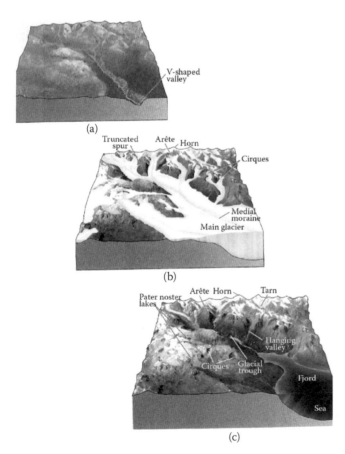

FIGURE 8.21   Glacial topography. (a) Before glaciation, (b) during glaciation, and (c) after glaciers have melted back.

Source: Courtesy of Oxford University Press.

the Alps is the most famous example, but there are horns in most glaciated mountains, including the Sierras.

As the glacier flows down the formerly V-shaped river valley, it grinds and plucks away the rocks on the valley walls until they form nearly vertical concave surfaces, making a **U-shaped valley** (vertical on the sides, nearly flat on the bottom). The gently sloping ridges are sliced off at the bottom to form **truncated spurs**. The smaller tributaries of the river in the old landscape once flowed smoothly downslope until they met the trunk stream. But when a thick glacier fills a valley and chops away at the old slopes of the river drainage, its tributary glaciers will come in at the top of the trunk glacier, not the base (Figure 8.21). When the glacier melts back, these side valleys come to a steep cliff, then drop off abruptly. This is called a **hanging valley**. The streams flow not down a smooth, continuous grade to the trunk stream but instead down the hanging valley and over a cliff to form a waterfall.

The Sierras exhibit most of the features known from glaciated mountains around the world. Nearly every part of the Sierras above 2,400 m (8,000 ft) in elevation was heavily glaciated at the peak of the last glacial episode, about 18,000–20,000 years ago (Figure 8.22). Most of these glaciers melted back during the glacial–interglacial transition from 18,000 to 10,000 years ago. However, a handful of small glaciers still hang on in the Sierras, as discussed below.

Nevertheless, the landscape of much of the Sierras shows the unmistakable imprint of glaciation. Perhaps one of the most famous glacial valleys in the world (and certainly in nearly every geology textbook) is Yosemite Valley (Figure 8.23). A glance down the axis of the valley shows the classic U-shaped profile, with the steep face of El Capitan showing the typical vertical surface of a glacially carved valley. There are numerous hanging valleys on each side, producing the many famous waterfalls of Yosemite. The sides of the higher peaks are all sculpted as cirques,

there are cirques on either side of a ridge, they will pluck away at it until it forms a knife-edged ridge with steep concave sides known as **arête** (French for "edge" or "ridge"). If glaciers have sculpted a mountain peak until it has concave cirques on all sides, it is called a **horn**. The Matterhorn in

FIGURE 8.22 Image showing the original ice extent in the Sierras. The red dots show the location of the current glaciers.

Source: Photo by H. Bagasic.

and there are numerous arêtes connecting them. On the floor of the valley are numerous glacial moraines, which are largely responsible for damming up the valley and forming Yosemite's famous lakes, such as photographers' favorite, Mirror Lake. The legendary peak known as Half Dome (Figure 8.23b) shows a cirque on the side that faces the valley, with the classic domed exfoliation surface on the opposite side. The U-shaped valleys can be seen in most of the larger canyons of the Sierras, such as Kings Canyon (Figure 8.23d).

Putting all this evidence together, we can reconstruct the history of how the Yosemite landscape evolved. Before the ice ages (Figure 8.24a—c), the landscape was carved into V-shaped valleys with gentle gradients, as is typical of normal river erosion around world. Then came the huge ice sheets, which buried the Yosemite landscape completely and scoured the gentle rolling topography of most of the Tioga Plateau. Eventually, the ice sheets melted back and were restricted to the ancient V-shaped stream valleys. These were sculpted into steep-walled U-shaped valleys. The gentle side valleys that once descended smoothly down to the main trunk stream were also glaciated and flowed into the top of the trunk glacier. When these melted back after the last ice age ended, the side valleys were left as hanging valleys dropping abruptly into the main U-shaped valley of Yosemite (Figure 8.24f).

FIGURE 8.23 Sierran glacial features: (a) The main U-shaped valley of Yosemite, with the vertical cliff of El Capitan on the left, the vertical face of Half Dome in the center distance, and the hanging valleys with their waterfalls. (b) Half Dome, a cirque on the west (left) side and exfoliated dome on the east (right) side. (c) Looking down from El Capitan into Yosemite Valley, with its U-shaped profile. Across the main valley is the hanging valley that feeds Yosemite Falls. (d) U-shaped valleys can be seen all throughout the Sierras, such as at Kings Canyon.

Source: (a–b) Courtesy of Wikimedia Commons. (c) Courtesy of US Geological Survey. (d) Photo by G. Hayes.

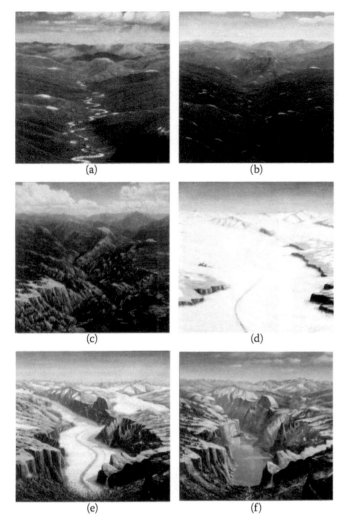

(a)   (b)

(c)   (d)

(e)   (f)

FIGURE 8.24 Reconstruction of the history of the topography of Yosemite. (a–c) Before glaciation, the landscape was gently sloping with V-shaped stream valleys that smoothly descended into the main trunk stream. (d) During the peak glaciations, the entire area was covered by glacial ice, with only a few peaks poking through. (e) As the ice melts back, large trunk glaciers carve the main valley, while side glaciers produced hanging valleys. (f) After the glaciers have vanished, the modern landscape has large U-shaped valleys, with hanging valleys coming in from the side.

Source: From Huber, N. K. The geologic story of Yosemite National Park. *US Geological Survey Bulletin*, 1595, 1–64, 1995, courtesy of US Geological Survey.

If you hike in the High Sierras, the signs of glaciation are everywhere. Many of the surfaces are polished by the grinding of ice masses, with the long parallel scratches known as **glacial striations** (Figure 8.25a) and **chatter marks** (Figure 8.25b), both signatures of glaciers the world over. The base of the mass of ice is full of large rocks that acted like the teeth in a rasp. If the weight of the ice dragged the "teeth" along the bedrock, they produced scratches and striations (Figure 8.25a). If the rocks were pushed down to form a gouge, then retracted, and then pushed down to gouge again, they

(a)

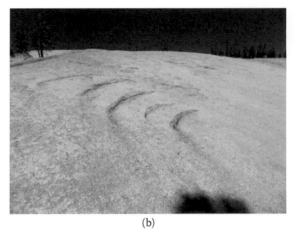

(b)

(c)

FIGURE 8.25 Glacial features on bedrock: (a) Striations, formed when rocks caught in the base of the glacier act like the teeth in a rasp. They drag along and scratch parallel grooves, here carved into the top of hexagonal columnar basalt at Devil's Postpile. (b) Chatter marks, arcuate-shaped grooves formed when rocks in the base of the glacier would gouge rock, then retreat, and then press down and gouge again. This example is from Pothole Dome, Yosemite National Park. (c) Glacial erratics are large boulders transported from long distances by glaciers and then just dumped where the glacier melts. This surface at Olmstead Point is covered with erratic boulders.

Source: (a) Photo by G. Hayes. (b–c) All other photos by the author.

formed chatter marks (Figure 8.25b). If the glacier carried finer-grained sediments, they may have smoothed the surface like a piece of sandpaper, creating **glacial polish** (Figure 8.25a).

Every once in a while in glaciated terrain, you might find large rocks or even huge boulders just sitting on the glaciated surface (Figure 8.25c), often in a balanced or precarious position. In many cases, these huge boulders are made of a rock not found anywhere in the area but derived from mountains many miles to the north. These are known as **glacial erratics**. The word comes from the Latin verb *erro*, "to wander." If you make an error, you wander from the truth, and erratic behavior is also like wandering aimlessly. Erratics are a common feature in much of the High Sierras (Figure 8.25c).

Another signature of glaciated bedrock is a feature known as a *roche moutonée*, French for "sheep rock" (presumably

because of their sheep-like shape in some cases). When ice overrides a hard knob or ridge of bedrock, it tends to flow up the slope of the knob and polish it flat. As it flows down the back side, however, it tends to pluck the rocks away from the base and form a steep slope (Figure 8.26a). Thus, the downstream side of the *roche moutonée* is steep, while the upstream side has a shallow slope. This can be seen clearly in many places, such as Pothole Dome and Lembert Dome in West Tuolumne Meadows near Yosemite Valley (Figure 8.26b and c).

Meanwhile, the downhill flow of the river of ice constantly carries a huge load of boulders, gravel, sand, and clay like a giant conveyer belt. As we have seen already, where the melting of the glacial snout stalls, this material piles up to form moraines (Figure 8.20). Moraines are composed of a completely unsorted, unstratified mix of every grain size of sediment (boulders, gravel, sand, and clay)

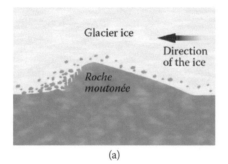

(a)

(b)

(c)

FIGURE 8.26 *Roche moutonée*: (a) Diagram showing how they are formed when the glacial ice grinds down a bedrock knob or ridge on the upstream side into a gentle slope and plucks the downstream side. (b) Pothole Dome, a famous roche moutonée in Tuolumne Meadows, Yosemite National Park. (c) Lembert Dome, a larger roche moutonée just east of Pothole Dome, Yosemite National Park.

Source: (a) Courtesy of Wikimedia Commons. (b–c) Photos by the author.

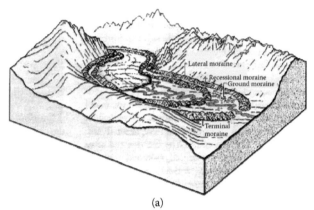

(a)

(b)

(c)

FIGURE 8.27 Glacial moraines: (a) Diagram showing lateral, recessional, and terminal moraines. (b) Lee Vining Canyon from the air, showing the lateral moraine from the Tahoe stage (21,000 years ago) (labeled *O*), the inner moraines (*I*) from the Tioga stage (70,000 years ago), and recessional moraines (*R*) in the middle of the valley. (c) Moraines below Convict Lake and Mt. Tom, showing the strong development of lateral moraines and a large terminal moraine. The moraines in the center and right are from the Tioga stage, while the lobes to the left are older Tahoe stage moraines covered by more recent glacial activity.

Source: Photo by M. B. Miller.

called **glacial till** or **diamictite**. The lack of size sorting and stratification reflects the process of glacial sedimentation, which dumps everything it is carrying indiscriminately in one place without regard to size and without beds being formed. Unstratified, unsorted piles of sediment are rarely found anywhere but in glacial moraines, so they are useful clues as to where glaciers have been. Not only do moraines form at the furthest point of a glacial advance before it retreats (**terminal moraine**), but also as the snout melts back, it may leave a smaller **recessional moraine** if it stalls in one place for a period of time. The sides of the glacier also accumulate sediment known as **lateral moraine**.

As you travel down the glacial valleys and out of the Sierras, there are numerous moraines formed by ice lobes that flowed out of the mountains. These are especially common in Owens Valley, where nearly every valley from the High Sierras has a large pair of lateral moraines that formed on the edge of the glacial lobe, and one or more terminal moraines where the snout stalled and melted (Figure 8.27). For example, from Lee Vining Canyon, an aerial view shows a series of moraines of different ages. There are lateral moraines from the Tahoe glacial stage (21,000 years ago, at the peak of the last glacial maximum). Inside these are moraines from the much older Tioga stage, about 70,000 years ago. In the middle of the valley are recessional moraines from times when the glacier snout paused during its retreat.

Like ice sheets all over the globe, the glaciers of the Sierras are now in rapid retreat as global warming melts all the world's ice. In 2006, a US Geological Survey (USGS) inventory showed that there were about 500 glaciers recorded in the Sierras, which covered an area of more than 50 km^2, and about 788 smaller ice bodies, which covered an area of about 13 km^2. All these are in retreat, and some are nearly gone (Figure 8.28). On average, the glaciers have retreated about 70% since 1900. At the rate they are retreating, they will all be gone by 2050.

This point is dramatically illustrated by before and after shots of glaciers, taken by early USGS geologists about a century ago, and then by more recent geological research in the past decade (Figure 8.28). For example, once-huge glaciers like Darwin Glacier (Figure 8.29a and b) are nearly gone. The same can be said for Lyell Glacier (Figure 8.29c and d) and Dana Glacier (Figure 8.29e and f). All these shots are consistent with research on glaciers and ice caps around the world. In less than 35 years from the publication of this book, there will be no traces of the glaciers that once sculpted the Sierras so dramatically.

8.10 SUMMARY

The oldest rocks of the Sierras (Figure 8.14) are early Paleozoic metasedimentary and metamorphic rocks of the Shoo Fly terrane and the Feather River ophiolite. These were accreted during the Devonian Antler orogeny. Late Paleozoic–Triassic rocks of the Northern Sierra terrane and the Calaveras terrane were accreted during the Sonoma

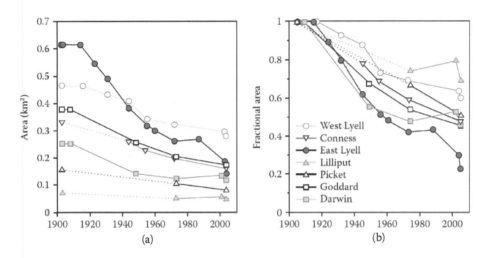

FIGURE 8.28 Plot showing the retreat of many Sierran glaciers, as indicated by (a) total area and (b) percentage loss of area (proportional area).

Source: Photo by H. Bagasic.

FIGURE 8.29 Before and after photo pairs of the retreat of Sierran glaciers. (a–b) Darwin Glacier on August 14, 1908, and on August 14, 2008. (c–d) Lyell Glacier on August 7, 1903, and on August 14, 2003. (e–f) Dana Glacier in August 1883, and on September 8, 2013.

Source: Courtesy of US Geological Survey and H. Bagasic. (a) photo by G. K. Gilbert. (b) Photo by H. Bagasic. (c) Photo by G. K. Gilbert. (d) Photo by H. Bagasic. (e) Photo by I.C. Russell. (f) Photo by H. Bagasic.

orogeny, followed by the Jurassic accretionary wedge rocks of the Foothills terrane.

Meanwhile, the Sierran arc began to intrude in the present Owens Valley during the Jurassic and then migrated to the western Sierra in the Early Cretaceous, followed by a shift to the eastern Sierras in the Late Cretaceous, possibly in response to steepening and then shallowing of the subducting plate. Many different magma bodies melted through the country rocks and forced their way upward, enclosing roof pendants and incorporating numerous xenoliths, as well as metamorphosing the older terranes in the western Sierra foothills. Only the Minarets Caldera remains of this ancient Cretaceous volcanic chain; all the older rock has been eroded away during uplift in the early Cenozoic, exposing the once-buried magma chambers. Igneous intrusion ended in the latest Cretaceous Laramide orogeny as the subducting plate went horizontal and was too shallow below the Sierras to reach melting depths.

Although the Sierras were somewhat elevated in the early Cenozoic, the much higher Nevadaplano plateau to the east formed rivers that cut across the Sierras, filling their channels with Eocene sediments bearing gold, and with Oligocene volcanics erupted from Nevada volcanoes. Additional late Cenozoic uplift is indicated by the filling of old river valleys with 9 Ma lava flows, which have since been uplifted and eroded to form inverted topography. Finally, the Sierras were extensively glaciated during the ice ages, forming the classic glacial features seen in Yosemite and many other places on the Sierra crest.

RESOURCES

Alt, D., and Hyndman, D.W. 2000. *Roadside Geology of Northern and Central California*. Mountain Press, Missoula, MT.

Bartley, J.M., Glazner, A.F., and Mahan, K.H. 2012. Formation of pluton roofs, floors, and walls by crack opening at Split Mountain, Sierra Nevada, California. *Geosphere, 8*, 1086–1103.

Bateman, P.C. 1981. Geological and geophysical constraints on models for the origin of the Sierra Nevada batholith, 71–86. In Ernst, W.G. (ed.), *The Geotectonic Development of California*. Prentice Hall, Englewood Cliffs, NJ.

Bateman, P.C., Clark, L.K., Huber, N.K., Moore, J.G., and Rinehart, C.D. 1963. *The Sierra Nevada Batholith—A Synthesis of Recent Work across the Central Part. U.S. Geological Survey Professional Paper 414-D*. U.S. Geological Survey, Reston, VA.

Bateman, P.C., and Eaton, J.P. 1967. Sierra Nevada batholith, California. *Science, 158*(3807), 1407–1417.

Burchfiel, B.C., and Davis, G.A. 1981. Triassic and Jurassic tectonic evolution of the Klamath Mountains-Sierra Nevada geologic terrane, 50–70. In Ernst, W.G. (ed.), *The Geotectonic Development of California*. Prentice Hall, Englewood Cliffs, NJ.

Cassel, E.J., Graham, S.A., and Chamberlain, C.P. 2009. Cenozoic tectonic and topographic evolution of the northern Sierra Nevada, California, through stable isotope paleoaltimetry in volcanic glass, *Geology, 37*, 547–550.

DeCelles, P.G. 2004. Late Jurassic to Eocene evolution of the Cordilleran thrust belt and foreland basin system, western U.S. *American Journal of Science, 304*, 105–168.

Fiske, R.S., and Tobisch, O.T. 1994. Middle Cretaceous ash-flow tuff and caldera-collapse deposit in the Minarets Caldera, east-central Sierra Nevada, California. *Geological Society of America Bulletin, 106*, 582–593.

Glazer, A.F., and Stock, G. 2010. *Geology Underfoot in Yosemite National Park*. Mountain Press Publishing, Missoula, MT.

Guyton, B. 1998. *Glaciers of California: Modern Glaciers, Ice Age Glaciers, the Origin of Yosemite Valley, and a Glacier Tour through the Sierra Nevada*. California Natural History Guides 59. University of California Press, Berkeley.

Henry, C.D. 2009. Uplift of the Sierra Nevada, California. *Geology, 37*, 575–576.

Hildebrand, R.S. 2013. *Mesozoic Assembly of the North American Cordillera*. Geological Society of America Special Paper 495. Geological Society of America, Boulder, CO.

Hill, M. 1975. *Geology of the Sierra Nevada*. California Natural History Guides 37. University of California Press, Berkeley.

Huber, N.K. 1995. The geologic story of Yosemite National Park. *U.S. Geological Survey Bulletin, 1595*, 1–64.

Huber, N.K., and Wahrhaftig, C. 1987. *The Geologic Story of Yosemite National Park*. U.S. Geological Survey Bulletin 1595. U.S. Geological Survey, Reston, VA.

Jones, C.H., Farmer, G.L., and Unruh, J. 2004. Tectonics of Pliocene removal of lithosphere of the Sierra Nevada, California. *Geological Society of America Bulletin, 116*, 1408–1422.

Konigsmark, T. 2007. *Geologic Trips: Sierra Nevada*. GeoPress, Mendocino, CA.

Meldahl, K.H. 2013. *The Rough-Hewn Land: A Geologic Journey from California to the Rocky Mountains*. University of California Press, Berkeley.

Memeti, V., Paterson, S.R., and Putirka, K.D. 2014. *Formation of the Sierra Nevada Batholith: Tectonic Processes and Their Tempos*. Geological Society of America Special Paper 000. Geological Society of America, Boulder, CO.

Muir, J. 1911. *My First Summer in the Sierra*. Dover Books, New York.

Mulch, A., Graham, S.A., and Chamberlain, C.P. 2006. Hydrogen isotopes in Eocene river gravels and paleoelevation of the Sierra Nevada. *Science, 313*, 87–89.

Prothero, D.R., and Dott, R.H., Jr. 2010. *Evolution of the Earth* (8th ed.). McGraw-Hill, New York.

Schweickert, R.A. 1981. Tectonic evolution of the Sierra Nevada Range, 87–131. In Ernst, W.G. (ed.), *The Geotectonic Development of California*. Prentice Hall, Englewood Cliffs, NJ.

Schweickert, R.A., and Lahren, M. 1999. Triassic caldera at Tioga Pass, Yosemite National Park, California: Structural relationships and significance. *Geological Society of America Bulletin, 111*, 1714–1722.

Schweickert, R.H., and Lahren, M. 2006. Geologic evolution of the Saddlebag Lake Pendant, eastern Sierra Nevada, California: Tectonic implications. *Pacific Section SEPM Book, 101*, 27–56.

Schweickert, R.A., and Snyder, W.S. 1981. Paleozoic plate tectonics of the Sierra Nevada and adjacent regions, 182–202. In Ernst, W.G. (ed.), *The Geotectonic Development of California*. Prentice Hall, Englewood Cliffs, NJ.

Stephenson, D.B., and Girty, G.H. 2006. Constraining the source of the Orocopia Schist, SE California: A comparative study of the major and trace element chemistry of Cretaceous sandstones deposited in forearc settings. *Pacific Section SEPM Book, 101*, 77–96.

Tobisch, O.T., Saleeby, J.B., Renne, P.R., McNulty, B., and Tong, W. 1995. Variations in deformation fields during development of a large volume magmatic arc, central Sierra Nevada, California. *Geological Society of America Bulletin*, *107*(2), 148–166.

Videos

Sierra Nevada Geology

www.youtube.com/watch?v=2Ym7cf44cCg

www.youtube.com/watch?v=rnaA84zepLk

www.youtube.com/watch?v=f5woxoWGLVU&list=PLMZCpLkCzAQXlFaSwHjn3O4_49AMPizW9&index=4

www.youtube.com/watch?v=x0DGpr4OrjQ&list=PLMZCpLkCzAQXlFaSwHjn3O4_49AMPizW9&index=8

www.youtube.com/watch?v=pErDEYc5X1A&index=9&list=PLMZCpLkCzAQXlFaSwHjn3O4_49AMPizW9

www.youtube.com/watch?v=U4PzefDPImY&list=PLMZCpLkCzAQXlFaSwHjn3O4_49AMPizW9&index=11

www.youtube.com/watch?v=ij7DU6VQBZIExfoliation:www.youtube.com/watch?v=yAZ1V_DJKV8

Sierran Glaciers

www.youtube.com/watch?v=7wUZLW-ahk4

www.youtube.com/watch?v=mgnzSTY5zRg

Glaciation

"The Miracle Planet: Riddles of Sand and Ice": www.youtube.com/watch?v=-NymNwrrZAg

Smartville Complex field trip video: www.youtube.com/watch?v=utRE6uQF93Q

9 Mantle Rocks and Exotic Terranes
The Klamath Mountains

Despite the picture-postcard aspect of the pines and peaks, it was the strangest landscape I had ever seen. The ridges were not particularly high or craggy, rather a succession of steep, pyramidal shapes that marched almost geometrically into the blue distance. The big conifers that crowned them enhanced an impression of regularity, almost of discipline. There was a tension in the ridges that departed radically from conventional notions of the irregularity and relaxation of wide-open spaces. It was almost at attention. I felt the hair stand up on the backs of my arms and legs. The faint sibilance of the wind in pine needles called attention to quiet so intense that I was reluctant to move, like a grouse chick crouched on the forest floor.

The ridges seemed not only vigilant, but reticent, as though hidden within them might be the most extraordinary things. Perhaps this impression was colored by my awareness that I was looking toward the Bluff Creek drainage, where giant humanoid footprints had been found in the dust of a road-building project in the early 1960s. The pyramidal ridges seemed to say "mystery" to my mind in the way that the shape or color of the parent bird's bill say "food" to its nestlings. Pyramids have a way of doing that, as evidenced by the lasting fascination of certain Egyptian tombs. The Siskiyou ridges might have been the vegetated remains of some prehistoric city, vast beyond comprehension. They did not seem altogether natural, at least, not with the insensate simplicity often associated with nature.

—David Rains Wallace
The Klamath Knot

9.1 DATA

Highest point: Mt. Eddy: 2,751 m (9,026 ft); Thompson Peak: 2,744 m (9,002 ft); many other peaks above 1,500–2,200 m (4,700–7,000 ft)

Major ranges: Marble, Salmon, Scott, Siskiyou, Trinity Alps

Predominant rocks: Paleozoic and Mesozoic metasedimentary and metavolcanic rocks, Mesozoic ultramafic seafloor rocks, Cretaceous granitic intrusions

Plate tectonic setting: Paleozoic and Mesozoic accretion of exotic terranes, followed by Cretaceous igneous intrusions

Geologic resources: Gold, nickel, chromium, cobalt, and water

Forests: Marble Mountains, Siskiyou Mountains, Salmon–Trinity Alps Wilderness Areas, Klamath National Forest, Rogue River National Forest

State parks: Castle Crags

9.2 GEOGRAPHY OF THE KLAMATH MOUNTAINS

The Klamath Mountains are a very distinctive region of California, both geographically and geologically (Figure 9.1). Actually, the Klamaths are a complex of many different mountain chains that form a big arc-shaped region in the northwestern corner of California, continuing into the southwestern part of Oregon (Figure 9.1). Most of the ranges are about 1,500–2,200 m in elevation at their crests, with the highest peak reaching 2,700 m (Figure 9.2). Deep canyons with large rivers cut between the ranges and sometimes right through them. The fact that the ranges and ridges are all roughly the same elevation, with rivers cutting through water gaps, has suggested to many geologists that it was once a low region with rivers crossing buried mountain ranges (discussed later in the chapter). Since those ranges rose into the sky, the rivers have been forced to cut down several thousand meters in whatever course they started with, even if it meant cutting through a mountain range rather than around it.

The Klamaths get a lot of rainfall from the Pacific each year (and snowfall in the higher elevations), so they are covered by some of the most densely forested places in California. The Klamaths are also one of the richest botanical regions in the West, with a full spectrum of plants adapted to the coastal fogs, the high mountain snows, and the relatively dry and sunny slopes of the interior canyons. The winters are mild enough, and the summers moist enough, that many different plant species can live in this relatively protected region. Part of this diversity comes from the fact that it is an **ecotone**—a transitional zone of overlap of the boundaries of many different plant ranges. Many plants that live in Alaska have their southernmost occurrences in the Klamaths, along with plants from Arizona at the northern limit of their ranges. There are more than 30 species of conifers in these mountains, along with huge stands of coast redwoods in the west. Russian Peak holds the world record for 17 different conifer species in a single square mile. The forests support an economy based largely on logging, with tourism (especially salmon fishing, hiking, and rafting). Grizzly bears,

DOI: 10.1201/9781003301837-9

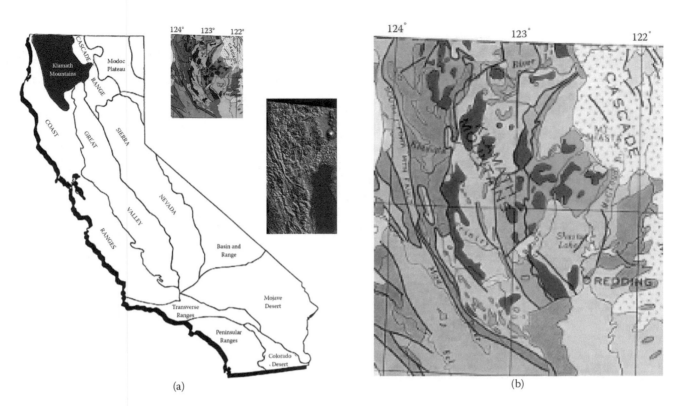

FIGURE 9.1 (a) Index map of California highlighting the Klamath Mountains. In the upper right is the geologic map of the region and the digital shaded relief map of northwestern California. (b) Enlarged geologic map of the Klamaths.

Source: (a) Courtesy of US Geological Survey. (b) Courtesy of US Geological Survey.

FIGURE 9.2 Typical Klamath landscape, with jagged pine-covered peaks and lakes and rivers. This is Bear Mountain and Devil's Punchbowl in the Siskiyou Wilderness. The mountains show classic U-shaped glacial profiles and cirques, and the lakes are dammed by glacial moraines.

Source: Courtesy of Wikimedia Commons.

cougars, gray wolves, Roosevelt elk, and allegedly, the legendary Bigfoot still roam these remote, isolated forests with few people. (For a more detailed account of why Bigfoot does not exist, see the book *Abominable Science* by Daniel Loxton and myself.)

9.3 GEOLOGY OF THE KLAMATH MOUNTAINS

The mountain ranges of the Klamaths do not follow the straight southeast–northwest trend of the axis of the Sierras, or the southeast–northwest features and topography dictated by the San Andreas fault to the south. Instead, they form a crescent-shaped curve concave to the east (Figures 9.1 and 9.3), a pattern seen nowhere else in the West. In fact, they represent a very ancient region of California that started out with as many as eight separate terranes (Figure 9.3) from far out in the ancient ancestor of the Pacific Ocean that have been added to California during a long series of accretion events. They are among the most ancient rocks in California, just like the accreted terranes of the western Sierra foothills, and only slightly younger than the Precambrian and lower Paleozoic rocks of Death Valley or the Mojave Desert.

A glance at the geologic map of California (Figures 1.1 and 9.4) shows that the Klamaths and the western Sierra foothills indeed look very similar, with a broad belt of Paleozoic rocks (pale blue on the map) on the east and a series of early Mesozoic accreted terranes (green on the map) to the west, all intruded by Cretaceous granitic rocks (red on the map). Even though they were originally mapped and named separately, in recent years it is clear that the Klamaths are just the northern continuation of the western Sierra foothills, now separated by large fault displacements (Figure 9.4). Let us look at each of the tectonic assemblages of the Klamaths in order.

9.3.1 Eastern Klamath Terrane

The oldest rocks in the Klamaths are lumped together into the Eastern Klamath terrane (Figures 9.3 through 9.5). As the name suggests, this terrane forms the basement rock of the entire eastern part of the province; it was the first terrane to be accreted onto North America. The rocks are nearly all Paleozoic in age, ranging from Cambrian to earliest Triassic. Thus, they are thought to be the equivalent of the Shoo Fly terrane in the Sierras (Figure 9.4) and possibly the Northern Sierra terrane as well.

The base of the rock sequence is a thick slice of oceanic crust known as the **Trinity ophiolite** (Figure 9.5a and b). It is composed of a suite of Cambrian (and possibly Precambrian) through Ordovician ultramafic magma chambers rich in peridotite and gabbro, many of which show the layered structure, suggesting a long history of fractional crystallization (see Chapter 2). In many places, the layered peridotites and gabbros are rich in chromite (chromium oxide), a typical mineral in ultramafic assemblages. For this reason, there has been a long history of chromium mining

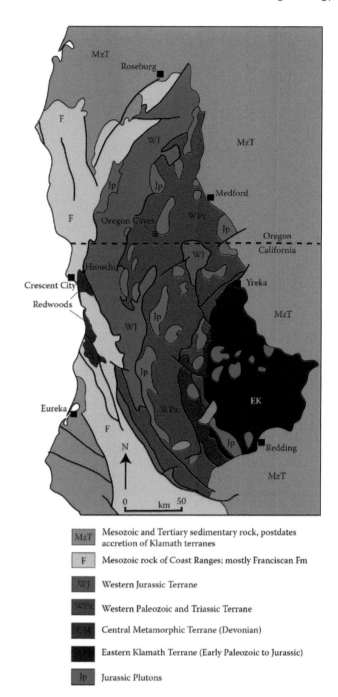

FIGURE 9.3 Tectonic terranes of the Klamaths.

Source: Photo by M. B. Miller.

in the Klamaths. There are also areas of extensive pillow lavas (Figure 9.5b) and sheeted dikes, all part of the classic ophiolite complex. Late in their history, this sliver of oceanic crust was intruded by Ordovician, and then Silurian and Devonian granitic rocks (Figure 9.5a), apparently produced by a more dacitic–rhyolitic volcanic arc complex that produced magmas melted from the overlying mantle that intruded into the pre-existing oceanic crust, or erupted as volcanic rocks on top of it. Currently, geologists interpret the Trinity ophiolite package as formed by back-arc spreading

Map legend:

MzT	Mesozoic and Tertiary sedimentary rock, postdates accretion of Klamath terranes
F	Mesozoic rock of Coast Ranges; mostly Franciscan Fm
WJ	Western Jurassic Terrane
WPz	Western Paleozoic and Triassic Terrane
CM	Central Metamorphic Terrane (Devonian)
EK	Eastern Klamath Terrane (Early Paleozoic to Jurassic)
Jp	Jurassic Plutons

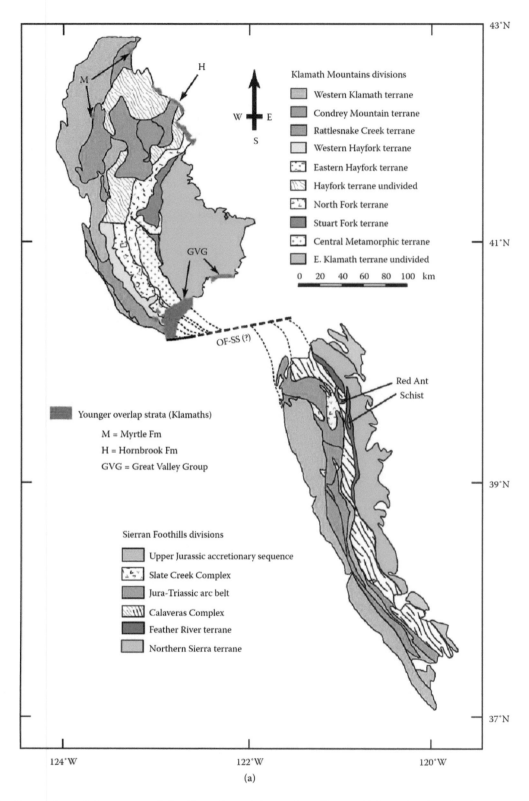

FIGURE 9.4 Comparison of the terranes of the Klamaths and the northern Sierras: (a) Map showing the similarity of the terranes. (b) Comparison of their similar tectonic cross sections.

Source: (a) With permission from the Geological Society of America. (b) Based on several sources; drawn by E. T. Prothero.

(producing the ophiolitic crust) behind a volcanic arc that was then crumpled as it accreted to the North American continent during the Late Devonian Antler orogeny.

Unconformably deposited above the Trinity ophiolite basement after its Antler collision is a sequence of late Paleozoic and early Mesozoic (Devonian through Jurassic) metasedimentary rocks known as the Redding section (Figure 9.5a and c). Sixteen formations are recognized, including andesitic volcanic and volcaniclastic units, along with mostly deep-water shales with turbidite sandstones.

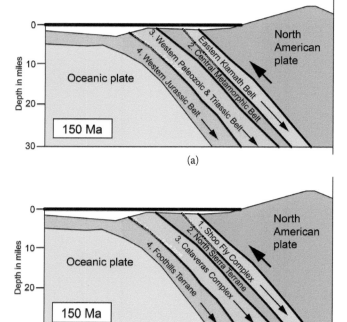

FIGURE 9.4 (Continued)

Among the remarkable rock units in this package is the McCloud Limestone (Figures 9.5a and 9.6), which is more than 800 m (2,600 ft) thick. It contains a classic Permian fauna of many different kinds of corals, crinoids, bryozoans, and brachiopods, all typical of shallow carbonate seas during the Permian (Figure 9.7). What is remarkable about this assemblage is that similar types of Permian limestones occur widely in the western Cordillera of North America, including Alaska, the Yukon, British Columbia, northern Washington, and in the northern Sierras (Figure 9.6). When paleontologists studied these Permian fossils more closely, they discovered that they match assemblages of fossils from *the other side of the predecessor of the modern Pacific Ocean* in the tropical southwestern Pacific region (the Tethys Seaway) during the Permian. Today, geologists recognize that this is evidence that these rocks are part of exotic terranes (Stikinia to the north and the Cache Creek terrane in British Columbia) that have been carried across the entire width of the Pacific since the Permian and slammed into North America during the Triassic or Jurassic (Figure 9.6c). We will see additional cases of exotic Southern Hemisphere Cretaceous rocks when we discuss the Laytonville Limestone in the Coast Ranges (Chapter 12).

9.3.2 Central Metamorphic Terrane

The next package, accreted to the west and beneath the East Klamath terrane, is the Central Metamorphic terrane (Figures 9.3, 9.7, and 9.8). It is composed of the Salmon Hornblende Schist and the Abrams Mica Schist, which are Devonian metasedimentary and metavolcanic rocks, plus serpentinized fragments of oceanic crust (Figure 9.8a). All these rocks were subjected to intermediate-grade (amphibolite-grade) metamorphism, so most of the rocks are now schists and slates. Both this terrane and the core of the

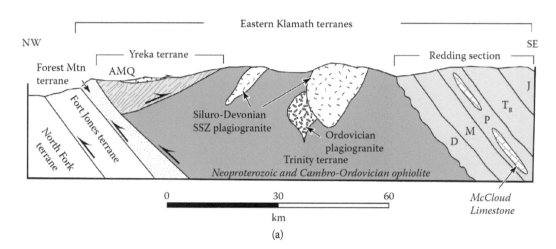

FIGURE 9.5 (a) Diagram showing a schematic cross section of the Eastern Klamath terrane. (b) Cabin Meadows pillow lavas from the Trinity ophiolite. (c) Carboniferous limestones of the Bragdon Formation on Interstate 5 south of Mt. Shasta, typical of the Redding section.

Source: (a) With permission from the Geological Society of America. (b) Courtesy of US Forest Service. (c) Photo by the author.

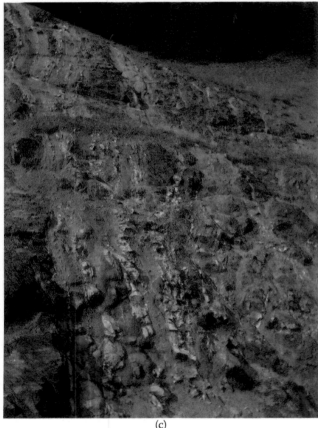

FIGURE 9.5   (Continued)

Eastern Klamath terrane were accreted to North America during the Late Devonian Antler orogeny (Figure 7.22).

9.3.3   WESTERN PALEOZOIC–TRIASSIC TERRANE

The next structural block to the west of the East Klamath and Central Metamorphic terranes is the Western Paleozoic–Triassic terrane (Figures 9.3, 9.4, 9.7, and 9.8b). It consists largely of deep-water Devonian to Triassic shales plus cherts full of siliceous plankton known as radiolarians,

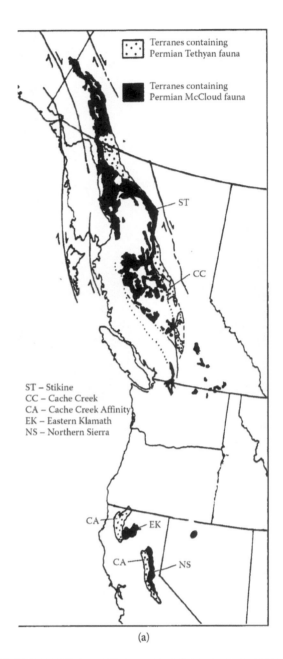

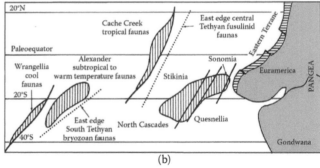

FIGURE 9.6   (a) Map showing the distribution of Permian limestones from exotic terranes in the western North American ranges. (b) Exotic terrane map showing the suggested origin of the major terranes that make up the western Cordillera of North America.

Source: From Prothero, D. R., and R. H. Dott Jr. *Evolution of the Earth* (8th ed.). McGraw-Hill: New York, 2010.

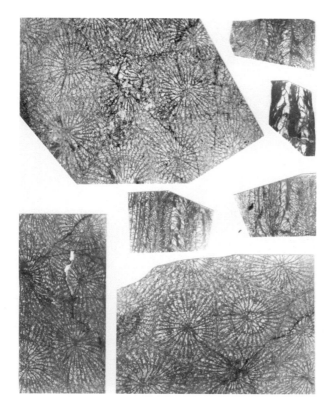

(a)

(b)

FIGURE 9.7 Cross sections of fossil corals from the McCloud Limestone.

Source: Photo by C. Stevens.

named the North Fork, Haystack, and Rattlesnake Creek assemblages.

The North Fork assemblage is the farthest east of the units. It is made of blocks of ophiolite (ultramafics, gabbro, basalt dikes, and pillow lavas) plus red radiolarian chert and shales that give a Triassic age based on their radiolarians. The Haystack assemblage is the largest and thickest unit of the three and situated in the middle. At the base is the thickest unit, the Hayfork Bally meta-andesite, which grades upward into shales, cherts, and minor limestones. In addition, there are numerous limestone blocks from the Permian Tethys (comparable to the McCloud Limestone), apparently part of the Stikinia terrane (Figure 9.6b and c). The Rattlesnake Creek assemblage is the farthest west of the three units. It is composed of a mix of dismembered ophiolite (ultramafics, gabbros, sheeted basalt dikes, and pillow lavas), shales, and cherts, with lenses and pods of limestone, phyllite, granitic rocks, sandstone, and conglomerate. The limestone pods also contain Permian Tethyan fossils, just like the McCloud Limestone, so much of this terrane was once part of Stikinia. The entire terrane is also highly shredded and sheared, consistent with the notion that it represents an early Mesozoic accretionary wedge. This is confirmed by large blocks of blueschist in many of the shredded rock units in this terrane.

The Western Paleozoic–Triassic terrane apparently correlates with rocks of the Calaveras Complex in the northern Sierras (Figure 9.4). Like the Calaveras, it was

FIGURE 9.8 (a) Schists and serpentinites of the Central Metamorphic terrane, Highway 299, near Big Bar, California. (b) Schists from the Western Paleozoic–Triassic terrane, Highway 299, east of Del Loma, California. (c) Dark green-black serpentinites from the Josephine ophiolite, Western Jurassic terrane, Highway 299, west of Willow Creek. (d) Castle Crags, a Cretaceous granitic intrusion that melted its way through the earlier rocks of the Klamaths.

an accretionary wedge that apparently accreted to North America during the Triassic Sonoma orogeny, when the entire northwestern corner of Nevada (Sonomia terrane) was added to the continent (Figure 7.22).

9.3.4 WESTERN JURASSIC TERRANE

Finally, the westernmost tectonic sliver of the Klamaths is known as the Western Jurassic terrane (Figures 9.3, 9.4, and 9.8c). It includes metasedimentary deep-water sandstones and shales called the Galice Formation, and meta-andesite

(c)

(d)

FIGURE 9.8 (Continued)

rocks of the Rogue Formation, plus oceanic crust rocks of the Josephine ophiolite (Figure 9.8c), and other metamorphic rocks such as the Briggs Creek Amphibolite and the Condrey Mountain Schist. All these Jurassic rocks are heavily sliced and shredded as part of a Jurassic accretionary wedge accreting to the edge of North America as the Jurassic arc volcanics began to erupt just east of the modern Sierras. In this respect, it is the equivalent of the Foothills terrane in the northwest Sierra (Figure 9.4).

9.3.5 CRETACEOUS GRANITIC INTRUSIONS

After all these blocks were tectonically accreted to the western edge of North America, they were intruded by Cretaceous granitic rocks. This is exactly as occurred in the Sierras. In many places in the Klamaths, such as Castle Crags just west of Mt. Shasta (Figure 9.8d), there are small intrusions and dikes of Cretaceous tonalite, quartz monzonite, and quartz diorite (Figure 9.3).

9.3.6 KLAMATH TECTONIC SUMMARY

Thus, the Klamaths have essentially the same tectonic history as the western Sierra foothills (Figure 9.4). Both regions are composed of an easternmost block of Paleozoic rocks that were accreted during the Devonian–Mississippian Antler orogeny (the Shoo Fly and Northern Sierra terranes in the Sierras, and the Eastern Klamath and Central Metamorphic terranes in the Klamaths), a belt of rocks accreted during the Triassic Sonoma orogeny (Calaveras terrane in the Sierras and Western Paleozoic–Triassic terrane in the Klamaths), and an accretionary wedge complex built up in front of the Jurassic predecessor of the volcanic arc (Foothills terrane in the Sierras and Western Jurassic terrane in the Klamaths). Finally, both regions were intruded by the major Andean-style volcanic arc that became the Sierra Nevada batholith and its southern and northern counterparts (Figure 8.7).

This leaves the last mystery: Why are the Klamaths offset so far to the west of the line of the Sierra Nevada batholith (Figures 8.7, 9.4, and 9.9)? Various tectonic models have been suggested, although there does not seem to be a consensus as to what tectonic forces located the Klamaths so far west. It seems clear that there must be a large left-lateral fault zone between the two blocks, based on the offset (Figures 9.4, 9.7, and 9.9). Several models have proposed some sort of extreme extension behind and to the east of the Klamaths. However, the extension behind the Sierras in the Basin and Range Province (Chapter 7) more than doubled the crustal width of Nevada in the late Cenozoic, so the extension behind the Klamaths would have to have been even more extreme.

Ernst et al. (2008) proposed a model (Figure 9.9) that postulates that oblique subduction and shearing are part of the story. Their model starts with the accretion of the Calaveras terrane to the south and the Eastern Hayfork block during the Early Jurassic (Figure 9.9a). As these terranes docked, they came in at an oblique angle in a transform plate boundary, causing **transpression** (compression plus transform faulting), which formed shear along the collision zone. By the mid-Jurassic (Figure 9.9b), the shear and rifting became more extreme, with the Rattlesnake Creek

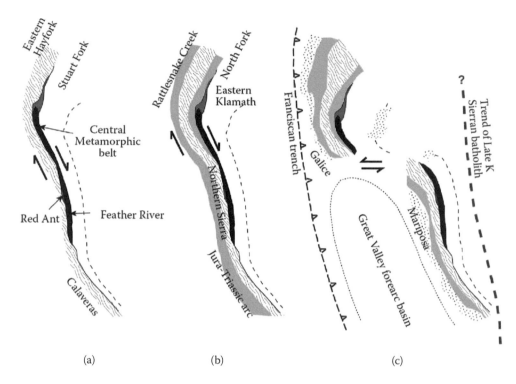

FIGURE 9.9 Tectonic model for the accretion and then dismemberment of the Klamath–Sierra terranes (see text for discussion): (a) Early Jurassic, (b) Middle Jurassic, (c) end of Jurassic.

Source: From Ernst W. G., et al., *Geological Society of America Bulletin*, 120, 179–194, 2008, Figure 8.

assemblage accreted on the western side and the North Fork assemblage accreted on the eastern side of the Hayfork block. Meanwhile, the Triassic–Jurassic arc and accretionary wedge begins to build in the western Sierras. Finally, at the end of the Jurassic (Figure 9.9c), a change in plate motions causes the shear forces to displace the Klamaths from the Sierras as a new subduction zone and accretionary wedge jumps out to the west (the Franciscan trench). Large transpressional basins opened up and filled with the Galice Formation sediments in the Klamaths and the Mariposa sediments of the Foothills terrane in the Sierras (Chapter 8). The Great Valley forearc basin (Chapter 10), on the other hand, formed in front of and to the west of the Sierras, so they did not shift to the west as much as the Klamaths did. The Sierra Nevada Andean-style arc then erupted in earnest. The rest of the Sierran eruptions north of the Sierras are apparently buried by the lavas of the Modoc Plateau (Chapter 6) to the north, although in a few places those intrusions also melted their way through the now-extinct tectonic blocks of the Klamaths.

9.4 CENOZOIC HISTORY OF THE KLAMATHS

Once the Sierran arc and Great Valley forearc basin (Chapter 10) were fully established and became the dominant tectonic features of Cretaceous California, the Klamath terranes became extinct and no longer showed any signs of volcanism or other tectonic activity. As early as 1902, geologists such as Diller noticed that the Klamath

ridges are all about the same elevation (Figure 9.10), suggesting that they formed a continuous surface (**peneplain**) that has since been uplifted and eroded. Subsequent mapping by Irwin (1981) has confirmed that these concordant ridges are real and documented a blanket of upper Miocene deposits (Wildcat Group, St. George Formation, and Wimer Formation) formed in a marine transgression that once buried the Klamaths. This suggests that the Klamaths sank well below sea level during the late Miocene, were eroded flat, and then were blanketed by shallow marine deposits. The geochemistry of these sediments suggests that they were eroded and transported from rivers that originated in the Idaho batholith to the east (Aalto, 2006). Dating of the erosional surfaces suggests that the uplift of the Klamaths to their present elevation, and the incision of their deep canyons, is very young, possibly early to middle Pleistocene. The fact that the rivers often cut directly through the mountains, creating canyons several thousand feet deep with water gaps, rather than going the easy way around them, is also consistent with the idea that the drainage network was developed on a flat peneplain and then forced to carve down through the underlying structural mountains as they were slowly uplifted beneath the drainages (a superposed drainage).

There are many possible tectonic reasons for this late Cenozoic uplift of the Klamaths: decreasing age of the subducting oceanic plate, decreasing dip angle of subduction, translation of the Sierra Nevada microplate northwestward as it impinges upon the Klamath block, and migration of

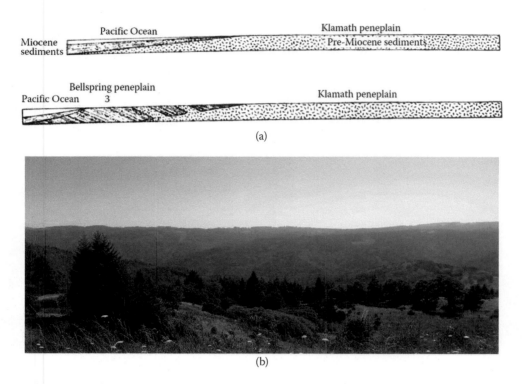

FIGURE 9.10 Klamath peneplain: (a) Diagram showing the erosional surface beveled into the pre-Miocene bedrock and the late Cenozoic transgressive deposits. (b) Photograph showing the flat ridges of the same elevation across the Klamath Mountains.

Source: Photo by the author.

the Blanco fracture zone northward along the continental margin (Aalto, 2006). However, no particular explanation is strongly supported at the moment.

9.5 KLAMATH GLACIATION

The highest peaks of the Klamaths, especially in the Trinity Alps, were above the snow line enough during the Pleistocene that they accumulated significant snowfields and were glaciated. Numerous U-shaped valleys can be found in the highest peaks, along with cirques, arêtes, and glacial moraines (Figures 9.2 and 9.11).

Today, about 35 small snowfields remain in the Trinity Alps, covering about 1.9 km². There are several small glaciers located in cirques on the north sides of Thompson Peak (2,742 m, or 8,994 ft) and Caesar Peak (2,720 m, or 8,920 ft). Most do not have enough snow on them to become fully flowing glaciers that slide downhill. In addition, these glaciers (like glaciers around the world) are rapidly melting back due to the effects of global warming. Like the Sierran glaciers (Chapter 8), they are vanishing rapidly and will probably all be gone by the end of the century.

9.6 SUMMARY

The Klamath Mountains have a long and complex history. The Eastern Klamath terrane was assembled during the early Paleozoic with ophiolitic seafloor rocks that were then accreted to North America during the Devonian Antler orogeny. The Central Metamorphic belt also consists of early Paleozoic metasedimentary and metavolcanic rocks accreted during the Antler orogeny. The Western Paleozoic and Triassic terranes were accreted in numerous smaller tectonic blocks, possibly as an accretionary wedge during the early phases of Sierran arc volcanism. Finally, the Western Jurassic terrane was an accretionary wedge complex built during the Jurassic phase of Sierran arc volcanism.

Complex tectonic forces then shifted the Klamaths to the west, and they became inactive as a new forearc basin in the Great Valley built during the Cretaceous Sierran volcanic arc eruptions, and a new accretionary wedge built to form the Coast Ranges. Small Cretaceous intrusions did melt their way through the crust in parts of the Klamaths, but the main Sierran arc was apparently where the Modoc Plateau now covers the landscape in late Cenozoic lavas.

During the Cenozoic, the Klamaths subsided and became eroded into a flat peneplain, covered with late Miocene marine sediments. These were uplifted and stripped away only during the Pleistocene, at which time the rivers carved the deep modern canyons, often cutting through the mountains to form water gaps, rather than flowing around the ridges the easy way. Finally, the highest peaks in the Trinity Alps were extensively glaciated in the Pleistocene, and about 35 small snowfields remain from those glaciers, all now rapidly retreating and melting due to global warming.

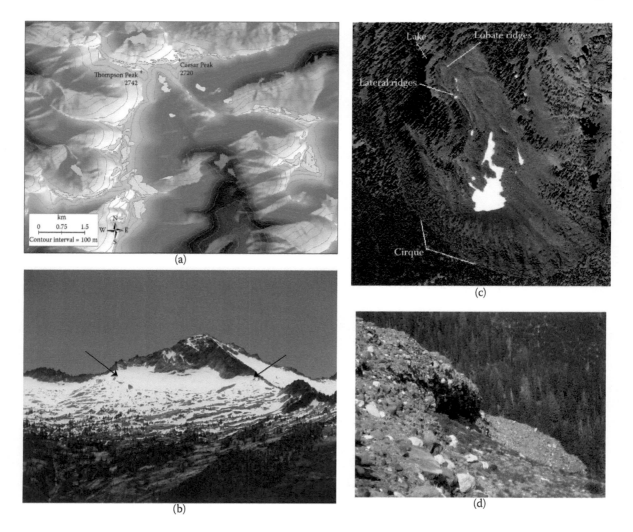

FIGURE 9.11 Glaciers of the Klamaths: (a) Shaded topographic relief map showing the remaining glaciers in the Trinity Alps in 1974. (b) Glacier in Caesar Peak Perennial Snowfield Geologic Area in the Klamath National Forest. Arrows point to the edges of the stable glacier ice that does not vanish in the summer. (c) Cirques and moraines in the Ash Creek Butte. (d) Glacial moraine in the Cement Banks Geologic Area.

Source: (a) Modified from US Geological Survey topographic map base. (b–d) Courtesy of US Forest Service.

RESOURCES

Aalto, K.R. 2006. The Klamath peneplain: A review of J.S. Diller's classic erosion surface. *GSA Special Paper*, *410*, 451–463.

Alt, D., and Hyndman, D.W. 2000. *Roadside Geology of Northern and Central California*. Mountain Press, Missoula, MT.

Burchfiel, B.C., and Davis, G.A. 1981. Triassic and Jurassic tectonic evolution of the Klamath Mountains-Sierra Nevada geologic terrane, 50–70. In Ernst, W.G. (ed.), *The Geotectonic Development of California*. Prentice Hall, Englewood Cliffs, NJ.

Ernst, W.G., Snow, C.A., and Scherer, H.H. 2008. Contrasting early and late Mesozoic petrotectonic evolution of Northern California. *Geological Society of America Bulletin*, *120*, 179–194.

Hildebrand, R.S. 2013. *Mesozoic Assembly of the North American Cordillera*. Geological Society of America Special Paper 495. Geological Society of America, Boulder, CO.

Irwin, W.P. 1981. Tectonic accretion of the Klamath Mountains, 29–49. In Ernst, W.G. (ed.), *The Geotectonic Development of California*. Prentice Hall, Englewood Cliffs, NJ.

Loxton, D., and Prothero, D.R. 2013. *Abominable Science: The Origins of the Yeti, Nessie, and Other Famous Cryptids*. Columbia University Press, New York.

McPhee, J. 1993. *Assembling California*. Farrar Straus Giroux, New York.

Prothero, D.R., and Dott, R.H., Jr. 2010. *Evolution of the Earth* (8th ed.). McGraw-Hill, New York.

Wallace, D.R. 1983. *The Klamath Knot: Explorations of Myth and Evolution* (20th anniv. ed., 2003). University of California Press, Berkeley.

Videos

Assembling California: Interview with John McPhee and Eldridge Moores: www.youtube.com/watch?v=vLwd_giY9dg

Klamath glaciers: www.youtube.com/watch?v=GvWQ5Umukws

www.youtube.com/watch?v=irHx0AwoDLE

www.youtube.com/watch?v=_MmYnGjLkHo

10 Oil and Agriculture
The Great Valley

From the Auburn suture of the Smartville Block, where you glimpse for the first time (westbound) the Great Central Valley of California, the immense flatland runs so far off the curve of the earth that its western horizon makes a simple line to the extremes of peripheral vision. In California's exceptional topography—with its crowd-gathering glacial excavations, its High Sierran hanging wall, its itinerant Salinian coast—nothing seems more singular to me than the Great Central Valley. It is far more planar than the plainest of plains. With respect to its surroundings, it arrived first. At its edges are mountains that were set up around it like portable screens. . . . The ground surface is so nearly level that you have no sense of contour. A former lakebed can be much the same, where sediments laid in still water have become a valley floor. Such valleys tend to be intimate, however, while this one is fifty miles wide and four hundred miles long. . . . The Great Central Valley has no counterpart on this planet.

—John Mcphee
Assembling California

10.1 DATA

Geography: Great Valley (also known as the Central Valley or the San Joaquin–Sacramento Valley)

Elevation: About 170 m (500 ft) at the edge of the foothills, to near sea level at the mouth of the Sacramento–San Joaquin Delta

Highest point: Sutter Buttes (647 m, or 2,122 ft)

Major rivers: Sacramento and San Joaquin

Predominant rocks: Jurassic Coast Range ophiolite basement, covered by the Cretaceous Great Valley Sequence, and Cenozoic marine and nonmarine sedimentary rocks

Plate tectonic setting: Forearc basin of the Sierra Nevada arc

Geologic resources: Oil, sand, gravel, and rich soils for agriculture

State parks: Colusa, Great Valley Grasslands, San Luis Reservoir, Turlock Lake, Shasta Dam

10.2 GEOGRAPHY OF THE GREAT VALLEY

One of California's most striking features (Figure 10.1) is its Central Valley, a long flat trough-shaped depression between the Sierras and the Coast Ranges. The term *Central Valley* is commonly used to refer to the topographic feature, but the term *Great Valley* is what geologists use to describe the geological basin (including the subsurface bedrock) that lies beneath the Central Valley. The valley is about 60–100 km (40–60 mi) wide and about 720 km (450 mi) long from northwest to southeast. Its total area is about 58,000 km^2 (22,500 mi^2), covering about 13.7% of California's total area. By itself, the Central Valley is bigger than nine other states, so if it were another state, it would be the 42nd largest state by area.

Across this landscape is about 15 million acres of farmland, the richest agricultural region in the history of the world. The harvest from 2014 alone topped $17 billion, more than the value of all the gold ever mined in California. It yields about 8% of the total US agricultural output on less than 1% of the land. More than 230 non-tropical crops are grown in the Central Valley, including corn, wheat, rice, and cotton, but it is the primary source of valuable fruits and vegetables like tomatoes, grapes, apricots, and asparagus. The western San Joaquin Valley is the nut capital of the world, producing 1,900 million pounds of almonds a year, 90% of the world's supply and 99% of the American supply. It also produces 99% of the nation's walnuts and pistachios, and 90% of American tomatoes. The second biggest industry in the Central Valley is oil, which is discussed in Chapter 17.

The Central Valley is drained by two huge river systems, the Sacramento River to the north (about 719 km, or 447 mi, long) and the San Joaquin River to the south (about 587 km, or 365 mi, long), so the entire valley can be subdivided into the Sacramento and San Joaquin Valleys (see Chapter 18). The Sacramento River drains the much wetter northern part of California, so the water volume passing out the mouth of the river is about 27 km^3 (22 million acre-feet) of water, compared to the drier drainage basin of the San Joaquin River, which yields only 7.4 km^3 (6 million acre-feet) of water. The two huge rivers then flow into a common delta, which opens into the Straits of Carquinez north of Martinez, California, and then into San Francisco Bay. Much of the valley is heavily irrigated, and the delta is heavily used for water-loving crops like rice, so only 21 km^3 (17 million acre-feet) of the Sacramento River and only 3.7 km^3 (3 million acre-feet) of the San Joaquin River reach the ocean. At one time, large lakes (such as Tulare Lake and Lake Buena Vista) covered parts of the landscape (relics of the last ice age), but most are now dry lakebeds since their water has been taken for irrigation. Tulare Lake has been plowed under and is now a large area of irrigated fields.

The Central Valley has a "hot Mediterranean climate" in the standard Köppen scheme of climate classification.

DOI: 10.1201/9781003301837-10

In summer, temperatures above 40°C (104°F) are typical for Bakersfield, Fresno, and Sacramento. However, the weather tends to be cool and damp in the winter, when dense "tule fogs" occur. Sometimes these cause huge chain-reaction car crashes on the highways. The Coast Ranges create enough of a rain shadow that the region is a relatively dry steppe grassland in the north, and even a semidesert with arid scrub vegetation in the south around Bakersfield.

Without irrigation, this rich farmland would revert to its natural state of grasslands and sage scrub, as can be seen in any uncultivated part of the landscape. Before humans began farming and introduced cattle, the region was inhabited by animals that roamed the grasslands, riparian forests, and tule marshes. These included tule elk, pronghorns, black-tailed deer, grizzly and brown bears, gray wolves, foxes, coyotes, ocelots, beaver, otter, minks, ferrets, weasels, musk rats, rabbits, ground squirrels, kangaroo rats, badgers, skunks, and many other small mammals. During the Pleistocene, there were also mammoths, mastodonts, bison, horses, camels, ground sloths, Ice Age lions, and saber-toothed cats. In addition, there are hundreds of species of birds (especially migrating waterfowl). Because of the mild climate and ample food and water, more than 100,000 Native Americans lived there, one of the largest indigenous populations in North America.

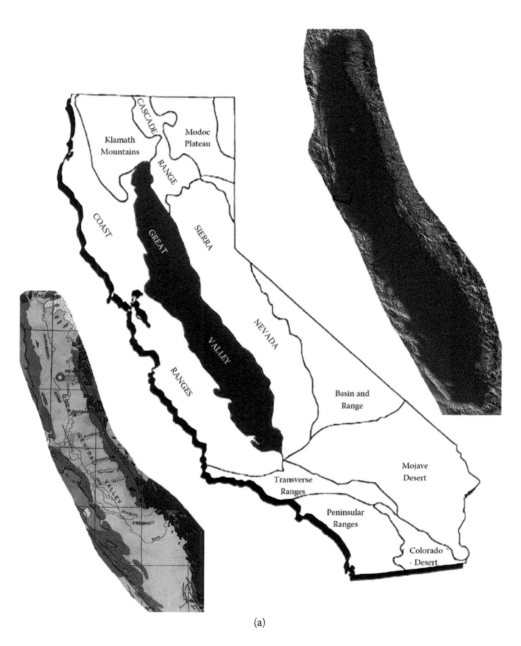

(a)

FIGURE 10.1 (a) Index map of California highlighting the Great Valley. In the lower left is the geologic map of the region, and in the upper right is the digital shaded relief map of the Central Valley. (b) Geologic map of the Great Valley and surrounding areas.

Source: (a) Courtesy of US Geological Survey. (b) Courtesy of California Division of Mines and Geology.

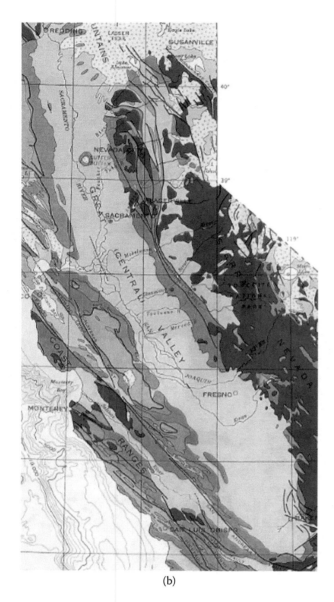

(b)

FIGURE 10.1 (Continued)

In addition to the huge area under cultivation, the Central Valley is also undergoing a population explosion due to the relatively cheap housing prices and low cost of living compared to the big cities along the coast. More than 6.5 million people live in the area now, and it is the fastest-growing region in California. More than 12 cities have populations in excess of 60,000: Sacramento (2.5 million), Fresno (930,000), Bakersfield (840,000), Stockton (700,000), Modesto (520,000), and seven more.

10.3 GEOLOGY OF THE GREAT VALLEY

The long trough-shaped configuration of the Great Valley is no accident. It has been a tectonic trough for a long time, since it started as the forearc basin of the great Andean-style arc of the Sierras (Figures 4.12a, 4.14, and 10.2), lying to the east of the accretionary wedge of the Coast Ranges (see Chapter 12). As such, it is comparable to other forearc

basins, such as the Willamette–Puget Sound lowlands in west side of the Cascades in Washington and Oregon, or the coastal valleys of Peru and Chile, which lie to the west of the Andean chain. Those forearc basins are largely emergent from the ocean and filling with nonmarine sediment today (as is the Great Valley), but during most of their history, these basins (Figure 10.3) were flooded with marine seawater, and nearly the entire sequence of sedimentary rocks is marine in origin (as are the ancient rocks of the Willamette Valley).

Forearc basins like the Great Valley, or the Willamette–Puget Sound lowland, tend to subside continuously through millions of years, accumulating thick sequences of mostly marine sediments (largely nearshore sandstones and off-shore shales). Thus, the cross section of these basins tends to have a lens-like or wedge-like shape, thinning on the edges and reaching enormous thicknesses in the center where the basin subsides the fastest. This geometry can be seen on seismic profiles (Figure 10.4) that show the cross section of the basin. Many such profiles have been run across the Great Valley, along with many deep wells in search for oil that is so abundant in certain places. They all demonstrate that the sedimentary fill thickens toward the center of the basin and thins out at the edges. The bottom of the basin is enormously deep, in some places more than 18 km (60,000 ft) thick. The thickness of the Cretaceous strata alone exceeds 12 km (40,000 ft) in many parts of the basin, and the Cenozoic sequence is almost as thick.

10.3.1 JURASSIC OPHIOLITE BASEMENT

So how did the Great Valley first form, and what lies beneath this enormous wedge of sedimentary rocks? Remember that during the Triassic and Jurassic, the predecessor of the Sierran arc (Chapter 8) was to the east of the present batholith, and some of the Triassic and Jurassic rocks of the Sierran foothills were accretionary wedges forming just west of the arc volcanoes in the modern Sierra foothills. Just to the west of these arc rocks would have been Jurassic oceanic crust in the process of subducting beneath the future Sierran arc. According to one interpretation, when the Cretaceous arc expanded, the subduction zone shifted westward to the present California coast and developed an accretionary wedge in the modern Coast Ranges, forming a forearc basin. Naturally, it was floored by the same ocean crust rocks that had been subducting in the Jurassic.

Indeed, this is what we find at the bottom of the basin wherever we can reach it. A handful of deep wells have drilled down to it, and it shows up clearly on the seismic profiles (Figure 10.4), where seismic waves were bounced off the subsurface layers and their two-way travel times for the reflected waves are recorded by small seismographs at the surface.

The best evidence comes from outcrops of what is called the Coast Range ophiolite. There is a strip of ophiolites and serpentinites along the western edge of the Sacramento Valley, where the basement comes to the surface. There are

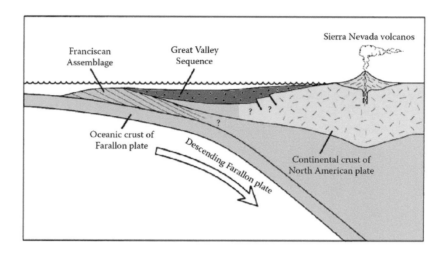

FIGURE 10.2 Tectonic model of an Andean-style forearc basin.

Source: Courtesy of Wikimedia Commons.

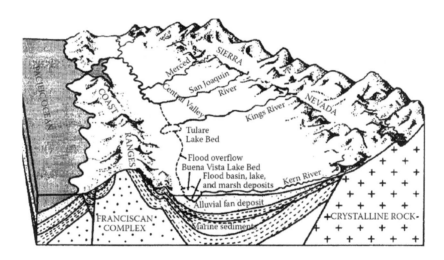

FIGURE 10.3 Simplified cross section of the Great Valley forearc basin.

Source: Courtesy of US Geological Survey.

also fragments of Jurassic ophiolite in the Coast Ranges, so we will look at them in more detail in Chapter 12 (Figure 10.5). In a few places, like Mt. Diablo east of the Bay Area and New Idria north of Coalinga, there are structural uplifts that bring serpentinized ophiolite to the surface of the basin itself. Thus, we can study outcrops of this deep basement rock, but we usually have to do so by leaving the Great Valley and going west into the Coast Ranges.

The other line of evidence comes from another type of remote geophysical sensing: gravity. Gravity measurements across the axis of the Great Valley (Figure 10.6) have been conducted many times, and they allow us to model the thickness and depth of the dense ophiolitic basement rock compared to the much less dense sedimentary rocks of the Great Valley basin fill. The gravity anomaly measured across the basin shows even stronger gravitational attraction as you move into the basin center, indicating an increasing thickness of the dense ophiolitic basement rock. The gravity anomaly model shown in Figure 10.6 suggests that the ophiolitic basement rock extends from depths of about 5 km to possibly as deep as 15–20 km into the upper part of the continental crust.

Putting all this evidence together gives a cross-sectional profile, something like that in Figure 10.7. The basement rock on the extreme eastern edge of the Great Valley is Sierran metamorphics and granitics, which crop out just beyond the east edge of the basin. To the west, however, the top surface of the Coast Range ophiolite basement begins, only 2–3 km down, but getting as deep as 5 km at the deepest part of the basin. On the western margin, faults and folding (possibly associated with the compression of the San Andreas fault farther west) bring these deep basin sedimentary rocks up in a steep western limb of a syncline and crop out at the surface of the western edge of

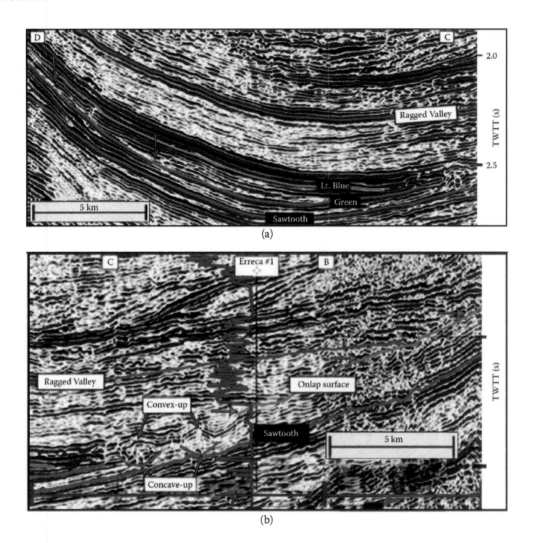

FIGURE 10.4   Seismic reflection profiles of the Great Valley subsurface. The reflection horizons (indicated by the yellow, red, and black bands over the seismic data) clearly show the trough-like folding of the beds in the basic center, interrupted by many faults. TWTT(s) = two-way travel time in seconds.

Source: Courtesy of US Geological Survey.

the basin, where they can be studied in outcrop. In some places, slices of the Coast Range ophiolite are also found at the surface, sitting on top of the Franciscan accretionary wedge complex of the Coast Ranges (see Chapter 12). If the seismic and gravity profiles are accurate, however, most of the top surface of the ophiolite continues to plunge to depths of 15–20 km. All these Franciscan accretionary wedges are sharply truncated by the San Andreas plate boundary, where different rocks from elsewhere are found on the opposite side of the plate.

10.3.2   Cretaceous Great Valley Group

The Jurassic ophiolite basement rocks apparently subsided quickly and were flooded by deep oceans, because above them the forearc basin filled with more than 12 km (40,000 ft) of Cretaceous marine sediments known as the **Great Valley Group** (or Great Valley Assemblage).

The lowest unit in the Great Valley Group (Stony Creek Formation) is actually Upper Jurassic (Figure 10.8), and this formation spans the interval from 150 to 120 Ma. In the Early Cretaceous, from about 120 to 95 Ma, the Lodoga Formation was deposited in the forearc basin (Figure 10.8). The Late Cretaceous (95–65 Ma) is represented by a number of different formations (Figure 10.8), lumped into units called the Boxer, Cortina, and Ramsey.

Nearly all the sediments of the Great Valley Group represent deep marine settings, so there are thousands of meters of deep-water shales, often punctuated with turbidite sandstones (Figure 10.9). These outcrops display long sequences of thin-bedded shales and sandstones, with great consistency and uniform thickness of beds for many kilometers. In many units (such as the Williams Canyon Formation and similar units), there are channels from submarine canyons incised into ancient submarine fan deposits. Where they crop out on the northern rim of the Sacramento Basin, they

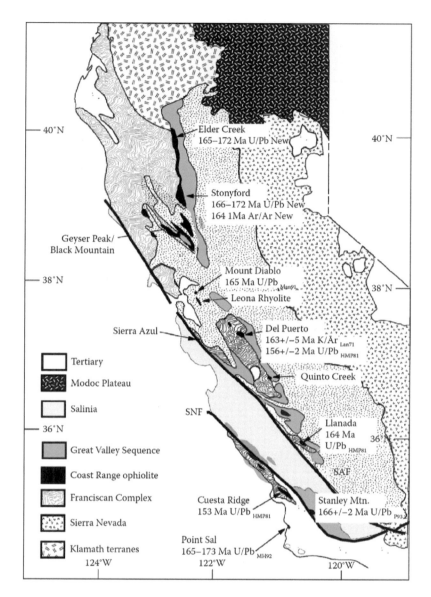

FIGURE 10.5 Map showing outcrops of Coast Range ophiolite in the western flank of the Great Valley and adjacent blocks.

Source: Modified from several sources.

are rich in ammonites and other Cretaceous marine fossils (see Chapter 20). Nearly all the sediment is derived from the eastern and western margins of the basin, and much of it is rich in volcanic debris and ash from the continuous eruption of the Sierran arc to the east. Although the Great Valley Group can be reached by drilling down into the basement of the center of the basin, it crops out on the western edge of the Great Valley, especially in the foothills of the western Sacramento Valley, and in the western San Joaquin Valley as far south as Coalinga (Figures 10.5 and 10.9).

10.3.3 CENOZOIC DEPOSITS OF THE GREAT VALLEY

The Cenozoic sedimentary package of the Great Valley is much thicker in the San Joaquin Basin, but much thinner

and less complete in the Sacramento Basin. The sequence of rocks is complex, with different formations and sedimentary environments from one side of the valley to the other. It is very thick in the subsurface but tends to be thinner and discontinuous, with more unconformities as the units from the middle of the basin pinch out toward the flanks of the basin (Figures 10.10 and 10.11). The best outcrops are on the west side of the Great Valley in the Coast Range foothills, where most of the formations plunge down to the east from surface exposures to much thicker deposits in the center of the basin. Despite the difficulty of sampling rocks from the deep central part of the basin, the complexity of the stratigraphy and structure has been intensely studied, because these rocks are the principal producers of oil in California.

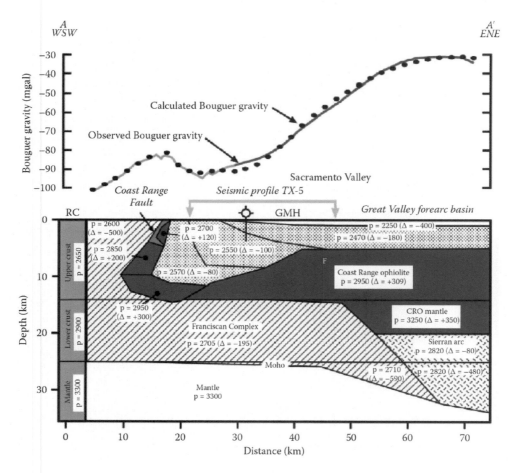

FIGURE 10.6 Gravity profile model across the Great Valley. The higher gravity readings indicate a denser-than-average unit (dark gray Coast Range ophiolite) must underlie the valley-fill sedimentary rocks (stippled units on the top).

Source: With permission from the Geological Society of America.

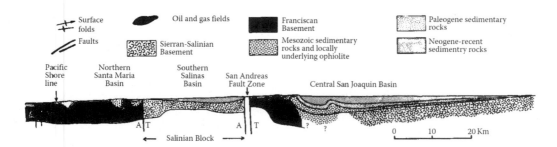

FIGURE 10.7 True-scale cross section of the Great Valley.

Source: Image by S. A. Graham.

10.3.3.1 Paleocene

The Paleocene (Figure 10.11) is poorly represented in the Great Valley. Most areas have Eocene deposits lying unconformably on Cretaceous deposits. However, the deep marine Martinez Formation (just east of the Bay Area in Martinez, California) is the basis for the "Martinez molluscan stage" in the standard California biostratigraphic molluscan zonation. The Moreno Formation in the Panoche Hills spans the Cretaceous–Paleocene boundary (see Chapter 20). Elsewhere, the San Francisquito Formation on the south flank of the Great Valley and in the Transverse Ranges spans most of the early and middle Paleocene.

All these Paleocene sequences were formed in deep ocean setting, much like the underlying Cretaceous beds.

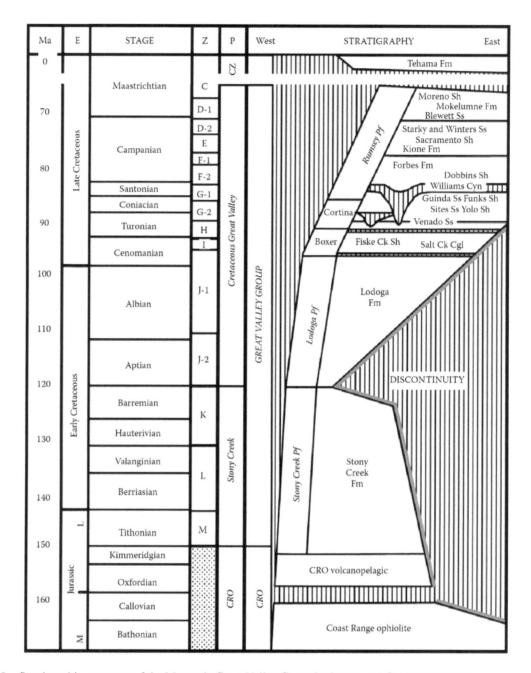

FIGURE 10.8   Stratigraphic sequence of the Mesozoic Great Valley Group in the western Sacramento Valley.

Source: Courtesy of SEPM.

They are largely deep-water shales with turbidite sandstones, although there are some well-developed submarine fan deposits in the central San Joaquin Basin. The different species and subspecies of the distinctive high-spired snail *Turritella* are used to zone the Paleocene beds (especially early Paleocene *T. peninsularis* and late Paleocene *T. infragranulata*). The benthic foraminifera of the late Paleocene Ynezian stage have also been used for telling time, but more recently, the best biostratigraphic indicators are the planktonic foraminiferans and calcareous nannofossils.

10.3.3.2   Eocene

The Paleocene is very thin or absent across the Great Valley, but the Eocene is very thick and complex. The basal Eocene near Sacramento and Martinez is represented by a submarine fan sequence and was the type area for the early Eocene Meganos stage. Overlying it is the early Eocene Capay molluscan stage, whose type area is near Winters, California. The late early and middle Eocene are represented by a number of nearshore sandstones in various parts of the basin, which were deposited as deltas and

(a) (b)

FIGURE 10.9 Outcrop images of the deep-water shales and thin turbidite sandstones of the Great Valley Group in the western Sacramento Valley: (a) Thin-bedded turbidites and shales of the Boxer Formation, near Williams. (b) Outcrop of Venado Sandstone near Lake Berryessa.

Source: Photos by S. A. Graham.

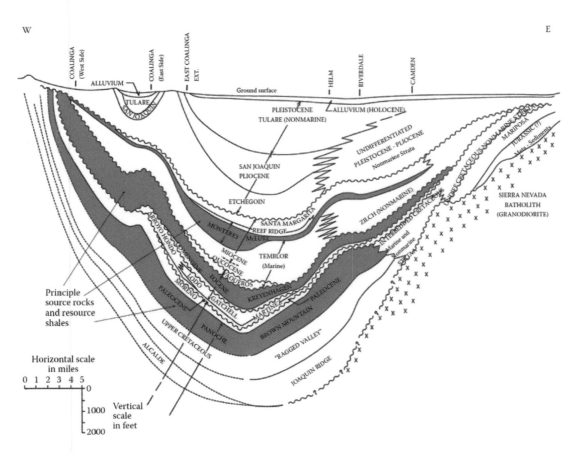

FIGURE 10.10 Geologic cross section of the San Joaquin Basin showing the oil-bearing units in green. Notice how the formations are thickest in the basin center and pinch out as they reach the edges.

Source: Courtesy of California Division of Oil and Gas Resources.

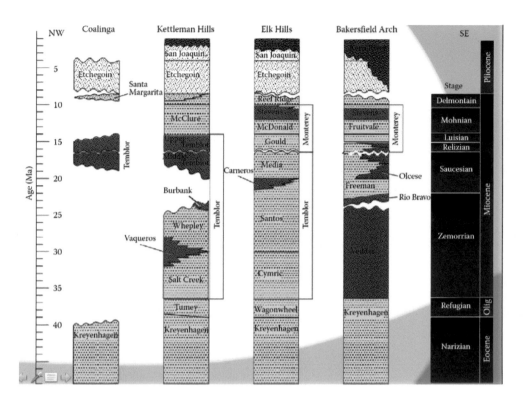

FIGURE 10.11 Stratigraphic column for parts of the Cenozoic of the Great Valley.

Source: Courtesy of Pacific Section SEPM.

submarine fans, building out from the edges of the basin. These include the Arroyo Hondo, Cantua, Gatchell, Yokut, and Domengine/Avenal Sandstones near Coalinga, plus the Ione Sandstone near Martinez and also in the Sierra foothills (see Chapter 20). In the middle Eocene, most of the central part of the San Joaquin Basin was filled by a very thick package of marine shales known as the Kreyenhagen Formation, which is a major oil source rock. Nearshore and deltaic sands of the Point of Rocks Formation build out from the western edge near the Temblor Range and the Famoso Sand from the eastern edge.

The upper Eocene is best represented by the enormously thick package of shales and turbidites of the Tejon Formation in the San Emigdio Mountains on the southern rim of the basin type area of the middle–late Eocene Tejon molluscan stage. The center of the San Joaquin Basin was filled by the Tumey shales overlying the Kreyenhagen Shales, with isolated shallow marine sand lenses of the Oceanic Sandstone in the subsurface only.

10.3.3.3 Oligocene

In most parts of the Great Valley, the Oligocene is very thin or absent entirely, and only a few areas have Oligocene formations. This is due to a major global sea-level drop in the middle Oligocene, which ended marine deposition in the Great Valley briefly, or even incised into and eroded away some of the older rocks.

In the southwestern San Joaquin Basin, there are short intervals of Oligocene deposits represented by the lower

members of the Temblor Formation: the Cymric Shale, Wygal Sandstone, and Santos Shale members in the Temblor Range. Thick wedges of the nearshore Vedder Sandstone built out from the southeastern edge of the basin. However, the vast majority of Oligocene time is not represented by any kinds of rocks at all.

10.3.3.4 Miocene

In contrast to the highly discontinuous and incomplete Oligocene, the Miocene section in the San Joaquin Basin is thick and complex.

In the southwestern San Joaquin Basin, the early Miocene is represented by additional members of the Temblor Formation: Santos Shale and Agua Sandstone, capped by the Carneros Sandstone, and Media Shale. The center of the basin was filled with the Freeman Silt and Santos Shale and Media Shale. The eastern edge of the basin near Bakersfield has numerous wedges and lenses of sandstones building out from deltas coming from the Sierras, including the Jewett Sand and the Olcese Sand.

In the middle Miocene, the deep-water Round Mountain Siltstone filled the San Joaquin Basin. It is famous for the amazing collection of marine fossils found at Sharktooth Hill (see Chapter 20).

The most famous and well-studied unit of the San Joaquin Basin, however, is the middle–upper Miocene Monterey Formation, which crops out in marine basins along the coast as well, from Monterey to Orange County. It is the richest oil producer in the state (see Chapter 17), so it has been the subject of much research interests by academic and industry geologists alike. In the San Joaquin Basin, this organic-rich deep-water shale is punctuated by many small sand bodies (such as the Stevens Sandstone lens) that are the main reservoir rocks that trap the migrating oil. On the southern rim of the basin in the Tejon Hills, the shallow marine Santa Margarita Sandstone and the nonmarine Chanac Formation (yielding fossil land mammals) built out from the edge of the basin into the deep-water Monterey shales. This sequence was the basis for the Cerrotejonian mammalian stage of the late Miocene.

10.3.3.5 Pliocene

The Pliocene is represented by the marine shales and sandstones of the Etchegoin Group (Jacalitos, Etchegoin, and San Joaquin Formations), which are 3,000 m (almost 9,000 ft) thick in the Coalinga–Kettleman Hills area. These units are famous for their abundant shallow marine mollusks (especially scallops) and sand dollars, as well as the peculiar teeth of the extinct hippo-like marine mammals known as desmostylians. In the southwestern basin near Bakersfield, nonmarine and shallow marine deposits of the Kern River Formation built out into the basin from rivers in the Sierras (just as the modern Kern River drains the southern Sierras today).

10.3.3.6 Quaternary

After a long history of entirely marine deposition since the Early Cretaceous, the seas finally retreated in the Pleistocene. This may be due to global sea-level drop due to the increase in ice volume, pulling water out of the ocean basins. All the Quaternary rocks of the Great Valley are nonmarine, represented mostly by the lake deposits and alluvium of the Tulare Formation that blankets most of the valley, and is the foundation for most of the rich soils. Most of this sediment apparently came down from the Sierras as they were uplifted and rapidly eroded by glaciers.

The southern San Joaquin Valley once had huge lakes, now represented by the dry lake basin of Tulare Lake and Buena Vista Lake. In the wet winter of 2023, Tulare Lake flooded and was once again a real lake, causing millions of dollars of damage to farms that had been built in the lake bed. Near McKittrick in southwestern Kern County, there are famous tar seeps that yield late Ice Age mammals much like those found in La Brea tar pits in Los Angeles, including dire wolves, saber-toothed cats and Ice Age lions, bison, deer, horses, pronghorns, peccaries, ground sloths, mastodonts and mammoths, and many rodents, rabbits, bats, and shrews. Near Chowchilla in Madera County, the Fairmead Landfill has also produced a large fauna of early Pleistocene mammals, including saber-toothed cats, ground sloths, mammoths, horses, camels, pronghorns, deer, dogs, and numerous rodents (see Chapter 20).

10.4 SUTTER BUTTES

Nearly the entire Mesozoic and Cenozoic history of the Great Valley is a tale of sedimentary basin fill. However, there is one striking igneous intrusion within the basin, an extinct volcanic plug called Sutter Buttes (also known as Marysville Buttes) (Figure 10.12), located just outside Yuba City. The eerie profile of this volcanic complex is an unmistakable feature sticking out of the flat plains around it and can be seen for many miles on a clear day. It is 647 m (2,122 ft) high, and about 16 km (10 mi) in diameter.

Some have nicknamed it "the world's smallest mountain range." The buttes were known as *los tres picos* ("the three peaks") to the Spanish and Mexican colonials. Their current name comes from pioneering Swiss settler John Sutter, who colonized the Sacramento Valley and owned all the land in the area in the 1840s, before the Gold Rush overran his claims.

Detailed studies of the crater show that it is an andesitic–dacitic cone which erupted in the early Pleistocene, beginning at 1.6 Ma and finishing at 1.4

(a)

(b)

FIGURE 10.12 Sutter Buttes: (a) aerial view and (b) view from ground level.

Source: Courtesy of Wikimedia Commons.

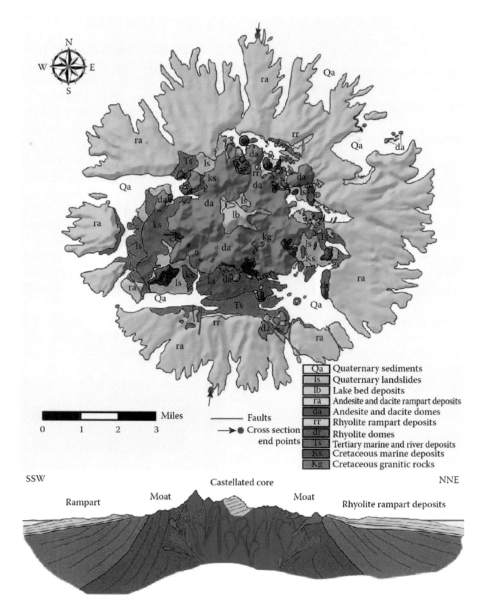

Qa	Quaternary sediments
ls	Quaternary landslides
lb	Lake bed deposits
ra	Andesite and dacite rampart deposits
da	Andesite and dacite domes
rr	Rhyolite rampart deposits
dr	Rhyolite domes
Ts	Tertiary marine and river deposits
Ks	Cretaceous marine deposits
Kg	Cretaceous granitic rocks

FIGURE 10.13   Geologic map and cross section of Sutter Buttes.

Source: Courtesy of US Geological Survey.

Ma. The center of the buttes is a volcanic plug made of andesite and dacite domes, called the "castellated core" (Figure 10.12). These were in the throat of the original volcano, because most of the original cone has eroded away since 1.4 Ma. There are lake sediments, and even layered lake terraces, in the center, showing that it was once filled with water.

Surrounding the castellated core is a "moat," a ring of valleys with relatively few volcanic rocks, and exposures of the older sediments through which the volcano erupted. The sedimentary layers span the Cretaceous and early Cenozoic since they were steeply tilted as the volcanic mass pushed up beneath them (Figure 10.12). Since these layers are not protected by a hard volcanic caprock,

the soft sandstones and shales erode rapidly, producing the valleys of the moat. The outer part of Sutter Buttes is called the "rampart," a low-sloping apron (Figure 10.13) that surrounds the entire volcano and covers most of its width. It is composed of volcanic flows, especially volcanic breccias (agglomerates), ignimbrites, and lahars that erupted outward from the volcano during its period of activity.

The tectonic interpretation of Sutter Buttes has been controversial. Some geologists think that it might be a Cascade volcano, the next one in the chain south of Shasta and Lassen, that became extinct as the Mendocino triple junction moved north (see Chapter 15). Other geologists have compared Sutter Buttes to similar-aged volcanoes in the Coast Ranges, although these seem to have a different chemical composition of their magmas. Some have argued that there might be a hidden rift in the floor of the Sacramento Basin which would allow magma to reach the surface. So far, there is no consensus about which tectonic model best explains Sutter Buttes.

10.5   SUMMARY

The Great Valley was formed when the subduction zone jumped from its Jurassic position near the western foothills of the Sierras to the California coast, leaving a large slice of Jurassic Coast Range ophiolite to floor a huge forearc basin. This basin then sank almost continuously for 140 m.y., filling with 12,000 m (40,000 ft) of the Cretaceous deep-water shales and turbidite sandstones of the Great Valley Group. The San Joaquin Basin (but not much of the Sacramento Basin) filled with about 6,000 m (20,000 ft) of Cenozoic marine sediments, consisting mostly of marine shales in the center of the basin and nearshore and deltaic sandstones building out from the margins. Many of these shales were organic-rich and became important oil producers, making the San Joaquin Basin the primary petroleum province of California. During the Pleistocene, sea level retreated and the valley filled with alluvium and large Ice Age lake beds. Between 1.6 and 1.4 Ma, Sutter Buttes intruded, producing the sole volcano in the valley.

RESOURCES

Dickinson, W.R. 1981. Plate tectonics and the continental margin of California, 1–28. In Ernst, W.G. (ed.), *The Geotectonic Development of California.* Prentice Hall, Englewood Cliffs, NJ.

Hausback, B.P., and Nilsen, T.H. 1989. *Sutter Buttes Field Trip Guide.* American Geophysical Union, Washington, DC.

Ingersoll, R.V., and Dickinson, W.R. 1977. *Field Guide: Great Valley Sequence, Sacramento Valley.* Geological Society of America, Boulder, CO.

Ingersoll, R.V., and Nilsen, T.H. (eds.). 1990. *Sacramento Valley Symposium and Guidebook.* Pacific of the SEPM (Society for Sedimentary Geology), Fullerton, CA.

Kuespert, J.G., and Reid, S.A. (eds.). 1990. *Structure, Stratigraphy and Hydrocarbon Occurrences of the San Joaquin Basin, California.* Guidebook No. 64. Pacific of the SEPM (Society for Sedimentary Geology).

Williams, H., and Curtis, G.H. 1977. *The Sutter Buttes of California.* University of California Publications in Geological Sciences 116. University of California, Berkeley.

Videos

Central Valley from space: www.esa.int/spaceinvideos/Videos/2015/06/Earth_from_Space_Central_California

Sharktooth Hill

www.kerngoldenempire.com/news/top-stories/a-dig-at-shark-tooth-hill-with-the-buena-vista-museum
www.youtube.com/watch?v=-tkpAx549D8
www.youtube.com/watch?v=sLQOspKgmRg

11 The San Andreas Fault System

I live a half mile from the San Andreas fault—a fact that bubbles up into my consciousness every time some other part of the world experiences an earthquake. I sometimes wonder whether this subterranean sense of impending disaster is at least partly responsible for Silicon Valley's feverish, get-it-done-yesterday work norms.

—Gary Hamel

In clear weather, a pilot with no radio and no instrumentation could easily fly those four hundred miles navigating only by the fault. The trace disappears here and again under wooded highlands, yet the San Andreas by and large is not only evident but also something to see—like the beaten track of a great migration, like a surgical scar on a belly. In the south, . . . it cuts through two high roadcuts in which Pliocene sedimentary rocks look like rolled up magazines, representing not one tectonic event but a whole working series of them exposed at the height of the action.

—John McPhee
Assembling California

11.1 CALIFORNIA FAULTS AND EARTHQUAKES

Although California is famous for its earthquakes, they happen in many states. The biggest quakes in American history occurred in 1811–1812 near the town of New Madrid, Missouri; they rang church bells as far as Boston. The 1886 Charleston, South Carolina, quake almost leveled the city. As we discussed in Chapter 7, the entire Basin and Range Province of Utah, Nevada, Arizona, and eastern California has active faults, especially in the zone in western Nevada and down the Wasatch fault running down the spine of Utah. Alaska suffers through constant quakes. The 1964 Alaska event was the biggest American quake ever measured by a seismograph.

But because California is the only state cut by three different types of plate boundaries, it is one of the most seismically active regions in the country. A glance at the fault map and seismicity map of California (Figure 11.1a—c) shows that the known and mapped faults occur almost everywhere, and it's virtually impossible to avoid them. Most of the faults are parallel to the plate boundary and the San Andreas transform system. Through their motion, part of the strain of the sliding between the two plates is taken up in subsidiary faults, reducing the amount of energy that builds on the San Andreas itself. Since the Pacific plate is moving northwest with respect to the North American

plate, the sense of shear on these strike-slip faults is always right-lateral (i.e., the block on the opposite side of the fault from where you stand appears to move to the right). More than 95% of the active strike-slip faults in California are right-lateral, with one glaring exception: the Garlock fault, which runs along the Tehachapi and Sierra Nevada mountains and out into the Mojave Desert, is left-lateral. As discussed in Chapter 7, the Garlock fault is the exception because it takes up the westward motion and rotation of the southern tip of the Sierras (Figure 7.13).

The San Andreas is not a single fault but an entire system of parallel faults that together take up the stresses of two plates grinding past one another. The San Andreas and related faults do not move smoothly along their entire length (Figure 11.2). In some places, such as the central Coast Ranges near Coalinga, or the Salton Trough and Imperial Valley, the fault appears to slip often, with many small earthquakes that release their stress gradually. In other regions, such as the Bay Area (which hasn't slipped since 1906) or the area from north of Los Angeles to down past San Bernardino (which hasn't moved since 1857), the fault tends to be locked and slips only rarely every few centuries, with big displacements and huge amounts of accumulated energy released when it does finally move. This is demonstrated by the map showing the fault probabilities of major segments of the San Andreas (Figure 11.3). The lowest probabilities of large earthquakes are in the central Coast Ranges and the Imperial Valley, because these regions are high probabilities of many small earthquakes. The highest risks of big quakes are in the two zones mentioned earlier, which appear to be long overdue.

However, many of the other faults in California besides the San Andreas have experienced significant quakes on them in historic times. Most of the big recent quakes in California have not happened on the plate boundary at all but on these subsidiary faults. These include major earthquakes, such as the 1872 Owens Valley quake ($M_w = 7.8$) and 1952 Kern County quake ($M_w = 7.5$), as well as smaller quakes that damaged urban areas, such as the 1933 Long Beach quake ($M_w = 6.3$), 1971 San Fernando–Sylmar quake ($M_w = 6.7$), 1987 Whittier quake ($M_w = 5.9$), and 1994 Northridge quake ($M_w = 6.7$). This last one occurred on a thrust fault buried deep beneath the surface and never mapped before (called a "blind thrust"); its existence was only revealed once it moved. California has at least hundreds of such quiet faults that we don't know about, so the fault map of the state underestimates the true number of active faults. Because legislation restricting new development is based on these fault maps, such laws also underestimate the risk of earthquake damage.

DOI: 10.1201/9781003301837-11

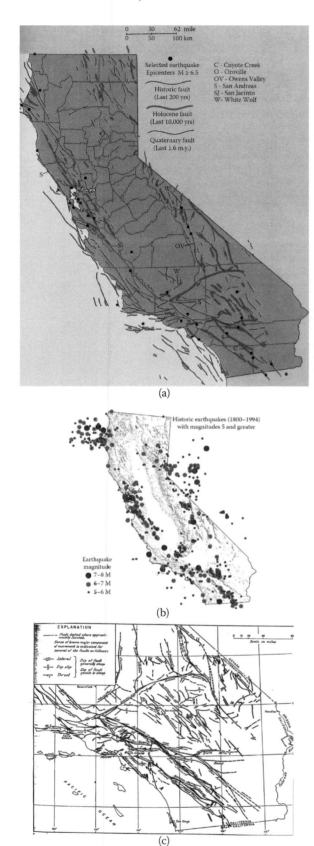

(a)

(b)

(c)

FIGURE 11.1 California seismicity: (a) Map of the major active faults of California. (b) Map showing the size and magnitude of the largest quakes in recent California history. (c) Map showing the major faults in the southern part of the state.

Source: Courtesy of California Division of Mines and Geology.

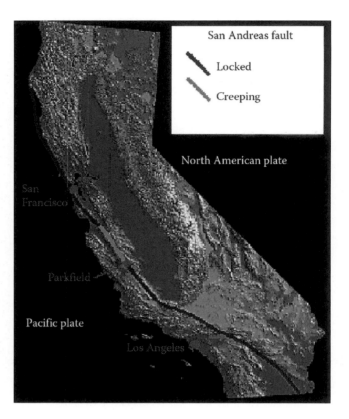

FIGURE 11.2 Areas of locked and active fault motion in California.

Source: Courtesy of US Geological Survey.

The biggest quakes have occurred on the San Andreas fault, and some of them have released enormous amounts of energy. The entire stretch of the San Andreas from down by Cajon Pass and then north of the Transverse Ranges and up to Parkfield in the central Coast Ranges was ruptured during the Fort Tejon earthquake (estimated $M_w = 7.9$) and has been locked ever since January 9, 1857. The rupture stretched 225 km (350 mi), and the Pacific plate lurched about 10 m (33 ft) northward relative to North America in just seconds in some places. There was only 4.5 m (15 ft) of slip in other areas.

Another famous recent quake along the San Andreas was the 1989 Loma Prieta quake on October 17, 1989 ($M_w = 7.9$), which occurred in another locked stretch of the San Andreas fault. It struck at 5:04 in the evening, postponing the first-ever World Series between the two Bay Area Major League Baseball teams, the San Francisco Giants and Oakland A's, just as the game was about to start. Damage was severe not only near the epicenter above Santa Cruz but also especially in the southern Bay Area, where the Bay Bridge broke (Figure 11.4a). In Oakland, the Nimitz Freeway collapsed, and one deck pancaked on top of another, killing most of the 63 victims (Figure 11.4b and c). There were also extensive damage and fires in the Marina District of San Francisco, where the ground liquefied into quicksand and houses sank and then caught fire when the gas mains broke and ignited.

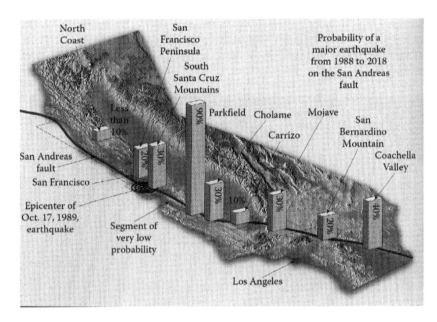

FIGURE 11.3 Earthquake probability for the different segments of the San Andreas fault. The highest probabilities are around Parkfield and the Coachella Valley, because of their frequent small quakes, since the fault creeps steadily in these areas (see discussion in text). But the remaining high-probability areas are "seismic gaps," where the fault is locked but overdue for an earthquake after a long quiet period. One such gap was the south Santa Cruz Mountains, which finally moved in the 1989 Loma Prieta earthquake.

Source: Courtesy of US Geological Survey.

But the biggest quake on the San Andreas fault in just over a century was the 1906 San Francisco quake. It occurred at 5:12 in the morning on April 18, 1906, with $M_w = 7.9$ (Richter M = 8.3). Most people were sound asleep, but the few people up that early saw the ground rise and fall in waves. The number of masonry buildings quickly collapsed, or their brick facades or chimneys fell.

Numerous fires were soon out of control, and the entire city began to burn over the next three days (Figure 11.5), destroying ten times as many buildings as did the initial quake (covering over 500 city blocks). San Francisco had burned before, so it had cisterns and water systems all over the city, but the quake ruptured and damaged these, so the firefighters were helpless. Instead, they began to dynamite buildings in the path of the fire to make a fire break but only made the problem worse. Martial law was declared, and looters (and even normal property owners protecting their houses) were shot on sight. Some of these ad hoc gangs, known as the "Specials," were appointed by Mayor Schmitz and took the opportunity to settle some old scores or steal with impunity in that lawless chaos.

After three days, the fires began to go out, and the city lay in ruins. The official death toll was only about 700 white people, because in those days the Chinese and Latinos were not considered worth counting, so the death toll was certainly much higher; most estimates place it around 3,000. Once the fires were out, there was a problem with sanitation and disease, with even 150 cases of the bubonic plague. Eventually, the city rebuilt, but the city fathers and boosters tried to cover up the real cause. They tried to censor any mention of the quake by calling it the "San Francisco fire" (something familiar in those days, when many cities burned often) to minimize the fears of people coming to rebuild. Fortunately, they were not able to completely prevent the publication of the two-volume expert report by the panel of distinguished geologists and seismologists, even though the authors of that report complained of censorship and attempts to hamper their work and prevent the news from getting out. Because of this, the San Francisco quake and the reports that followed marked the birth of modern seismology.

11.2 HOW OFTEN DO QUAKES OCCUR?

The few major historic quakes just discussed only give a general sense of which faults are currently active, what their recent activity is like, and how big their energy and displacement can be. To better understand the long-term likelihood of a quake along a particular stretch of a fault, we turn to another approach to earthquake prediction: **paleoseismology**, or studying the record of ancient earthquakes to predict future ones. Geologists dig a trench across ancient deposits that sit across active fault lines and then map all the exposures in the trench to analyze the stratigraphy of the marker beds.

For example, in a classic area called Pallett Creek near Devil's Punchbowl and Littlerock, California, geologist Kerry Sieh pioneered this technique back in the late 1970s and 1980s. He mapped and analyzed all the layers in his trenches and then noted where older beds had been cut by an ancient fault that was then buried by younger beds (Figure 11.6). The Pallett Creek deposits formed in an

(a)

(b)

(c)

FIGURE 11.4 Effects of the 1989 Loma Prieta quake. (a) Collapse of the upper deck of the Bay Bridge. (b) Collapse of the upper and lower deck of the Nimitz Freeway in Oakland. (c) A car crushed by the collapsed upper deck of the Nimitz Freeway.

Source: Courtesy of US Geological Survey.

(a)

(b)

(c)

FIGURE 11.5 Damage after the 1906 San Francisco quake: (a) View down Sacramento Street showing the advancing fire. (b) Troops watching for looters near the San Francisco City Hall, which was totally destroyed in the quake. (c) Broken pavement, warped streets, and bent trolley tracks after the quake.

Source: Courtesy of Wikimedia Commons.

ancient sag pond sitting on the fault, so they were full of small layers of fossilized plant material called lignite that could be radiocarbon-dated.

After dating which layers had been cut by different faults, Sieh was able to establish that the deposits covered the last

FIGURE 11.6   (a) The Pallett Creek trenches when they were fully excavated in the late 1970s. (b) Close-up of one of the trenches. The fault that breaks the AD 1225 peat layer is visible in the center. Kerry Sieh is pointing to a peat layer that is radiocarbon-dated at AD 1470 (shortly before Columbus's voyage), which is not broken by that same fault. Thus, the age of that particular fault can be bracketed between 1225 and AD 1470. (c) Recurrence interval at Pallett Creek. Except for the historic 1857 Fort Tejon quake, the earthquakes are given alphabetical labels going back from X. The vertical bars on each point indicated the plus or minus error of the radiocarbon date.

Source: Image by K. Sieh.

2,000 years of activity on this stretch of the San Andreas. More importantly, he showed a recurrence interval of about 137 ± 8 years; a later study suggested a recurrent interval of 145 ± 8 years. Consider that in 2017 it will have been 160 years since the 1857 Fort Tejon quake, and you can see why seismologists are very worried about this seismic gap. When it does slip, it will probably be offset about 10 m (33 ft) in seconds, just as it did in 1857. It will be the "big one" that all Southern Californians have awaited for so long.

11.3   WHAT DOES THE SAN ANDREAS FAULT LOOK LIKE?

As discussed in Chapter 5, the faults of California are not giant chasms opening up into the deep earth, with lava at the bottom, like the movies show them. Instead, the best clue to a fault line in California is a long straight valley, because the fault grinds up and pulverizes the rock on its slip plane and makes the fault line the easiest place for water to erode. Indeed, this is so common in California that geologists often suspect that any straight valley is a fault line until proven otherwise.

The features of the strike-slip faults are often quite different from other kinds of fault features (Figure 11.7). Not only is there a straight valley above the fault line, but there may also be sharp fault scarps or benches, or cutoff ridges on the edge of the fault zone. The fault zone often serves to trap water migrating through the subsurface against the fault plane; many fault planes are marked by springs, or by lines of vegetation that are nurtured by this trapped groundwater. In some cases, the water pools at the surface of low spots in the fault valley. These are known as **sag ponds** (Figure 11.7).

The most striking feature, however, are places where a stream approaches the fault and does an abrupt right-hand turn and then turns left again to its original course. This pattern of **offset drainage channels** (Figure 11.7) is found all up and down the San Andreas fault system and is a handy indicator of how much motion has occurred. At one time, the stream valley cut straight across the fault, but it has been gradually offset and become more and more sinuous through time as the lower part of the stream moved away from the headwaters. If a geologist can find a way to date when the stream was straight before the offset began, then it is possible to calculate the rate of slip.

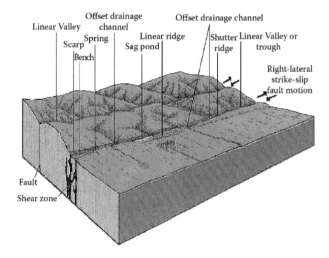

FIGURE 11.7 Physiographic features of a typical strike-slip fault zone.

Source: Courtesy of US Geological Survey.

Another common feature is compression of the fault zone in the rocks on either side, forming what are known as pressure ridges (Figure 11.8). When a roadcut slices through one of these, the beds are often intensely folded, crumpled, and even thrust-faulted (Figure 11.8b).

11.4 TOUR OF THE FAULT ZONE

Let's follow the San Andreas fault system from south to north and see some of its classic exposures. You can trace the entire fault line in satellite image using Google Earth or any similar satellite map software (such as the US Geological Survey [USGS] Map Viewer, http://viewer.nationalmap.gov/viewer/), switching back and forth from the map image to the aerial photo image.

The southern end of the San Andreas starts in the Gulf of California (also called the Sea of Cortez), where the spreading ridge between Baja California and mainland Mexico changes into a transform fault. This region is called the Salton Trough, and we'll discuss it in greater length in Chapter 14.

11.4.1 SALT CREEK

On the east side of the Salton Sea near the Durmid Hills is a striking example of the San Andreas offset. Salt Creek drains out of the hills to the east, and then just east of the highway and the railroad tracks, it does a sharp right-hand turn, then a sharp left, and flows beneath State Highway 111 and the railway trestle bridge (Figure 11.9). The offset here is about 800 m (2,500 ft). Trenches dug into the ancient deposits of Glacial Lake Cahuilla, which once filled the Salton Trough, give the history of past motions of this stretch of the San Andreas fault. Almost 1,200 years are recorded in the trenches, and the data from the dated breaks in the layers in the trench wall give a recurrence interval of about 180 years.

Many other places in the Imperial Valley show signs of offset, especially after earthquakes. For example, the orange trees in Figure 11.9c were originally planted in straight rows. They were offset during the May 1940 quake.

11.4.2 COACHELLA VALLEY AREA TO CAJON PASS

From the Salton Sea north, the fault curves to the northeast of Palm Springs, Palm Desert, Indio, and the other cities of the Coachella Valley. It cuts off the western edge of the Orocopia Mountains and the Mecca Hills. In the Mecca Hills especially, the relatively young late Miocene Mecca Hills Formation and the Pliocene–Pleistocene

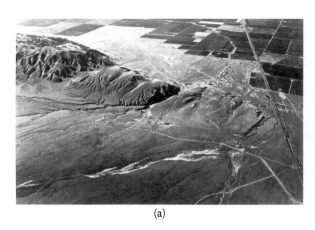
(a)

(b)

FIGURE 11.8 Pressure ridges form when crust on each side of a fault is buckled up into folds and anticlines. (a) Wheeler Anticline, a fold in the southern rim of the San Joaquin Basin just west of Interstate 5, and just north of the San Andreas in the San Emigdio Mountains to the south. It has buckled upward so rapidly that several small water gaps and wind gaps have carved down through its crest as it rose. (b) Roadcuts in Highway 14 just south of Palmdale, where the Pliocene Anaverde Formation was formed by lake beds on the San Andreas only 3 Ma, then intensely folded and crumpled.

Source: (a) Photo by J. Shelton. (b) Photo by G. Hayes.

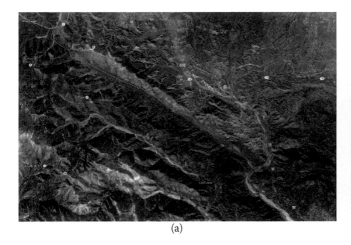

(a)

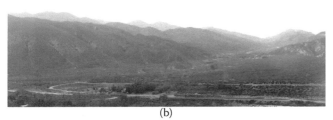

(b)

(c)

(d)

FIGURE 11.12 (a) Satellite image of the San Andreas fault from Cajon Pass to Wrightwood. The fault follows the trace of Lone Pine Canyon Road. (b) View west from Interstate 15, looking up the trace of the San Andreas fault in Lone Pine Creek. Beyond the notch in the distance is Wrightwood. (c) The Blue Cut with the San Andreas Valley on the extreme left. The left-most outcrop is Pelona Schist, with dark-green San Francisquito Formation and light tan Cajon Pass Formation on the right. (d) Outcrops of the early–middle Miocene Cajon Formation at Mormon Rocks in the Cajon Pass. Although these rocks were once thought to be the offset match for the Punchbowl Formation on the southwest side of the fault (see Figure 4.4), in fact their fossil mammals and the mismatch of their rock types show they are much older.

Source: (a) Courtesy of US Geological Survey. (b–d) Photos by the author.

FIGURE 11.13 Scar on the hillside where the Wrightwood debris flow began.

Source: Photo by the author.

studies were done that dated the recurrence interval of this strand of the San Andreas (Figure 11.6, discussed earlier).

From Devil's Punchbowl, Highway N4 (Fort Tejon Road) runs parallel and just north of the San Andreas fault valley through Littlerock. Barrel Springs Road then follows the fault valley to the northwest. The fault valley then continues up to Palmdale, where it crosses Highway 14, the Antelope Valley Freeway. Here you can see the intense folding in the pressure ridge just north of the fault valley (Figure 11.8b). Lake Palmdale is a natural sag pond that has been enhanced by a dam at the mouth of the creek. You can also see the California Aqueduct bringing water from the northern Sierras (Chapter 18), which crosses the fault and the highway here. Since the aqueduct must cross the fault somewhere, it does so here, where it can be monitored. The canal is surprisingly deep and runs very fast—cars have plunged into it and vanished without being found for years. It also has a series of seismically triggered gates along this stretch, so when the inevitable earthquake happens again here, only one segment of the aqueduct will break and leak its water. The rest of the water would be held back by closed gates until repairs are made.

11.5.1 PALMDALE TO CARRIZO PLAIN

Going west out of Palmdale, the San Andreas Valley trends northwest until it joins Elizabeth Lake Road, where it passes through Leona Valley, and Elizabeth Lake, before it changes to Pine Canyon Road as it passes through Lake Hughes. Many of these lakes were also sag ponds on the fault valley that have been enlarged by dams. The fault follows Pine Canyon Road as it continues through Three Points and Quail Lake, another reservoir that started as a sag pond. Here it joins Interstate 5, and the interstate follows the fault valley all the way from here to Gorman to Frazier Park. At this point, the interstate plunges down the canyon called "the Grapevine" past historic old Fort Tejon (the nearest early settlement and thus source of the name of the Fort Tejon quake of 1857), and then down to the San Joaquin Valley at the bottom, and the pressure ridges of Wheeler Anticline (Figure 11.8a).

Meanwhile, the San Andreas fault follows the central valley of the little ski resort town of Frazier Park and Lake of the Woods. Here it switches to the Cuddy Valley Road up to the resort community of the Pine Mountain Club. To the north are the San Emigdio Mountains, some of the highest (2,282 m, or 7,487 ft), most rugged, and remote mountains anywhere near Los Angeles. They rest on the North American plate side of the San Andreas fault, and their north slope goes down to the San Joaquin Valley, so they are a major tectonic boundary as well as a boundary between physiographic provinces.

The geology of these mountains is very interesting as well. They are built of a 5,000 m (15,000 ft) thick sequence of Eocene–Miocene sedimentary rocks sitting on granitic basement rocks from the Sierra Nevada arc and Tehachapi Mountains to the northeast. The sequence is exposed on the west wall of the Grapevine Canyon and also up the mountaintops to the west. They include 1,220 m (4,000 ft) of the Eocene Tejon Formation (basis for the Eocene Tejon stage in California), then 270 m (800 ft) of the Eocene–Oligocene San Emigdio Formation and 1,000 m (3,300 ft) of the Oligocene Pleito Formation, followed by 2,400 m (7,500 ft) of the late Oligocene–Miocene shallow marine Temblor Formation. These rocks are also intensely folded by the pressure along the San Andreas fault just to the south, so they show tremendous shortening due to this compression. They represent one of the best Eocene–Oligocene sequences in California, but they are difficult to reach in most places with limited road access, plus the steep cliffs and steep mountain valleys and peaks.

The fault continues to the northwest, where stretches of Mil Potrero Highway and Hudson Ranch Road follow its valley and eventually meet with Highway 33 from Ventura. At this point, the fault travels through miles and miles of private ranchland, with only limited private dirt road access, until you reach the Carrizo Plain.

11.5.2 CARRIZO PLAIN

To the northeast of the Carrizo Plain, on the North American side of the fault, is the Temblor Range. It gets its name from the Spanish word for earthquake, *temblor*. (Often you'll hear TV reporters making a weird amalgam of two words and calling an earthquake a "tremblor.") It is composed of the same Cenozoic rocks typical of the entire San Joaquin Basin (Chapter 10), but here they are tilted as they come up from the subsurface and are compressed and folded by the pressure of the nearby San Andreas fault system. The Temblor Range has some of the best surface exposures of classic type sections of certain biostratigraphic stages.

As the fault enters the Carrizo Plain National Monument, you are traveling along the most photographed stretch of the entire San Andreas. Almost every time you see aerial footage flying over the huge scar of the fault line (Figure 11.14a), it is the Carrizo Plain. One of the best-studied places is Wallace Creek.

Wallace Creek (Figure 11.14b) is a legendary place in the annals of geology and especially seismology. Along with Pallett Creek near Devil's Punchbowl south of Palmdale, it is a place where you can not only see the fault very well but also see its effects on the landscape. By digging trenches, paleoseismologists have found out about prehistoric earthquakes along that stretch of the San Andreas fault. At Wallace Creek, you can hike up to the top of the main overlook and look back to the northwest up the trail. There you can see how Wallace Creek flows straight west out of the Temblor foothills, then does a sharp right turn, and then turn left so its course resumes flowing west (Figure 11.4c). At one time, this old creek bed used to flow straight across the fault line, but as the fault has moved, it caused the lower part of the creek bed to creep farther and farther north until the entire creek did this dramatic dogleg, with the middle offset segment sitting right on the San Andreas fault.

This huge offset of about 500 m (1,600 ft) is the product of numerous earthquakes on the San Andreas over the past few thousand years, including 10 m (33 ft) caused by the Fort Tejon earthquake of 1857. Farther to the southeast is a much older stream drainage that is offset by 20 m (65 ft), which all occurred in seconds during the Fort Tejon quake. Digging trenches for paleoseismology, Kerry Sieh has analyzed these ancient earthquakes and dated the slip on different offset streams. He calculated that the fault moved on average about 33.9 ± 2.9 mm/year over the past 3,700 years, and over the longer span of 13,250 years, its average slip rate was 35.8 ± 5.4 mm/year. Of course, these are averages. In reality, the fault appears to be dormant here between 250 and 450 years between quakes, and then it moves at least 10 m each time. By contrast, Pallett Creek appeared to have moved more often, with only 130–150 years between major quakes. Thus, the fault sticks along its length in zones like this that are locked and overdue, and it slips different amounts and in different segments at different times.

If you look to the southwest from Wallace Creek, you can see the Caliente Range (Figure 11.14c), which has an amazing sequence of rocks ranging from the Paleocene Pattiway Formation at the bottom, capped by Oligocene red beds of the Simmler Formation, and then a thick Miocene sequence of marine Vaqueros Sandstone interbedded with the nonmarine, mammal-bearing Caliente Formation. Magnetic stratigraphy (Prothero, 2001) dated the Pattiway Formation between 58 million and 60 million years in age, and the Soda Lake Shale Member of the Vaqueros Formation at 22–23 million years in age.

11.5.3 CARRIZO PLAIN TO PARKFIELD

From Wallace Creek, the San Andreas forms a series of gentle valleys, often covered by grasslands and stands of California live oaks, so typical of the central Coast Ranges. It runs up the Elkhorn Road in the Carrizo Plain, and then across private ranchland near the Carrisa Highway to the south. For long stretches, it is not accessible by public road until you reach Whale Rock Reservoir Road, which runs parallel to it and to the south of the fault valley. Then it runs parallel

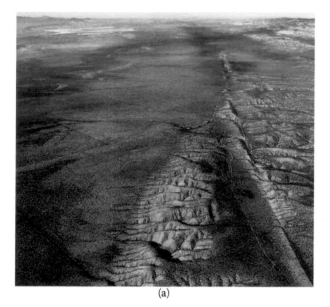

(a)

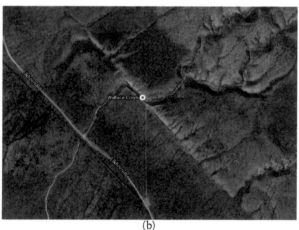

(b)

(c)

FIGURE 11.14 Carrizo Plain: (a) Aerial view showing the dramatic scar of the San Andreas fault valley. (b) Satellite view of the offset drainage at Wallace Creek. (c) The dogleg of Wallace Creek from ground level, looking north. Wallace Creek drains down from the hill off to the right of the photo, then does a right-angle jog (trench in the middle of the shot), and then turns left by 90° and heads out into the Carrizo Plain. The Caliente Range is in the background.

Source: (a) Courtesy of Wikimedia Commons. (b) Courtesy of US Geological Survey. (c) Photo by the author.

to Soda Lake Road, followed by Bitterwater Road for many kilometers, until it finally reaches Highway 46 and Cholame Creek (near the spot where iconic 1950s actor James Dean died in a car crash). From there it follows Cholame Road until it reaches the town of Parkfield, California.

Virtually every seismologist and geologist in California has heard of Parkfield, although only few have made a pilgrimage to this remote little town of only 18 residents. There are hundreds of scientific instruments at this site, monitored by the USGS in Menlo Park via satellites and Internet links. Parkfield is in a stretch of the fault that is not locked but seems to move at fairly regular intervals. Since this is where the most consistent fault activity on the San Andreas occurs, it is the site of many experiments attempting to catch an earthquake in action. Moderate earthquakes (magnitude about 6) have occurred on the Parkfield section of the San Andreas fault at fairly regular intervals: 1857, 1881, 1901, 1922, 1934, and 1966 (Figure 11.15).

The 1857 Fort Tejon earthquake was apparently started by a fault rupture in the central Coast Ranges in Parkfield, which then propagated southward. It appeared to move every 22 years on average since 1857 (Figure 11.15a and b). Based on the last known earthquake in Parkfield in 1966, seismologists predicted that it would move again in 1988. In 1985, they invested huge amounts of time and money and energy installing every instrument and experiment they could imagine and making Parkfield the most closely watched seismic zone in the world. The year 1988 came and went, and the years continued to tick by, but with no earthquake. Eventually, many of the experiments were shut down; the rest are remotely monitored by the USGS in Menlo Park near Stanford. Finally, on September 28, 2004, the long-awaited quake finally came, only 16 years late—and it did not trigger any movement of the old Fort Tejon segment of the San Andreas. Although the experiments produced excellent results, the unpredictability of this "regular" segment of the San Andreas fault severely chastened seismologists, who were hoping for earthquake prediction.

Nevertheless, Parkfield continues to interest seismologists. In June 2004, scientists began an NSF-sponsored drilling project (Figure 11.15c) right into the San Andreas known as the San Andreas fault Observatory at Depth (SAFOD). For three years, the drill hole got deeper and deeper, until they stopped in December 2007, after having drilled to a depth of more than 10,000 ft. The scientific results of SAFOD are still being analyzed, although it proved to be a very productive experiment.

11.5.4 Parkfield to San Francisco

From Parkfield, the fault trends up mostly private ranchland along Slacks Canyon Road, then Peach Tree Road past Highway 198, and then further northwest on Peach Tree Road (which also becomes Airline Highway). At Lonoak it crosses Lewis Creek, then follows Airline Highway (Highway 25) to the northwest, and eventually reaches the town of Hollister, which has experienced many small quakes over the years.

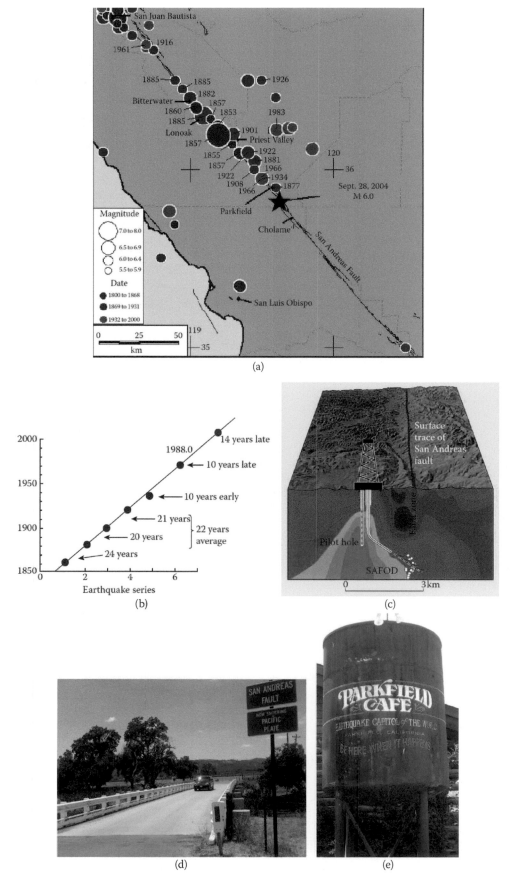

FIGURE 11.15 Parkfield: (a) Location of recent quakes near Parkfield. (b) Plot of the recurrence interval of quakes at Parkfield. (c) Diagram of the SAFOD experiment. (d) Bridge over the San Andreas fault at Cholame Creek in Parkfield. Each side is marked by a sign indicating which plate you are crossing into, so the other side has a sign indicating that you are entering the North American plate. (e) The tiny town of Parkfield has whimsical slogans on several of its structures, including this iconic water tower.

Source: (a) Courtesy of California Division of Mines and Geology. (b) Redrawn from several sources by E. T. Prothero. (c) Courtesy of US Geological Survey. (d–e) Photos by the author.

| (a) | (b) |

FIGURE 11.16 Frequent slow creep along the fault in Hollister has created numerous offsets and broken buildings and pavements: (a) Cracked and twisted house in Hollister. (b) Offset culvert near the DeRose Winery.

Source: Photos by G. Hayes.

Hollister's small quakes show constant creep in the town, which can be seen in broken pavement and even buildings being torn apart by each small temblor (Figure 11.16).

At Gilroy ("Garlic Capital of the World"), the San Andreas and Highway 25 merge with US Highway 101, which passes Mission San Juan Bautista (damaged by many of the quakes on the fault here), and then Highway 101 travels up to the Bay Area. But the San Andreas cuts across the crest of the Santa Cruz Mountains (site of the 1989 Loma Prieta quake already discussed) and then plunges back down at Los Gatos to form the foothills of the mountains south of all the Silicon Valley towns in San Mateo County. It passes through Portola Valley and runs beneath parts of Interstate 280. Then it travels beneath Crystal Springs Reservoir and Lake San Andreas (both are dammed sag ponds) near San Bruno; Lake San Andreas gave its name to the fault (Figure 11.17). Finally, it cuts to the northwest right across Daly City, passes through the San Francisco Zoo, before plunging off into the ocean just at the southern boundary of the city limits of San Francisco.

There is one additional place where the San Andreas fault cuts across land again, the Gualala block, starting at Point Reyes north of San Francisco. It is discussed in Chapter 12.

11.6 BACKTRACKING THE PLATE BOUNDARY

Paleoseismology can find evidence of ancient earthquakes along the transform fault boundary that happened a few thousand years ago and see evidence of movements along the fault of less than a kilometer. But the transform plate boundary has been active for more than 30 m.y. (see Chapter 15). Can we reconstruct the past movements of the plate boundary and how far the Pacific plate has moved relative to the North American plate over the geologic past? Such a restoration of rocks to their original position is called a **palinspastic** reconstruction in geology.

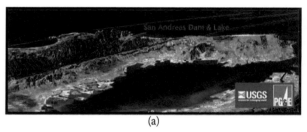

(a)

(b)

FIGURE 11.17 Lake San Andreas: (a) Diagram showing the trace of the San Andreas across the San Francisco Peninsula. (b) Aerial photo of Lake San Andreas and Lower Crystal Springs Reservoir. The fault valley runs down the center of these lakes.

Source: Courtesy of US Geological Survey.

The best way to do this is to find matching rock units that once straddled the fault zone but are now offset by long distances. These displaced links between two sides of the fault are sometimes called **piercing points**. If you know their age, you can calculate the rate at which the plate has moved since those rocks were formed.

Take for example the spectacular outcrops at Pinnacles National Park, in the spine of the central Coast Ranges

(Figure 11.18a and b). These deeply eroded volcanic rocks erupted 23 Ma, but they are only on the Pacific plate.

They are chopped off on the northeast side by the San Andreas fault, and there is no match across on the other side. However, if you go 313 km (195 mi) to the south on the San Andreas fault, you will find the other half of these rocks. This matching unit, known as the Neenach volcanics, lies in the western end of the Mojave Desert (Figure 11.18a). They are the same age and same rock type. Thus, these units were once the same volcanic event, chopped in half by the San Andreas, and moved 313 km (195 mi) in 23 Ma, or 13.6 mm/year (a bit slower than typical plate motions on other boundaries).

There are many other piercing points that give us the rate of fault movement further back in the geologic past (Figure 11.19). For example, the Gualala block (see Chapter 12) north of Point Reyes has a comparable set of outcrops down in the Transverse Ranges. The tropical reef clams known as rudistids in the Gualala block are far to the north of where they would have lived naturally. This shows that this block has moved more than 514 km (320 mi) in about 70 Ma (7.3 mm/year).

The Eocene Butano Sandstone north of Santa Cruz has its counterpart on the other side of the fault in the Point of Rocks Sandstone in the Temblor Range. This gives about 354 km (220 mi) of translation in about 40 Ma, or about 8.8 mm/year. A similar offset applies to the match between the Orocopia Schist east of the Salton Sea and the Pelona Schist in the San Gabriel Mountains (Chapter 13). The Paleocene rocks of the La Panza Range in the central Coast Ranges have a match with the San Francisquito Formation in the Transverse Ranges.

Thus, we can backtrack the Pacific plate side of the San Andreas back in time at least as far as the Cretaceous with confidence. If the contact between the Sierran basement on the east side and the end of the Sierran basement north of the Gualala block is the same piercing point, that gives 560 km (350 mi) since the Jurassic (about 150 Ma), or 3.7 mm/year. It also implies that almost the entire block west of the San Andreas was at least as far south as Southern California in the Jurassic, and most of what is now coastal California would have been south of the Mexican border.

Such huge offsets have staggering implications. One problem is that plate tectonic models (Chapter 15) indicate that the modern San Andreas didn't form until about 30 Ma at the earliest, so all the earlier offset must have been caused by shear from some other source. The source of that transport is still controversial, although some geologists have proposed a "proto–San Andreas" to explain it. In recent years, geologists have come to realize that only the northern portion of the San Andreas needs to have displacements prior to 30 Ma. The southern portion has no piercing points offset by that much and probably didn't open until after 30 Ma. More recently, a number of geologists have argued that the Salinian block (see Chapter 12) is the only portion of the fault zone that shows pre–30 Ma displacements. Some geologists identify the San Gregorio–Hosgri fault system as the proto–San Andreas that caused the huge displacements of Cretaceous through Eocene rocks, while the post–30 Ma San Andreas eventually took over that job

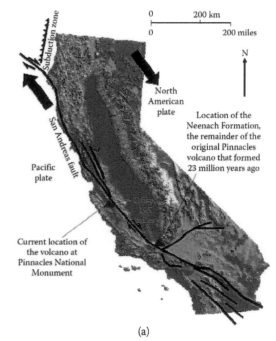

(a)

(b)

FIGURE 11.18 (a) The Pinnacles–Neenach piercing point. The two volcanic units erupted in the southern Mojave, where the Neenach volcanics are located about 23 Ma, and then the Pinnacles volcanics were carried north along the San Andreas since then. (b) Photograph of Pinnacles National Monument.

Source: (a) Courtesy of US Geological Survey. (b) Courtesy of Wikimedia Commons.

and ruptured all the way to the southern part of California. These topics will be discussed further in Chapter 15.

11.7 SUMMARY

California is legendary for its earthquakes, and a lot has been learned about seismology by studying earthquakes and faults in California. In particular, they have a set of distinctive surface features, and by trenching into them and dating the layers, paleoseismologists can study their prehistoric movements. The San Andreas fault is the largest and most important fault in the state, since it is a transform fault

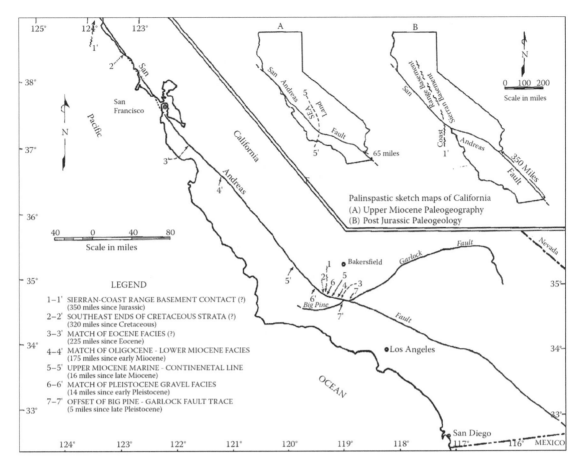

FIGURE 11.19 Piercing points that restore the original position of the blocks west of the San Andreas. Point 1 is the offset of Jurassic rocks in the Gualala block, which have moved 350 mi since 150 Ma (or 2 mi/m.y.). Point 2 is the contact with the Cretaceous basement (which has moved 320 mi since 65 Ma, or 5 mi/m.y.) on both sides of the fault. Point 3 is the match of the Eocene Butano Sandstone on the west with the Point of Rocks Sandstone on the east side of the fault (225 mi since 37 Ma, or 6 mi/m.y.). Point 4 is the match of Oligocene–lower Miocene rocks across the fault (175 mi since 20 Ma, or 8.75 mi/m.y.). Point 5 is the late Miocene match (65 mi since 8 Ma, or 9 mi/m.y.). Point 6 is the offset of Pleistocene gravels (14 mi since 2 Ma, or 7 mi/m.y.). Point 7 is the offset of the young Big Pine and Garlock faults (5 mi in 100,000 years). In the inset diagram in the upper right is a cartoon showing the magnitude of the offset of the western block (if treated as one coherent unit) since the late Miocene and Late Jurassic. In reality, the western block is not a single coherent terrane but a composite of many smaller blocks. For example, much of the Mesozoic offset (points 1 and 2) may have taken place on the San Gregorio–Hosgri fault, since the modern San Andreas did not develop until about 30 Ma (see Chapter 15).

Source: From California Division of Mines and Geology Bulletin, 170, 1954.

that marks the boundary between the North American and Pacific plates. A tour of the San Andreas along most of its length allows the geologist or seismologist to witness much evidence of how and when it has moved.

RESOURCES

Collier, M. 1999. *A Land in Motion: California's San Andreas Fault.* University of California Press, Berkeley.

Dvorak, J. 2014. *Earthquake Storms: The Fascinating History and Volatile Future of the San Andreas Fault.* Pegasus, New York.

Hough, S.E. 2004. *Finding Fault in California: An Earthquake Tourist's Guide.* Mountain Press, Missoula, MT.

Iacopi, R. 1971. *Earthquake Country: How, Why, and Where Earthquakes Strike in California.* Sunset Publishing, Oakland, CA.

Lynch, D.K. 2015. *The Field Guide to the San Andreas Fault.* Sunbelt Publications, El Cajon, CA.

Powell, R.E., Weldon, R.J., II, and Matti, J.C. 1993. *San Andreas Fault System: Displacement, Palinspastic Reconstruction, and Geologic Evolution.* Geological Society of America Memoir 178. Geological Society of America, Boulder, CO.

Prothero, D.R. 2001. *Magnetic Stratigraphy of the Pacific Coast Cenozoic.* Pacific Section SEPM Special Publication, 91, 394. https://doi.org/10.1669/0883-1351(2002)017<0527:BR>2.0.CO;2

Winchester, S. 2006. *A Crack in the Edge of the World: America and the Great California Earthquake of 1906.* Harper Perennial, New York.

Yeats, R.S., Sieh, K.E., and Allen, C.R. 1997. *Geology of Earthquakes.* Oxford University Press, Oxford.

VIDEOS

http://video.nationalgeographic.com/video/101-videos/earthquake-101

www.youtube.com/watch?v=vbLjzZg79ZU

www.youtube.com/watch?v=3owcJB0x6m0

www.youtube.com/watch?v=JfbNWFBr5g8

12 Mélanges, Granitics, and Ophiolites
The Coast Ranges

The Franciscan mélange contains rock of such widespread provenance that it is quite literally a collection from the entire Pacific basin, or even half of the surface of the planet. As fossils and paleomagnetism indicate, there are sediments from continents (sandstones and so forth) and rocks from scattered marine sources (cherts, graywackes, serpentines, gabbros, pillow lavas, and other volcanics) assembled at random in the matrix clay. Caught between the plates in the subduction zone, many of these things were taken down sixty-five thousand to a hundred thousand feet and spit back up as blueschist. This dense, heavy blue-gray rock, characteristic of subduction zones wherever found, is raspberried with garnets.

—John McPhee
Assembling California

This morning I saw a coyote walking through the sagebrush right at the very edge of the ocean—next stop China. The coyote was acting like he was in New Mexico or Wyoming, except that there were whales passing below. That's what this country does for you. Come down to Big Sur and let your soul have some room to get outside its marrow.

—Richard Brautigan
A Confederate General From Big Sur

12.1 DATA

Geography: Coast Range Mountains (from north to south: Yolla Bolly, King, Diablo, Santa Cruz, Gabilan, Santa Lucia, Sierra Madre, and San Rafael)

Elevation: Solomon Peak, Trinity County: 2,312 m (7,583 ft); Mt. Tamalpais, Marin County: 784 m (2,571 ft); Cone Peak, Monterey County: 1,572 m (5,155 ft); Mt. Linn, Yolla Bolly Mountains: 2,468 m (8,096 ft)

Major watercourses: Klamath River, Eel River, Russian River, Salinas River, Humboldt Bay, San Francisco Bay

Predominant rocks: Mesozoic Franciscan Complex, Cenozoic sedimentary basins

Plate tectonic setting: Accretionary wedge complex of the Sierra Nevada arc

Geologic resources: Water, sand, gravel; mercury; natural gas and oil

National parks: Redwood National Park, Point Reyes National Seashore, Golden Gate National Recreation Area

State parks: Jedediah Smith Redwoods, Prairie Creek, Humboldt Redwoods, Big Sur, Guadalupe Dunes

12.2 GEOGRAPHY OF THE COAST RANGES

The Coast Ranges are not impressively high mountain ranges, like the Cascades or the Sierra Nevada mountains. In most cases, they form low rolling hills or smaller mountains that rise from the Pacific Ocean. Only a few are more than a mile in elevation. Most are about 1,000 m (3,300 ft) or lower. The entire Coast Ranges extend over 640 km (400 mi) from the Oregon border to the north edge of the Transverse Ranges (Figure 12.1). The width of the range in the east–west direction varies from place to place, but typically it is about 160 km (100 mi) wide from the Pacific Ocean to the Great Valley. The Coast Ranges merge with the Klamaths at the north side and are bounded by the Central Valley on the east and the Transverse Ranges to the south.

The Coast Ranges have a strong northwest–southeast orientation because they are products of the San Andreas transform boundary, plus all the faults and related structures parallel to the San Andreas (see Chapter 11). Many of the valleys are carved along fault lines and cut through mountains thanks to faulting. Much of the bedrock is a wide variety of different accreted blocks of different ages and different rocks types, so there is very little consistency of landscape or geology from one range to the next.

Their ecology and climate are just as variable as their geology. In the northern Coast Ranges, lots of moisture blows off the Pacific Ocean, and the climate tends to be wetter, producing lush vegetation. This is the realm of the California Coast Redwoods, the tallest trees in the world. The slightly drier inner valleys are cloaked in grasslands and big stands of California live oaks. In Napa, Lake, Mendocino, and Sonoma Counties, the microclimate of places like the Napa Valley produces some of the best wine grapes in the world. The central Coast Ranges are famous for the rugged coastline at Big Sur, with huge redwood groves just inland. As you move to the southern Coast Ranges, the climate becomes drier. The coastal area gets the most rain, and fog is cloaked in coastal sage scrub, which gets much of its moisture from fog since the rains are less frequent. Inland, the landscapes of the southern Coast Ranges are covered in drier sagebrush and chaparral plants,

DOI: 10.1201/9781003301837-12

FIGURE 12.4 Outcrops of the pinkish-gray Laytonville Limestone, north of Laytonville, on Highway 101. Paleomagnetic data show they come from the Southern Hemisphere near the present latitude of Peru in the Early Cretaceous.

Source: Photo by the author.

Many of the cherts in these belts have radiolarian fossils (Hagstrum and Murchey, 1993) that indicate they came from exotic terranes that were once in the tropical Pacific (see Figure 9.6b). In addition, there are isolated limestone blocks in these terranes that have Permian fossils from the southern tropics in the Tethys Seaway, across the Pacific over 250 Ma (just like the McCloud Limestone in the Klamaths—Chapter 9). Another one of these is the Laytonville Limestone (Figure 12.4). Not only does it have Early Cretaceous (101–88 Ma) foraminiferans from the tropical regions, but it has also been analyzed by at least two independent paleomagnetic studies (Alvarez et al., 1980; Tarduno et al., 1986). All the paleomagnetic data place the Laytonville Limestone in the Southern Hemisphere in the Cretaceous, at a latitude of about 14°–17° ± 7° south (about the modern-day latitude of Peru). This is a long distance (over 50° of latitude!) to travel north in only 30 m.y. and implies a very fast rate of seafloor spreading to get this Southern Hemisphere limestone block all the way to Northern California by no later than the Eocene (50 Ma) and incorporated into an accretionary wedge in only about 30 m.y. The implications of this extremely fast plate motion of the Laytonville block are still debated, but there is no doubting its original latitude since it is based on both the paleomagnetic data and the Southern Hemisphere fossils.

We can't describe every part of this large and complex region in detail, so we will look at two areas that show the full spectrum of geological phenomena typical of the Central Franciscan belt: Mt. Diablo and San Francisco.

12.6 MT. DIABLO

Mt. Diablo (Figures 12.3 and 12.5a and b) is in the Coast Ranges just east of San Francisco Bay and the Berkeley

Hills. It towers 1,173 m (3,849 ft) above the landscape, and on a clear day, you can see the Sierras from the top. The forty-niners traveling between San Francisco and the Mother Lode used it in navigation. In addition, on a very clear day, you can see the mountain from the high spots in the eastern part of the Bay Area. It got its name not because people thought it was devilish but because some Chupcan prisoners escaped from the Spanish in the willow thickets to the north, so the Spaniards called the thicket *Monte del Diablo* ("thicket of the devil"), and then the name was later applied to the peak to the south.

Mt. Diablo shows a complete spectrum of Mesozoic and Cenozoic rocks, all buckled upward into a faulted and slightly recumbent dome-like structure, with the oldest rocks in the north and center of the dome, and younger rocks dipping away on the flanks to the south. The north flank of the uplift exposes Jurassic Coast Range ophiolite (sometimes called the Mt. Diablo ophiolite), and pillow lavas are widely exposed on the summit and on its north flanks (Figure 12.5a and c). In other places, there are large areas of serpentinite. Structurally, beneath this is a thick package of Franciscan mélange, intensely folded and faulted. As you drive up the roads from the North Gate, you encounter many outcrops of deformed and folded ribbon cherts (Figure 12.5a and d) and classic shredded mélange (Figure 12.5e). On the east flank of the range, the Cretaceous Great Valley Sequence is more than 20,000 m (60,000 ft) thick, although it thins rapidly toward Mt. Diablo.

Overlying these Mesozoic basement rocks are exposures of the Paleocene Martinez Formation (mostly to the north in Concord and Martinez), and the early Eocene near-shore Domengine Sandstone (Figure 12.5f). This unit is very fossiliferous and forms the rugged outcrops of Rock City on the south slope of the summit. On the north side of the range, the purity of the white quartz sandstones of the Domengine was an important source of silica for glass. The old Black Diamond Mine near Antioch, California, is now an important historic landmark. In Nortonville and four other nearby ghost towns, the Domengine Formation also had small seams of coal, which formed in Eocene coastal swamps. Even though these low-grade coals were not especially good for furnaces, they were mined from the time of the Gold Rush until 1906, producing 4 million tons of coal, by far the largest coal fields in California.

The Eocene rocks were capped by the upper Oligocene Kirker Tuff, and then on the south flank of the range (near Danville and San Ramon) along Blackhawk Ridge and Fossil Ridge is a thick sequence of middle–upper Miocene to Pliocene rocks, famous for their fossils. The lowest units are the middle Miocene marine sandstones and shales of the Briones, Neroly, and San Ramon Formations (12.5–10 Ma). These rocks interfinger with the middle to upper Miocene Sycamore Formation, which yields the legendary Blackhawk Ranch fauna, the type of fauna for the Montediablan stage (see Chapter 20). Based on magnetic stratigraphy (Prothero and Tedford, 2000), it dates between 8 and 9 Ma. Overlying the Sycamore Formation in the Bay

FIGURE 12.5 Mt. Diablo: (a) Map of the geology. (b) View of the peaks from the northwest showing the multiple summits. (c) Outcrop of pillow lavas from the Jurassic Coast Range ophiolite, near the top of the summit road. (d) Outcrop of folded ribbon cherts from the Cretaceous Franciscan Complex on the north flank of the range, along the North Gate Road. (e) Typical outcrop of highly sheared and deformed Franciscan mélange, North Gate Road. (f) Close-up of the fossiliferous Eocene Domengine Sandstone, here full of scallop shells, which overlies the Mesozoic rocks.

Source: Photos (a) and (f) from Wikimedia Commons; all other photos by the author.

Area and on the edge of Mt. Diablo is the Miocene–Pliocene Pinole Formation, with tuffs dated at 5.3–5.5 Ma, and the Pliocene Tassajara Formation, with the Lawlor tuffs dated 4.1–4.6 Ma. The fact that these tuffs were laid across the area in a flat sheet shows that the uplift of the Berkeley Hills and Mt. Diablo is very recent (late Pliocene at the earliest).

12.7   SAN FRANCISCO BAY AREA

The "All-American City by the Bay" is underlain by a combination of exotic terranes from many different places across the Pacific Ocean, all accreted together as part of the Franciscan mélange (Figure 12.6). The area south of Daly City in the San Francisco Peninsula is part of another tectonic block, the Salinian block, which will be discussed in the next section. But the rest of San Francisco is part of the Central Franciscan belt, and it displays the chaotic composite nature of all the bedrock in this region.

From east to west, the major terranes that make up San Francisco are as follows (Figure 12.6):

1. The **Yolla Bolly terrane**, already discussed, best known from the Yolla Bolly Mountains farther north. It makes up most of Angel Island and the southeastern part of the Tiburon Peninsula.

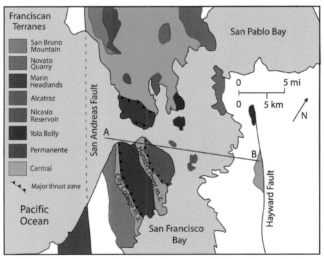

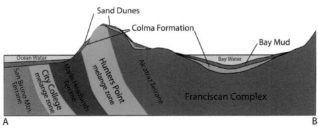

FIGURE 12.6 Simplified geologic cross section and geologic map of the terranes of the San Francisco Bay region. (Figure 12.7a). It has been overthrust on top of the Hunter's Point mélange zone.

Source: Modified from Blake, 1984; courtesy of W. Elder and US Geological Survey.

2. The **Alcatraz terrane**, which makes up not only Alcatraz Island but also Treasure Island (which forms the midway point of the Bay Bridge) and the entire northeast part of the city of San Francisco, especially the Presidio, North and South Beach, the Embarcadero, the Financial District, Nob Hill and Russian Hill, and the Marina District. It consists almost entirely of a thick sequence of deformed Lower Cretaceous (based on fossil mollusks dating between 130 and 140 Ma) turbidite sandstones and shales (Figure 12.7a). It has been overthrust on top of the Hunter's Point mélange zone.

3. The **Hunter's Point mélange zone**, which makes up another strip of central San Francisco, trending northwest to southeast from the south end of the Golden Gate Bridge, down through Haight-Ashbury and the Castro and Mission Districts, down to Potrero Hill and Hunter's Point on the east end (Figure 12.6). As the name implies, it is a thick sequence of Jurassic–Cretaceous mélange with abundant ophiolites and serpentine (Figure 12.7b).

4. The **Marin Headlands terrane** occupies the central strip of San Francisco, from Golden Gate Park down through the mountains at the center of the peninsula (Mt. Sutro, Twin Peaks, Mt. Davidson, Diamond Heights, and Bernal Heights) and down to Candlestick Hill, and east across the Bay into the Coyote Hills. As the name implies, it also occupies most of the Marin Headlands on the north end of Golden Gate Bridge. It consists largely of oceanic rocks, especially folded ribbon cherts (Figure 12.7c) and ophiolites (Figure 12.7d), both of which are easy to study on the Marin Headlands, and deep-water turbidite sandstones. The fossils date the rocks from Lower Jurassic to Upper Cretaceous, but the microfossils (especially the radiolarians) in the cherts show that the Marin Headlands are another terrane that came from the other side of the tropical southern Pacific in the Jurassic–Cretaceous.

5. The **City College mélange** occupies another strip to the southwest of the Marin Headlands terrane. It runs northwest to southeast in a narrow strip from Land's End, down to beneath San Francisco City College, and then east to Bayshore. Like the Hunter's Point mélange, it consists of highly shredded Jurassic–Cretaceous ophiolites, cherts, and deep-water shales and sandstones (Figure 12.7e).

6. The **San Bruno Mountain terrane** is the next block to the west of the previous units. It runs from Sunset District on the Pacific Coast in the northwest, down through Lake Merced, Ingleside, Oceanview, and San Bruno Mountain State Park, and down to Brisbane on the bay. Another slice of San Bruno Mountain terrane can be found up in Marin County up by Mt. Tamalpais. It is composed

FIGURE 12.7 Typical rock exposures representative of the different terranes of San Francisco. (a) Sandstones from the Alcatraz terrane just below the legendary prison on Alcatraz Island. (b) Serpentinites from a landslide north of the Presidio from the Hunter's Point mélange zone. (c) Crumpled ribbon cherts from the Marin Headlands terrane. (d) Pillow lavas from Point Bonita, also on the Marin Headlands terrane. (e) Crumpled ribbon cherts from Glen Canyon Park, part of the City College mélange zone. (f) Turbidite sandstones and shales at the summit of San Bruno Mountain, part of the San Bruno terrane.

Source: (a) Courtesy of Wikimedia Commons. (b, d) Photos by W. Elder. (c) Photo by the author. (e–f) Courtesy of Wikimedia Commons.

of a thick sequence of Upper Cretaceous turbidites and deep-water shales (Figure 12.7f).

7. Just to the east of Mt. Tamalpais is the **Nicasio Reservoir terrane**, which underlies Nicasio and its reservoir, down to Woodacre and San Geronimo.

8. Undifferentiated **Central Franciscan terrane** makes up most of the Marin County area north of the Headlands, along with the strip between the San Bruno Mountain terrane and the San Andreas fault down in Daly City.

9. In addition to the terranes that make up the San Francisco Peninsula, the **Novato Quarry terrane** occupies a strip along the western coast of Richmond and the southwestern shore of San Pablo Bay (especially McNear's Beach, Gallinas, and China Camp State Park, and parts of San Rafael and Novato). It is composed mostly of upper Cretaceous turbidite sandstones.

This brief summary only begins to scratch the surface of the complexity of the geology of the Bay Area. For more information, there is Clyde Wahrhaftig's excellent guidebook *A Streetcar to Subduction* (cited at the end of the chapter) and professional papers by Blake, Howells, Murchey, Wakabayashi, and others who have deciphered this complex puzzle.

12.8 CENOZOIC ROCKS OF THE EASTERN FRANCISCAN BELT

Most of the rocks that make up the Eastern Franciscan belt are Paleozoic and Mesozoic exotic terranes (such as the Laytonville Limestone and the Marin Headlands terrane), plus mélange and ophiolite, and the deep-water shales and turbidites of the Great Valley Group on the eastern edge of the Coast Ranges. However, there are a number of important Cenozoic beds, such as the extensive Paleocene through Miocene beds of the Temblor Range east of the Carrizo Plain, or the thick Cenozoic sequence surrounding Mt. Diablo (discussed earlier). Cenozoic sediments from the Great Valley also cover much of the older Mesozoic Franciscan basement rocks on the eastern edge of the Coast Ranges.

There are also important Cenozoic volcanics as well. Chief among these is the Clear Lake volcanic field (or Sonoma volcanic field), near Clear Lake in Lake and Sonoma Counties, the northern Coast Ranges. These very young eruptions of basalt, andesite, and rhyolite from multiple magma chambers began about 2.1 Ma, with numerous eruptions in the Pleistocene, most recently about 10,000 years ago. In the vicinity are abundant young cinder cones and flows, with bombs (Figures 6.3e) and pyroclastics around Clearlake Oaks. The biggest volcanoes are Mt. Konocti (1,306 m, or 4,285 ft) and Cobb Mountain (1,449 m, or 4,724 ft) (Figure 12.8a).

The volcanoes seem to be extinct, but hot magma is not far below, as proven by the numerous hot springs in the area (especially the Geysers, a historic hot spring resort, and now an important geothermal plant (Chapter 23); Figure 12.8b) and the famous Calistoga Hot Springs. In some places, the magmas have fossilized ancient tree trunks, such as at Calistoga Petrified Forest State Park (see Chapter 20).

The magmas of the Clear Lake volcanics intruded through Jurassic ophiolites in the area, so they mobilized rare elements and scavenged and concentrated them into veins. In addition, the hot groundwater in the area created hydrothermal deposits that dissolved rare elements, like mercury and sulfur, and concentrated them in the surface. Consequently, there are large mercury mines in the region, such as the Sulphur Bank mercury mine (Figure 12.8c), discovered in 1856, and one of the world's largest mercury deposits. From 1865 to 1869, it produced more than 900,000 kg (2 million pounds) of mercury, chiefly as mercury sulfide or the mineral cinnabar. The New Almaden mine down by San Jose was also an important deposit during the Gold Rush years. Mercury was important to Gold Rush miners, who used this deadly element to separate the gold from the worthless ore rock.

The mining district also produced huge deposits of sulfur from the young volcanoes, borax in the lakes nearby, and about 3.4 million ounces of gold (about $1 billion worth) from the McLaughlin mine (Figure 12.8d) from 1985 to 2002. Today, most of the historic mercury mines like Sulphur Bank and New Almaden are so full of toxic mercury that they are closed and are being cleaned up as part of the Superfund project for the most toxic waste sites in the United States.

The position of the Clear Lake volcanic field just a few dozen miles south of the Cascades (Mt. Shasta and Mt. Lassen) has suggested to some geologists that they are an extinct part of the once-longer Cascade chain. The same has been suggested about Sutter Buttes, which are east of the Clear Lake volcanoes and the same age. Although this is controversial, it fits with plate tectonic models that suggest that the Cascade arc has been shutting down from south to north as the Mendocino triple junction moves north (see Chapter 15).

12.9 SALINIAN BLOCK

Once you go west across the San Andreas fault in the Coast Ranges, you're in a completely different world that came up from far to the south. This terrane that is bounded by the San Andreas on the east and the Sur-Nacimiento fault on the west (Figure 12.9) is known as the **Salinian block** (named after the Salinas River and Salinas Valley, which is the heart of the block). It extends from the Transverse Ranges in the south (see Chapter 13) all the way up to the south part of San Francisco Peninsula, then the fault slice at Point Reyes, and finally a sliver of rocks in the Gualala block just south of Cape Mendocino.

For decades, geologists puzzled over the mismatch of the rocks across the fault, contrasting the ophiolites and mélange and Great Valley sedimentary package east of the fault with what appeared to be Sierran-style granites

FIGURE 12.8 Clear Lake–Sonoma volcanic field: (a) Mt. Konocti, overlooking Clear Lake. (b) Geysers geothermal field. (c) Sulphur Bank Mine on the shores of Clear Lake, source of mercury, borax, and sulfur. (d) McLaughlin gold mine.

Source: (a–b) Courtesy Wikimedia Commons. (c–d) Courtesy of US Geological Survey.

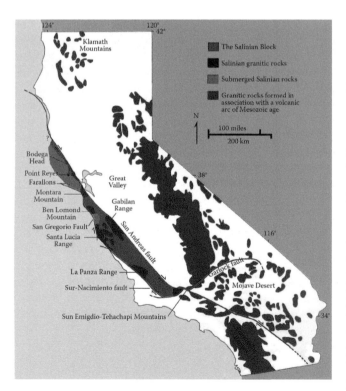

FIGURE 12.9 Map showing the location of the Salinian block. The Cretaceous granitic batholiths are indicated in red; the rest of the block is in green.

Source: Courtesy of US Geological Survey.

to the west. But as more and more piercing points (see Chapter 11) were found, it was clear that the rocks of the Salinian block had originated at least 200 km south of their present location, at the southern tip of the Sierra Nevada mountains in what is now Mexico and Southern California (Figure 12.10). The granitic rocks of the Salinian block are indeed the southern continuation of the Sierras that have moved north since 60 Ma. The granitic rocks that are currently in Southern California (such as the Peninsular Ranges batholith, discussed in Chapter 14) were once even farther south below the present-day Mexican border.

The most distinctive rocks of the Salinian block are the numerous intrusions of Cretaceous granitics that resemble the Sierra Nevada batholith (Figures 12.9 and 12.11). These include granitics in the La Panza Range, Santa Lucia Range, and Gabilan Range in central California; the Ben Lomond granitics north of Santa Cruz; the Montara Mountain granitics just south of San Francisco; and the granitic rocks on the edge of Point Reyes north of the Bay Area, and their northernmost occurrence at Bodega Head. Most of these rocks are tonalites, quartz monzonites, and granodiorites, just like the rocks of the Sierras; there are no true granites (with more than two-thirds of their total feldspar as potassium feldspar) as igneous petrologists define the term. These rocks exhibit many of the same features seen in the Sierras, including numerous xenoliths, dikes, and evidence of multiple magma chambers intruding one

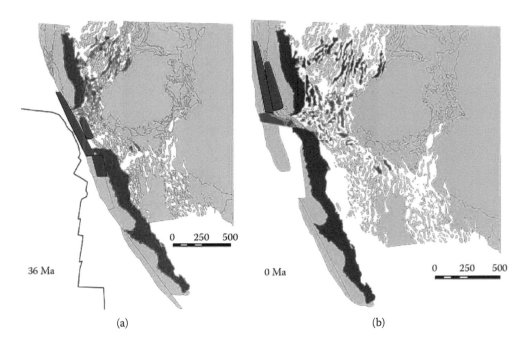

36 Ma

0 Ma

(a)   (b)

FIGURE 12.10   Reconstruction of the Salinian block in its original position south of the Sierras and then transported north to its present position: (a) At 36 Ma, the Salinian block (green) lies just adjacent to the Transverse Ranges block, which was then oriented in a north–south position, and filled the gap between the Sierran arc and the Peninsular Ranges arc (both in red). (b) Today, the Salinian block has moved north parallel to the Sierras, the Transverse Ranges rotated 90° clockwise, and the Peninsular Ranges have moved up behind them.

Source: Modified from McQuarrie and Wernicke (2005.)

after another. Indeed, if you didn't notice the difference in vegetation (many of these rocks are beneath coastal redwood forests), you would swear that you were looking at the Sierras when you hike in the La Panza Range, or the Santa Lucia Range, or the Gabilan Range, or the Ben Lomond or Montara Mountains.

12.9.1   LA PANZA RANGE

At the south end of the Salinian block is the Carrizo Plain and the Caliente Range (Figure 11.14), with a thick sequence of Paleocene through late Miocene sedimentary rocks exposed in the slopes of Caliente Mountain (see Chapter 11). Just north of the Caliente Range, you reach the first major batholith in the Salinian basement, the La Panza Range. It is about 50 km (30 mi) in length and due east of the Temblor Range (on the other side of the San Andreas fault) and due south of the Santa Lucia Range. The range itself (Figure 12.11a) is low and heavily vegetated with oak woodlands and coastal sage scrub and chaparral, so there are few dramatic domes of granitic rocks like the Sierras.

12.9.2   SANTA LUCIA RANGE

The Santa Lucia Range is by far the largest, highest, and most rugged of the Salinian ranges, forming much of the coast of Big Sur, the bedrock of Monterey and Carmel, and the mountains inland from the coast. It includes many peaks about a mile in elevation, including Junipero Serra

Peak (1,785 m, or 5,857 ft, in elevation), Ventana Double Cone Peak (1,480 m, or 4,856 ft), and Mt. Carmel (1,347 m, or 4,420 ft). Cone Peak (1,572 m, or 5,158 ft), which is the steepest coastal mountain in the lower 48 states, rises from the ocean to nearly a mile (1609 m) above sea level over only 5 km (3 mi) from the ocean. As a consequence of these high elevations and the closeness to the ocean, the Santa Lucias are heavily forested, with big groves of coast redwoods (their southernmost occurrence), Douglas fir, and Ponderosa pine on the western slopes facing the ocean, an area that receives as much as 150 cm (60 in) of rain in some years. The east side, however, is in a rain shadow and is vegetated by California live oaks and coastal sage scrub.

To the east of the Santa Lucia Range is the Salinas Valley, which is a deep sedimentary basin filled with mostly Cenozoic formations. These include the type areas of the nearshore marine lower Miocene Vaqueros Formation, the middle Miocene Reliz Canyon Formation, and a thick sequence of the middle–upper Miocene Monterey Formation. In a few places (such as San Ardo), the Salinas Basin produces significant oil and gas.

12.9.3   BEN LOMOND AND MONTARA MOUNTAINS

Ben Lomond Mountain forms the core of the Santa Cruz Mountains that separate the coast of Santa Cruz from the interior valley of San Juan Bautista and Gilroy. Most of it is heavily vegetated by dense stands of coast redwoods, so exposures and outcrops are very limited, largely restricted

(a)

(b)

(c)

(d)

FIGURE 12.11 Typical granitic rocks of the Salinian block: (a) La Panza Range granitics, Pozo Summit. (b) Granitic rocks of the eastern Santa Lucia Ranges. (c) Granitic rocks in the coastal cliffs of Garrapata State Park. (d) Quartz diorites of Bodega Head on Bodega Bay to the north of Point Reyes, part of the Salinian block.

Source: (a) Courtesy of US Forest Service. (b) Photo by S. Graham. (c) Photo by G. Hayes. (d) Photo by C. Heptig.

to fresh roadcuts, quarries, and stream bottoms during the dry season. Above the granitic rocks and especially in the La Honda Basin between Ben Lomond Mountain and Montara Mountain is a thick package of more than 12,000 m (40,000 ft) of Cenozoic sedimentary rocks. The package is highly discontinuous, with thin sequences of upper Paleocene Locatelli Formation, Eocene Butano Sandstone, Oligocene San Lorenzo Formation, lower Miocene Vaqueros Formation, and middle–upper Monterey Formation; most of these units are separated by large unconformities. The Locatelli Formation has been paleomagnetically dated at 61–58 Ma (Prothero, 2001) and shows paleomagnetic directions that indicate hundreds of kilometers of northward transport. The lower Eocene Butano Sandstone closely matches the Point of Rocks Sandstone down in the Temblor Range in the southern San Joaquin Basin, so it is one of the piercing points that suggest 354 km

(220 mi) of translation since 40 Ma (see Chapter 11). The middle Eocene–Oligocene San Lorenzo Formation and lower Miocene Vaqueros Formation discontinuously span the interval from 38 to 25 Ma, based on paleomagnetic dating (Prothero, 2001).

The Montara Mountain block (Figure 12.11c) rises to 579 m (1,898 ft) just west of San Mateo and many of the cities of the Silicon Valley area and continues down into the Santa Cruz Mountains. On the east, the block abuts the San Andreas fault, which runs right under Crystal Springs Reservoir and San Andreas Lake (see Chapter 11). On the west, the block extends from Half Moon Bay on the south through Moss Beach and north to Point San Pedro. The granitic rocks in Montara Mountain have been dated from 81 to 90 Ma. On the flanks of this block is Franciscan Complex mélange just east of the San Andreas fault in the San Mateo area, and on the west, tiny fragments of the Western Franciscan terrane

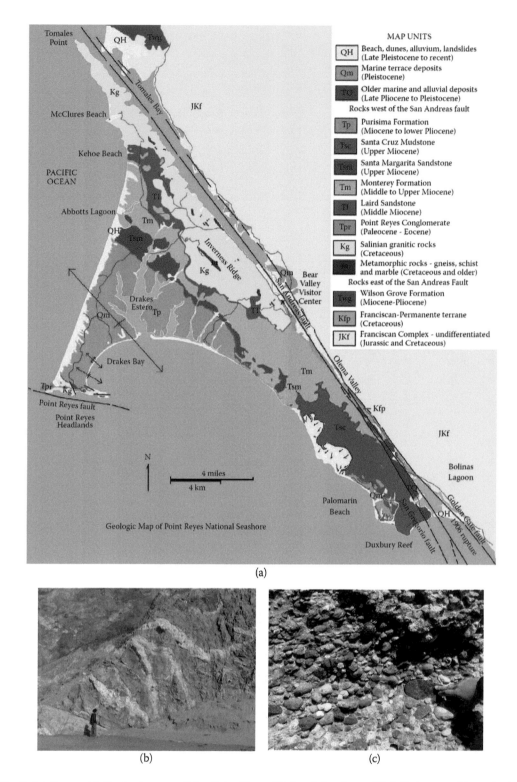

(a)

MAP UNITS

QH	Beach, dunes, alluvium, landslides (Late Pleistocene to recent)
Qm	Marine terrace deposits (Pleistocene)
TQ	Older marine and alluvial deposits (Late Pliocene to Pleistocene)

Rocks west of the San Andreas fault

Tp	Purisima Formation (Miocene to lower Pliocene)
Tsc	Santa Cruz Mudstone (Upper Miocene)
Tsm	Santa Margarita Sandstone (Upper Miocene)
Tm	Monterey Formation (Middle to Upper Miocene)
Tl	Laird Sandstone (Middle Miocene)
Tpr	Point Reyes Conglomerate (Paleocene - Eocene)
Kg	Salinian granitic rocks (Cretaceous)
	Metamorphic rocks - gneiss, schist and marble (Cretaceous and older)

Rocks east of the San Andreas Fault

Twg	Wilson Grove Formation (Miocene-Pliocene)
Kfp	Franciscan-Permanente terrane (Cretaceous)
JKf	Franciscan Complex - undifferentiated (Jurassic and Cretaceous)

(b)

(c)

FIGURE 12.12 Point Reyes: (a) Geologic map of the Point Reyes National Seashore. (b) Salinian granitic rocks in eastern Point Reyes, abutting the San Andreas fault, at Kehoe Beach. Most of the rock is granodiorite, with white aplite dikes cutting through it. (c) Close-up of the Point Reyes Conglomerate at Point Reyes, made of cobbles of eroded Salinian granodiorites. (d) Outcrops of Monterey Formation near Kehoe Beach access trail. (e) Light gray siliceous mudstones of the Miocene–Pliocene Purisima Formation at Drakes Beach, near the Kenneth C. Patrick Visitor Center. (f) Blocks of blueschist from the Franciscan rocks east of San Andreas fault, on the east shore of Tomales Bay. (g) Famous photo shot by pioneering geologist Grove Karl Gilbert of the offset of a fence line during the 1906 San Francisco earthquake, here in the Olema Valley.

Source: (a–c) Courtesy of US Geological Survey. (d–g) Courtesy of US Geological Survey.

(d)

(e)

(f)

(g)

FIGURE 12.12 (Continued)

(such as the Cretaceous turbidites at Point San Pedro) sit on the other side of the bounding faults to the southwest.

12.9.4 POINT REYES

North of Montara Mountain, the San Andreas fault and Salinian block plunge offshore, only to resume again at Point Reyes (Figure 12.12). Point Reyes is now a National Seashore (part of the National Park Service) and a popular destination for beachcombers from the Bay Area. At Drake's Bay on the south shore, Sir Francis Drake allegedly landed and made repairs to his ship, the *Golden Hinde*. He and his crew stayed on that beach while recovering from his 1579 around-the-world voyage raiding the Spanish Main. (The evidence is limited, and historians have long argued about the exact landing place.) This was the first and, for a long time, only visit by English to the California shores, which was predominantly Spanish territory (with some later Russian outposts at Fort Ross and up the northern coast).

Along its eastern edge, the San Andreas fault forms a distinctive trough running from Bolinas Lagoon in the southeast through Olema Valley and Bear Valley, and then down beneath the long narrow inlet of Tomales Bay (Figure 12.12a). Immediately to the west of the fault are large areas of Salinian granitic rocks running from

Inverness Ridge to Tomales Point (Figure 12.12b). The Point Reyes headlands, on the western edge of the area, is also held up by Salinian granitics (Figure 12.12a), as well as by the Paleocene–Eocene Point Reyes Conglomerate (Figure 12.12c), which was eroded from the Salinian granitic after early Cenozoic uplift. The next youngest units are the middle Miocene Laird Sandstone, overlain by middle–upper Miocene Monterey Shales (Figure 12.12d), upper Miocene Santa Margarita Sandstone and Santa Cruz Mudstone, and uppermost Miocene–Pliocene Purisima Formation, another deep-water shale (Figure 12.12e).

The eastern shore of Tomales Bay is within the National Seashore, but geologically it is east of the San Andreas fault and therefore on the Eastern Franciscan block. It contains typical Franciscan rocks, including mélange and especially blueschist (Figure 12.12f). Much of the offset of the 1906 San Francisco earthquake was shown by the area around Point Reyes, especially with Grove Karl Gilbert's iconic photo of the fence offset by the quake in Olema Valley (Figure 12.12g).

Just to the north of Tomales Bay, Bodega Head protrudes out into the Pacific Ocean. It, too, is a piece of Salinian granite, with the San Andreas fault cutting through the neck of the promontory. The Bodega Head quart diorite (Figure 12.11d) looks much like granitic rocks at Point

Reyes, Montara Mountain, Ben Lomond Mountain, the Santa Lucia Range, and the La Panza Range (Figure 12.9).

12.9.5 GUALALA BLOCK

After Bodega Head, the San Andreas fault plunges back into the Pacific Ocean for many miles off the California coast. It comes back ashore in one final fault slice, known as the Gualala block (Figure 12.13a). This narrow slice of rocks west of the San Andreas fault runs from Fort Ross in the south to Point Arenas in the north (Wentworth et al., 1998). It has no Cretaceous granitic rocks, so it may or may not be a part of the Salinian block. The tectonic slice of the Gualala block is very narrow (only a few tens of kilometers wide), and most of the land portions of the terrane are covered in dense redwood forests and coastal vegetation. Geology has to be conducted largely with beach cliff exposures, the only place where erosion produces fresh rocks.

The oldest rocks exposed on these beaches are pillow lavas of the Upper Cretaceous Black Point ophiolite, which probably represents a slice of oceanic crust onto which the rest of the Gualala rocks are deposited. Overlying this unit is the Upper Cretaceous Gualala Formation (Figure 12.13b and c), which is about 3,000 m (10,000 ft) thick and has two members. The lower Stewart's Point Member (Figure 12.13b) consists of shallow marine sandstones, shales, and minor conglomerates that have light-colored granitic pebbles in them, suggestive of a Salinian granitic source area. Overlying this is the Anchor Bay Member (Figure 12.13c), which is also made of dark-colored sandstones and shales but has dark-colored mafic pebbles derived from ophiolites, with no Salinian granite clasts. The Anchor Bay Member also yields fossils for the weird conical oysters known as rudistids, which formed the reefs of the Cretaceous. Most are found in only tropical–subtropical Cretaceous paleolatitudes, so the presence of this warm-water group as far north as the Gualala block suggests it has gone a long way since the Cretaceous.

Unconformably overlying the Cretaceous Gualala Formation is the Paleocene–Eocene German Rancho Formation (Figure 12.13d), which may be up to 6,000 m (20,000 ft) thick in some places. It consists mostly of light-gray to yellow thick-bedded sandstones with minor shales, ranging from shallow marine settings to deep marine turbidites. Many geologists believe that the distinctive mineral assemblage of the German Rancho Formation matches Eocene rocks down in the Mojave Desert, another piercing point that places the Gualala block many hundreds of kilometers south of its present location during the Eocene.

On Iversen Point are outcrops of the Oligocene Iversen basalt, which occurs with a late Oligocene fossil assemblage and gives radiometric dates of 22.8–23.4 Ma (latest Oligocene). It is overlain by outcrops of the middle Miocene Skooner Gulch Formation, which has a lava flow in it dated 15.6 Ma. On Bowling Ball Beach, one layer in the Miocene Galloway Formation was partially cemented and forms spectacular cannonball concretions (Figure 12.13e)

in layers that go for many tens of meters (see front cover). Finally, up by Point Arena, there are classic upper Miocene exposures of Point Arena Formation (Figure 12.13f), equivalent to the Monterey Shale. These have been intensively studied, because the offshore basins have produced significant oil in the past.

Geologists have argued at length about how far away the Gualala block has come from. Although it has no Salinian granite outcrops, it appears to have Salinian granitic pebbles in the lower Gualala Formation. Its rudistid clams suggest at least a subtropical location in the Late Cretaceous. In addition, it has rock units (especially German Rancho Formation sandstones) that are a good match for the Butano Sandstone on the Salinian block north of Santa Cruz. The provenance of both of these Eocene sandstones also suggests that they were near the eastern Mojave Desert about 50 Ma (Figure 12.13a). Most geologists estimate the displacement of the Gualala block to be at least 125 km, and some suggest as much as 430 km farther south.

12.10 WESTERN FRANCISCAN TERRANE

On the west side of the Salinian terrane is yet another block of Franciscan mélange and associated rocks. Known as the Western Franciscan terrane, it forms a narrow triangular block (Figures 12.2 and 12.14) that goes from the northern boundary faults of the Transverse Ranges (such as the Big Pine and Pine Mountain faults) in the south and then up to San Simeon and Hearst Castle to the north. It is separated from the Salinian block by the Sur-Nacimiento fault zone, and the San Gregorio–Hosgri fault zone bounds it on the west (Figure 12.14). Since we know the Salinian block came up from Southern California and Mexico as the southern extension of the Sierras, most geologists think the Western Franciscan block was the accretionary wedge complex that originally formed to the west of the Salinian arc volcanics. This would make it the southern extension of the Eastern Franciscan block, which has been shifted northward until it lies to the west instead of the south of its equivalent rocks in the Eastern Franciscan block (Figure 12.10).

As you travel from Hearst Castle and San Simeon down Highway 1 to Cambria, and then to Morro Bay and San Luis Obispo, there are typical Franciscan rocks everywhere (Figure 12.15). The roadcuts in the region are often composed of bright-green serpentinites and greenschist-grade metamorphics. There are numerous pillow lavas and ophiolites as well (Figure 12.15a and b). At the famous Point Sal ophiolite on Vandenbergh Air Force Base, there are legendary beach cliff exposures of a classic ophiolite, including pillow lavas, sheeted dikes, and layered gabbros (Figure 4.9d–f). In addition, there are many exposures of mélange in the region, especially in the beach cliffs south of Hearst Castle and San Simeon (Figure 12.15c), and classic exposures of ribbon chert in many places (Figure 12.15d). All these rocks closely resemble those of the Eastern Franciscan belt and indeed represent the southern extension of that belt, now moved to the west. They also indicate

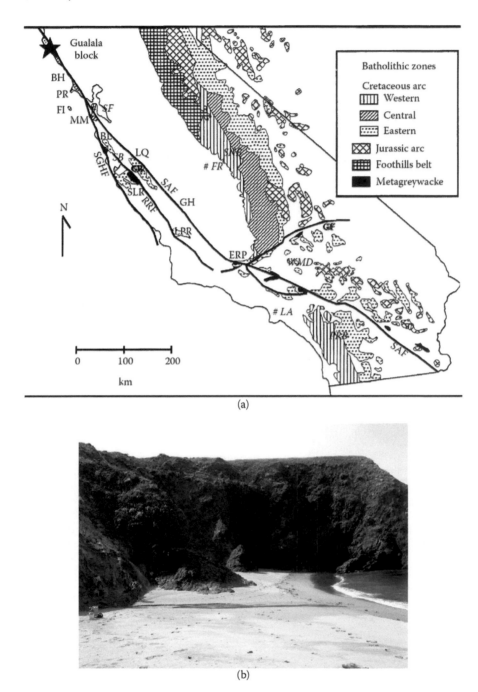

FIGURE 12.13 Gualala block. (a) Index map showing the location of the Gualala block. Also shown are matches to similar rocks in the Salinian block and Mojave Desert. Major tectonic blocks: SB = Salinian block; SNB = Sierra Nevada batholith; WMD = Western Mojave Desert. Cretaceous Salinian granitic outcrops: BH = Bodega Head; BL = Ben Lomond; FI = Farallon Islands; GR = Gabilan Range; LPR = La Panza Range; MM = Montara Mountain; PR = Point Reyes; SLR = Santa Lucia Range. Neogene faults: GF = Garlock fault; RRF = Reliz–Rinconada fault; SAF = San Andreas fault; SGHF = San Gregorio–Hosgri fault. Jurassic ophiolitic basement rocks: ERP = Eagle Rest Peak; GH = Gold Hill; LQ = Logan Quarry. (b) Typical deep marine sandstones of the Upper Cretaceous Stewart's Point Member, Gualala Formation, Pebble Beach. (c) Massive sandstones of the Upper Cretaceous Anchor Bay Member, Gualala Formation, Anchor Bay. (d) Deep-water turbidite sandstones and shales of the Paleocene–Eocene German Rancho Formation, Walk-On Beach. (e) Spectacular "bowling ball" concretions in layers of the upper Miocene Galloway Formation, Bowling Ball Beach. (f) Exposures of the upper Miocene Point Arena Formation, a Monterey equivalent, in the cliffs below Point Arena Lighthouse. (g) Tentative correlation of the Gualala block (left) to the rocks of the Santa Cruz Mountains (right).

Source: (b–d) Photos by the author. (e–f) Courtesy of Wikimedia Commons. (a, g) Maps by Wentworth et al. (1998).

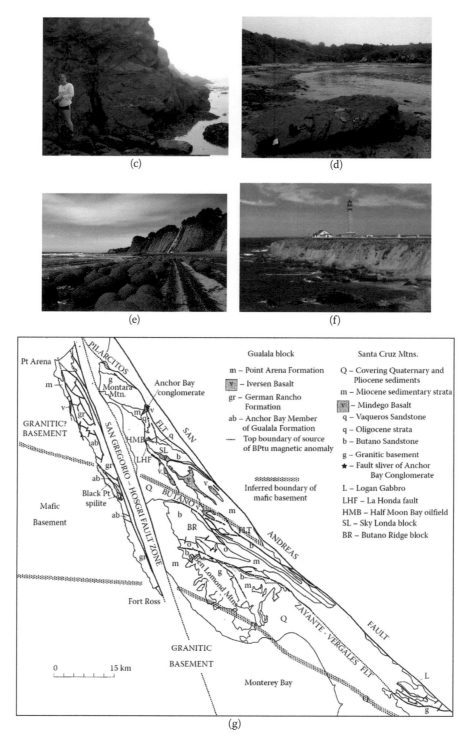

FIGURE 12.13   (Continued)

the activities of an immense accretionary wedge complex forming west of the Salinian arc volcanoes, originally down in Southern California or Mexico before the Salinian block slid north.

As was true of the Eastern Franciscan block, these rocks are often hosts for unusual mineral deposits, especially mercury, which was important in gold mining. The town of Cambria was founded in 1869 and named after the Roman name for Wales (which also led Adam Sedgwick to name the "Cambrian Period" in the 1830s for its exposures in western Wales). The town grew from the old Mexican rancho when quicksilver (mercury) was found in 1862, and it became a mining boomtown. Mercury deposits are common in the Franciscan all up and down the coast of California, because the processes that make accretionary wedges are ideal for concentrating mercury. For the several years that Cambria

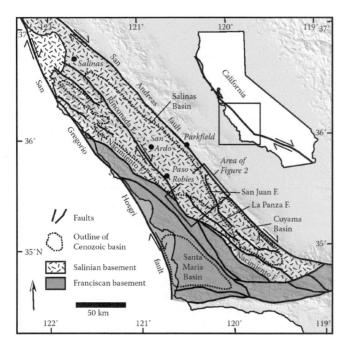

FIGURE 12.14 Index map showing the bedrock of the Salinian block and the Western Franciscan block, with the major sedimentary basins.

Source: With permission from the Geological Society of America.

was a booming mine town, prospectors flooded the area. More than 150 claims were filed in the early 1870s. The most successful of these claims, the Oceanic Quicksilver Mining Company, at one time employed 300 and was the largest mine in the area and the sixth largest in the world. During the boom of 1876, $282,832 worth of quicksilver was produced; four years later, production had decreased to only $6,760. A devastating fire in 1889 virtually ended the mercury business, and Cambria became a quiet dairy community. Today, Cambria is primarily an artists' colony (there are many galleries all over town) and a resort town, with numerous beach houses and nice hotels, especially along famous Moonstone Beach (where moonstone agates are found).

On top of these Cretaceous Franciscan accretionary wedge rocks are younger Cenozoic sedimentary rocks, especially down in the Santa Maria Basin in northern Santa Barbara County (Figures 12.14 and 12.16). Like the other basins in Southern California (San Joaquin Basin, Los Angeles Basin, and Ventura Basin), the Santa Maria Basin is an important producer of oil. It is entirely on the Western Franciscan block, so the basement here is Franciscan mélange and ophiolites. It was sliced up by numerous large faults that run parallel to the San Andreas fault and helped assemble these slivers and slices of crustal rock as they slid north with the edge of the Pacific plate (Figure 12.16). Around the margins of the basin, Great Valley Sequence sedimentary rocks are present beneath upper Oligocene to lower Miocene near-shore sandstone (Vaqueros Formation), deep-water mudstone (Rincon Formation), and volcanic rocks (Tranquillon

and Obispo Formations). In the center of the basin, lower Miocene nonmarine sediments (red sandstones and shales of the Lospe Formation) shed from uplifted blocks along basin-forming faults are present locally, resting in part on Franciscan rocks interpreted to have been exposed in the footwalls of early Miocene extensional faults. Very deep-water marine, mostly Miocene sedimentary rocks (Point Sal and Monterey Formations), blanket the region, filling in early Miocene extensional lows and onlapping ancient structural high spots. Uppermost Miocene to Quaternary marine and nonmarine sandstones and shale units (Sisquoc, Foxen, Careaga, and Paso Robles Formations) record filling of the basin and emergence of flanking uplifts (Figure 12.16).

Large folds in the central part of the basin (Casmalia–Orcutt Anticline, San Antonio–Los Alamos Valley Syncline, Lompoc–Purisima Anticline, and Santa Rita Syncline) are associated with north-trending reverse faults of the Pliocene and Quaternary age. The thickness of the

(a)

FIGURE 12.15 Characteristic mélange rocks of the Western Franciscan block: (a) Pillow lavas, west of Point San Luis pier, near Avila Beach. (b) Pile of pillow lavas on the beach in south Cayucos. (c) Shredded mélange exposures, San Simeon State Beach. (d) Contact between ribbon cherts (vertical maroon-colored deformed beds at bottom left) and weathered pillow basalts (yellowish-brown rocks on the right), Estero Bay, north of Cayucos.

Source: Photos by the author.

(b)

(c)   (d)

FIGURE 12.15   (Continued)

deformed basin fill in the central Santa Maria Basin probably approaches 5,000 m (more than 15,000 ft) in synclines in the footwalls of reverse fault systems. Oil accumulations, typically relatively heavy and sulfur-rich, are mostly trapped in fractured Monterey Formation anticlines, in fractured Monterey truncated by an unconformity along the northeastern flank of the basin, or in Pliocene sandstone lenses above an unconformity on the northeastern basin margin. The petroleum industry has had a large presence in the area since oil was first discovered at the Orcutt Oil Field in 1902. By 1957, there were 1,775 oil wells in operation in the Santa Maria Valley, producing more than $640 million worth of oil.

Finally, there is also significant Cenozoic volcanic activity in the region as well. In Pismo Beach, there are numerous outcrops of the Obispo Tuff, which was dated at 19 Ma. As you travel Highway 1 between San Luis Obispo and Morro Bay, you pass a chain of mountains known as the Morros (Figure 12.17). They are all extinct volcanoes, making up the Morro Rock–Islay Hill intrusive complex. They

originally erupted 20–26 Ma but have long since cooled, and most of the original volcano is long since eroded away. Only the hard, congealed magma in the throat of the volcano, which was resistant to erosion, remains.

In Morro Bay is the largest and youngest of the volcanoes, El Morro (or Morro Rock), which forms its harbor (Figure 12.17c). It is 576 ft tall and used to be a popular hike for tourists and rock climbers. Morro Bay is filled with fishing boats and usually has a population of sea otters living there.

12.11   THE MONTEREY TERRANE

The combination of the Salinian block and the Nacimiento block in the Western Transverse Ranges, along with a third unit, the Sierra de Salinas block, has been considered all parts of a single larger block known as the Monterey terrane (Ducea et al., 2009) (Figure 12.18a). These three blocks are separated by large-scale thrust faults, such as the Sur fault and the Salinas shear zone (Figure 12.18b).

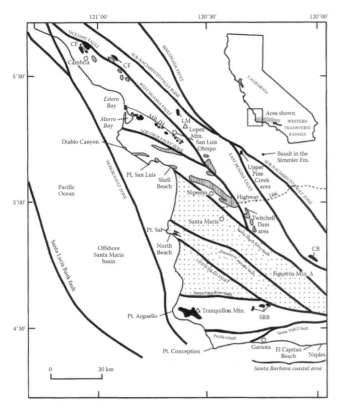

FIGURE 12.16 Geologic map of the Santa Maria Basin and vicinity, showing the location of the Morro Rock–Islay Hill volcanic complex.

Source: Courtesy of Pacific Section SEPM.

They are believed to have been assembled together after the Late Cretaceous, because they are capped by undeformed Cenozoic rocks. As we saw earlier in this chapter, the Salinian block is mostly the original southern extension of the Sierra Nevada arc, which moved from its position south of the Sierras to almost due west of them as it was carried along the San Andreas fault in the Cenozoic. The Nacimiento and Sierra de Salinas blocks are basement rocks from the Franciscan complex of the accretionary wedge. This trench and wedge material was tectonically underplated beneath the Salinian block and shut off any further subduction that created the Salinian arc volcanoes. Tectonic evidence suggests that these blocks were assembled and sutured together as the Monterey terrane over about 6 m.y. during the latest Cretaceous, probably when the Laramide orogeny produced horizontal subduction of the Farallon plate (Figure 7.21) beneath the western edge of North America. This is analogous to how the Cocos plate is undergoing shallow subduction and tectonic underplating beneath Central America and South America today.

12.12 SUMMARY

The Coast Ranges are a complex of many distinct rock units accreted at different times or moved to their present position by faults. When the Sierras erupted in the

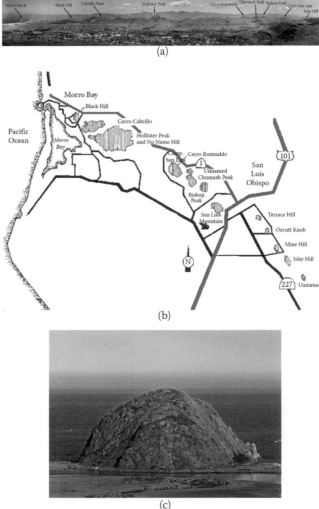

FIGURE 12.17 Morro Rock–Islay Hill volcanic complex: (a) Panoramic aerial view of the chain of extinct volcanoes, from El Morro on the left (west) to Islay Hill on the east (right). (b) Map showing the major named volcanic plugs in the chain. (c) El Morro, the largest and youngest of all the volcanic plugs, at Morro Bay.

Source: (a) Courtesy of US Geological Survey. (b) Redrawn by E. T. Prothero. (c) Photo by the author.

Cretaceous, they formed a huge accretionary wedge complex that became the Franciscan mélange, ophiolites, and other subducted trench rocks that were plastered onto North America over the entire Cretaceous. These formed the Eastern Franciscan block and its southern continuation, the Western Franciscan block, which has since slid north to lie to the west rather than the south of the Eastern Franciscan. The central Salinian block was once the southern continuation of the Sierra Nevada batholith but also slid north hundreds of kilometers to lie to the west of the Sierras.

After their Cretaceous history as accretionary wedges, these rocks became basement for extensive Cenozoic basins, especially the Salinas Basin, the Santa Maria Basin, and smaller basins in Northern California. In many places,

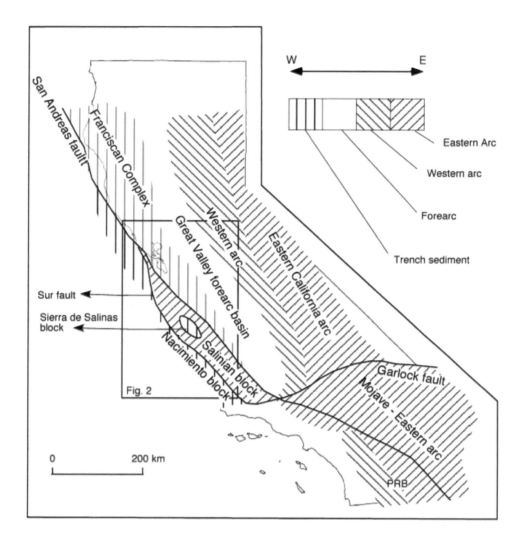

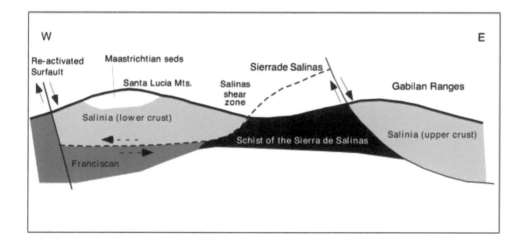

FIGURE 12.18 The Monterey Terrane: (a) Map showing the location of the major tectonic blocks in California. The Salinian, Nacimiento, and Sierra de Salinas blocks together are considered to be parts of a larger block, the Monterey Terrane. PRB = Peninsular Ranges Batholith. (b) Cross-section showing how the Franciscan rocks and the Sierra de Salinas are underplated between the arc rocks of the Salinian terrane in both the Santa Lucia Mountains and the Gabilan Ranges.

Source: From Ducea et al. (2009).

they were intruded by Cenozoic volcanics, such as the late Oligocene–early Miocene Morro–Islay Hill complex in the San Luis Obispo–Morro Bay area and the Pliocene–Pleistocene Clear Lake volcanic field in Sonoma County. Finally, during the late Cenozoic, the wedge complex and its basins have all been uplifted to form the distinctive mountains known as the Coast Ranges.

RESOURCES

Aalto, K.R., and Harper, G.D. 1989. *Geologic Evolution of the Northernmost Coast Ranges and Western Klamath Mountains, California*. American Geophysical Union Field Trip Guidebook T308. American Geophysical Union, Washington, DC, 1–82.

Alt, D., and Hyndman, D.W. 2000. *Roadside Geology of Northern and Central California*. Mountain Press, Missoula, MT.

Alvarez, W.A., Kent, D.V., Premoli-Silva, I., Schweickert, R.A., and Larson, R.A. 1980. Franciscan complex limestone deposited at 17° south paleolatitude. *Geological Society of America Bulletin, 91*, 476–484.

Bailey, E.H., Irwin, W.P., and Jones, D.L. 1964. Franciscan and related rocks and their significance in the geology of western California. *California Division of Mines and Geology Bulletin, 183*.

Blake, M.C. (ed.). 1984. *Franciscan Geology of Northern California*. Pacific Section SEPM Field Trip Guidebook 43. Pacific Section SEPM, Fullerton, CA.

Blake, M.C., Jr., and Jones, D.L. 1981. The Franciscan Assemblage and related rocks in Northern California: A reinterpretation, 306–329. In Ernst, W.G. (ed.), *The Geotectonic Development of California*. Prentice Hall, Englewood Cliffs, NJ.

Dickinson, W.R. 1970. Relation of andesite, granites, and derivative sandstones to arc-trench tectonics. *Review of Geophysics and Space Physics, 8*, 813–860.

Ducea, M.H., Kidder, S., Chesley, J.T., and Saleeby, J.B. 2009. Tectonic underplating of trench sediments beneath magmatic arcs: The central California example. *International Geology Review, 51*, 1–26.

Elder, W.P. (ed.). 1998. *Geology and Tectonics of the Gualala Block, Northern California*. Pacific Section SEPM Book 84. Pacific Section SEPM, Fullerton, CA.

Ernst, W.G. 1970. Tectonic contact between the Franciscan mélange and the Great Valley Sequence, crustal expression of the late Mesozoic Benioff zone. *Journal of Geophysical Research, 75*, 886–902.

Ernst, W.G., Snow, C.A., and Scherer, H.H. 2008. Contrasting early and late Mesozoic petrotectonic evolution of Northern California. *Geological Society of America Bulletin, 120*, 179–194.

Hagstrum, J.T., and Murchey, B.L. 1993. Deposition of Franciscan Complex cherts along the paleoequator and accretion to the American margin at tropical paleolatitudes. *GSA Bulletin, 105*, 766–768.

Hamilton, W.B. 1969. Mesozoic California and the underflow of Pacific mantle. *Geolological Society of America Bulletin, 80*, 2409–2430.

Hildebrand, R.S. 2013. *Mesozoic Assembly of the North American Cordillera*. Geological Society of America Special Paper 495. Geological Society of America, Boulder, CO.

Hopson, C.A., Mattinson, J.M., and Pessagno, E.M., Jr. 1981. Coast range ophiolite, western California, 418–510. In Ernst, W.G. (ed.), *The Geotectonic Development of California*. Prentice Hall, Englewood Cliffs, NJ.

Konigsmark, T. 1998. *Geologic Trips: San Francisco and the Bay Area*. GeoPress, Gualala, CA.

McQuarrie, N., and Wernicke, B.P. 2005. An animated tectonic reconstruction of southwestern North America since 36 Ma. *Geosphere, 1*, 147–172.

Moores, E.M. 1970. Ultramafics and orogeny, with models of the U.S. Cordillera and the Tethys. *Nature, 228*, 837–842.

Page, B.M. 1981. The southern coast ranges, 329–417. In Ernst, W.G. (ed.), *The Geotectonic Development of California*. Prentice Hall, Englewood Cliffs, NJ.

Prothero, D.R., and Tedford, R.H. 2000. Magnetic stratigraphy of the type Montediablan Stage (Late Miocene), Black Hawk Ranch, Contra Costa County, California: Implications for regional correlations. *Paleobios, 20*, 1–10.

Sloan, D., and Karachewski, J. 2006. *Geology of the San Francisco Bay Region*. University of California Press, Berkeley.

Stevens, C.H. (ed.). 1983. *Pre-Jurassic Rocks in Western North America Suspect Terranes*. Pacific Section SEPM (Society of Sedimentary Geologists), Fullerton, CA.

Stoffer, P.W., and Gordon, L.C. (eds.). 2001. *Geology and Natural History of the San Francisco Bay Region: A Field Trip Guidebook*. USGS Bulletin 2186. U.S. Geological Survey, Reston, VA.

Tarduno, J.A., McWilliams, M., Sliter, W.V., Cook, H.E., Blake, M.C., Jr., and Premoli-Silva, I. 1986. Southern hemisphere origin of the Cretaceous Laytonville Limestone of California, *Science, 231*, 1425–1428.

Wahrhaftig, C. 1984. *A Streetcar to Subduction, and Other Plate Tectonic Trips by Public Transport in San Francisco*. American Geophysical Union, Washington, DC.

Wahrhaftig, C., and Sloan, D. 1989. *Geology of San Francisco and Vicinity*. American Geophysical Union Field Trip Guidebook T105. American Geophysical Union, Washington, DC, 1–69.

Wakabayashi, J. 1992. Nappes, tectonics of oblique plate convergence, and metamorphic evolution related to 140 million years of continuous subduction, Franciscan Complex, California. *Journal of Geology, 100*, 19–40.

Wentworth, C.M., Blake, M.C., Jr., Jones, D.L., Walter, A.W., and Zoback, M.D. 1984. Tectonic wedging associated with emplacement of the Franciscan Assemblage, California Coast Ranges, 163–173. In Blake, M.C. (ed.), *Franciscan Geology of Northern California*. Pacific Section SEPM Field Trip Guidebook 43. Pacific Section SEPM, Fullerton, CA.

Wentworth, C.M., Jones, D.L., and Brabb, E.E. 1998. Geology and regional correlation of the Cretaceous and Paleogene rocks of the Gualala Block, California. *Pacific Section SEPM Book, 84*, 3–26.

VIDEOS

Wernicke and McQuarrie posted a short video showing the rotation of the northward movement of the Salinian block: www.youtube.com/watch?v=rzVU0PoZLE4

Calistoga Petrified Forest

www.youtube.com/watch?v=0jMz7s1AXt0
www.youtube.com/watch?v=Cm69aiwK5Dg
www.youtube.com/watch?v=WGmaeMcl7NQ
Clear Lake volcanoes: www.youtube.com/watch?v=kxORAntid1I

Geology of San Francisco

www.youtube.com/watch?v=dAul4-vE5TM
www.youtube.com/watch?v=3owcJB0x6m0
www.youtube.com/watch?v=r74L1OL5uE4

www.youtube.com/watch?v=lv8lUOd20vU
www.youtube.com/watch?v=4qM2bjNrg60

Point Reyes

www.youtube.com/watch?v=VAwZ1iYJbhk
www.youtube.com/watch?v=GcjcmnDQqBo
www.youtube.com/watch?v=BptP5BfAOBI
www.youtube.com/watch?v=kxORAntid1I

Mt. Diablo

www.youtube.com/watch?v=9EKhk84pgws
www.youtube.com/watch?v=rIcRLJd0MdE&list=PLr2AidRIEM
oGHlF4rafmKYHIHrdAgsPYa

Coast Ranges

www.youtube.com/watch?v=aBIXBbxF11Y
www.youtube.com/watch?v=TQ1aGcrADBo

13 Compression, Rotation, Uplift
The Transverse Ranges and Adjacent Basins

An afternoon drive from Los Angeles will take you up into the high mountains, where eagles circle above the forests and the cold blue lakes, or out over the Mojave Desert, with its weird vegetation and immense vistas. Not very far away are Death Valley, and Yosemite, and Sequoia Forest with its giant trees which were growing long before the Parthenon was built; they are the oldest living things in the world. One should visit such places often, and be conscious, in the midst of the city, of their surrounding presence. For this is the real nature of California and the secret of its fascination; this untamed, undomesticated, aloof, prehistoric landscape which relentlessly reminds the traveller of his human condition and the circumstances of his tenure upon the earth.

—**Christopher Isherwood**
Exhumations

13.1 DATA

Geography: Transverse Ranges and associated tectonic basins (Los Angeles Basin, Ventura Basin, San Fernando Basin, Ridge Basin, and Soledad Basin)

Elevation: Mt. San Gorgonio: 3,505 m (11,501 ft); San Bernardino Peak: 3,246 m (10,649 ft); Mt. San Antonio: 3,068 m (10,064 ft); Cucamonga Peak: 2,700 m (8,859 ft); Mt. Pinos: 2,962 m (8,831 ft); White Ledge Peak: 1,414 m (4,640 ft)

Mountain ranges: Santa Ynez, San Rafael, Santa Susana, Santa Monica, San Gabriel, San Bernardino, Little San Bernardino, Orocopia Mountains

Major rock types: Precambrian metamorphics and gabbros, Mesozoic granitic plutons and marine sedimentary rocks, Cenozoic marine and nonmarine sedimentary rocks

Plate tectonic history: Mesozoic subduction, accretion; Cenozoic transpression, rotation, and transform motion

Famous faults: San Andreas, Santa Ynez, Pine Mountain, Big Pine, Oak Ridge, Malibu Coast–Santa Monica–Hollywood, San Gabriel, Sierra Madre–Cucamonga, Raymond, Newport–Inglewood, Whittier, plus many unnamed blind thrusts

Resources: Oil and gas, diatomite, limestone, gold

National parks: Joshua Tree National Park, San Gabriel Mountains National Monument, Channel Islands National Park, Santa Monica Mountains National Recreation Area

National forests: San Bernardino, Angeles, Los Padres

13.2 GEOGRAPHY OF THE TRANSVERSE RANGES

A glance at the map of California (Figure 13.1) shows a striking feature. Most of the ranges along the coast are oriented northwest–southeast, parallel to the San Andreas fault that produced them. But in Southern California, the trend of the ranges curves to almost due east–west, roughly transverse to the northwest–southeast trend of the other mountains. These are the Transverse Ranges. They run from the North American plate in the Mojave Desert (Orocopia, Little San Bernardino Mountains, and San Bernardino Mountains), cross the San Andreas fault at Cajon Pass, then west through the San Gabriel Mountains, Santa Monica, and the Santa Susana Range, and finally to the San Rafael and Santa Ynez Ranges to the coast at Point Conception.

Other than the Sierras, the Transverse Ranges include some of the highest mountains in California, with one peak (Mt. San Gorgonio) more than 3,500 m (11,503 ft), four peaks more than 3,000 m (10,000 ft), and ten individual ranges more than 1,400 m (4,500 ft) in elevation. They form an imposing barrier or wall to the north of the huge Los Angeles Metropolitan Area, as well as many other large cities, like those of the "Inland Empire" (San Bernardino–Riverside), Ventura–Oxnard, and Santa Barbara, and hundreds of smaller cities. Their extreme relief and elevation mean that the higher peaks in the San Gabriel and San Bernardino Mountains get significant snow, above 1,800 m (6,000 ft), in the winter as the cold Pacific storms come in from the ocean. They also produce a strong rain-shadow effect in the Mojave Desert behind them. The individual valleys in the ranges are among the deepest and most steep-walled in all of California. Some canyons in the San Gabriel Mountains have almost 3,000 m (10,000 ft) of vertical relief over very short distances, twice as deep and much steeper than the Grand Canyon. This is because they are among the fastest-rising mountains in the world, with average uplift rates of several millimeters per year, exceeded only by the rates of uplift of rapidly rising ranges such as the mountains of New Zealand or the Himalayas. The extremes in elevation and rainfall mean big differences in vegetation and ecology as well. The coastal side of the valleys and

DOI: 10.1201/9781003301837-13

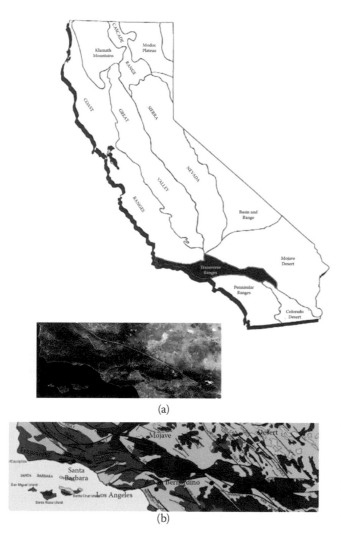

(a)

(b)

FIGURE 13.1 (a) Index map of California highlighting the Transverse Ranges. In the lower left is the satellite image of the Transverse Ranges. (b) Simplified geologic map of the Transverse Ranges.

Source: (a) Courtesy of US Geological Survey. (b) Courtesy of California Division of Mines.

foothills is dominated by chaparral plants, which survive on a few heavy rainstorms during the winter and then endure drought with no measurable precipitation during the spring, summer, and fall. Near the ocean, the vegetation is coastal sage scrub, which adapted to absorbing moisture from the coastal fogs that roll landward during the spring and early summer before the region goes completely dry. As you climb the higher ranges like the San Rafael Mountains, the San Gabriels, and the San Bernardinos, the elevation is high enough to support the middle-elevation piñon-juniper forests and Joshua trees on the dry desert slopes to the north. The highest part of these ranges supports dense pine forests, dominated by drought-tolerant conifers like the Ponderosa pine and the Jeffrey pine. The top of the highest peak (Mt. San Gorgonio) is above the tree line (3,400 m, or 11,000 ft) and has an alpine tundra climate.

Although they have similar topographic features and vegetation, the bedrock of the Transverse Ranges and the geologic history are radically different from one mountain to the next. Each range has its own distinct history, with few rock units that can be found in any other mountain range. They have little in common except their east–west orientation, so we discuss them one range at a time and cannot generalize much about their geologic history. In the past 40 years, detailed studies have shown that their tectonic history is even more complex and incredible than anyone could have ever imagined. Each range is an independent tectonic block that behaves in a different way from the other ranges. This peculiar tectonic history explains why the mountains are transverse to the general northwest–southeast trend of most of the coastal mountains in California. We will look at this complex tectonism later in the chapter.

In addition to extremely rapidly rising ranges, the fault-bounded sediment-filled basins around the ranges are among the deepest, narrowest, and fastest-sinking basins in the entire world. They are important not only for the huge populations that now live on their surface but also for their subsurface resources, especially groundwater, oil, and gas. The Transverse Ranges and their surrounding basins are intimately connected by common tectonic forces: huge faults that cause both rapid uplift and subsidence, as well as horizontal transform motion and even tectonic rotation.

13.3   THE EASTERN RANGES

Let us begin with the relatively simple geologic history of the ranges in the eastern half of the Transverse Range system. All these ranges are on the North American plate, so they have experienced relatively little complex tectonism, especially compared to the ranges on the Pacific plate.

13.3.1   SAN BERNARDINO MOUNTAINS

The San Bernardino Mountains (Figures 11.11 and 13.2) are the highest of all the Transverse Ranges, capped by peaks such as Mt. San Gorgonio at 3,505 m (11,501 ft) and San Bernardino Peak, reaching 3,246 m (10,649 ft). Their southern front is an extremely steep fault scarp (Figure 13.2a and b), since it is chopped off by the San Andreas fault and associated faults, such as the Mission Creek fault (Figure 11.11). The northern boundary of the range by Victorville and Apple Valley is a thrust fault, pushing the range up and northward over the Mojave Desert and away from the San Andreas fault zone every time there is a major earthquake in the area.

The oldest rocks in the San Bernardino Mountains are Proterozoic metamorphics, mostly amphibolites and gneisses (Figure 13.3a). These rocks have been dated at 1.75 Ga. They are found in roof pendants and as xenoliths in the Cretaceous granitics that make up the bulk of the range. Most are so highly metamorphosed that it is not clear whether they are metasedimentary gneiss or metamorphosed granitics or both. They are very similar to

(a)   (b)

FIGURE 13.2 San Bernardino Mountains. (a) Aerial image of the south flank of the range, looking east. The sharp break at the San Andreas fault at the foot of the mountains, and the steep relief of the mountain range above the cities of Highland and San Bernardino, is very clear. (b) Aerial image of the southeast flank of the range, looking west. Mt. San Gorgonio is the peak in the center. The traces of the San Andreas and related faults cut across the base of the mountains in the foreground.

Source: Courtesy of US Geological Survey.

(a)   (b)

(c)   (d)

FIGURE 13.3 Typical rocks of the San Bernardino Mountains. (a) The oldest rocks are 1.7-billion-year-old gneisses, similar to the Mendenhall Gneiss of the San Gabriel Mountains, the Pinto Gneiss of the Little San Bernardino Mountains, and similar gneisses in the Orocopia Mountains. (b) Early Paleozoic quartzites of the roof pendant on the north flank of the range, Chicopee Canyon. (c) Limestones of Furnace Formation, Cushenberry Canyon. (d) Cretaceous granitic rocks, which make up most of the range, here shown just west of Big Bear Lake.

Source: Photos by the author.

the Proterozoic gneisses found in the metamorphic core complexes of the Basin and Range (Chapter 7), as well as the Mendenhall gneisses of the San Gabriel Mountains. Geologists consider them to be a part of the ancient Mojave terrane, also dated around 1.7 Ga, which forms the oldest basement rock in the central and southwestern United States (Figure 7.15).

The next youngest unit in the ranges is roof pendant made up of a thick stack of metasedimentary rocks on the north flank of the mountains (Figure 13.3b). The lower part of this sequence is a 620 m (1,900 ft) thick stack of quartzites, variously called the Saragossa Quartzite, the Arrastre Formation, or the Chicopee Canyon Formation. It crops out mainly in the hills north of Baldwin Lake. The quartzites have many well-preserved sedimentary structures, such as ripple marks, cross-bedding, pebble conglomerates, and other stratification types, along with trace fossils, such as the vertical tubular burrows known as *Skolithos*. Unfortunately, the rocks are too highly metamorphosed to preserve body fossils, so there are no biostratigraphic data to date the sequence. The sequence is usually correlated to the thick late Proterozoic–Cambrian sequence found across the Basin and Range (especially in Death Valley, the Providence Mountains, and the Marble Mountains). Stewart and Poole (1975) correlated this sequence to the Wood Canyon Formation (latest Proterozoic to Cambrian) and the Lower Cambrian Zabriskie Quartzite (rich in *Skolithos*), Carrara Formation, and Bonanza King Formation in the Death Valley area (Figure 13.4).

Lying above these early Paleozoic quartzites in the same roof pendant is a 1,300 m (4,000 ft) thick stack of limestones

(Figure 13.3c) originally known as the Furnace Limestone, which are mostly metamorphosed to marble. Brown (1991) mapped the rocks more carefully and found that it is the equivalent of limestones found throughout the Basin and Range. The lower part of this package appears to correlate with the Cambrian Bonanza King, Nopah, and Carrara Formations (Figure 13.4), but the upper part has Devonian and Mississippian index fossils, including diagnostic species of brachiopods, bryozoans, and crinoids. These rocks have been correlated with the Devonian Sultan Limestone, and the Mississippian Monte Cristo and Bird Springs Limestones, widely exposed in southern Nevada and in the Mojave Desert. Thus, the limestone sequence is actually two sedimentary packages of very different ages, separated by an unconformity that wipes out the Upper Cambrian, Ordovician, Silurian, and some of the Devonian (spanning from 350 to 510 Ma, or about 160 m.y. in duration). This is similar to the unconformity found in much of the Paleozoic cratonic rocks in the southwest, for example, in the Grand Canyon, where Mississippian Redwall Limestone lies disconformably on Cambrian Muav Limestone, with channel fill deposits of Devonian Temple Butte Limestone caught between the erosional surfaces.

The vast majority of the rock outcrops in the San Bernardino Mountains, however, are Mesozoic granitic rocks (Figure 13.3d) related to not only those in the Little San Bernardino and San Gabriel Mountains but also the granitics of the Sierra Nevada mountains and the Peninsular Ranges. The oldest intrusions are quartz monzonites dated around 215 Ma (Triassic), comparable to the Lowe Granodiorite in the San Gabriel Mountains. Intruded into these early plutons are Jurassic tonalites, quartz diorites, and diorites dated between 127 and 158 Ma. These rocks have been compared to the Upper Jurassic plutonics in the Independence area of the Owens Valley, which preceded the building of the Cretaceous portion of the Sierra Nevada batholith. The bulk of the range, however, was intruded by quartz monzonites and granodiorites (Figure 13.3d) that are dated between 86 and 112 Ma, or Cretaceous in age. These plutons are comparable to the Cretaceous granitics found in the Sierras and the Peninsular Ranges and are thus part of the dismembered arc volcano chain that once ran from the Coast Range batholith in British Columbia down to the Peninsular Ranges in Baja California (Figure 8.7).

13.3.2 Little San Bernardino Mountains

To the east of the San Bernardino Mountains across the Pinto Mountain fault and Morongo Valley are the Little San Bernardino Mountains, so named because they are much lower in elevation than the bigger range to the west. This range is also bounded by strands of the San Andreas fault (the Mission Creek fault) on the south side (Figure 11.11) and the Pinto Mountain fault on the north side. Most of the range is within the bounds of Joshua Tree National Park, along with the Twenty-Nine Palms Marine Base on the northern side.

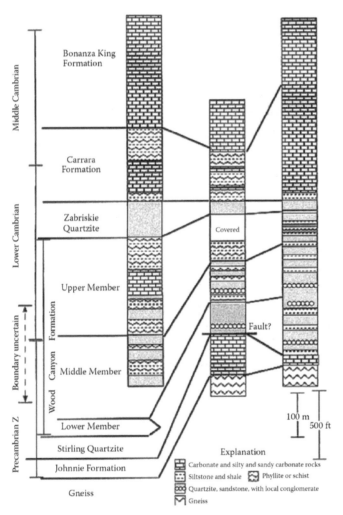

FIGURE 13.4 Correlation of the San Bernardino Mountain Paleozoic rocks (central column) with the sequence at Quartzite Mountain near Victorville (left column) and the Providence Mountains in the central Mojave Desert (right column).

Source: Modified after Stewart, J. H., and Poole, F. G. *Geological Society of America Bulletin*, 86: 205–212, 1975; with permission from the Geological Society of America.

The Little San Bernardinos have a much simpler geology than most of the other Transverse Ranges. The oldest rocks are the Proterozoic (1.7 Ga) Pinto Gneiss, found as xenoliths and roof pendants in the eastern and in the western part of the range (Figure 13.5a). This hard gneiss often forms the higher ranges and the rims of the valleys, such as Queen Valley and Pleasant Valley, Hexie Mountain, and the rim of Pinto Basin. In the Upper Jurassic (151 Ma), these gneisses were intruded from below by the granitic rocks, especially the White Tank monzogranite (a rock intermediate between a true granite and a monzonite). The rocks forming the

exposures at Joshua Tree are classic examples of spheroidally weathered jointed granitic rock (Figure 13.5b) and make up the bulk of the outcrops in the park. The granitic rocks were domed upward after their intrusion, and as they reached the surface, they broke into a series of crisscrossing joints that promoted the spheroidal weathering of the rock units. Finally, Malapai Hill is a deeply weathered mass of basalt (Figure 13.5c) which has been dated at 15.93 Ma (middle Miocene). This basalt shows some columnar jointing and may be similar to other Miocene basalts in the Basin and Range Province that were formed when crustal extension began in the Miocene.

13.3.3 OROCOPIA MOUNTAINS

The Orocopia Mountains are the easternmost of the Transverse Ranges (Figure 13.6). They are bounded to the west by the San Andreas fault and the Coachella Valley, with the folded and faulted late Cenozoic beds of the Mecca Hills (see Chapter 11) caught between them. The Salton Wash fault separates them from the Chocolate Mountains to the south. The fault blocks and grabens of the Basin and Range Province lie to the north, and the Chuckwalla Mountains are in the east. They are about 10.2 km (18 mi) wide in the east–west direction and slightly narrower in the north–south direction.

The geology of the Orocopia Mountains was poorly understood until pioneering fieldwork in the 1950s by John C. Crowell mapped the units and interpreted them. In the 1960s and 1970s, Lee Silver of Caltech not only studied the rocks in detail but also advised the National Aeronautics and Space Administration when it used the Orocopia Mountains to train the Apollo astronauts. Today, nearly the entire range is within the Orocopia Mountains Wilderness area, protecting it from mining, development, and too many people exploiting the land and destroying its natural beauty.

The oldest rocks in the Orocopias (Figure 13.6) are a band of rock in the core of the range, uplifted and exposed by the downward sliding of the overlying block on the Orocopia Mountains detachment fault. They include 1-billion-year old gabbros and anorthosites that resemble the San Gabriel anorthosite, and 1.7-billion-year-old gneisses (Figure 13.7a) that resemble the Mendenhall Gneiss of the San Gabriels, the Pinto Gneiss in the Little San Bernardino Mountains, and the gneisses and amphibolites of the San Bernardino Mountains. Intruded into the northeastern part of the range are Mesozoic granitic rocks, especially in the Hayfield Mountains (Figure 13.6). These are part of the widespread belt of granitic plutons that goes from the Coast Range batholith in British Columbia down through the Sierra Nevadas and then on to the Peninsular Ranges batholith (Figure 8.7).

In the western half of the range, overlying the Orocopia Mountains detachment fault, is one of the most remarkable units in the range, the Orocopia Schist (Figures 13.6, 13.7b, and 13.8a). Its protolith is Cretaceous and Paleogene arc rocks that have undergone blueschist-grade metamorphism,

(a)

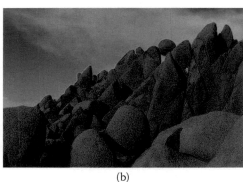

(b)

(c)

FIGURE 13.5 Typical outcrops of rocks in the Little San Bernardino Mountains: (a) The black-and-white banded Pinto gneiss, dated about 1.7 Ga. (b) Jurassic White Tank monzogranites with crisscrossing joints forming spheroidally rounded boulders, Joshua Tree National Park. (c) The Malapai Hill basalt, dated at 16 Ma, showing columnar jointing.

Source: (a) Photo by T. Savage. (b–c) Photos courtesy of Wikimedia Commons.

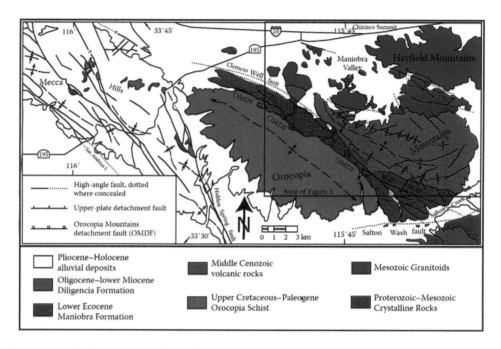

FIGURE 13.6 Simplified geologic map of the Orocopia Mountains.

Source: With permission from the Geological Society of America.

so most of their original fabric and mineralogy have been destroyed. As mentioned in Chapter 11, the Orocopia Schist has long been correlated with the Pelona Schist of the San Gabriel Mountains, and also the Rand Schist in the central Mojave Desert (Figure 13.7c). Mason Hill and Tom Dibblee (1953) first noticed the similarity between the Orocopia and Pelona Schists and realized that it was a critical piercing point that indicates about 210–250 km (125–155 mi) of displacement along the San Andreas fault since the early Paleogene. Since then, the Orocopia Schist has been studied in greater detail (Haxel et al., 1985; Haxel and Dillon, 1978; Grove et al., 2013). In addition to these units, the Rand Schist in the northern Mojave Desert, the Portal Ridge schist just north of the Pelona schists in the Transverse Ranges west of Palmdale, and the Sierra de Salinas schist in the Coast Ranges are considered isolated remnants of this terrane. The protolith for the Orocopia–Pelona schist is thought to be the Jurassic–Cretaceous sequence of deep marine turbidite sandstones and shales, with some metavolcanic rocks and even oceanic crustal rocks (Figure 13.8), although the Rand Schist, Portal Ridge Schist, and Sierra de Salinas Schists have a protolith that was derived from Sierran arc rocks (Jacobson et al., 2011; Grove et al., 2013). As such, it probably represents a highly metamorphosed accretionary wedge, similar to the Franciscan rocks of the Coast Ranges in both age and lithology. It was probably connected to the ancient forearc and accretionary wedge of the plutons and arc volcanoes associated with the Hayfield Mountain granitics (Figure 13.8a). By 75 Ma, the growth and accumulation of the subduction complex and forerarc basin had ceased, and the arc became inactive. This was true of most of the Sierran arc rocks throughout the region, thanks to the

Laramide orogeny (see Chapter 7). Then, 65 Ma, the entire ancient arc complex underwent shearing and compression. The ancient arc rocks were forced deeper into the subduction zone as the Orocopia detachment fault (equivalent to the Vincent thrust in the San Gabriel Mountains) pushed thick slices of rocks on top of them and caused blueschist-grade metamorphism (Figure 13.8b). Recent work has suggested that there may also be Pelona–Orocopia Schist equivalents in the Western Franciscan mélange, west of the Sur-Nacimiento fault as well.

The final stage (Figures 13.6 and 13.8c) of evolution of the Orocopia Mountains is the development of Paleogene sedimentary basins on top of the Mesozoic metamorphic and granitic rocks. The Maniobra Basin to the north and the Diligencia Basin to the south (Figure 13.6) apparently opened up in the Paleocene and Eocene, just as the other Laramide tectonic basins did further to the west in the modern Rocky Mountains. The first basin to open was the Maniobra Basin, which filled with 1,560 m (5,120 ft) of Paleocene–lower Eocene deep marine sandstones and shales, mostly part of a large submarine fan complex. Some geologists have correlated these rocks to Paleocene sandstones, such as the San Francisquito Formation, seen in the western San Gabriel Mountains, as well as Eocene sandstones in the Simi Valley and elsewhere, suggesting another possible piercing point. Unconformably overlying these lower Eocene rocks is the upper Oligocene–lower Miocene Diligencia Formation, which is about 1,500–2,000 m (5,000–7,000 ft) of mostly continental sandstones, conglomerates, and evaporites, with extensive lava flows (dated between 19 and 23 Ma) filling the bottom of the basin and interbedded with the rest of the Diligencia sequence. Some

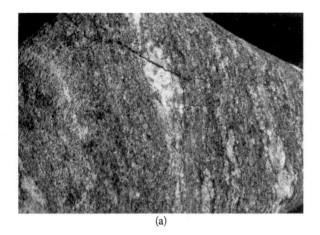

(a)

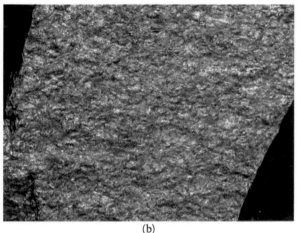

(b)

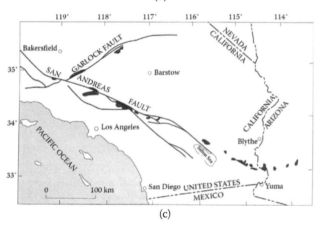

(c)

FIGURE 13.7 Typical rocks of the Orocopia Mountains: (a) 1.7-billion-year-old gneiss, much like the Pinto Gneiss in the Little San Bernardino Mountains. (b) Shiny muscovite-rich Orocopia Schist. (c) Map showing the match between the Orocopia Schist (red patches) in the Orocopia Mountains to the east (and its Arizona equivalents) and the transported remnants of the same body, known as Pelona Schist, in the San Gabriel Mountains.

Source: (a–b) Photos by the author. (c) With permission from the Geological Society of America.

geologists connect this unit with the Vasquez Formation (also dated by a 23-million-year-old old basalt at the base; see later in this chapter) in the Soledad Basin, and also the Plush Ranch Formation up in the Tejon Basin, suggesting

yet another piercing point, although this has been challenged (Spittler and Arthur, 1973).

The entire Orocopia Range has been subjected to intense structural deformation (Figure 13.6) during the development of both the Orocopia Schist and Orocopia detachment fault in the Mesozoic, and then compressional events that down-warped the Maniobra and Diligencia Basins. Since the Miocene, all these rocks have been intensely folded and thrust-faulted. Late Cenozoic extension on detachment faults unroofed the oldest rocks in the core of the range.

13.4 THE WESTERN RANGES

The ranges to the west of the San Andreas fault have very different tectonic histories and only few rock types that resemble those on the eastern side. This is because they have traveled north from regions much farther south during the motion of the San Andreas fault. The oldest rocks in the region almost certainly formed south of the present-day Mexican border. In addition, there is very little similarity in rock type from one range to the next, showing that they had very different tectonic and sedimentary histories. The fact that they are discussed together is purely an accident of the fact that today they happen to align with the other ranges in the area.

13.4.1 SAN GABRIEL MOUNTAINS

The San Gabriel Mountains (Figure 13.9a) tower above the Los Angeles Basin and have a strong effect on the weather and climate of the region, and thus the lives of millions of people. They trap the clouds from big winter storms off the Pacific that create rainfall over the city and also block the foggy marine layer coming in from the ocean so that it does not cross over into the deserts. Under certain conditions, this forms an inversion layer, trapping the smog against the mountains as well. These smoggy conditions used to be very common in Los Angeles, although decades of environmental efforts have hugely reduced the serious air pollution problems that used to be common before the 1980s.

The highest peaks include Mt. San Antonio (3,068 m, or 10,064 ft) and Cucamonga Peak (2,700 m, 8,859 ft), but the entire mountainous region of the Angeles National Forest and San Gabriel Mountains National Monument is very high, with 11 peaks more than 2,440 m (8,000 ft) in elevation. Covered mostly by Ponderosa pine forest, it is visited by millions of people on the spectacularly steep and winding Angeles Crest Highway every year. The snow on the mountains every winter can be significant, so thousands of skiers and snowboarders drive up from Los Angeles to the ski resorts on the northwest flank of the range near Wrightwood and Mountain High. Despite the presence of millions of people nearby, the San Gabriel Mountains are quite rugged and wild, with large populations of cougars, bears, deer, and coyotes that prowl the campgrounds and occasionally wander down into the suburbs below. Finally, Mt. Wilson Observatory, with three century-old telescopes,

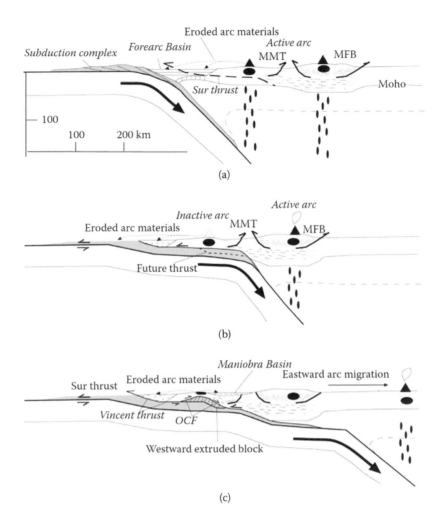

FIGURE 13.8 Tectonic evolution of the Orocopia Schist and the later structural and tectonic features: (a) stage 1, 80–75 Ma; (b) stage 2, 75–65 Ma; (c) stage 3, 65–50 Ma.

Source: From A. Yin, *Geology*, 30, 2, 183–186. With permission from the Geological Society of America.

sits on Mt. Wilson peak. It was the site where Hubble and Humason discovered the expanding universe, which led to "Big Bang" cosmology.

The geology of the San Gabriels is the most complex of the entire Transverse Ranges (Figures 13.9b and 13.10). In map view, the range has a lens-like or football-like shape (Figure 13.9b), because it is formed between the straight line of the San Andreas fault along the north boundary and several curved fault lines along the south boundary. The most important of these southern faults is the Sierra Madre–Cucamonga fault zone, a huge thrust fault that plunges north into the foothills of the range and then flattens out (Figure 13.10). The entire range sits on top of this thrust zone, and every time there is an earthquake along this fault, the range is pushed both up and south over the Los Angeles Basin by several centimeters. The last such quake was an M5.8 event in 1991, and before that was the Sylmar quake (M = 6.7) of 1971.

Within this enormous fault slice is complicated geologic history. The oldest rocks in the ranges are the 1.7 Ga Mendenhall gneisses and schists (Figure 13.11a).

These rocks are equivalent to the other late Proterozoic high-grade metamorphics of the Mojave terrane found in the other ranges, such as the Pinto Gneiss at Joshua Tree National Park and the gneisses of the Orocopia Mountains and San Bernardino Mountains. These metamorphic rocks occur as xenoliths and roof pendants surrounded by all the younger intrusive rocks that make up most of the range. There are large areas of them just north of Mt. Wilson, in the south flanks of Mt. San Antonio and the north side of Mt. Waterman (Figure 13.9b).

The next major event was the intrusion of the San Gabriel Anorthosite magma chamber, about 1.2–1.0 Ga. This pluton is found in the western part of the range, especially along the Angeles Forest Highway just north of Big Tujunga River (Figure 13.9b). Anorthosite (Figure 13.11b) is a gabbro made almost entirely of bluish-gray calcium plagioclase (the mineral anorthite), so it is a very peculiar rock. In the San Gabriel Mountains, these mafic magmas intruded to form a huge layered mafic magma chamber (see Chapter 2). Their minerals crystallized out, settled down on the floor of the magma chamber, and accumulated centimeter-thick

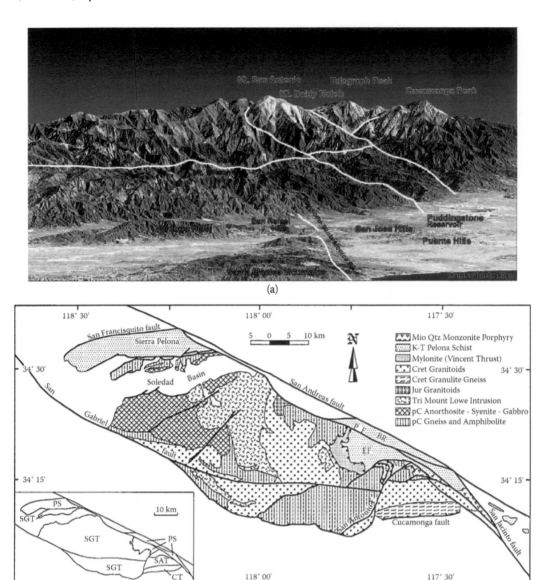

FIGURE 13.9 (a) Exaggerated vertical relief image of the south flank of the San Gabriel Mountains, showing the major peaks and the red fault lines. (b) Simplified geologic map of the San Gabriel Mountains.

Source: (a) Courtesy of US Geological Survey. (b) Modified from Ehlig, P. L., in Ernst, W. G. (ed.), *The Geotectonic Development of California*, Prentice Hall, Englewood Cliffs, NJ, 1981, 258–283.

alternating layers of pure black pyroxene and gray calcium plagioclase. Later, they were cooled and tilted into a vertical orientation of the layers today (Figure 13.11c and d). In some places, the settling crystals on the floor of the magma chamber formed remarkable features, such as the teepee-shaped mounds of calcium plagioclase crystals, known as crescumulate structures, that piled up in some places (Figure 13.11c).

Anorthosites are very special and peculiar rocks. The vast majority intruded at the same time in the late Proterozoic (1.2–1.0 Ga), and few formed or intruded before or afterward. Anorthosites occur not only in the Transverse Ranges but also in the Beartooth Plateau of Montana, the Adirondacks of New York, and in Quebec and Labrador, as well as Scotland,

Scandinavia, and parts of western Europe. A billion years ago, they formed a single belt in the Rodinia supercontinent that has since been torn apart. Igneous petrologists have puzzled over and debated for years over the reasons for this distinctive "anorthosite event." Whatever the cause, it was a unique event of melting and intrusion of peculiar magma that happened no other time in geologic history. Finally, when the Apollo 15 astronauts brought back crustal moon rocks, they turned out to be anorthosites as well.

The next youngest rocks in the San Gabriels are a series of Mesozoic granitic intrusions which incorporated the Mendenhall metamorphics and the San Gabriel Anorthosite magma chamber as roof pendants (Figure 13.11d). The first of these was the Triassic Lowe Granodiorite, dated between

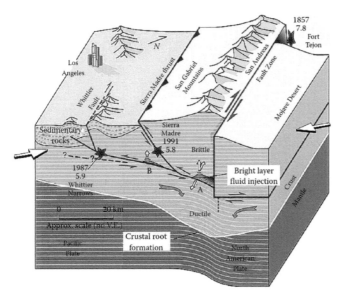

FIGURE 13.10 Cartoon block model of the San Gabriel Mountains.

Source: Courtesy of US Geological Survey.

220 and 208 Ma. This pluton lies mostly in the north-central part of the range (Figure 13.9b), north of Mt. Wilson and Mt. Lowe and west of Mt. Waterman. Next came a series of Cretaceous granitic intrusions, much like those in the Sierras, the Peninsular Ranges, and the Salinian block, and also in the San Bernardino Mountains and in Joshua Tree (Figures 13.3c and 13.5b). The largest of these is the Upper Cretaceous (80 Ma) Josephine Granodiorite, which crops out in the center of the range around Mt. Waterman, as well as the south flank near Mt. Josephine. The complex relationships of all these different plutons are greatly simplified in Figure 13.9b, and the outcrop in Figure 13.11d captures some of their relative ages through their crosscutting relationships.

If that story wasn't complicated enough, the fault block that makes up the San Gabriel Mountains above the Sierra Madre–Cucamonga thrust (Figure 13.10) is actually a composite of two major blocks. Within this fault block, the entire mass of Proterozoic to Mesozoic rocks of the central and western San Gabriels is pushed over the top of a lower block, made of the Pelona Schist along the Vincent thrust in the eastern San Gabriels (Figures 13.9b and 13.12). As discussed earlier, the Pelona Schist is the offset equivalent of the Orocopia Schist, and the Vincent thrust is the other portion of the Orocopia detachment fault (Figure 13.7c). The Vincent thrust is a huge shear surface, with a major zone of rocks that have been crushed and pulverized by the thrusting into a rock known as mylonite, which is metamorphosed by shearing rather than by great heat and pressure. As already discussed, the Pelona–Orocopia Schist is thought to be the remnants of a Cretaceous arc complex and accretionary wedge that has undergone blueschist-grade metamorphism as it was pushed into the subduction zone after it formed (Figure 13.8). As mentioned already, the presence of so many similar rocks in both the San Gabriel Mountains and the Orocopia Mountains (the Pelona–Orocopia Schists, the Vincent–Orocopia faults, and the occurrence of the same 1.7-billion-year-old gneiss, metamorphics, anorthosites, similar Paleocene, Eocene, and Miocene sedimentary rocks, and similar Mesozoic plutons) shows that they are dismembered parts of the same range that have separated ever since the San Andreas fault cut them in half.

The final surprise is that the San Gabriel Mountains are a very young range. Not only have they moved 250 km (150 mi) from their Miocene location just west of the Orocopia Mountains, but also their very rapid uplift is very recent. Studies of the Miocene sandstones in the region (Woodburne, 1975), such as the Punchbowl Sandstone in Devil's Punchbowl (Figure 11.13), show no evidence of the characteristic and distinctive rock types (especially gabbros and anorthosites), eroding down from the San Gabriels in the region today. Instead, it is full of basalt and rhyolite pebbles that have been traced to the Sidewinder volcanics between Barstow and Victorville. Thus, the uplift of the range occurred rapidly in the late Neogene as it slid north along the San Andreas fault and was pushed up on the Sierra Madre–Cucamonga thrust (Figure 13.10).

13.4.2 Santa Monica Mountains

The Santa Monica Mountains are nowhere near as high as the San Gabriel Mountains to the north, but they are important in very different ways. Malibu Peak at 948 m (3,111 ft) in elevation is the highest spot, although the canyons, such as Malibu Canyon, are very steep, with almost 800 m (2,500 ft) of vertical relief. The Santa Monica Mountains run almost due east–west from the central part of Los Angeles on the east and west to Point Mugu and the Oxnard–Ventura Basin. Their eastern end forms the Hollywood Hills, home to the entertainment industry and the famous Hollywood sign. These hills separate the Los Angeles Basin on the south from the San Fernando Basin to the north. This mountain range in the center of a huge urban area creates traffic bottlenecks, as there are only a handful of freeways and roads that cross the mountains. The eastern terminus of the range is Griffith Park, a wild area of steep mountains, as well as beautiful parks, playgrounds, the Griffith Planetarium, and the Los Angeles Zoo. Griffith Park is the largest, wildest, and most rugged urban park in the middle of any city in the world.

Further west, the south flank of the range drops into the Pacific Ocean in the movie star colonies of Pacific Palisades and Malibu. On the north side of the western Santa Monica Mountains are the communities that have grown up in eastern Ventura County, including Calabasas, Tarzana, Thousand Oaks, Westlake Village, and Agoura Hills. The rugged heart of the range is remote and scenic enough that many movie ranches are located there. Thousands of movies and TV shows have been shot in the Santa Monica Mountains, which have substituted for locations from the Wild West to Korea in the *M*A*S*H* television series.

(a)   (b)

(c)   (d)

FIGURE 13.11 Typical rocks of the San Gabriel Mountains: (a) The 1.7-billion-year-old gneiss, showing intense folding. (b) Hand samples of the San Gabriel Anorthosite showing reaction rims around the plagioclase in the more pyroxene-rich parts (right sample) and calcium plagioclase crystals of pure anorthite (left sample). (c) Roadcut through the San Gabriel Anorthosite pluton, just south of Baughman Springs, Angeles Forest Highway. The vertical stripes were originally horizontal layers of crystals that accumulated on the floor of the magma chapter about 1 Ga. The white triangles are crescumulate structures that formed at the bottom of the magma chamber. These mounds of plagioclase crystals built up from the floor of the chamber and have since been tilted vertically (top to the right). (d) Roadcut on the north end of Upper Big Tujunga Road showing the typical San Gabriel Mountain rock types and their crosscutting relationships. The black bodies are Mendenhall schists and gneisses. These are intruded by the white San Gabriel Anorthosite (with a spotted gabbro layer also visible). Two brownish dikes of the Triassic Lowe Granodiorite cut across horizontally. These are intruded by the pink granitic rocks across the top of the roadcut, probably from the Upper Cretaceous Josephine Granodiorite. Finally, a large vertical fault cuts through the right half of the outcrop, probably from recent uplift.

Source: (a, c) Photos by the author except. (c) Photo by B. Carter.

Since they are not as close to the San Andreas fault or as highly deformed or uplifted as the San Gabriel or San Bernardino Mountains, the Santa Monica Mountains do not expose ancient basement rock or the complex geology of those other ranges that border the San Andreas fault (Figure 13.13). The oldest rock unit in the range is the Santa Monica Slate (Figure 13.14a), composed of at least 2,460 m (8,000 ft) of shales metamorphosed into slates and phyllites. It crops out in the eastern Santa Monica Mountains north of Westwood and Malibu (Figure 13.13). The Santa Monica Formation yields Late Jurassic ammonites and bivalves, suggesting that it is a forearc basin equivalent of the Bedford Canyon Formation down in the Peninsular Ranges, or the Jurassic forearc basin rocks of the Foothills terrane in the Sierras and the Western Jurassic terrane in the Klamaths.

Intruded into the Jurassic slates are Cretaceous granitic rocks (Figure 13.14b), known as the Feliz granodiorite, the Vermont quartz diorite, and the Lar biotite quartz diorite. These crop out in the southeastern part of the range near Hollywood and Griffith Park (Figure 13.13). They are dated at 102 Ma, so they are older than the Josephine Granodiorite of the San Gabriel Mountains at 80 Ma or the granitics of the Verdugo Mountains at 83–94 Ma.

The Santa Monica Mountains are capped by a thick sedimentary sequence that makes up the bulk of the exposures (Figures 13.13 and 13.15a). The base of this sequence is about 1,860 m (6,050 ft) of Upper Cretaceous (Campanian–Maastrichtian) deep marine turbidite sandstones, shales, and submarine fan complexes known as the Chatsworth Formation in the Simi Hills (especially Rocky Peak and Stoney Point) and western San Fernando Valley

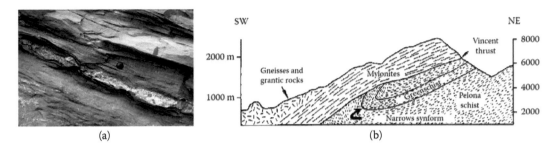

(a)   (b)

FIGURE 13.12 (a) Outcrop of the Pelona Schist from the eastern San Gabriels. (b) Simplified cross section showing how the rocks of the central–western San Gabriels are pushed over the top of the Pelona Schist in the eastern San Gabriels, along the mylonite zone formed on the Vincent thrust.

Source: (a) Courtesy of US Geological Survey. (b) Courtesy of US Geological Survey.

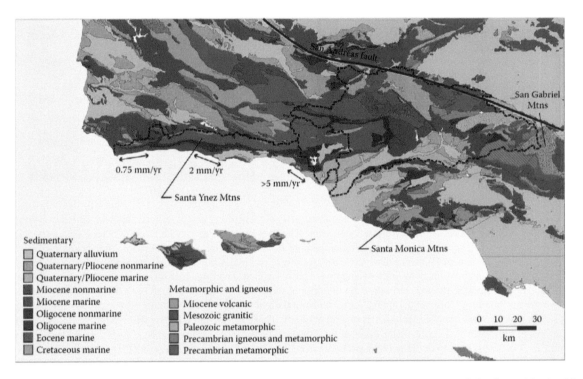

FIGURE 13.13 Simplified geologic map of the western Transverse Ranges, showing the geology of the Santa Monica Mountains (lower right), the Santa Ynez Range (lower left), and the more inland ranges.

Source: With permission from the Geological Society of America.

(Figure 13.14c). These are correlative with the Trabuco Conglomerate (Figure 13.14d) and the arkosic marine sandstones of the Tuna Canyon Formation in the central Santa Monica Mountains (Figure 13.15a).

Above this Cretaceous sequence is a thick package of sediments that spans much of the Cenozoic (Figure 13.15a). In the Simi Valley area lies the Paleocene Simi Conglomerate; this intergrades with and is overlain by the 540 m (1,700 ft) thick Santa Susana Shale, which yields late Paleocene marine fossils. In the central Santa Monica Mountains, the arkosic conglomeratic sandstone known as the Coal Canyon Formation is equivalent in age. These rocks are overlain by the middle Eocene Llajas Formation, composed of shales and sandstones with minor conglomerate.

As is true of much of the Transverse Ranges and Peninsular Ranges, most of the Oligocene is missing. However, in the western Simi Valley and western Santa Monica Mountains, there is a thick sequence of middle Eocene Sespe Formation red beds, overlain by upper Oligocene Sespe red beds (Figure 13.14e), with a 10-million-year unconformity between the two packages that wipes out the entire late Eocene and early Oligocene. The Sespe Formation in this area is famous for middle Eocene mammal fossils and a sparse fauna of late Oligocene mammals as well.

By far, the thickest portion of the Santa Monica Mountain sedimentary record is a long complex Miocene sequence. It begins with the gradual intergradation of upper Oligocene Sespe red beds into the shallow marine tan sandstones of the

FIGURE 13.14 Representative rocks of the Santa Monica Mountains and Simi Valley area. (a) Santa Monica Slate, Franklin Canyon Park. (b) Feliz granodiorite, forming the bedrock at Griffith Planetarium and the tunnel behind it. (c) Upper Cretaceous turbidite sandstones of the Chatsworth Formation, north side of Highway 118, Santa Susana Pass. (d) Cretaceous Trabuco Conglomerate, Stone Canyon. (e) Thick sequence of Sespe, Vaqueros, and Topanga Formations, Piuma Ridge. (f) Pillow lavas of the Conejo Volcanics, on the west side of Kanan Road just south of Highway 101. (g) Folded Monterey Shale.

Source: All photos by the author except (g), photo by R. Behl.

lower Miocene Vaqueros Formation, full of early Miocene mollusks such as the index fossil *Turritella inezana*. These are capped by a 670 m (2,200 ft) thickness of greenish-brown sandstones of the Topanga Canyon Formation. This unit is legendary for its rich fauna of middle Miocene mollusks, such as *Turritella ocoyana*, as well as abundant oysters and dozens of species of clams (Figure 20.16). Generations of young fossil collectors in Southern California first cut their teeth at localities along Old Topanga Canyon Road. In the western Santa Monica Mountains, the shallow marine rocks of the Topanga Canyon Formation interfinger with

contemporaneous volcanics of the Conejo volcanic field, dated between 13 and 15 Ma (Figures 13.13 and 13.15). In many places, you can find pillow lavas erupted underwater (Figure 13.14f), numerous dikes of Conejo volcanics cutting older rocks, and flows interfingered with marine sandstones. On the south side of Camarillo, the extinct Miocene volcanic cones are still visible.

The upper part of the Miocene is represented by 1,000 m (3,400 ft) of deep-water Monterey shales and cherts (some of which are mapped as the Modelo Formation), often intensely folded (Figure 13.15g). Deep marine shales and turbidite sandstones of the Puente and Repetto Formations are also found, although they are much thicker down in the adjacent Los Angeles and Ventura Basins. Finally, the Pliocene and Pleistocene rocks were tightly folded into a series of anticlines and synclines and numerous faults as they were deformed and uplifted into the modern Santa Monica Mountains (Figure 13.16). We will look at the tectonic reasons for this compression at the end of the chapter.

13.4.3 SANTA YNEZ RANGE

Space does not permit a complete review of all the complicated geology of every part of the Transverse Ranges (Figures 13.13 and 13.15b), including the fascinating geology of the Channel Islands or the San Rafael Mountains in the rugged hills north of Santa Barbara. Instead, we will look at one last mountain range that shows both similarities and differences to the ones we have already covered.

The Santa Ynez Range is the western extreme of the Transverse Range province, running from central Ventura County north of Ventura and Ojai all the way out to Point Concepcion. It forms a continuous crest along this entire distance, with a steep slope to the north in rural Santa Barbara County, and another steep slope down to the ocean in the coastal region of Santa Barbara County (Figure 13.17). There, cities like Ventura, Carpinteria, Goleta, and Santa Barbara and Highway 101 lie along its foothills and narrow coastal plain. Its highest point is 1,438 m (4,864 ft) in elevation, and it has numerous peaks almost that high, including Divide Peak at 1,435 m (4,707 ft), La Cumbre Peak at 1,215 m (3,985 ft), and Santa Ynez Peak at 1,310 m (4,298 ft).

Most of the range is underlain by Cenozoic sedimentary rocks (Figures 13.13 and 13.15b) which dip south toward the ocean on the south flank of the range. On the north slope of the range lies the Santa Ynez fault, which uplifted most of the range and tilted it to the south on the south flank, and to the north in the Santa Ynez Valley–Solvang–Buellton area.

The sedimentary sequence in the western Santa Ynez Range has a cumulative thickness of about 13,000 m (40,000 ft), almost all of it Cenozoic in age (Figure 13.15b). On the north flank of the range are limited exposures of Franciscan mélange, remnants of the Western Franciscan terrane just to the north (Chapter 12). In the far western Santa Ynez Range are outcrops of the Espada Formation, a dark-gray thin-bedded shale unit of probable Cretaceous age. In the area near Jalama Beach lies the Upper Cretaceous Jalama

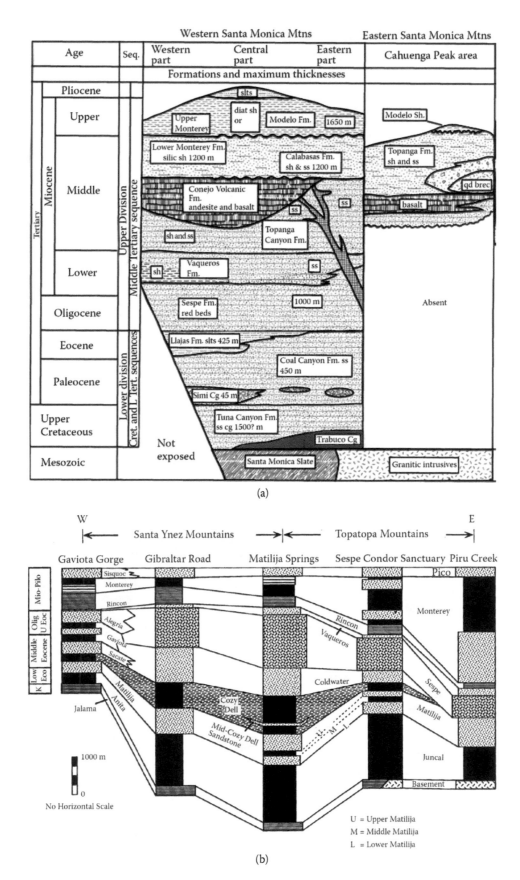

FIGURE 13.15 (a) Sequence of geologic formations in the Santa Monica Mountains. (b) East–west cross section of the Cenozoic stratigraphy of Ventura County (Topatopa Mountains) and the Santa Ynez Range, showing the facies change from the Sespe nonmarine red beds in the Oligocene of Ventura County (right) to the marine Gaviota and Alegria Formations in Gaviota Gorge (left).

Source: Courtesy of Pacific Section SEPM.

Formation, a thinly bedded shale and siltstone unit up to 900 m (2,800 ft) thick.

The Paleocene is entirely missing from the Santa Ynez Range, but the Eocene is incredibly thick, spanning almost 2,500 m (8,000 ft) of strata (Figure 13.15b). It begins with the deep-water Juncal Shale (1,300 m, or 4,000 ft, thick), dated between 47 and 50 Ma. The Juncal shallows upward into the nearshore marine Matilija Sandstone (650 m, or 2,000 ft, thick), also early–middle Eocene in age (44–46

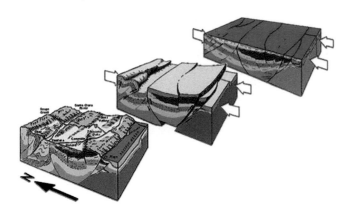

FIGURE 13.16 Geologic block diagram showing the history of the late Cenozoic folding, rotation, and faulting typical of the Santa Monica Mountains.

Source: Image by T. Atwater.

Ma), based on magnetic stratigraphy (Prothero, 2001). It is not only exposed in Matilija Gorge on Highway 33 north of Ojai but also forms prominent sandstone hogbacks along the spine of the Santa Ynez Range above Santa Barbara (Figure 13.18a). The Matilija Sandstone depositional setting transitions from deep marine turbidites to shallow marine sandstones and then grades upward into completely non-marine red beds (which can be seen at the bridge at Matilija Gorge on Highway 33), and then there is another transgressive sequence of shallow marine Matilija Sandstone changing to the deep marine Cozy Dell Shale (Figure 13.18b). The Cozy Dell Shale is poorly exposed as covered intervals on the slopes between sandstone ridges, except along Highway 33 north of Ojai, where it occurs in roadcuts. About 650 m (2,000 ft) thick, the Cozy Dell Shale has been dated paleomagnetically as late middle Eocene, between 44 and 42 Ma (Prothero, 2001).

The top of the Cozy Dell Shale shallows upward into the upper middle Eocene Coldwater Sandstone (Figure 13.18c). Full of shallow marine mollusk fossils, it is about 760 m (2,500 ft) thick and has been paleomagnetically dated at 42–40 Ma (Prothero, 2001). It is well exposed on Highway 33 in northern Ojai, along Coldwater Gorge just north of Santa Barbara, and forms a second prominent sandstone hogback running to the south and below the ridge-forming Matilija Sandstone along the south flank of the Santa Ynez Range all the way to Gaviota Gorge.

FIGURE 13.17   Oblique image of the Santa Ynez Range, looking from Gaviota to the northeast.

Source: Courtesy of US Geological Survey.

FIGURE 13.18 Typical outcrops of the Santa Ynez Range: (a) Middle Eocene Matilija Sandstone, which forms prominent vertically dipping sandstone hogbacks along the spine of the Santa Ynez Range. (b) Vertically tilted Cozy Dell Shale, Highway 33 south of Matilija Gorge. (c) Contact between white Coldwater Sandstone and red Sespe Shales, Highway 33 north of Ojai. (d) Red beds of the Sespe Formation north of Lake Casitas. (e) Gaviota Sandstone at Gaviota Gorge, dipping vertically with the top surface facing south (toward camera). The tunnels on Highway 101 cut through the Gaviota Sandstone. Alegria Shales form the valley lying above the Gaviota Formation (visible to the right of the shot). (f) Monterey Formation at Gaviota State Beach at low tide. (g) Hills built of an enormous thickness of turbidite sandstones, tan shales, and white diatomaceous shales of the Monterey Formation and Pico Formation, Balcom Canyon Road, south of Fillmore.

Source: (a–e, g) Photos by the author. (f) Photo by R. Behl.

Above the Coldwater Sandstone, there is a dramatic east–west facies change in the upper Eocene–lower Miocene rocks of the Santa Ynez Range (Figure 13.15b). At the eastern end in Ventura County along Highway 33, the Coldwater grades upward from shallow marine tan sandstones to the striking red beds (Figure 13.18d) of the Sespe Formation (already encountered widely in the Transverse Ranges and Peninsular

Ranges). Rare mammal fossils plus magnetic stratigraphy establish that the lower Sespe is late middle Eocene in age (40–38 Ma), and then an unconformity wipes out 9 m.y. of early Oligocene, and the upper Sespe is entirely upper Oligocene (29–27 Ma) in age (Prothero, 2001). However, if you trace the Sespe red beds from above Lake Casitas and then west along the Santa Ynez Range from one canyon to the next, you see a dramatic lateral facies change. The terrestrial mammal-bearing red beds are gradually replaced by shallow marine mollusk-bearing formations, the upper Eocene–Oligocene Gaviota sandstones and Alegria Shales, exposed best in the tunnels at Gaviota Gorge (Figure 13.18e).

Overlying the Sespe/Alegria throughout the range is the ubiquitous upper Oligocene–lower Miocene shallow marine Vaqueros Sandstone, found all the way from Monterey to San Diego County. On the south slopes of the Santa Ynez Range above Santa Barbara and Gaviota Gorge, it forms a third resistant sandstone ridge (beneath the Matilija at the top and the Coldwater just below it), separated by slopes of brush-covered shale (Cozy Dell between the Matilija and Coldwater, Alegria between the Coldwater/Sespe and the Vaqueros). The next shale slope above the Vaqueros is the lower Miocene Rincon Shale, another deep-water shale about 760 m (2,500 ft) thick. Finally, the southernmost slopes and the beach outcrops of the Santa Ynez Range are underlain by the middle–upper Miocene Monterey Formation. On the beaches at Gaviota and along the coast, there are excellent outcrops of Monterey Shale, held up by silicified layers that are highly wave-resistant (Figure 13.18f).

In some places north of the Santa Ynez Range, even younger units occur. North of Sespe Creek along Highway 33 and Chorro Grande Creek, the Monterey is capped by the upper Miocene Santa Margarita Formation, a marine unit that starts with deep-water phosphatic shales full of shark teeth at the base and then grades upward into shallow marine sandstone and, finally, thick deposits of gypsum deposited in saline lagoons. It has been paleomagnetically dated between 8 and 10 Ma (Prothero, 2001). In Ventura County along Santa Paula Creek and in the Topatopa Mountains (Figure 13.18g), there is an amazing sequence of Pliocene deep-water turbidites and shales of the Pico Formation that is more than 3,800 m (12,500 ft) thick, which is part of the extraordinary basin fill of the Ventura Basin (see following section).

13.5 SEDIMENTARY BASINS AROUND THE TRANSVERSE RANGES

Not only did the Transverse Ranges rise rapidly many thousands of meters in the late Cenozoic, but even more remarkable is the history of sedimentary basins trapped between the ranges. These sedimentary basins are very different from typical basins of the Midwest or the Great Plains. There the basin is usually hundreds of kilometers wide but barely a kilometer or two deep—somewhat like a shallow dish or bowl. Instead, the sedimentary basins of Southern California are remarkably deep but narrow. Some are as deep as they are wide, shaped like a trough rather than a dish.

Even though you can drive all over Southern California today and think that most of the city is quite flat, this is just an illusion. Beneath your feet the bedrock sinks down to make an enormous sedimentary basin (Figure 13.19) that happens to be completely filled with very young sediment, so the top of the basin forms a flat surface. This remarkable discovery became apparent when oil companies drilled down in many places around Los Angeles. Instead of drilling a few hundred meters to hit bedrock (as in most sedimentary basins in the world), they drilled beneath the downtown civic center of Los Angeles through more than 9,000 m (30,000 ft, or nearly 6 mi) of very young sediment (mostly deposited in the last 6 m.y.) and *still* did not hit basement rock! Further drilling ended up revealing the complex three-dimensional geometry of the Los Angeles Basin (Figure 13.19b). It's not a shallow dish or bowl as in most sedimentary basins in the world. Instead, it is divided into numerous subbasins, each of which moved at a different rate and depth, because each basin is a separate fault-bounded block. Most of the blocks sank down extremely rapidly during the last 6 m.y. As they did so, they were filled with thousands of meters of deep marine shales and turbidite sandstones, so the sedimentary basin was covered by a

topographic basin filled with deep seawater. That works out to a sediment accumulation rate of about 1,500 m (5,000 ft, or about a mile) of sediment per million years, one of the fastest subsiding basins ever documented. These deep marine deposits of the Miocene Monterey Formation, the Pliocene Pico and Repetto Formation, and the Pleistocene San Pedro Formation are widely exposed in the uplifted areas around the Los Angeles Basin, such as the Palos Verdes Peninsula, or in Elysian Park around Dodger Stadium. In these hills, it is typical to see resistant sandstone ledges formed by turbidites, alternating with shales that are deeply weathered and poorly exposed. Only during the last million years or so has the Los Angeles Basin filled with nonmarine sediments, and now it is uplifted and exposed to subaerial erosion.

A similar story can be seen in the Ventura Basin. If you drive around the Oxnard Plain through Camarillo, Oxnard, and Ventura today, you see a flat area that was once dominated by agriculture but is now becoming a sprawling suburb. But beneath this flat surface is another deep basin (Figure 13.20), almost 7,600 m (25,000 ft, or nearly 5 mi). It, too, is filled with deep marine sediments that are 5 Ma or less in age, again sinking down and filling with sediment at the extraordinarily rapid rate of about 1,500 m (5,000 ft) per million years (Figure 13.20b). In the case of the Ventura Basin,

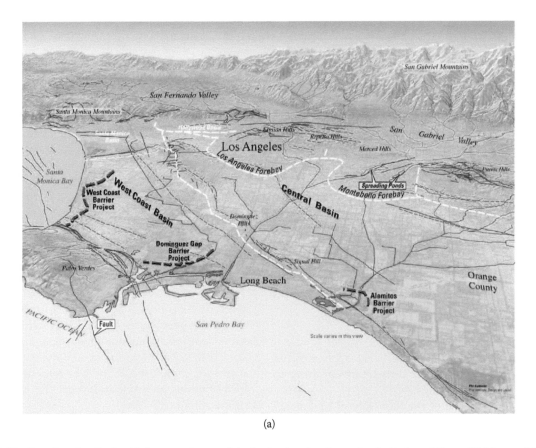

(a)

FIGURE 13.19 Los Angeles Basin: (a) Structural map of the basin showing the major faults in red lines. (b) Map and three-dimensional diagram showing the basement surface beneath the basin. (c) Stratigraphic sequence of rocks in the Los Angeles Basin.

Source: (b) From Geology of Southern California. *California Division of Mines and Geology Bulletin* 170, 1954.

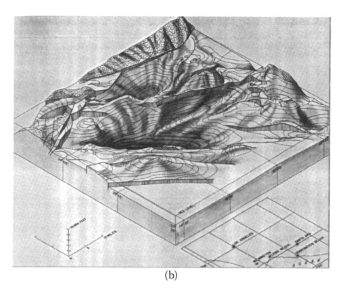

(b)

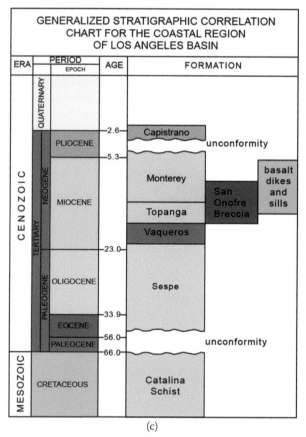

(c)

FIGURE 13.19 (Continued)

the south and north flanks of the basin are steep and even overturned, since there are thrust faults pushing up from the north and south sides.

On the northwest flank of the San Gabriel Mountains and south side of the Sierra Pelona is yet another remarkable sedimentary basin. Known to geologists as the Soledad Basin (Figure 13.9b), it was once wild country but, since the 1990s, has experienced almost complete urbanization and is now covered by the city of Santa

Clarita. The Soledad/Santa Clarita Basin is another deep fault-bounded trough that sank down almost 6,000 m (20,000 ft, or about 4 mi) and completely filled with sediment. Unlike the other basins, however, this one began much earlier. Its oldest units are 23 Ma basalt flows that poured across the bottom of the basin as it began to subside. These are then capped by 7,600 m (25,000 ft) of lower Miocene Vasquez Formation, a thick sequence of alluvial fan conglomerates and river sandstones that have since been tilted and faulted (Figure 13.21). Vasquez Rocks County Park is the most famous area in this formation, because it has been a legendary filming location for movies, TV shows, and commercials since the silent movie days. Because it is close to the Hollywood studios, it is cheaper to film on location here, and thus it has seen hundreds of Westerns and science fiction productions, as well as modeling shoots and car commercials. Overlying the Vasquez Formation are the Miocene Mint Canyon and Tick Canyon Formations, followed by the Pliocene–Pleistocene Saugus and San Fernando Formations. Altogether, this basin is almost as steep and narrow and filled with as much sediment as the Los Angeles or Ventura Basin, except that it began to open and subside at 23 Ma, not 6 Ma. Also, it is further inland, so it filled with mostly nonmarine sediments, rather than the deep marine deposits found in the more coastal basins that were below sea level.

Finally, travelers taking Interstate 5 between Castaic and Gorman encounter another remarkable sedimentary basin along the Old Ridge Route Highway that has carried travelers over the Transverse Ranges since the days of stagecoaches. The Ridge Basin is a peculiar sedimentary basin, because it is extremely narrow (about 4.5 km, or 15 mi, wide) but very deep (12 km, or 40,000 ft, or about 8 mi, of sediments deposited in just the past 15 million years). This basin is just the opposite of the shallow dish-shaped basins of the Midwest. It is deeper than it is wide (Figure 13.22). It is a classic example of a **transtensional** basin that formed as the two blocks of the irregular edge of the San Gabriel fault (the predecessor of the San Andreas plate margin here) moved past one another. As they slid by, gaps opened in their edges, and very steep-sided deep fault-bounded basins opened up in these gaps. Sediment poured into these basins as fast as the bottom dropped down, trying to keep up with the rate of subsidence. Since most of the lowest part of the basin is deep-water Castaic Formation deposits, the bottom was dropping down too fast for sediments to completely fill it. Most of the sediment poured in from the eastern edge, even at the western edge of the basin, which contains only one unit from the west, the Violin Breccia (Figure 13.22b).

Even more convenient for geologists who want to study it, you don't need to drill down to see the sediments at the bottom of the basin. Instead, once the basin finished filling up, it was tilted to the north by tectonic activity, so it was uplifted and eroded and exposed, with the oldest units

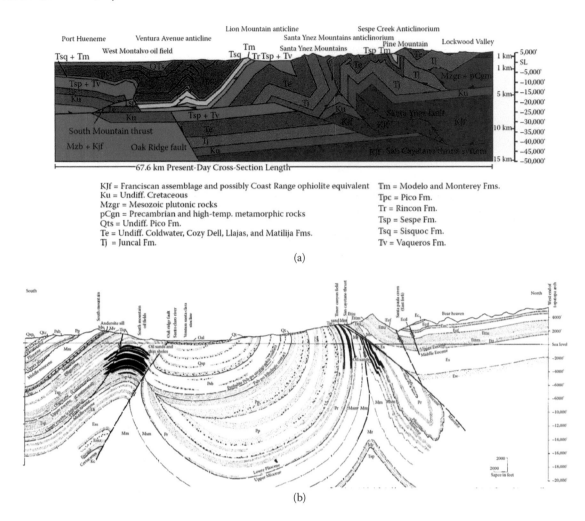

FIGURE 13.20 (a) Geologic cross section of western Ventura County, showing the intense folding of the ranges north of Ventura and Ojai. (b) Geologic cross section of the Ventura Basin.

Source: From *California Division of Mines and Geology Bulletin*, 170, 1954.

(Castaic Formation) exposed in the south, and the youngest unit (Hungry Valley Formation) in the north. Thus, as you drive north along Interstate 5, you climb the stack of rocks into younger and younger beds.

The lowest units are the upper Miocene (12–10 Ma) Castaic Formation (Figures 13.22 and 13.23a and b), a series of deep-water sandstones formed by turbidity currents and deep-water shales. This formation alone is more than 9,000 m (30,000 ft) thick. At the north end of Castaic Lake, you can actually see the outcrops of the bottom of the Castaic Formation. They show how sediments filled the basin as it sank, first with gravels and sandstones formed in alluvial fans, before being drowned by shallow marine sands (full of scallop shells). These were eventually completely drowned by deep-water deposits.

The next eight units are part of the Ridge Route Formation (Figure 13.22b), which is more than 9,000 m (30,00 ft) thick and ranges from 10 to 5 Ma in age (late Miocene). The ridges going east–west across the highway near the Templin Highway exit to Interstate 5 are the Marple Canyon Sandstone Member (Figure 13.23c) of the Ridge Route Formation (Figure 13.22),

which represents a shallow marine and nearshore deposit capping the Castaic Formation as the basin went from deep marine to shallow marine. The transition can be seen in roadcuts at the junction of Templin Highway and the Old Ridge Route (Figure 13.23c). Above this is the Ridge Route Formation, which has a complex stratigraphy (Figure 13.23d and e). In the center of the Ridge Basin, it was a deep-water lake (the shale members) alternating with episodes of alluvial fans building into the center from both edges (the sandstone members). At the eastern edge (Figures 13.22b and 13.24), the lake shale units disappear as it gets shallow, and you just get fan sandstones and conglomerates of the undifferentiated Ridge Route Formation. If you go to the western edge approaching the boundary fault (San Gabriel fault), you get a very distinctive unit, the Violin Breccia (Figures 13.22b and 13.24), which is made of sharp angular pebbles and cobbles dumped into the western edge of the basin during flash floods and debris flows from the uplifted San Gabriel fault scarp to the west. Thus, these outcrops represent a great example of lateral facies change, where the sedimentary environments change from east to west. What is Violin Breccia on the west

FIGURE 13.21 The famous tilted Vasquez Formation sandstones at Vasquez Rocks County Park, one of the most popular filming locations in movie history.

Source: Photo by the author.

transitions into the equivalent deep-water lake shale in a very short distance—and then as you continue east, the lake shale shallow and then disappear, and you just have alluvial fan sandstones and gravels from sources to the west.

The ridges of Marple Canyon Sandstone gradually vanish to the north, and next there is a broad shale valley along old Highway 99. This is the first of the deep-water lake shale units, the Paradise Ranch Shale (Figure 13.22). Then there is another east–west ridge of sandstone, the Fisher Springs Sandstone (Figure 13.23c). It is a lake delta deposit that built out from the east and west edges (Figure 13.24). It is full of ripple marks, mud cracks, fossil wood, fossil camel tracks, and other signs of shallow-water deposition, and fossil mammal bones have come from here as well.

Above the Fisher Springs Sandstone is the Osito Canyon Shale Member (Figure 13.23e). In the Piru Gorge Narrows, the Osito Canyon Shale abruptly transforms right into Violin Breccia (Figure 13.22), right at the west edge of the basin. Above the Osito Canyon Shale is the Frenchman

Flat Sandstone Member (Figures 13.22b and 13.23e). This is another lake delta sandstone that built into the center of the basin from its edges. Above it is the Cereza Peak Shale Member, the third lake shale up the section through the basin. Above this is the Piru Gorge Sandstone Member (Figure 13.22b), the fourth shallow delta sandstone that built out from the edge of the basin. Down where old Highway 99 runs, there are amazing roadcuts of the Piru Gorge Sandstone, with mud cracks, ripple marks, large petrified logs still in place, and other remarkable features that clearly tell what sedimentary environment it represented. It was apparently a lake delta, building out into deeper-water shale.

The Piru Gorge Dam is built in the natural narrows in the Piru Gorge Sandstone Member to block the old valley where Highway 99 used to run. Now it is flooded by Pyramid Lake, a huge reservoir that is part of the California Aqueduct system, bringing water all the way from the Feather River in the northern Sierras. Water covers most of the exposures of the next lake shale up (Figure 13.22), the Posey Canyon Shale Member (visible on the cliffs just north of the dam). At the north end of the reservoir are roadcuts in the uppermost members of the Ridge Route Formation, the Alamos Canyon Siltstone Member, and the Apple Canyon Sandstone Member (Figure 13.22).

Above last outcrops of the Apple Canyon Sandstone is the next formation that overlies the Ridge Route Formation. Called the Hungry Valley Formation, it is dated between 4.5 and 2 Ma (Pliocene). It is mostly deposits from ancient rivers and deltas that filled the Ridge Basin as the entire basin was nearly completely filled. Today, it is exposed on both sides of Interstate 5 but is part of a designated off-road vehicle area, so the outcrops are all chopped up by dirt bikes and ATVs.

13.6 TECTONICS OF THE TRANSVERSE RANGES

In the Transverse Ranges, the mountain ranges rose 3,000 m (10,000 ft) in just the past few million years. These mountains are aligned transverse to the trend of the San Andreas and its related faults, even though each mountain range has a different history. There are gigantic narrow fault-bounded sedimentary basins that are up to 10 km (33,000 ft) deep, filled with sediment less than 5 million years in age. There is evidence of enormous shallow thrust faults that have stacked one block of ancient rocks on top of much younger ones and push mountain ranges up in the sky after each quake. In addition, there is intense folding of sedimentary layers, some of which are less than 5 Ma. These are the remarkable features of the Transverse Ranges, and they have few counterparts in any other mountain range in the world.

For decades, geologists struggled to explain the tectonics of these peculiar mountains. They usually attributed it to the stresses of the "big bend" of the San Andreas fault, where the fault changes direction from southeast–northwest to

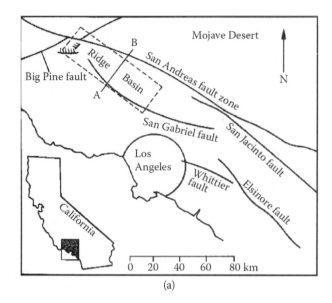

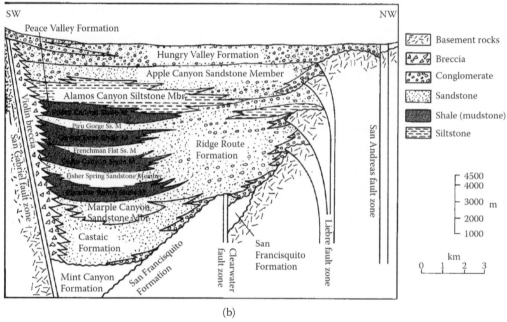

FIGURE 13.22 (a) Location map of the Ridge Basin. (b) Cross section of the Ridge Basin showing the major stratigraphic units and their relationships.

Source: Courtesy of Pacific Section SEPM.

east–west over part of the Transverse Ranges. According to these older models, all that was needed was stress to build up where the irregular edge of the fault boundary pressed against the other side, making transtensional basins (Figure 11.11).

Then in the late 1970s and 1980s, Bruce Luyendyk and his graduate students at the University of California, Santa Barbara, began to make paleomagnetic measurements on rocks throughout the Transverse Ranges. What they found was astounding. Almost every rock unit in the Transverse Ranges that was older than 16 Ma (middle Miocene and older) had a paleomagnetic direction that pointed east, not north, like the magnetic direction was supposed to

(Figure 13.25). This was true in the Santa Ynez Range, the Channel Islands, the Santa Monica Mountains, the Vasquez Rocks in the San Gabriel Mountains, and even in the Orocopia Mountains. (The San Bernardino and Little San Bernardino Mountains do not have rocks of the appropriate age or lithology.) As the 1980s progressed, these easterly directions were found in every pre-Pliocene rock in the Transverse Ranges, ruling out the possibility that it was some anomalous behavior of one or two rock units. In addition, rocks in the Mojave Desert or south of the Transverse Ranges showed no eastward directions, so the magnetic pole of the Earth didn't do something weird.

FIGURE 13.23 Typical outcrops of the Ridge Basin: (a) Deep-water Castaic Formation, near the top of the formation at the northeast junction of Templin Highway and Old Ridge Route. This outcrop shows a huge soft-sediment gravity fold. (b) The Castaic Formation is full of deep-water slump folds, including these S-folds. (c) Transition from the marine Castaic Formation (left foreground) to the deltaic and river deposits of the Marple Canyon Sandstone (at the top the road), northwest junction of the Old Ridge Route and Templin Highway. (d) Outcrops of the Fisher Springs Sandstone on old Highway 99. The top surfaces of the sandstone beds have abundant ripple marks, mud cracks, and even Miocene camel tracks. (e) Osito Canyon Shale (a lake shale) at the bottom, with deltaic Frenchman Flat Sandstone at the top of the cliff, on old Highway 99 just south of Frenchman Flat.

Source: All photos by the author.

The only explanation, then, is that the pre–middle Miocene rocks of the Transverse Ranges must have rotated or turned about 90° clockwise since they were formed. This kind of tectonic rotation has been found in many mountain belts in the last 40 years, but it is still staggering in its implications. It makes tectonic sense, however. The San Andreas fault creates a right-lateral shear in the area, so blocks caught between these shearing plates either slide smoothly northwest–southeast or are caught

like ball bearings between two sliding plates and rotate in a clockwise sense.

The original models proposed by Luyendyk, Kamerling, Hornafius, and others in the 1980s proposed a rather startling reconstruction (Figure 13.26). If you restored the rotated blocks to their original position prior to 16 Ma, then the Transverse Range blocks were originally oriented north–south, not east–west. (Figure 13.26a). Over the course of the later Cenozoic, the ranges began to turn clockwise, almost

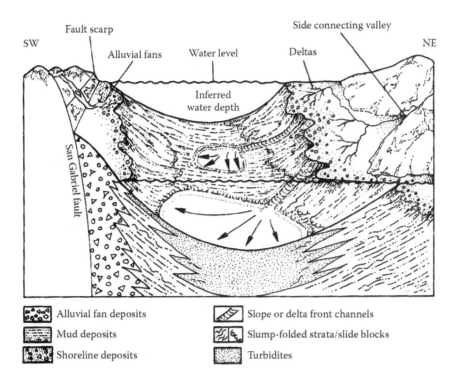

FIGURE 13.24 Reconstruction of the depositional setting of the Ridge Basin, with deep-water lake shales forming in the center and deltas and rivers building out from the eastern edge to form the interfingering sandstone bodies. On the western rim, the Violin Breccia is formed by coarse alluvial fans eroding from the San Gabriel fault scarp.

Source: Courtesy of Pacific Section SEPM.

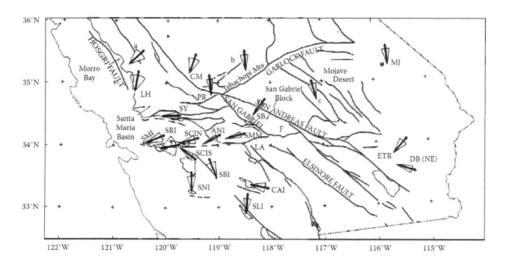

FIGURE 13.25 Paleomagnetic directions from the Transverse Ranges.

Source: From Luyendyk, B. P., and Hornafius, J. S., *Cenozoic Basin Development of Coastal California*, 259–283, 1991. Prentice Hall, Englewood Cliffs, NJ.

like a giant teeter-totter swinging around (Figure 13.26b). As they did so, the fault blocks would pull apart, opening up huge fault-bounded chasms in the crust between them (Figure 13.26a). This explains why the Los Angeles, Ventura, and other basins ripped open along fault lines and dropped so rapidly and filled with so much sediment.

Today, the ranges are oriented east–west, and the blocks west of the San Andreas have been chopped off and moved

north along the San Andreas fault (Figure 13.26a). The Orocopia–San Gabriel blocks, which were once a single mountain range, are separated by about 250 km (150 mi). This means that in the original position of the mountains, Santa Barbara would have been *south* of San Bernardino; now it is hundreds of kilometers to the *west*.

When the model was first proposed, it was hard to accept, because it turned all the conventional thinking on its head.

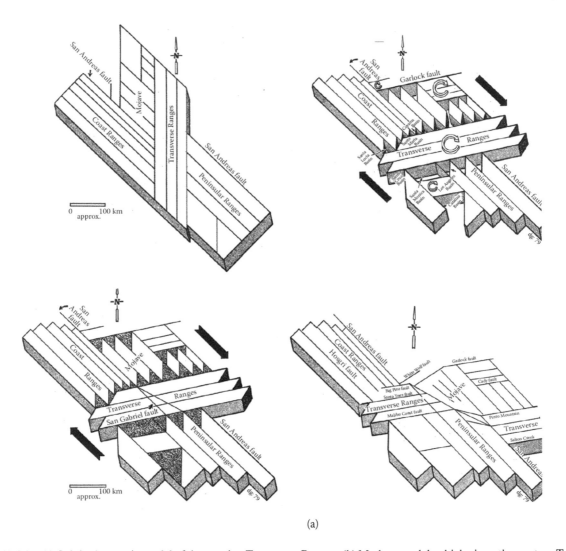

(a)

FIGURE 13.26 (a) Original tectonic model of the rotating Transverse Ranges. (b) Modern model, which views the western Transverse Range block (WTRB) as a swinging door, with the Santa Barbara end swinging clockwise and northwesterly from the south of the region to the east of the offshore bedrock.

Source: (a) From Luyendyk, B. P. *Geological Society of America Bulletin*, 103: 1528–1536, 1991. With permission from the Geological Society of America Bulletin. (b) Image by T. Atwater.

It was as jarring as plate tectonics must have seemed to the early geologists who could not fit it into their assumptions and framework of understanding. But the data were irrefutable, and gradually the model has been tweaked and modified to reflect geologic constraints.

For one thing, it is not necessary to have a giant tee-ter-totter (Figure 13.26) turning within the heart of the continental crust of Southern California. Instead, a more reasonable model views the Transverse Ranges block as a huge swinging "door" (Figure 13.26b). At one time, the block was oriented north–south parallel to the coast, with Santa Barbara at the southern tip. As the rotation began, the door swung open to the west, until the range axis was perpendicular to the California coast, and Santa Barbara was at the western tip. In 1994, Nicholson and others postulated that there was a small "Monterey microplate" in the edge of the subducting Farallon plate (see Chapter 15). When this

microplate was subducted and clogged up the subduction zone, it might have triggered this peculiar rotation in the block adjacent to it, according to Nicholson et al. (1994).

And this remarkable story doesn't end there. For decades, it was thought that the rotated blocks only included the main axis of the Transverse Ranges south of the Big Pine and Pine Mountain faults (Figures 13.25 and 13.27), and the transition to the southern Coast Ranges was supposedly unrotated. In the original models (Figure 13.26), the southern Coast Ranges blocks were supposed to slide smoothly northward parallel to the San Andreas fault with no rotation. Yet early paleomagnetic data on the volcanic rocks of the early Miocene Morro–Islay Hill Complex (Figure 12.17) between San Luis Obispo and Morro Bay also showed clockwise rotation (Figure 13.27). The mistaken impression was due to bad results on just a few data points north of the Big Pine–Pine Mountain faults, where

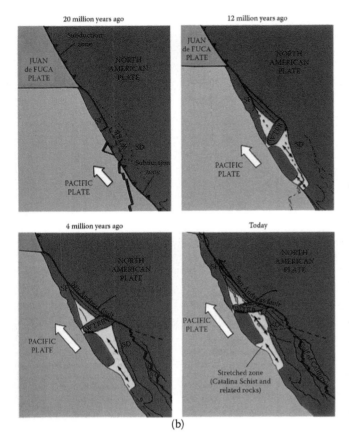

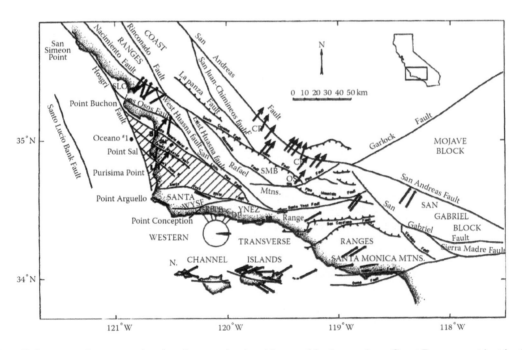

FIGURE 13.26 (Continued)

young normal overprints from the modern Earth's magnetic field had not been properly removed during demagnetization. Renewed research on a variety of pre-Miocene rocks from all over the southern Coast Ranges and northern part of the Transverse Ranges showed that they all had clockwise tectonic rotation, typically only about 45–60° (Prothero, 2006). In addition, in some sedimentary basins such as the Cuyama Basin, it was possible to look at the entire Miocene–Pliocene paleomagnetic history of the rocks and determine whether they rotated smoothly or in jerks and stops. It turns out that most rotated quickly and then were stable over long periods of time before rotating rapidly again (Prothero, 2006).

In addition to the rotation, there is tremendous horizontal compression and folding of the blocks as the plates push hard against each other and turn and grind into one another. This is demonstrated by the incredibly tight folding in places like the Santa Monica Mountains (Figures 13.14g and 13.16) and the mountains north of Ojai (Figure 13.20a) and the overturned folding and thrust faulting along the margins of the Ventura Basin (Figure 13.20b). This compression is still an important component today, since the thrust faults in the region are still very active. For example, the 1994 Northridge earthquake occurred on a deeply buried "blind thrust" whose existence was unknown until it moved.

In the more than 30 years since these results began to be published, we have learned a tremendous amount about

FIGURE 13.27 Paleomagnetic vectors showing that rotation is widespread in the southern Coast Ranges, not just in the Transverse Ranges south of the Big Pine–Pine Mountain faults.

Source: From Prothero, D. R., *Paleomagnetism and Tectonic Rotation of the Southern Coast Ranges, California*, Pacific Section SEPM Special Publication 101, Pacific Section SEPM, Fullerton, CA, 2006, 215–236.

the remarkable tectonics of the Transverse Ranges. But the tectonic rotation shown by the paleomagnetic data explains not only why the ranges arose so quickly and showed such intense compression, folding, and thrust faulting but also why enormously deep but narrow basins ripped open and their bottoms dropped out along steep faults. Both are explained by the rotation model.

13.7 SUMMARY

Although the Transverse Ranges have long been treated as a single mountain range and province, they are actually a string of unrelated fault-bounded blocks with very little in common from one range to the next. The east of the San Andreas fault ranges (Orocopias, San Bernardinos, and Little San Bernardinos) are on the North American plate and have not moved with the Pacific plate. They have relatively simple geology of late Proterozoic Mojave terrane metamorphics, intruded by Mesozoic granitics, and some have Cenozoic sedimentary fill as well.

In the middle, the San Gabriel Mountains are extremely complex, with Proterozoic metamorphics intruded by 1.2 Ga anorthosites, which were then intruded by Triassic and Cretaceous granitic plutons. These rocks of the western San Gabriels were overthrust on the Vincent thrust on top of late Mesozoic metasedimentary rocks of the Pelona Schist in the eastern San Gabriels. All these rocks have counterparts in the Orocopia Mountains, especially the Orocopia Schist beneath the Orocopia detachment fault, showing they were once a single range that was sliced in half by the San Andreas fault. The entire range is sliding on the Sierra Madre–Cucamonga thrust, which pushes the range up and to the south during each earthquake.

The western Transverse Ranges (Santa Monica Mountains, Santa Ynez Range, and others) are not nearly as high, old, or complex. Their oldest rocks are usually Mesozoic metamorphics (sometimes with Cretaceous intrusions), and all have a thick stack of Cenozoic marine sedimentary rocks, intensely folded and faulted since the late Miocene. Between these ranges are remarkable narrow fault-bounded sedimentary basins that are more than 10 km (6 mi) deep, filled with upper Miocene and younger marine sediments.

Paleomagnetic data have shown that all the Transverse Ranges (and some of the southern Coast Ranges) have rotated clockwise 45–90° since the middle Miocene. This is accomplished by swinging the entire block (which once trended north–south, with Santa Barbara at the southern tip) outward like a door on a hinge away from the continent as dextral shear of the adjacent crustal blocks is driven by San Andreas transform motion. This model explains not only why the ranges are rotated, faulted, compressed, and uplifted so rapidly but also why huge fault-bounded chasms opened up between the blocks and dropped so rapidly, filling with deep-water marine sediments as fast as the basin sank.

RESOURCES

Brown, H.J. 1991. Stratigraphy and paleogeographic setting of Paleozoic rocks in the San Bernardino Mountains, California. *Pacific Section SEPM Book*, 67, 193–207.

Crowell, J.C. 1981. An outline of the tectonic history of southeastern California, 583–500. In Ernst, W.G. (ed.), *The Geotectonic Development of California*. Prentice Hall, Englewood Cliffs, NJ.

Ehlig, P.L. 1981. Origin and tectonic history of the basement terrane of the San Gabriel Mountains, Central Transverse Ranges, 258–283. In Ernst, W.G. (ed.), *The Geotectonic Development of California*. Prentice Hall, Englewood Cliffs, NJ.

Fife, D.L., and Minch, J.A. (eds.). 1982. *Geology and Mineral Wealth of the California Transverse Ranges*. South Coast Geological Society, Santa Ana, CA.

Grove, M., Jacobson, C.E., Barth, A.P., and Vucic, A. 2013. Temporal and spatial trends of Late Cretaceous–early Tertiary underplating of Pelona and related schist beneath southern California and southwestern Arizona. *Geological Society of America Special Paper*, 374, 1–26.

Haxel, G.B., and Dillon, J.T. 1978. The Pelona-Orocopia schist and Vincent-Chocolate Mountain thrust system, southern California. *Pacific Section SEPM, Pacific Coast Paleogeography Symposium*, 2, 453–469.

Haxel, G.B., Tosdal, R.M., and Dillon, J.T. 1985. Tectonic setting and lithology of the Winterhaven Formation: A new Mesozoic stratigraphic unit in southeasternmost California and southwestern Arizona. *U.S. Geological Survey Bulletin*, 1599, 1–19.

Hill, M.L., and Dibblee, T.W., Jr. 1953. San Andreas, Garlock, and Big Pine faults, California—a study of the character, history, and tectonic significance of their displacements. *Geological Society of America Bulletin*, 64, 443–458.

Jacobson, C.E., Grove, M., Pedrick, J.N., Barth, A.P, Marsaglia, K.M., Gehrels, G.E., and Nourse, J.A. 2011. Late Cretaceous-early Cenozoic tectonic evolution of the southern California margin inferred from the provenance of trench and forearc sediments. *Geological Society of America Bulletin*, 123, 485–506.

Jahns, R. (ed.). 1954. Geology of southern California. *California Division of Mines and Geology Bulletin*, 170.

Kamerling, M.J., and Luyendyk, B.P. 1979. Tectonic rotations of the Santa Monica Mountains region, western Transverse Ranges, California, suggested by paleomagnetic vectors. *Geological Society of America Bulletin*, 90, 331–337.

Luyendyk, B.P., and Hornafius, J.S. 1991. Neogene crustal rotations, fault slip, and basin development in Southern California, 259–283. In Ingersoll, R.V., and Ernst, W.G. (eds.), *Cenozoic Basin Development of Coastal California*. Prentice Hall, Englewood Cliffs, NJ.

Nicholson, C., Sorlien, C.C., Atwater, T., Crowell, J.C., and Luyendyk, B.P. 1994. Microplate capture, rotation of the western Transverse Ranges, and initiation of the San Andreas transform as a low-angle fault system. *Geology*, 22, 491–495.

Prothero, D.R. (ed.). 2001. *Magnetic Stratigraphy of the Pacific Coast Cenozoic*. Pacific Section SEPM Special Publication 91. Pacific Section SEPM, Fullerton, CA.

Prothero, D.R. 2006. *Paleomagnetism and Tectonic Rotation of the Southern Coast Ranges, California*. Pacific Section SEPM Special Publication 101. Pacific Section SEPM, Fullerton, CA, 215–236.

Sharp, R.P., and Glazner, A.F. 1993. *Geology Underfoot in Southern California*. Mountain Press, Missoula, MT.

Spittler, T.E., and Arthur, M.A. 1973. Post-early Miocene displacement along the San Andreas fault in southern California. In Kovach, R.L., and Nur, A. (eds.), in *Proceedings of the Conference on Tectonic Problems of the San Andreas Fault System*, Stanford University Publications in the Geological Sciences, Stanford, CA, *13*, 374–382.

Stewart, J.H., and Poole, F.G. 1975. Extension of the Cordilleran miogeosynclinal belt to the San Andreas fault, southern California. *Geological Society of America Bulletin, 86,* 205–212.

Woodburne, M.O. 1975. *Cenozoic Stratigraphy of the Transverse Ranges and Adjacent Areas, Southern California.* Geological Society of America Special Paper 162. Geological Society of America, Boulder, CO.

Yin, A. 2002. Passive-roof thrust model for the emplacement of the Pelona-Orocopia Schist in Southern California. *Geology, 30,* 183–186.

VIDEOS

Wernicke and McQuarrie posted a short video showing the rotation of the Transverse Ranges and the northward movement of the Salinian block: www.youtube.com/watch?v=rzVU0PoZLE4

Animations

Tanya Atwater maintains a website (http://emvc.geol.ucsb.edu/1_DownloadPage/Download_Page.html) with excellent animations of the tectonics of the western United States, especially California. These are highly recommended for comprehending the three-dimensional motions of the tectonic blocks of California.

14 Granitics, Gravels, and Gems
The Peninsular Ranges and Salton Trough

Thank God I arrived the day before yesterday, the first of the month, at this port of San Diego, truly a fine one, and not without reason called famous.

—**Junipero Serra, 1769**

14.1 DATA

Geography: Peninsular Ranges batholith, Salton Trough

Elevation: Santiago Peak, Santa Ana Mountains: 1,733 m (5,687 ft); San Jacinto Peak: 3,296 m (10,804 ft); Mt. Palomar: 1,871 m (6,140 ft); Cuyamaca Peak, San Diego Mountains: 1,985 m (6,512 ft); the Salton Sea 69 m (227 ft) below sea level

Mountain ranges: Santa Ana, San Jacinto, Santa Rosa, Agua Tibia, Palomar, San Diego, Laguna

Predominant rocks: Mesozoic granitics and mafic plutonic rocks (Peninsular Ranges batholith), Mesozoic volcanic rocks, Mesozoic and Cenozoic marine and nonmarine sedimentary rocks

Plate tectonic events: Mesozoic Andean-style subduction zone volcanism and forearc basin sedimentation, Cenozoic transform tectonics

Famous faults: San Andreas, San Jacinto, Banning, Elsinore, Imperial

Resources: Gemstones (especially tourmaline), gold, water, sand, gravel

National parks: Cabrillo National Monument

14.2 GEOGRAPHY OF THE PENINSULAR RANGES

The region of Southern California south of the Transverse Ranges (Chapter 13) all the way down to the tip of Baja California Peninsula is known as the Peninsular Ranges Province (Figure 14.1). The US–Mexican border arbitrarily cuts through the middle of the province, so only the northern 240 km (150 mi) of the range is in California. Since this book focuses on California geology, it will not consider the 1,200 km (745 mi) of the Peninsular Ranges south of the border, although they are an important part of the story and well documented elsewhere. The entire province is dominated by the Cretaceous Peninsular Ranges batholith, sometimes called the "Southern California batholith." It runs down from the southern Mojave Desert to Riverside, south through the Santa Ana Mountains in Orange County and the San Jacinto Mountains east of Riverside, and then down through the numerous ranges in central San Diego County (Figure 14.2), and then down the spine of Baja California.

As mentioned in previous chapters, these ranges were once part of a continuous chain of Andean-style volcanic arcs and their magma chambers that ran from the Coast Ranges batholith in British Columbia to the Idaho batholith, to the Klamaths, and then down the Sierras, then the Salinian block, and finally the Peninsular Ranges batholith (Figure 8.7). The Peninsular Ranges batholith formed south of the Mexican border but has since moved north along the San Andreas over the past 40 million years (Chapter 11). When you restore the batholith to its original Cretaceous location, it was just the southern extension of the Sierra Nevada batholith and its southern half, the Salinian batholith (Figure 12.10). Since that time, the Salinian block has moved from south of the Sierras to west of them, and the Peninsular Ranges batholith has slid north to replace it and align with the south end of the Sierras.

The Peninsular Ranges batholith shares many characteristics with its Sierran counterpart. It is steeper on the eastern side, where it is truncated by large normal faults like the Elsinore fault or San Jacinto fault, and gradually sloping down to the ocean on its western side. Thanks to the rain shadow effect, the western slopes get most of the rainfall and vegetation. However, it is not covered by giant Sierra-style pine forests but mostly coastal sage scrub and piñon-juniper forest, since even the wettest parts of the area are semidesert. Only in the highest elevations do you find Ponderosa pine forests, like those of the more northerly ranges. Like the Sierras, the Peninsular Ranges create a big rain shadow behind them, forming the deserts of the Coachella Valley and Salton Trough and even affecting the areas to the east, including southern Arizona. The entire range is also chopped up by a series of strike-slip faults parallel to the San Andreas transform plate boundary, especially the Elsinore fault system, the San Jacinto fault system, and numerous smaller faults.

The rest of the province is influenced by the batholith in the center as well. Almost the entire coastal region of San Diego and Orange Counties was once a Cretaceous forearc basin, with thick deposits of Cretaceous and Cenozoic rocks still preserved there. In other places, the Cretaceous batholiths trap remnants of wall rocks and roof pendants of rocks that were there before the intrusions (Figure 14.3). Thanks to the dry climate and rapid uplift along faults, many of these units are now exposed at the surface and can be seen in the mountains and deserts of Orange and San Diego Counties today.

DOI: 10.1201/9781003301837-14

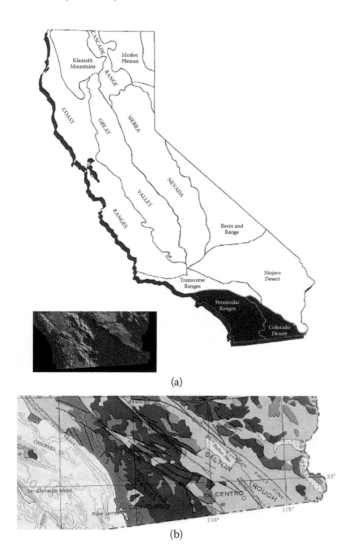

(a)

(b)

FIGURE 14.1 (a) Index map of California highlighting the Peninsular Ranges and Colorado Desert. In the lower left is the satellite image of the Transverse Ranges. (b) Geologic map.

Source: (a) Courtesy of US Geological Survey. (b) Courtesy of California Division of Mines.

14.3 PREBATHOLITHIC ROCKS OF SAN DIEGO COUNTY

The oldest rocks in the Peninsular Ranges are the Julian Schist and some other metamorphics, caught up as roof pendants and xenoliths in the Cretaceous intrusions (Figure 14.3). The Julian Schist (Figure 14.4a) is a Permo-Triassic metasedimentary unit thought to have started out as a marine sandstone–shale unit during the Permian and Triassic. Even though the unit is foliated, in outcrop you can still see the original bedding of thin sandstones and shales (now quartzites and schists). Careful analysis of the mineral grains in the Julian Schist pinpoints the source of the sediments as the Precambrian rocks now found in northern Mexico and southern Arizona. At one time, these sands and muds may have covered a large area of coastal California (then down in Mexico), but most were uplifted and heavily

metamorphosed and eroded so that only tiny roof pendants of the Julian Schist remain (Figure 14.3).

During metamorphism, the Julian Schist became host for many quartz veins from the intruding granitic rocks, which scavenged rare elements, including gold. Indeed, the town of Julian in the San Diego Mountains was originally a gold mining town. The Julian Gold Rush began in 1869, when a former slave, A. E. "Fred" Coleman, found gold flakes in a creek now named after him. Soon, hundreds of miners rushed to the area, founding numerous towns, including Julian (along with now-defunct places like Branson City, Eastwood, and Coleman City). On February 22, 1870, the first "lode" vein was discovered in Julian, and soon hard-rock mining became very productive. However, the mines were eventually exhausted, and most of the towns became ghost towns or vanished altogether. Julian today is famous as an artist's colony, and also for the many fruit orchards (especially apples) that were planted nearby shortly after the Gold Rush began.

In addition to the Julian Schist, there are other metamorphic roof pendants around the Cretaceous batholith. Near the border between San Diego and Imperial Counties, there are bodies of white marble, quartzite, and mica schist. These are thought to be similar to the Paleozoic limestones, sandstones, and shales found in the Mojave Desert, possibly as old as Ordovician in age. In other places, there are outcrops of amphibolite, caused by medium-grade metamorphism of oceanic crust. Thus, there must have been a significant Paleozoic and Triassic history in the region, although most of it is now reduced to metamorphic rocks without fossils that are tiny remnants of the original country rock and are trapped as small roof pendants.

In the Cuyamaca–Laguna Mountain (CLM) area, there are granitic gneisses thought to be remnants of a Jurassic (or older) volcanic arc complex (Figures 14.3 and 14.4c). These rocks are rich in quartz, feldspar, hornblende, muscovite, and especially platy biotite micas that give them their foliation. Based on the mineral composition, they are thought to have undergone temperatures approaching 600°C (1,100°F) and were buried about 13 km (8 mi) in the crust after they formed. Despite their intense metamorphism, their geochemistry resembles the Jurassic granitic rocks of the early Sierran arc up in the Owens Valley (Figure 8.5a). In addition, they give dates of 160–170 Ma, so they are Middle-Late Jurassic in age. Thus, before the intrusion of the Cretaceous arc rocks, the CLM gneiss was a Jurassic precursor in this region that then underwent intense metamorphism, followed by intrusion Cretaceous magmas after 120 Ma.

14.4 CRETACEOUS INTRUSIONS IN SAN DIEGO COUNTY

Just as in the Sierras to the north, the bulk of the igneous activity in the Coast Range–Idaho–Sierra–Salinian–Peninsular Ranges batholith (Figure 8.7a) occurred in the Cretaceous. These rocks include not only the batholiths made from former magma chambers, which dominate the

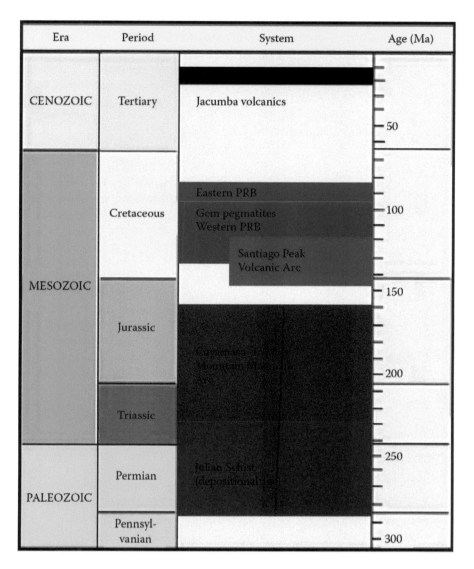

Era	Period	System	Age (Ma)
CENOZOIC	Tertiary	Jacumba volcanics	50
MESOZOIC	Cretaceous	Eastern PRB / Gem pegmatites / Western PRB / Santiago Peak Volcanic Arc	100
	Jurassic	Cuyamaca-Laguna Mountain Magmatic Arc	150 / 200
	Triassic		
PALEOZOIC	Permian	Julian Schist (depositional age)	250
	Pennsyl-vanian		300

FIGURE 14.2 Geologic history of the San Diego region.

Source: Redrawn after M. Walawender.

range, but also the volcanic rocks that once were fed by the magma chambers and stood above them. Most of these volcanics have been eroded away, but on the western foothills (Figure 14.3) of the Peninsular Ranges are scattered outcrops of their last remnants, known as the Santiago Peak volcanics (Figure 14.5). Most are andesite–dacite in composition, although some are closer to basalts. Much of the unit is composed of volcanic sediment fed into a deep-water forearc basin setting. Although a few parts are Upper Jurassic, most of them date to the Early Cretaceous (128–110 Ma), so they are the volcanics that were fed by the first generation of magma chambers to intrude the Peninsular Ranges, known as the Western Peninsular Ranges batholith (Figure 14.2). At one time, they must have extended across the entire Peninsular Ranges terrane, but as early as 95 Ma, uplift and erosion have removed all but a tiny remnant lying in the flanks to the west of the range.

The next phase of intrusion occurred in the Early Cretaceous in the western part of the Peninsular Ranges batholith (Figures 14.2, 14.3, and 14.6). Many different types of intrusions occurred. They include rocks much like the Sierran granitics, including abundant tonalite, quartz diorite, and quartz monzonite (Figure 14.6b). The intrusions can be separated into tonalite plutons that date between 102 and 110 Ma and quartz monzonites that date between 110 and 120 Ma. These form the characteristic spheroidally weathered boulder outcrops that are so typical of most of the Peninsular Ranges (Figure 14.4c).

Finally, the Lower Cretaceous intrusions included not only the familiar suite of granitic rocks found in the Sierras but also large intrusions of mafic gabbros as well. The most famous of these is the San Marcos Gabbro, in northern San Diego County, around Vista, San Marcos, and Escondido. It is a classic gabbro, rich in pyroxene and calcium plagioclase, with some olivine in parts of it. Although radiometric dating of the San Marcos Gabbro has been a problem, based on field relationships, they all appear to be Lower Cretaceous intrusions. The San

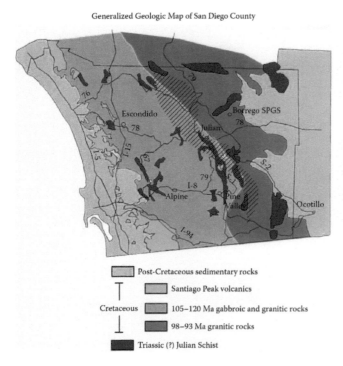

Generalized Geologic Map of San Diego County

Post-Cretaceous sedimentary rocks

Santiago Peak volcanics

Cretaceous — 105–120 Ma gabbroic and granitic rocks

98–93 Ma granitic rocks

Triassic (?) Julian Schist

FIGURE 14.3 Simplified geologic map of the San Diego region. Diagonally hachured area indicates outcrops of the Jurassic Cuyamaca–Laguna Mountain (CLM) gneisses.

Source: Redrawn after M. Walawender.

FIGURE 14.4 (a) Typical outcrop of the Julian Schist along Sunset Highway S-1 south of Julian. (b) Outcrop of the Cuyamaca-Laguna Mountain (CLM) Jurassic arc rocks (now gneisses) along Highway S-1, intruded by a Cretaceous quartz dike. (c) Spheroidally weathered outcrops of the 105 Ma (Lower Cretaceous) Cibbetts Flat Pluton on Highway S-1, Western Peninsular Ranges batholith. (d) Contact between the Lower Cretaceous Western Peninsular Ranges batholithic rocks (dark rocks in foreground) and the Upper Cretaceous La Posta Pluton (lighter rocks on the ridge beyond the saddle), Storm Canyon. (e) Outcrops of the La Posta pluton along old Highway 80.

Source: Photos by the author.

Marcos Gabbro has been a popular building stone for years, where it is misidentified as "black granite"—many large quarries still remain.

(To stone dealers, all rocks fall into two categories. If it is soluble in acid, it is "marble," even if it's really the unmetamorphosed parent rock of a marble, which is a limestone. All other igneous and metamorphic rocks, and even some sedimentary rocks, are called "granite." This is clearly different from how geologists use those words.)

These Western Peninsular Ranges intrusions and volcanics all occurred in the Early Cretaceous, no later than 110 Ma. By the Late Cretaceous (100–95 Ma), the plate apparently shifted to a shallower plunge angle, and the newer intrusions were focused in the Eastern Peninsular Ranges batholith (Figures 14.2, 14.3, 14.4d and e, and 14.6). This shallower, much more easterly zone of melting in the plate produced an intrusion known as the La Posta pluton (Figure 14.6). The La Posta intrusion forced its way to the east of the old extinct Jurassic arc of the CLM rocks and the Early Cretaceous intrusions of Western Peninsular Ranges batholith. Such an eastward shift of the intrusive belt is also true of the Sierras (Figure 8.5), suggesting that the plate dip was becoming shallower all across the Late Cretaceous arc over all of western North America. As we saw in Chapters 7 and 8, this was followed by the latest Cretaceous–middle Eocene (70–40 Ma) Laramide orogeny, when the plate began to slide horizontally beneath the overlying continental crust and formed no arc magmatism at all for 30 million years (Figure 7.21).

Between 97 and 93 Ma, the La Posta pluton produced huge mountains of granodiorite and quartz monzonites, which form most of the exposures in the Eastern Peninsular Ranges (Figures 14.3 and 14.4e). It covers more than 1,700 km² of eastern San Diego County, and another 1,200 km² occurs up in Riverside County or down in Baja California (Figure 14.6c). As these magma chambers rose, they caused uplift of the older rocks on top (especially the Santiago Peak volcanic arc) and caused most of the older material to erode away. They were probably the source of much of the heat (at least 1,000°C, or 1,830°F) that metamorphosed the Jurassic CLM arc rocks and the older Julian Schist.

In the very final stages of cooling, the melt was rich in quartz and a host of relatively rare elements that cannot be crystallized into the lattice of the minerals found in granitics like quartz monzonite, tonalite, and quartz diorite. This residue of leftovers included rare light elements, like lithium, boron, and beryllium, and heavier elements, like gold and silver. Some of these rare elements were scavenged from the metamorphic wall rocks (Julian Schist and Jurassic

FIGURE 14.5 Outcrop of porphyritic Santiago Peak volcanics.

Source: Photo by the author.

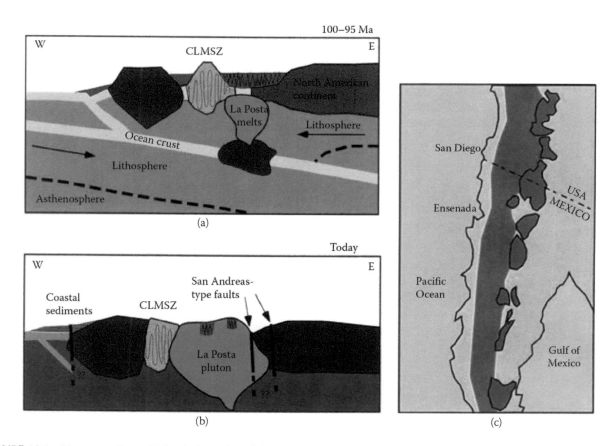

FIGURE 14.6 Plate tectonic model for the intrusion of the La Posta pluton in the Eastern Peninsular Ranges batholith due to shallowing subduction. It melted its way to the east of the Jurassic CLM Shear Zone (CLMSZ) and the Lower Cretaceous Western Peninsular Ranges batholith. (a) During the intrusion in the Late Cretaceous. (b) Today. (c) Map showing the location of the La Posta pluton (pink) in relation to the older Western Peninsular Ranges batholith (blue).

Source: Redrawn after M. Walawender.

gneisses) as the magmas melted their way through them. Others were concentrated by hydrothermal activity, as heated groundwater from just above the plutons dissolved rare elements out of the country rock and concentrated them in liquid silica.

The veins intruded much older rocks and then had centuries to cool slowly, making huge perfect crystals of minerals in what are called **pegmatite** veins and dikes. These produced not only enormous crystals of common igneous minerals, such as quartz, feldspars, and micas, but also rare minerals that were formed from the unusual chemistry of the melt. These include such gemstones as boron-rich tourmaline. In these pegmatite mines (Figure 14.7), there is a pink lithium-rich tourmaline (elbaite) that often has pink and green "watermelon" color zonation (Figure 14.8a and b). There is abundant pale lavender lithium-rich mica known as lepidolite (Figure 14.8c). Other gems include the pink lithium-rich form of pyroxene known as kunzite (Figure 14.8d), plus pink beryl (morganite) (Figure 14.8e) and blue beryl (aquamarine), topaz, and blood-red almandine garnet. The most famous district for these pegmatite minerals is near Pala in northern San Diego County, where more than a century of mining has produced incredibly large and beautiful gems, especially of "watermelon tourmaline" (a signature gem of this mining district). Between 1902 and 1911, the dowager empress of China, Cixi, was so fond of pink tourmaline (elbaite) that she created a demand that launched the mining district. The mining produces not only gems but also

FIGURE 14.7 Pegmatite veins in the mines of the Pala District, showing zones of crystals in the mine shaft.

Source: Courtesy of Wikimedia Commons.

(a)   (b)   (c)

(d)   (e)

FIGURE 14.8 Gem minerals of the Pala pegmatites. (a–b) Watermelon tourmaline, a form of pink elbaite tourmaline with a green "rind" around it reminiscent of a watermelon slice. (c) Lavender lithium-rich mica known as lepidolite. (d) Pink form of beryl known as morganite. (e) Purple mineral kunzite.

Source: (a, c) Photos by the author. (b, d, e) Photos courtesy of Wikimedia Commons.

minerals that provide ores of such rare elements as lithium (lepidolite) and boron. There is still considerable mining in the Pala pegmatites today. It is possible to arrange a visit to some of these mines and sort through the tailing piles for crystals of quartz, tourmaline, and many other gems.

14.5 MESOZOIC ROCKS OF ORANGE COUNTY

The rock sequence in the Santa Ana Mountains in Orange County has many similarities to its counterparts to the south (Figure 14.9). Their oldest rocks include early Mesozoic blueschists known as the Catalina Schist, which comes from the accretionary wedge complex of the Mesozoic arc rocks. At the base of the stratigraphic sequence on the western flanks of the Santa Anas in places like Silverado Canyon, the oldest unit is the Bedford Canyon Formation (Figures 14.9 and 14.10a). It is composed largely of sheared blocks of deep-water turbidite sandstones and shales, often intensely folded. The Bedford Canyon is dated by its ammonite fossils as Late Triassic to Jurassic in age and is the bedrock for most of the peaks of the northern Santa Ana Mountains. It is thought to have formed in an accretionary wedge in front of the early Andean-style volcanic arcs that formed in the Peninsular Ranges. In this regard, it is equivalent to the Foothill terrane in the northern Sierras, or the Western Jurassic terrane in the Klamath Mountains, which are Jurassic accretionary prisms for their respective arc volcano complexes.

Unconformably overlying the Bedford Canyon Formation is the Santiago Peak volcanics, the blanket of volcanic ashes and volcaniclastic sediments derived from the Early Cretaceous arc volcanoes that erupted up and down the Peninsular Ranges. The Santiago Peak volcanics (Figure 14.10b) are found only as remnants on the western slopes of the Santa Ana Mountains, as they are in the San Diego ranges as well. However, at one time the volcanic arc would have covered the entire region. Meanwhile,

there were intrusions of numerous Cretaceous plutons (Figure 14.10c) of different ages, now exposed on the crest of the Santa Ana Mountains (especially along the Ortega Highway), roughly corresponding to the plutonic activity of the Peninsular Ranges batholiths of San Diego County.

In Orange County, there is a thick Cretaceous forearc basin sequence that developed on top of the Santiago Peak volcanics, unlike the sequence in San Diego County. The lowest unit is the middle Cretaceous Trabuco Formation (Cenomanian, about 95 Ma), which is a conglomeratic sandstone (Figure 14.10d) deposited by alluvial fans on top of the eroded surface of Santiago Peak volcanics as the forearc basin began to subside. The Trabuco Conglomerates are also at the base of the Cretaceous sequence in the Santa Monica Mountains (Chapter 13). Overlying the Trabuco is a thick sequence of the Turonian–Santonian (94–84 Ma, Upper Cretaceous) Ladd Formation (Figures 14.9 and 20.11a), which represents shallow- to deep-water marine sandstones and shales as the forearc basin became deeper marine. The Holz Shale Member of the Ladd Formation in Silverado Canyon area is particularly famous for its rich fossil fauna of deep marine bivalves and gastropods. The collections are especially rich in the heavily ribbed shells of *Pterotrigonia*, large inoceramid clams, and the high-spired gastropod *Anchura* with its broad pointed flange on its aperture, as well as rare ammonites (see Chapter 20). The Ladd Formation is capped by the Upper Cretaceous (Campanian) Williams Formation, which is composed mostly of shallow marine sandstones and some conglomerates, representing a shallowing upward of the forearc basin fill as the Cretaceous came to a close. Although there may be a few latest Cretaceous (Maastrichtian) fossils in the upper part of the Williams Formation, the Cretaceous–Paleogene boundary is erased by a big erosional unconformity.

14.6 CENOZOIC ROCKS OF ORANGE COUNTY

In the Santa Ana Mountains in Orange County, the oldest Cenozoic unit above the Cretaceous–Paleogene unconformity is the upper Paleocene Silverado Formation, which is a distinctive reddish-tan unit (Figure 14.10e). It demonstrates a transgression from nonmarine sandstones at the base to shallow marine sandstones at the top, which yield late Paleocene mollusks and foraminifera. It reaches a total thickness of more than 550 m (1,700 ft). The Silverado Formation has been dated by magnetic stratigraphy as 56.6–57.7 Ma in age (Prothero, 2001). It is best exposed in Silverado Canyon and in its type section in the northern Santa Ana Mountains east of Irvine Lake.

Above the Silverado Formation is another unconformity, and then the next youngest unit in the Santa Ana Mountains and northern San Diego County is the Santiago Formation. This unit consists mostly of early Eocene (Domengine molluscan stage and Ulatisian foraminiferal stage) shallow marine deposits in the type area east of Irvine Lake. In northern San Diego County (especially around Vista and

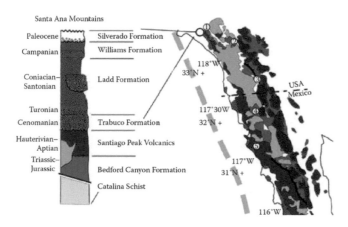

FIGURE 14.9 Geologic map and stratigraphic column of the Santa Ana Mountains, Orange County, California.

Source: Courtesy of US Geological Survey.

FIGURE 14.10 Typical Mesozoic and Cenozoic rocks of the Santa Ana Mountains, Orange County, California: (a) Deep-water Triassic–Jurassic turbidite sandstones and shales from the Bedford Canyon Formation. (b) Weathered Santiago Peak volcanics, Silverado Canyon. (c) Granitic rocks at the crest of Ortega Highway between Elsinore and San Juan Capistrano. (d) Conglomerates of the Cretaceous Trabuco Formation in Trabuco Canyon. (e) Reddish-tan sandstones of the upper Paleocene Silverado Formation, Silverado Canyon. (f) Brick-red sandstones and conglomerates of the upper Oligocene Sespe Formation, Irvine Regional Park. (g) Debris flow conglomerates of the San Onofre Breccia, made of huge blueschist boulders carried by debris flows. These deposits show inverse grading, typical of debris flows. Photo from the cliffs at Dana Point. (h) Cliffs of shallow marine Capistrano Formation sandstones and conglomerates, Dana Harbor.

Source: Photos by the author.

Oceanside), the Santiago Formation is considerably younger and different in facies. It consists of river and floodplain sandstones and siltstones, which were dated by magnetic stratigraphy between 41 and 46 Ma, or late middle Eocene (Prothero, 2001). In recent years, rapid development and excavation of the bedrock for housing and roadcuts has yielded a rich middle Eocene mammalian fauna in the Santiago Formation of northern San Diego County (see Chapter 20).

Another unconformity above the Santiago Formation removed most of the upper Eocene and lower Oligocene beds in Orange County. Lying above this unconformity is a thick sequence of upper Oligocene nonmarine red beds of the Sespe Formation (Figure 14.10f) so familiar from the Transverse Ranges (Chapter 13). At the top of this thick red bed package, the Sespe nonmarine deposits interfinger with and grade into the shallow marine sandstones of the Vaqueros Formation, with its typical early Miocene mollusks. In Orange County, however, the Sespe–Vaqueros sequence is much younger (early Miocene, 17–19 Ma) than up in Ventura County, where it is middle Eocene to late Oligocene in age.

Moving east from the Sespe exposures in the Santa Ana Mountains, the coastal plain of Orange County is entirely underlain by Miocene and Pliocene marine beds, often full of marine fossils. Vaqueros exposures are found not only in the Santa Ana Mountains but also in the Laguna Hills area, where they have yielded a large (and mostly unstudied) fauna of not only marine mollusks but also whales, manatees, seals and sea lions, and many other kinds of Miocene marine mammals that were found when housing subdivisions excavated the area. There are exposures of a middle Miocene debris flow deposit known as the San Onofre Breccia (Figure 14.10g) in the coastal cliffs of Dana Point and Laguna Beach, interfingering with deep marine sandstones and shales of the Capistrano Formation (Figure 14.10h). In Newport Back Bay, there are excellent exposures of the Monterey Formation, which are rich in diatomaceous earth, as they are throughout California. Finally, the Tertiary beds of Orange County are capped by a Plio-Pleistocene unit known as the Niguel Formation, and then the coastal areas have marine terraces capped with Pleistocene deposits.

14.7 CRETACEOUS AND CENOZOIC OF SAN DIEGO COUNTY

Southern San Diego County also preserves a thick wedge of Cretaceous forearc basin sediments as well as Cenozoic beds eroded off the Peninsular Ranges and affected by sea-level changes on the coastal plain (Figure 14.11). Up in the Rancho Santa Fe area north of San Diego are outcrops of the earliest unit, a reddish-brown boulder conglomerate known as the Lusardi Formation. Some of the boulders in the unit are 10 m (33 ft) in diameter. It is unconformably deposited on top of the irregular surface of Cretaceous Santiago Peak volcanics and batholithic rocks of the Peninsular Ranges

basement and represents Upper Cretaceous alluvial fans that formed as the Peninsular Ranges began to erode in the Late Cretaceous.

Along the coastal cliffs in Point Loma are rocks of the Upper Cretaceous Rosario Group, which is named for similar rocks near El Rosario in Baja California. The lowest unit is the 400 m (1,300 ft) thick Upper Cretaceous Point Loma Formation (Figure 14.12a). Its lower half is composed of yellow shallow marine sandstones and olive-gray clay shales deposited in deeper waters; the upper part is massive gray-black siltstones from deep waters. In addition to mollusks and microfossils, the Point Loma Formation also yields fragmentary specimens of duck-billed dinosaurs and armored ankylosaurs, one of the few places that yield dinosaur fossils in all of California (see Chapter 20). Overlying Point Loma is the uppermost Cretaceous (Maastrichtian) Cabrillo Formation. It is made largely of conglomerates in the lower part and nearshore sandstones in the upper part, with only a few shark teeth among the fossils found in the unit (Figure 14.11).

A major unconformity that wiped out the entire Paleocene and lower Eocene was incised into the top of the Cabrillo Formation. By the middle Eocene, the coastal basin began to subside again and accumulated a thick sequence of rocks ranging from nonmarine to deep marine environments, known as the La Jolla Group and Poway Group. The lowest unit is a conglomeratic sandstone known as the Mt. Soledad Formation, formed by alluvial fans on the eroded surface of the older rocks in the area. As you go west and seaward, the Mt. Soledad Conglomerates interfinger with and are overlain by shallow marine deposits of the Delmar Formation and Torrey Sandstone, exposed in the sea cliffs around La Jolla (Figure 14.12b). The Delmar Formation greenish-yellow mudstone and siltstones, with thick beds of oysters, were likely deposits of an ancient lagoon (Figure 14.12b and c). The Delmar Formation has been dated by magnetic stratigraphy at 48–49 Ma.

These, in turn, are overlain by deeper marine deposits of the Ardath Shale (Figure 14.12d), which eventually transgressed over much of the coastal region. Eventually, the seas regressed and there were deposits of the nearshore Scripps Formation (which yields terrestrial mammal fossils). Above the Scripps Formation is the mostly nonmarine Friars Formation, composed primarily of white river sandstone deposits and greenish floodplain mudstones (which is famous for its mammal fossils—see Chapter 20). The entire Ardath–Scripps–Friars sequence has been paleomagnetically dated at 45.5–47.5 Ma.

Above these rocks of the La Jolla Group is a second package of middle Eocene sedimentary rocks, the Poway Group (Figure 14.11). The lowest units are the boulders and gravels of the Stadium Conglomerate, exposed in the walls of Mission Valley near the stadium that hosts the San Diego Chargers (Figure 14.12e). To the south and west, it interfingers with and is overlain by the Mission Valley Formation, composed of shallow marine sandstones with abundant

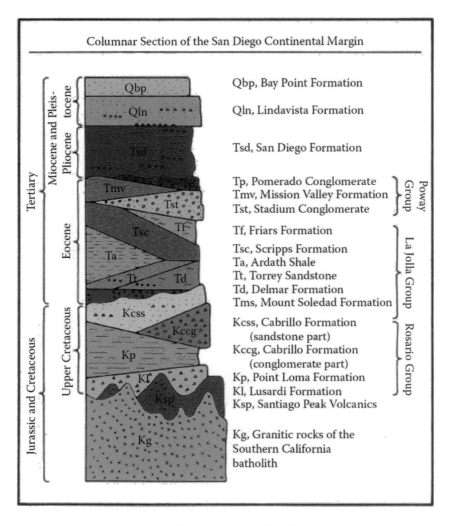

Columnar Section of the San Diego Continental Margin

Qbp, Bay Point Formation

Qln, Lindavista Formation

Tsd, San Diego Formation

Tp, Pomerado Conglomerate
Tmv, Mission Valley Formation
Tst, Stadium Conglomerate
} Poway Group

Tf, Friars Formation

Tsc, Scripps Formation
Ta, Ardath Shale
Tt, Torrey Sandstone
Td, Delmar Formation
Tms, Mount Soledad Formation
} La Jolla Group

Kcss, Cabrillo Formation
(sandstone part)
Kccg, Cabrillo Formation
(conglomerate part)
Kp, Point Loma Formation
Kl, Lusardi Formation
Ksp, Santiago Peak Volcanics
} Rosario Group

Kg, Granitic rocks of the
Southern California
batholith

FIGURE 14.11 Stratigraphic column for Mesozoic and Cenozoic rocks of the San Diego coastal region.

Source: Modified from Kennedy, M. P., and Moore, G. W. *American Association of Petroleum Geologists Bulletin*, 55: 709–722, 1971.

mollusks and exposed in many places along Mission Valley and in the roadcuts of Highway 163 through Balboa Park. The Mission Valley Formation yields many fossil mollusks, as well as a large fauna of late middle Eocene mammals (see Chapter 20).

After the Mission Valley transgression, another regression occurred, producing the Pomerado Conglomerate. Both the Pomerado and Stadium Conglomerates have been studied intensively and yield an unusual type of purplish-pink rhyolitic volcanic rock among their clasts. Known as "Poway clasts," these volcanic cobbles and pebbles have been traced to a source in northern Mexico, further confirming that the entire San Diego region was well south of the border in the middle Eocene. The Stadium–Mission Valley–Pomerado package has been dated paleomagnetically between 45 and 41 Ma.

A large unconformity wipes out the late Eocene, the entire Oligocene, and most of the Miocene in the San Diego region (Figure 14.11), suggesting that the coastal plain was uplifted and exposed to erosion. Down by the Mexican border on Otay Mesa, there are exposures of the upper Oligocene Otay Formation, which yield numerous fossil mammals (see Chapter 20), but otherwise, the late Eocene to late Miocene (37–5 Ma) is completely missing from most of the San Diego region.

Unconformably overlying these Eocene rocks are shallow marine rocks formed when rising sea levels drowned the coastal plain (Figure 14.11). The thickest unit is the San Diego Formation, which is Pliocene (2–3 Ma) in age. It yields a rich fauna of marine fossils, including many mollusks, seabirds, and marine mammals, especially porpoises, whales, walrus, fur seals, and an extinct relative of the Steller's sea cow (see Chapter 20). Finally, the entire San Diego region is underlain by Pleistocene terrace deposits of the Linda Vista Formation. Although it is a very young deposit, it is often cemented with red iron oxides, making a hard caprock that tops nearly all the mesas and flat terrace areas in the San Diego region.

14.8 THE BAJA–BC HYPOTHESIS

In the 1970s, paleomagnetic studies of rocks in coastal British Columbia and Alaska (the Insular superterrane and

FIGURE 14.12 Outcrops of Cretaceous and Cenozoic rocks of the San Diego area. (a) Outer submarine fan deposits of the Cretaceous Point Loma Formation at La Jolla Bay. (b) Beach cliffs composed of Torrey Sandstone (top) and Delmar Formation (bottom), Torrey Pines State Park. (c) Fossil oysters from ancient lagoonal deposits of the Delmar Formation. (d) Ardath Shale (bottom) and Scripps Formation (top) at Black's Beach. (e) Outcrop of Stadium Conglomerate near Lakeside.

Source: Photos by P. Abbott.

Coast Mountains terrane) produced some startling results. The paleomagnetic directions in Cretaceous rocks were much too shallow, suggesting that they had formed at a latitude around Baja California and the San Diego area today. Since the Cretaceous, they had slid north along faults with as much as 3,000 km of northward translation. Although there is no question that coastal California and the Salinian terrane and Monterey terrane (Chapter 12) have moved from a position south of the Sierras to their present location (a slip of possibly 1,000 km), it seemed pretty outrageous to suggest that outboard of this slip were additional blocks from even further south that travelled much further north.

The main doubts about the Baja–BC hypothesis centered on the lack of any clear-cut fault zones that might accommodate such a huge amount of slip, as well as any rock units that might act as "piercing points" to confirm that at one time the rocks of coastal British Columbia once lay adjacent to Mexico. Another source of doubt was the fact that much of the paleomagnetic data came from igneous plutons, and it is hard to establish the paleohorizontal in units which don't have horizontal bedding in them, like sedimentary rocks do. Some tilting of those same igneous plutons could make their paleomagnetic inclinations unusually shallow without having to move them horizontally. However, not all the shallow inclinations came from plutons. The thick sequence of Cretaceous marine rocks of the Nanaimo Group on Vancouver Island also produced shallow inclinations, consistent with the Baja–BC model.

It has now been more than 40 years since the Baja–BC model was first proposed, and there have been dozens of papers published on it, and even a special Penrose Conference in 2000 and again in 2022 to focus on just this controversy. Although numerous attempts to reconcile the differences have been proposed, there still seems to be no consensus on what model explains all the data. The current competing hypotheses include the following:

1. *Long-distance transport over 1,000 km.* Paleomagnetists (e.g., Housen and Beck, 1999) have tested and retested their data, and many are adamant that there is no alternative explanation for long-distance transport over 2,000 km. This is especially true when sedimentary rocks like the Nanaimo Group, whose horizontal orientation can be established, still give such shallow inclinations. Housen and Beck (1999) and Matthews et al. (2017) have provided additional evidence in the form of sedimentary zircons found in the Nanaimo Group, which have their source in Southern California or even Baja California. This provides some possibility of a "piercing point" that physically connects the two regions.

2. *Some combination of translation plus tilting.* Others (Butler et al., 2001) argued that the shallow paleomagnetic inclinations must suggest some horizontal movement (like we know occurred with the Salinian block), but that if the British Columbia rocks have undergone just a few degrees of tilting as well, you would not require more than 2,500 km of horizontal movement. This is a form of compromise but doesn't completely explain the shallow inclinations in the Nanaimo Group, whose tilting can be corrected for since we have its paleohorizontal orientation.

3. *Large-scale tectonic movements across the entire Cordillera.* Hildebrand (2015) has argued that British Columbia has moved at least part of that 2,500 km, but that large-scale sliding and rotation across the entire southern Cordillera region can also account for some of the translation of that terrane. This is another way to avoid the extremes of the original hypothesis and provides some additional possible piercing points, but it is also a broad sweeping hypothesis of movement over a huge area of the southwestern United States, which still has yet to be confirmed with additional independent data.

The Baja–BC debate is now over 40 years old and still not resolved. Some geological controversies are like that.

14.9 THE SALTON TROUGH

The Peninsular Ranges batholith is chopped off on its eastern edge by the Elsinore fault zone and also by the San Jacinto fault zone (Figure 14.13). The other side of these fault zones is the Salton Trough, formed by the transition between the rifting of the spreading ridge beneath the Gulf of California and the area where it transitions to a strike-slip transform boundary (see Chapter 11). It is a huge stretched-out fault graben formed by transtension that began to rift about 30 Ma when a seafloor spreading ridge pulled Baja California away from Mexico and began moving it 300 km northward (Figure 14.14). This created a huge extensional zone, with both stretched-out normally faulted crust in large detachment faults (Figure 7.7) and extreme shear due to the northward motion of the Pacific plate along the San Andreas fault. Most of this basin is below sea level, analogous to deep subsiding rift basins like the Dead Sea and Death Valley. The lowest point is in the center of the Salton Sea, which is 69 m (227 ft) below sea level, and the bottom of the sea is another 16 m (52 ft) deeper, only 1.5 m (5 ft) higher than the deepest point in Death Valley.

The consequence of all this stretching and shear is the immense basin known as the Salton Trough and Imperial Valley, where the crust is very thin. The mantle material, just below the surface as an active spreading ridge boundary in the Gulf of California, changes to a transform fault boundary where the San Andreas fault begins at the north end of the Gulf of California. The result is big "pull-apart" transtensional basins that stretch out and rip open, and their bottoms drop out so they are rapidly filled with sediment. Almost all the Imperial Valley is in a pull-apart basin, as is the Salton Sea.

Naturally, such thinned-out continental crust with the hot mantle not far below has lots of effects. First, the region is very seismically active, with dozens of medium-sized earthquakes in a single year. These go from at least as far back as the M7.8 earthquake in 1892, which leveled the entire region and toppled buildings all over San Diego County, to a M6.9 event in 1940, in which the crust jumped 4.5 m (15 ft) north in seconds; to the big earthquakes of April 4–5, 2010; to the M6.9 event in the Gulf of California on October 21, 2010; to the quakes of December 15, 2010, which included a swarm of tremors up to M4.4 in the Brawley Seismic Zone.

The good news is that this segment of the San Andreas is continuously moving in many small earthquakes and only rarely experiences a giant earthquake in the M6–7 range or larger. This is in contrast to the San Andreas fault in the Transverse Ranges, which has been locked since the Fort Tejon earthquake of 1857 and is overdue for a huge event. The bad news is that people living in the Imperial Valley experience several smaller earthquakes a year and several earthquakes per decade that do significant damage and kill people.

In addition to the geothermal fields, there are five small volcanic domes, known as Obsidian Buttes (Figure 14.15), southwest of Mullet Island. They form a northeasterly trend, perpendicular to the line of the San Andreas fault. The buttes rise 30–50 m (100–160 ft) above the valley floor. They are made of rhyolite and pumice, with large amounts of obsidian.

14.10 DEVELOPMENT OF THE SALTON TROUGH

Anza-Borrego State Park covers more than 2,420 km² (600,000 acres), and 20% of San Diego County is within

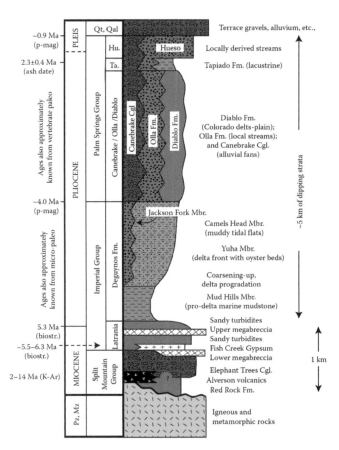

FIGURE 14.16 Stratigraphic section of the Miocene–Pleistocene deposits in the Salton Trough–Anza-Borrego Desert area.

Source: Image by R. Dorsey.

These magnetostratigraphic dates, in turn, allowed geologists to precisely date the time ranges of all the mammal fossils recovered from the area. This is now the "gold standard" by which paleontologists can date the time range of any of these mammals anywhere in North America. Further details can be found in Chapter 20.

14.11   SALTON SEA

The Salton Trough has long alternated between flooding by deep lakes and exposed barren dry desert. During the ice ages, a huge body of water known as Lake Cahuilla filled most of the Salton Trough. Its shorelines can be seen along the slopes of the Chocolate Mountains and as far north at Palm Springs. However, by 1900, the Salton Sink was again a desert. Today it is filled with a stagnant body of water known as the Salton Sea, which is an entirely human-made body of water.

The surface of the Salton Sea is 69 m (227 ft) below sea level, and the deepest area of the sink is only 1.5 m (5 ft) higher in elevation than the lowest point in Death Valley. It covers 970 km² (376 mi²), making the Salton Sea the largest lake in California. Its maximum depth is 16 m (52 ft) in the center of the basin, so it contains about 9.25 km³ (7.5 million acre-feet) of water. Its salinity is about 44 g/L of salts, which is saltier than the ocean (35 g/L), but not as salty as the Great Salt Lake.

Even though it is an artificial lake, it is maintained by the runoff from the surrounding mountains and drainage from agricultural areas, so it receives about 1.68 km³ (1.36 million acre-feet) of inflow each year. The basin is closed, or **endorheic**, with no rivers flowing out to the sea, so all the loss of water occurs through evaporation. In recent years, excessive use of the water in the region for agriculture has diminished the amount going into the Salton Sea, and the lake level is dropping. This threatens the huge bird sanctuaries at the south end of the lake, and there have been big efforts to restore the flow into the basin and keep its lake level from dropping further. One reason the Salton Sea bird sanctuary is so important is that nearly all their old habitats of coastal lagoons and marshes have been destroyed by development.

Before 1905, however, there was no water at the bottom of Salton Sink, and its main use was salt mining from the dried-up Ice Age lake basin. Yet the Colorado River was temptingly nearby, a source of water that could turn the dry desert into a cropland. An entrepreneur named Charles Rockwood became obsessed with the idea of such a canal system in 1892 and eventually got financial support to build a small canal system in 1900 to divert some of the flow of the Colorado River down into the Salton Sink, which started a boom in agriculture. The problem was that the Colorado River was (in the words of Marc Reisner from his outstanding book *Cadillac Desert*)

> like a forty-pound wolverine that can drive a bear off its dinner, it is unrivaled for sheer orneriness. The virgin

the walrus *Valenictis*, whales and dolphins, manatees, and a large variety of seabirds, including pelicans and flamingos (see Chapter 20). The "Elephant Knees" along Fish Creek is made of a thick shell bed composed mostly of oysters that lived when the water was brackish, rather than normal marine in salinity. This marine fauna is particularly interesting because it lived during the time just before the Isthmus of Panama closed and cut off the Caribbean from the Pacific. Thus, there are many Caribbean tropical mollusks among the fossils in the Imperial Formation.

About 4.2 Ma (Figure 14.16), the sea retreated from the Salton Trough, and the marine Imperial Formation was replaced by the nonmarine Palm Springs Formation (Figure 14.17). This impressive pile of sediments more than 3,000 m (10,000 ft) thick is exposed in many places in the Anza-Borrego badlands (see Chapter 20). It records the middle and late Pliocene (Blancan land mammal age; 4.2–1.8 Ma) and most of the ice ages (Pleistocene; Irvingtonian land mammal age) from 1.8 Ma to about 90,000 years ago. The dating on this sequence is very precise. Its magnetic pattern can be calibrated because there are dated volcanic ashes, such as the Bishop Tuff from near Mammoth Mountain (758,000 years old) interbedded with the sediments that allow precise calibration.

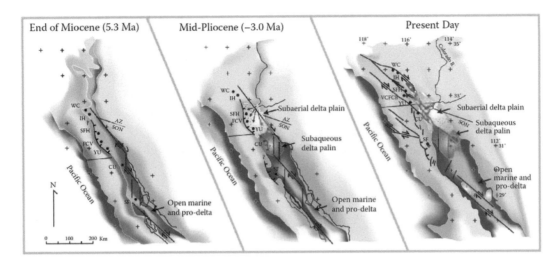

FIGURE 14.17 Tectonic history of the Salton Trough.

Source: Map by R. Dorsey.

(a)   (b)

FIGURE 14.18 Outcrops of the Anza-Borrego badlands: (a) Split Mountain Gorge–Fish Creek Canyon, with gigantic megabreccias of the Elephant Trees Conglomerate exposed in the walls. (b) Split Mountain Gorge–Fish Creek Canyon showing a gigantic gravity-slide fold due to soft-sediment deformation in the Latrania Formation.

Source: Photos by D. Patton.

Colorado was tempestuous, willful, headstrong. Its flow varied psychotically between a few thousand cubic feet per second and a couple of hundred thousand, sometimes within a few days. Draining a vast barren watershed whose rains usually come in deluges, its sediment volume was phenomenal. If the river, running high, were diverted through an ocean liner with a cheesecloth strainer at one end, it would have filled the ship with mud in an afternoon. The silt would begin to settle about two hundred miles above the Gulf of California, below the last of the Grand Canyon's rapids, where the river's gradient finally moderated for good. There was so much silt that it raised the entire riverbed, foot by foot, year by year, until the Colorado slipped out of its loose confinement of low sandy bluffs and tore off in some other direction, instantly digging a new course.

By 1904, the Imperial Canal became clogged with silt from the river, reducing its capacity to handle the flow of the Colorado. Then in 1905, heavy rainfall and snowmelt caused flooding of the Colorado River, which then overran the gates on the Alamo Canal. The flood poured down the canal, breached a dike in the Imperial Valley, and eroded two rivers, the New River in the west and Alamo River in the east. The river kept eroding deeper into the desert sand and soon cut a huge waterfall that started at 4.6 m (15 ft) in height but grew to 24 m (80 ft) tall before the breach was finally closed. As Reisner described it:

> In February [1905] the great surge of snowmelt and warm rain spilled out of the Gila River, just above the Mexican channel, and made off with the control gate. For the first time in centuries, the river was back in its phantom channel, the Alamo River, heading for its old haunt, the Salton Sink. As the surge advanced across the Imperial Valley, it cut into the loamy soil at a foot-per-second rate, forming a waterfall that marched backward toward the main channel. Even as their fields were being eaten and as their homes swam away, the valley people came out in the hundreds to this apparition, a twenty-foot falls moving backward at a slow walk. By summer, virtually all the Colorado River was out of its main channel, and the Salton Sink had once again become the Salton Sea.

Rockwood's company went bankrupt, so the Southern Pacific Railroad bought it out in order to protect their rail lines that were threatened. They now put their huge resources and energy into stopping the flood. The Southern Pacific repeatedly brought huge trainloads full of rocks and gravel to dump into the breach, only to fail over and over. For two more years, the Colorado experienced record rainfall, and these new rivers carried the entire Colorado River into the Salton Sink. In 1907 alone, the river had a record flow of 25 million acre-feet, or 8 quadrillion gallons. Finally, in late 1907, the rainfall and the flow of the river went into a lull, and the temporary dikes and weirs held. Engineers breathed a sigh of relief, because if the erosion had cut upstream much further, it would have reached the main channel of the Colorado River and diverted its course away from the Pacific Ocean. As the basin filled, the old town of Salton, an old railroad siding, and some Torres–Martinez Reservation land were submerged. The flood of water and the lack of drainage had produced a new sea where there was none before.

After the flooding stopped and the Salton Sea was now permanent, the residents accepted the new lake as a benefit. It became a tourist attraction in its early phases, with numerous beach resorts, especially where natural hot springs added to the amenities. The resorts of Bombay Beach, Desert Beach, Salton City, and North Shore were once boomtowns in the early twentieth century for people seeking warm waters to swim and bathe in. Many Hollywood celebrities of the Age of the Silver Screen would vacation at the Salton Sea, making it a trendy place to visit. But as the lake went from fresh water to a salty brine and became more

(a)

(b)

FIGURE 14.19 (a) Typical decrepit ghost towns remain of the once-thriving tourist industry around the Salton Sea, here shown in Bombay Beach. (b) Tilapia die-offs kill millions of fish when the winds stir up the waters and cause algal blooms, depleting the oxygen as the algae rot.

Source: Courtesy of Wikimedia Commons.

stagnant, it was no longer fun to swim in and most of the resorts have gone out of business or are just barely hanging on (Figure 14.19a).

Today, it is a stagnant salty basin that supports only salt-tolerant algae, brine shrimp, and a few other organisms, plus seabirds that feed on them during their migrations. It occasionally gets a large population of tilapia fish, which are tolerant of most of the extremes in the water and provide good fishing conditions when the water is suitable. They were stocked in the 1950s to promote tourism, and for a while, they were a reliable source for fishermen. However, in recent years, there have been huge die-offs of fish (Figure 14.19b), leading to millions of dead fish on the lakeshore and sand that is composed mostly of fish bones. The die-offs are thought to be the result of several factors. During the summer, the water temperatures reach 52°C (125°F) and the water is extremely salty and low in oxygen, so even hardy fish like tilapia can no longer survive. In addition, summer winds churn up nutrients, leading to huge algal blooms that then die off, consuming more of the oxygen in the water as they rot. Then the fish die, and the rotting stench can be detected for miles, even as far as Los Angeles.

There were few die-offs until fairly recently, and then they began to happen on a small scale each summer when the winds blew. But in 1999, more than 8 million fish died, and another die-off almost as big happened in 2013. Many scientists think that this is a sign that the Salton Sea is reaching its last stages of dying. Soon it will be too hot and salty and toxic to support all but the most minimal of living things.

14.12 SUMMARY

The Peninsular Ranges Province started with Permian–Triassic rocks of the Julian Schist, followed by Jurassic arc rocks that are now gneisses in roof pendants. These rocks were intruded in the Lower Cretaceous Western Peninsular Ranges batholith (and the Santiago Peak volcanics that erupted from above the magma chambers). In the Late Cretaceous, the locus of magmas shifted east to the Eastern Peninsular Ranges batholith as the down-going subducting slab reached a shallower angle.

Meanwhile, Jurassic accretionary wedges (Bedford Canyon Formation) and Cretaceous and Cenozoic forearc basins formed, which trapped an enormous thickness of mostly deep marine sediments. Some of these Cretaceous rocks include dinosaur fossils, and the Eocene sequence of San Diego is rich in mammal fossils. Limited deposition occurred in the Oligocene–early Miocene, and then extensive marine deposits covered coastal Orange and San Diego Counties in the late Miocene–Pliocene and Pleistocene.

Meanwhile, in the late Miocene, the rift zone between the San Andreas transform and the spreading ridge beneath the Gulf of California formed as Baja California ripped away and moved north from Mexico. In the Salton Trough, an enormous thickness of Miocene–Pleistocene marine and nonmarine sediments was deposited, rich in fossils. The Salton Sea is the last body of water in this down-dropped basin, which is well below sea level. It was accidentally formed in 1905 by a spill from canals from the Colorado River and is now drying up and dying off as the lake becomes hotter, saltier, and more depleted in oxygen.

RESOURCES

Abbott, P. 1999. *The Rise and Fall of San Diego: 150 Million Years of History Recorded in Sedimentary Rocks.* Sunbelt Publications, El Cajon, CA.

Bauer Morton, J. 2014. *Coast to Cactus: Geology and Tectonics, San Diego to Salton Trough, California.* Sunbelt Publications, El Cajon, CA.

Butler, R.F., Gehrels, G.E., and Kodama, K.P. 2001. A moderate translation alternative to the Baja British Columbia hypothesis. *GSA Today, 11,* 4–10.

California Bureau of Mines. 2014. *Mines of San Diego and Imperial Counties, California.* California Division of Mines, Sacramento.

Crowell, J.C. 1981. An outline of the tectonic history of southeastern California, 583–500. In Ernst, W.G. (ed.), *The Geotectonic Development of California.* Prentice Hall, Englewood Cliffs, NJ.

Gastil, G., Morgan, G.J., and Krummenacher, D. 1981. The tectonic history of Peninsular California and adjacent Mexico, 284–305. In Ernst, W.G. (ed.), *The Geotectonic Development of California.* Prentice Hall, Englewood Cliffs, NJ.

Hildebrand, R.S. 2015. Dismemberment and northward migration of the Cordilleran orogen: Baja-BC resolved. *GSA Today, 25*(11), 4–11.

Housen, B.A., and Beck, M.E., Jr. 1999. Testing terrane transport: An inclusive approach to the Baja B.C. controversy. *Geology, 27*(12), 1143–1146.

Jefferson, G.T., and Lindsay, L. 2006. *Fossil Treasures of the Anza-Borrego Desert: The Last 7 Million Years.* Sunbelt Publications, El Cajon, CA.

Kennedy, M.P., and Moore, G.W. 1971. Stratigraphic relations of Upper Cretaceous and Eocene formations, San Diego coastal area, California. *American Association of Petroleum Geologists Bulletin, 55,* 709–722.

Matthews, W., Guest, B., Coutts, D.S., Bain, H., and Hubbard, S.M. 2017. Detrital zircons from the Nanaimo Basin, Vancouver Island, British Columbia: An independent test of Late Cretaceous to Cenozoic northward translation. *Tectonics, 38*(5), 854–876.

Prothero, D.R. (ed.). 2001. *Magnetic Stratigraphy of the Pacific Coast Cenozoic.* Pacific Section SEPM Special Publication 91. Pacific Section SEPM, Fullerton, CA.

Reisner, M. 1987. *Cadillac Desert: The American West and its Disappearing Water.* Penguin Books, New York.

Remelka, P., and Lindsay, L. 1992. *Geology of Anza-Borrego: Edge of Creation.* Sunbelt Publications, El Cajon, CA.

Spear, S.G., and Burns, D. 1997. *Geology of San Diego County: Legacy of the Land.* Sunbelt Publications, El Cajon, CA.

Walawender, M.J. 2011. *Roadside Geology along Sunrise Highway.* Sunbelt Publications, El Cajon, CA.

VIDEOS

Santa Ana Mountains: www.youtube.com/watch?v=aKr6Pm-RfFQ
Julian gold mining: www.youtube.com/watch?v=shGSZv6YfZo

Pala Gem Mines

www.youtube.com/watch?v=1eZhevQsx3Q
www.youtube.com/watch?v=b1DEFX16T6M
www.youtube.com/watch?v=SNenyAAH4w8
www.youtube.com/watch?v=B6DaP4clbSw
www.youtube.com/watch?v=-6BPMALFdkk
www.youtube.com/watch?v=oed2t-sv5Ss

Geology of San Diego County

www.youtube.com/watch?v=FZp0i_hIkWM
www.youtube.com/watch?v=oy1VI9tX0bk
www.youtube.com/watch?v=oy1VI9tX0bk

Anza-Borrego State Park

https://vimeo.com/119833344
www.youtube.com/watch?v=ev3vPUhr_oM
www.youtube.com/watch?v=8uZ8oQQpEyY
www.youtube.com/watch?v=DU5-VvXDqdg
www.youtube.com/watch?v=sBkFY5RuuVg

Salton Trough

https://vimeo.com/132132710
https://vimeo.com/128414770
https://vimeo.com/125478882
www.youtube.com/watch?v=9HrH5cB_cCc
www.youtube.com/watch?v=BKsK13NDkpg
www.youtube.com/watch?v=8TjGAWxL23c
www.youtube.com/watch?v=Bj3icM-6xpI
www.youtube.com/watch?v=d9mMd31FdZY

15 Assembling California
A Four-Dimensional Jigsaw Puzzle

The far-out stuff was in the Far West of the country—wild, weirdsma, a leather-jacket geology in mirrored shades, with its welded tuffs and Franciscan mélange (internally deformed, complex beyond analysis), its strike-slip faults and falling buildings, its boiling springs and fresh volcanics, its extensional disassembling of the earth.

—John McPhee
Basin and Range

In previous chapters, we looked at California geology on a local scale, province by province. Throughout the entire discussion, however, common threads and similarities have emerged. In this summary chapter, we review all the rock units and events from the different provinces in chronological order so that similar events in different provinces going on at the same time are connected. Hopefully, this will help the reader synthesize the common events that occurred across the entire state and understand how many provinces followed similar tectonic and sedimentary histories.

As we saw in Chapter 1, the story of California is very complex. Unlike the relatively simple "layer-cake" geology of some states of the Midwest and Great Plains, with undeformed rocks that cover almost the entire state like blankets, nearly every part of California is a complex contorted piece of a broken puzzle. The individual parts were formed and arrived at different times, and many are exotic terranes stuck together by accretionary tectonics. Many rock units do not persist any great distance at all. Even now, California is one of the most tectonically active states in the United States, with an active plate boundary running from spreading to transform to subduction right in the heart of the state. In California, mountain ranges rise and deform in real time, and most of the tectonics we can see are very young. In this chapter, we try to assemble the pieces of this complex four-dimensional jigsaw puzzle (three dimensions of space, plus the fourth dimension, time) and try to clarify and capture the complexity and challenges of California's geologic and tectonic history.

15.1 IN THE BEGINNING: THE ARCHEAN AND PROTEROZOIC

The first thing to remember is that most of what we now know as California is a very young state, and most of it was not here even before 250 Ma. Only some areas of the Mojave Desert, Death Valley, and the White and Inyo Mountains have any *in situ* bedrock older than the Cambrian. In addition, there are a few slices of those same rocks incorporated into the roof pendants of the Transverse Ranges, but most came up from what is now Mexico. A few late Proterozoic rocks are also known from the Trinity ophiolite in the Klamaths, the Shoo Fly Complex in the Sierras, and a few other tiny remnants of Proterozoic metamorphics, but these come from exotic terranes that were not originally part of California.

The oldest dated rocks in California are early Proterozoic rocks dated 1.78–1.68 Ga. These rocks occur mostly in the Proterozoic cores of metamorphic core complexes in the Mojave Desert (Whitmeyer and Karlstrom, 2007), as well gneisses in the Transverse Ranges, like the Mendenhall Gneiss in the San Gabriels, the Pinto Gneiss in Joshua Tree National Park, and similar gneisses of the Orocopia Mountains and San Bernardino Mountains. There are no actual rocks older than the early Proterozoic anywhere near California, although some of these early Proterozoic gneisses have geochemistry suggesting a component that is reworked from an older, probably Archean source (older than 2.5 billion years). Geologists label this ancient predecessor of the Proterozoic rocks the "Mojavia" terrane, and it crops out mostly in the eastern Mojave Desert (Figure 7.15a).

In the western part of Arizona, there are gneisses dated 1.84 Ga, which are part of an ancient volcanic arc complex known as the "Elves Chasm arc." These accreted to the eastern edge of the Mojavia block (Figure 7.15a). Between 1.76 and 1.72 Ga, another block accreted to the southeastern edge of this growing terrane. Known as the Yavapai terrane (Figure 7.15a), it accreted to the east of the Elves Chasm block, mostly in central Arizona.

Sometime between 1.72 and 1.68 Ga, all these blocks were intruded by granitic plutons, especially the Mojavia and Yavapai blocks (Figure 7.15b). Many of these granitic rocks can be found in the Mojave Desert and adjacent areas in the Basin and Range of California, as well as in the roof pendants of the Transverse Ranges. In addition, units like the Zoroaster Granite in the bottom of the Grand Canyon are part of this magmatic event. Finally, between 1.65 and 1.62 Ga, the last major block was added to the southwestern United States, with the accretion of the Mazatzal terrane to the older terranes (Figure 7.15b).

By about 1 Ga, the North American continent (known as Laurentia) was part of a bigger supercontinent known as Rodinia. As discussed in Chapter 7, one of the first signs of this breakup was the appearance of rift valley systems and abandoned rifts (aulacogens), such as the latest Proterozoic Amargosa aulacogen in Death Valley (Figure 7.17).

DOI: 10.1201/9781003301837-15

15.2 PALEOZOIC HISTORY OF CALIFORNIA

Just as there are few Californian Proterozoic or Archean rocks outside the Mojave Desert and Transverse Ranges, there are relatively few *in situ* early Paleozoic rocks in California either, except in the Mojave Desert area of the Basin and Range (Chapter 7). The Paleozoic rocks of the Shoo Fly and Feather River Complex in the Sierras, and the Trinity ophiolite and the other Paleozoic rocks of the Eastern Klamath terrane, are all in exotic terranes and not originally from California. Most of California did not yet exist in the early Paleozoic.

The best record of the Cambrian through Silurian (Figure 15.1) comes from the thick passive margin wedge deposits that are found in the Death Valley–Nopah Range area, and in some other ranges in the Mojave, such as the Marble Mountains and Providence Mountains (Figure 7.19). This region was covered by shallow tropical seas in the Early Cambrian, as evidenced by the nearshore and beach sandstones of the Wood Canyon and Zabriskie

Formations, the trilobite-rich shales of the Carrara and Latham Formations, and the Chambless and Cadiz Limestones (see Chapter 20). By the middle Cambrian, the Mojave region was drowned by shallow marine carbonate shoals that were very similar to the modern Bahamas, but on a much larger scale. These deposited the Middle Cambrian Bonanza King and Upper Cambrian Nopah Limestones and shales.

These Cambrian seas were very different from anything we have on Earth today. Back then, North America straddled the equator, and the present-day "West Coast" was the north coast, and the modern "East Coast" was the south coast. North America was drowned by shallow limey seas so that only tiny bits of land in the upper Midwest were above the waves. The moon was much closer then, so the rush of the tides must have been enormous in many places, much like the true tidal waves that occur in places like the Bay of Fundy today. These shallow waters were filled with hundreds of species of trilobites, plus the reef-building

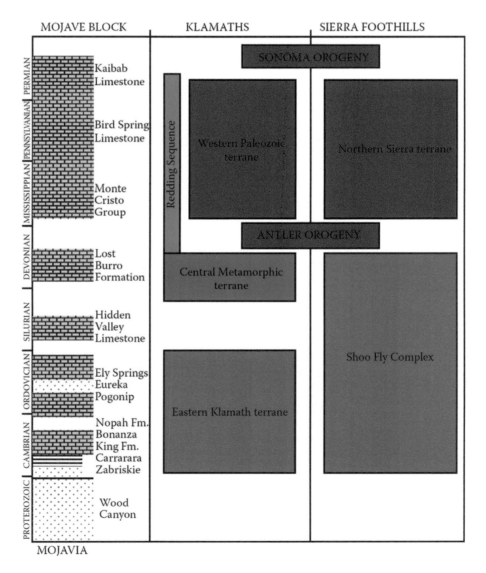

FIGURE 15.1   Summary of Paleozoic events and terranes of California.

sponge-like archaeocyathans, primitive brachiopods, and archaic members of the echinoderms (known as eocrinoids) and primitive cap-shaped mollusks. All these fossils are abundantly preserved in shales like the Carrara and Latham, and in limestones like the Chambless Limestone in the Marble Mountains.

If Stewart and Poole (1975) are right, the roof pendants in the northern San Bernardino Mountains are also remnants of this sedimentary package, with Cambrian Saragossa Quartzite that is equivalent to the Wood Canyon–Zabriskie–Carrara–Bonanza King of Death Valley (Figure 13.4). This would mark the southernmost remnant of this great Cambrian blanket of sediment, now largely eroded away and chopped up by faults. This tiny corner of the modern Mojave Desert was the only stable permanent piece of California from 550 Ma to at least 300 Ma. It remained a tropical seaway through the Ordovician (Pogonip Group limestones and Eureka Quartzite in Death Valley), Silurian (Ely Springs Formation and Hidden Valley Formation, both limestones), and Devonian (Lost Burro Formation, a limestone) (Figure 7.19). Almost all these rocks are shallow marine tropical limestones (most turned into dolostones by later action of magnesium-rich water), so the gigantic Bahama-style shallow carbonate seas persisted for at least 250 m.y. The shallow marine shelf of what is now the Mojave Desert remained a tropical seaway in this area, depositing fossil-rich limestones even during the Mississippian (Monte Cristo Group) and Pennsylvanian (Bird Spring Group). Once again, there appears to be Monte Cristo–Bird Spring equivalents trapped as roof pendants on the north slope of the San Bernardino Mountains, especially the Furnace Limestone at Cushenberry Grade (Figure 13.3c).

Meanwhile, as the shallow tropical seas drowned this stable, slowly subsiding margin of the continent that is now the Mojave Desert, other events were converging on California in the early Paleozoic (Figure 15.1). As discussed in Chapters 8 and 9, Cambrian through Devonian rocks were forming far to the west of California as exotic terranes, drowned by seas and full of shallow marine fossils. These exotic blocks from the west finally arrived in California and collided during the Late Devonian Antler orogeny (Chapters 7 through 9). Antler collisions accreted the Shoo Fly Complex in the western Sierran foothills and the Eastern Klamath and Central Metamorphic terranes in the Klamaths (Figure 15.1). These rocks were once fossiliferous shallow marine shales and limestones and sandstones (with some volcanics), much like those of Death Valley at the same time, but most of them have undergone at least low-grade metamorphism due to their collision and shearing as they accreted to the rest of North America (Figure 9.4). Today, they are slates and phyllites and quartzites, found in many areas in the western Sierras and the eastern Klamaths. In addition, many of the Cambrian through Devonian rocks that originally formed in Northern California were pushed over the Roberts Mountain thrust and now occur as a belt in central Nevada

(overthrust on top of early Paleozoic rocks that were originally in Nevada) (Figure 7.20).

After the Antler collision ended in the Early Mississippian, more exotic blocks from the west were not far behind (Figure 15.1). Huge exotic terranes docked in the Late Permian–Triassic during the Sonoma orogeny to become the Sonomia terrane (Figure 15.1) of northwest Nevada (Figure 7.20). Meanwhile, Carboniferous–Permian deep marine rocks that became the Northern Sierra terrane, plus the slivers of oceanic crust known as the Feather River ophiolite (Figure 8.13), slammed into the Sierras during the same orogeny. Likewise, the thick package known as the Western Paleozoic–Triassic terrane in the Klamaths (Figure 9.4) docked in the future location of the Klamath Mountains during the Sonoma orogeny.

15.3 EARLY MESOZOIC

By the end of the Paleozoic, only small parts of the Mojave Desert, roof pendants in the Transverse Ranges, and several exotic belts in the future Sierras and Klamaths were actually part of California. This changed dramatically during the Mesozoic (Figures 15.2 and 15.3), when multiple exotic terranes arrived in the Triassic and Jurassic. By the Cretaceous, much of California was a huge Andean-style volcanic arc (Figure 4.14), complete with a deep forearc basin (the Great Valley—Chapter 10) and accretionary wedge complex (the Franciscan of the Coast Ranges—Chapter 12). By the end of the Mesozoic, most of the basement rocks of California were in place, although there was still much more tectonism yet to come.

The first we see of Triassic–Jurassic predecessors of this great Andean-style Cretaceous arc are limited areas of volcanism and intrusions to the east of the Sierras, where there are a number of small Triassic–Jurassic plutons (Figure 8.5). There were also equivalent plutons to the south, such as the Triassic granitic intrusions in the San Gabriel Mountains (Lowe Granodiorite). There are numerous examples of rocks that might have been part of the Triassic–Jurassic accretionary wedge in front of this volcanic arc complex. These include the Triassic Julian Schist and the Bedford Canyon Formation in the Peninsular Ranges (Chapter 14) and the rocks of the Calaveras Complex in the western Sierras (Chapter 8).

This trend continues, with an even larger volcanic arc–forearc basin–accretionary wedge complex in the Jurassic. Most of the plutons of this age are to the east of the Sierras (Figure 8.5) and may also include the highly metamorphosed Jurassic gneisses of Cuyamaca–Laguna Mountain Shear Zone in the Peninsular Ranges (Chapter 14). Meanwhile, the large accretionary wedge known as the Foothills terrane was accreting on the western edge of the Sierran block (Chapter 8), as was the Western Jurassic terrane in the Klamaths (Chapter 9). By the end of the Jurassic, California was getting larger, with a small arc complex and numerous accretionary wedges. But things were about to become much more dynamic in the Cretaceous.

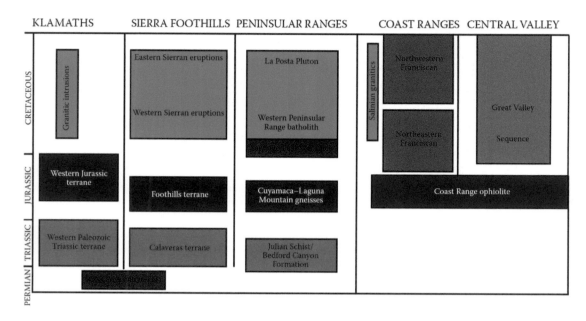

	KLAMATHS	SIERRA FOOTHILLS	PENINSULAR RANGES		COAST RANGES	CENTRAL VALLEY

FIGURE 15.2 Summary of Mesozoic events and terranes of California.

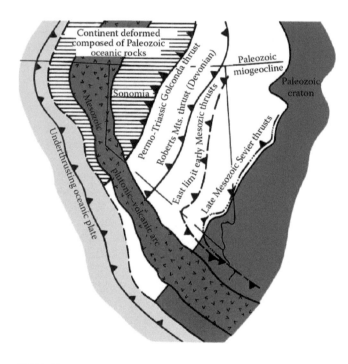

FIGURE 15.3 Paleogeographic reconstruction of the north-west–southeast trend of the Mesozoic arc volcanics in California and adjacent states, overprinting the earlier accreted terranes of the Antler and Sonoma Orogenies, which formed sutures along a continental margin that trended northeast–southwest.

Source: From Prothero, D. R., and Dott, R. H. Jr. *Evolution of the Earth* (8th ed.). McGraw-Hill: New York, 2010.

15.4 CALIFORNIA IN THE CRETACEOUS

In the Early Cretaceous (about 120 Ma), the full-blown Sierran volcanic arc began to erupt in earnest, mostly in the western part of the range (Figure 8.5). This arc complex

included not only the western Sierran plutons but also the Coast Range batholith and Idaho batholith to the north and the Cretaceous intrusions in the Klamaths (Figure 8.7). Other parts of this Andean-style volcanic chain and its magma chambers included the Salinian block to the south end of the Sierras, the Cretaceous granitics in the eastern Transverse Ranges (San Bernardino Mountains, Little San Bernardino Mountains, and San Gabriel Mountains), and at the southern end of this chain, the Western Peninsular Ranges batholith and its volcanic edifice, the Santiago Peak volcanics (Figures 14.3 and 14.8).

By the Late Cretaceous (100–80 Ma), there is evidence from many of the ranges that the angle of the down-going slab was shallowing. Eruptions in many of the volcanic arcs shifted eastward in response to the shallowing dip of the slab, displacing the melt zone of the plate to the east as well. This is true of the main body of the eastern Sierran arc (Figure 8.5), as well as the La Posta pluton of the Eastern Peninsular Ranges batholith (Figure 14.6). Most of the plutons of the eastern Transverse Ranges are also Late Cretaceous in age, such as the Josephine Granodiorite in the San Gabriels and most of the granitics of the San Bernardino and Little San Bernardino Mountains. The huge volcanic edifice that once erupted above these deep plutons has long since eroded away, with the exception of the Minarets Peak volcanics (about 100 Ma) on top of the Sierras west of Mammoth Mountain Ski Resort (Figure 8.6).

As this immense Andean-style arc erupted and built up for the 80 m.y. of the Cretaceous (65–145 Ma), the giant forearc basin of the Great Valley was down-warping and accumulating thick piles of deep marine shales and tur-bidite sandstones of the Great Valley Group (Chapter 10). Remnants of that Cretaceous forearc can also be found in front of the Peninsular Ranges with the Rosario Group beneath parts of San Diego County (Chapter 14), and the

Cretaceous sequence of Trabuco, Ladd, and Williams Formations in Orange County (Figure 14.10). In the Transverse Ranges, there are Cretaceous rocks, such as the Chatsworth Formation and the Trabuco–Tuna Canyon Formations in the eastern Santa Monica Mountains, and the Jalama and Espada Formations in the western Santa Ynez Range.

Finally, the most striking feature of the California Cretaceous is the construction of the giant accretionary wedge known as the Franciscan Complex of the Coast Ranges (Figure 4.14). As discussed in Chapter 12, the blocks of the northern part of the Franciscan get younger from east to west. The first to form were in the Eastern Franciscan block and the youngest in the western part of the north Franciscan rocks of the north Coast Ranges. Many of the rocks in the Franciscan are much older, such as the exotic blocks of Permian limestone from the Southern Hemisphere on the other side of the world, found widely not only in the Cretaceous Franciscan slices but also in the Western Jurassic terrane of the Klamaths (such as the McCloud Limestone). In addition to these old exotic blocks, the Franciscan yields such remarkable rocks as the Laytonville Limestone, which came from near the latitude of Peru in the Early Cretaceous but ended up in the northern Coast Ranges by the Late Cretaceous.

The Eastern Franciscan Complex in the northern Coast Ranges was the main accretionary prism for the Sierran arc, and an impressive wedge of tectonically accreted rocks it is. But its southern half, the Western Franciscan Complex from San Simeon and San Luis Obispo to Santa Maria, was once much farther south and completed the accretionary wedge. Other rocks that might once have been part of this prism even farther south were the Orocopia–Pelona Schist of the Transverse Ranges (Chapter 13), which is now much more highly metamorphosed than the Franciscan rocks.

15.5 THE CENOZOIC IN CALIFORNIA

The last part of the Cretaceous (70–65 Ma) was marked by a complete shutoff of all the Sierran-style arc volcanoes as the subducting plate went horizontal during the Laramide orogeny and transferred its stresses to the Rocky Mountain region (Figure 7.21). No more volcanic eruptions occurred in California. Consequently, the Sierras were gradually uplifted and beginning to erode, even as the Laramide orogeny continued through the Paleocene and early–middle Eocene (65–40 Ma).

15.5.1 PALEOCENE

Even though the arc volcanoes ceased erupting and the Sierras were uplifting and eroding, their forearc basins did not stop accumulating marine sediments (Figures 15.4 and 15.5a). The Paleocene Martinez Formation was deposited in the Great Valley basin in the eastern Bay Area, the Locatelli Formation in the Santa Cruz area, the Pattiway Formation in the southern San Joaquin Basin, the San

Francisquito Formation in the basins around the Transverse Ranges, the Silverado Formation in the western flank of the Peninsular Ranges, and the Maniobra Formation in the Orocopia Mountains. However, the Paleocene is missing from many of the Cenozoic basins of California, and even where there are Paleocene beds, they are relatively thin and only span a small portion of the Paleocene (Prothero, 2001), so there must have been widespread uplift in much of California in the Paleocene.

15.5.2 EOCENE

The 20 Ma of the Eocene (34–54 Ma) is represented by abundant sedimentary deposits in nearly all the regions of California (Figures 15.4 and 15.5b). Most important is the thick package of Eocene forearc basin sediments in the Great Valley, from the nearshore Yokut/Domengine/Avenal Sandstones of the early and middle Eocene, to the immensely thick deep-water Kreyenhagen Shales of the middle Eocene, to the middle Eocene Point of Rocks Sandstones in the southwestern Great Valley (Figure 10.11). As we saw in Chapter 14, there is also a thick lower–middle Eocene succession in the coastal plain of San Diego County, representing several cycles of transgression and regression (Figure 14.11). Even more remarkable is the thick Eocene sequence in the western Transverse Ranges (especially the Santa Ynez Range), including the deep-water Juncal Shale, the shallow marine Matilija Sandstone, the deep-water Cozy Dell Shale, the shallow marine Coldwater Sandstone, and the middle–upper Eocene Sacate, Gaviota, and Alegria Formations (Figure 13.15). The Eocene rocks of the Diligencia Basin in the Orocopia Mountains are also remarkably thick.

Even though arc volcanism had ceased and the arc was no longer active, it still had effects outside the Great Valley forearc basin. Some areas outside the major sedimentary basins accumulated sediments as the extinct Sierra volcanoes began to erode down and the "auriferous gravels" accumulated in their paleovalleys (Figure 8.15). Some of the youngest of the accretionary prism slices in the westernmost northern Coast Ranges were also slapped onto California during the Eocene.

About 40 Ma, the Laramide style of flat subduction came to an end (Figure 7.21). The subducting plate began to peel down from its horizontal orientation and sink down into the mantle again, where it could melt and form magmas (Figure 7.22). This generated a renewal of the volcanic arc, although it was located far to the east of the old extinct Sierra arc. It ran mainly from central Oregon and Washington across the narrow, pre–Basin and Range extension Nevada and Utah, across the San Juan volcanic field in southwest Colorado, and down into western New Mexico and adjacent Arizona.

15.5.3 OLIGOCENE

This huge volume of eruptions, especially in Nevada, created a high Nevadaplano volcanic plateau in Nevada,

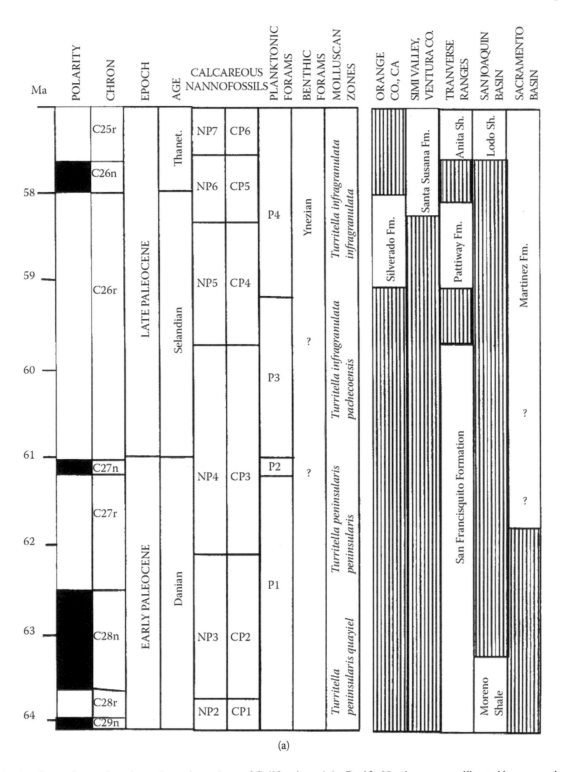

(a)

FIGURE 15.4 Cenozoic stratigraphy and geochronology of California and the Pacific Northwest, as calibrated by magnetic stratigraphy. The vertically lined boxes indicate time gaps; names of formations and their time extent are indicated by the unshaded boxes: (a) Paleocene of California, (b) Eocene of California and the Pacific Northwest, (c) Oligocene of California and the Pacific Northwest, and (d) Miocene of California and the Pacific Northwest.

Source: From Prothero, D. R. (ed.), *Magnetic Stratigraphy of the Pacific Coast Cenozoic*, Pacific Section SEPM Special Publication 91, Pacific Section SEPM, Fullerton, CA, 2001.

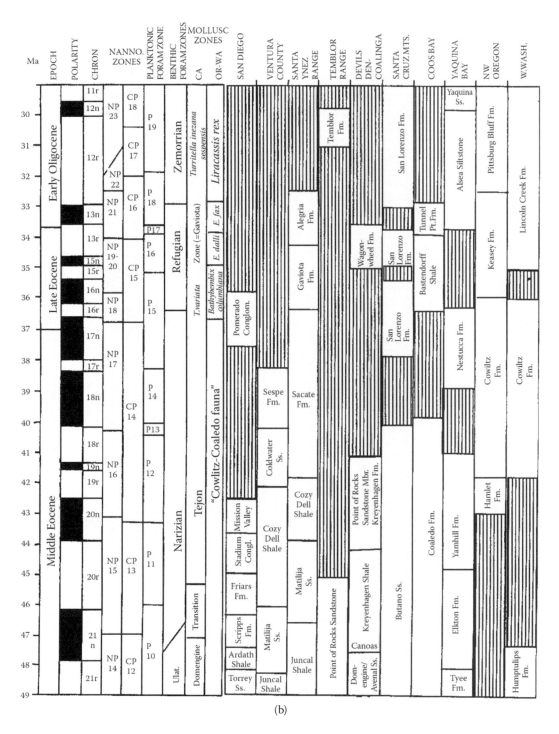

FIGURE 15.4   (Continued)

which funneled some of these volcanic ashes down the paleovalleys in the uplifting Sierras (Figures 8.15 and 8.17). Uplift was widespread across the entire region, ending the marine deposition of the Eocene and replacing it with the widespread Oligocene floodplain and river deposits (Figures 15.4 and 15.5c). These include the ubiquitous Sespe red beds in Santa Barbara, Ventura, Los Angeles, and Orange counties: the Simmler Formation red beds in the southern San Joaquin Basin, the Lospe Formation red beds of the Santa Maria Basin, and the Otay Formation in

southern San Diego County. A striking feature of these red beds is that they are either middle–late Eocene in age or upper Oligocene. Lower–middle Oligocene rocks are completely missing throughout the entire region, partly due to the general uplift and erosion of the area. It was also caused by the enormous global sea-level drop in the middle Oligocene that must have eroded whatever lower Oligocene deposits were formed.

Not all the Oligocene deposition in California is represented by nonmarine red beds. The San Joaquin Basin

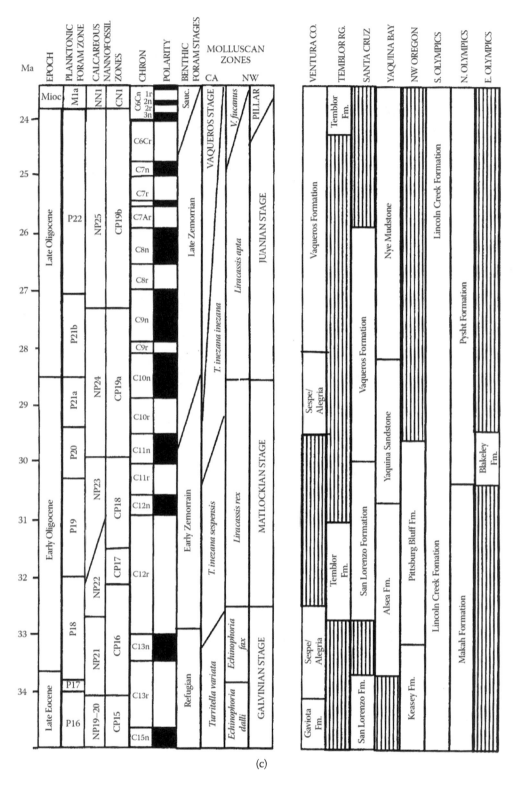

FIGURE 15.4 (Continued)

was deep enough in the Oligocene that marine deposition continued, even though it is very discontinuous and full of unconformities. There are thin wedges of lower Oligocene Temblor Formation in the Temblor Range, capped by an upper Oligocene–lower Miocene wedge of Temblor, with an 8-million-year unconformity between them. In the Santa Cruz area, the San Lorenzo and Vaqueros Formations are shallow to deep marine shales and sandstones that span most of the late Eocene, portions of the early and late Oligocene, and some of the early Miocene.

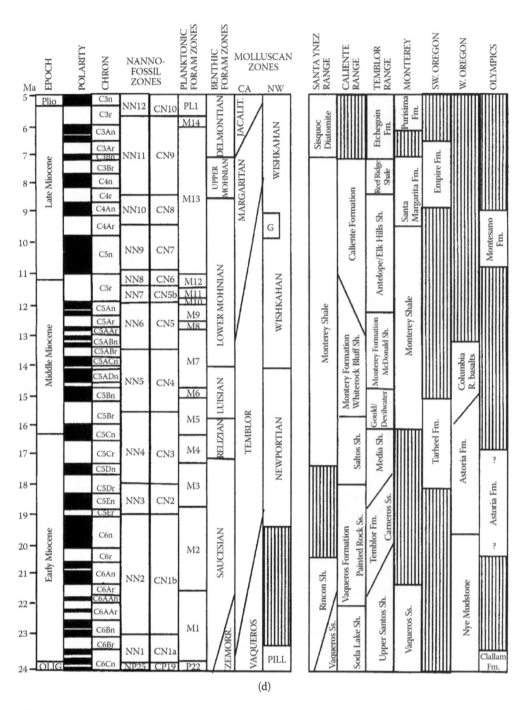

(d)

FIGURE 15.4 (Continued)

15.5.4 MIOCENE

By the later Oligocene–early Miocene, the shallow marine Vaqueros Sandstone became widespread over the entire coastal region, with outcrops running from its type area in the Salinas Valley down through the Transverse Ranges, and even down to Orange County. In the western San Joaquin Basin, the lower Miocene was represented by the upper members of the shallow marine Temblor Formation, with many of the same fossils as are found in the Vaqueros Formation.

By the middle Miocene, deep marine deposition was widespread in many parts of California. The youngest part of the Temblor Formation was deposited on the western flank of the San Joaquin Basin, while the Olcese Sand and Round Mountain Formations of Sharktooth Hill were accumulating in the southeastern San Joaquin Basin. By the late–middle and late Miocene, ocean basins were opening up along fault lines in many parts of California, accumulating huge thicknesses of deep marine shales and cherts of the Monterey Formation. These occur from the type area near Monterey in Northern California, through the immense thickness of Monterey deposition in the San Joaquin Basin, to Monterey deposits in the basins around the Transverse Ranges (Los Angeles, Ventura, and other basins), and all the way down

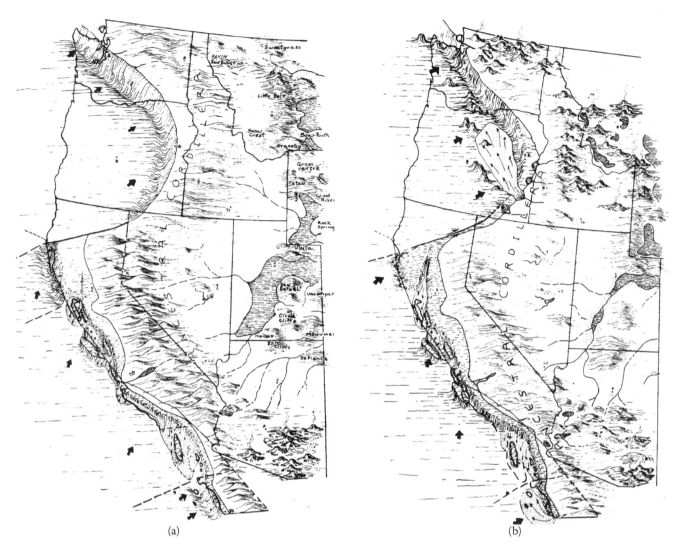

(a)   (b)

FIGURE 15.5   (a) Paleocene paleogeography of the western United States. The basement rocks of oceanic crust that make up western Oregon, Washington, and northern California had not arrived yet. Although there was a remnant forearc basin in front of the Sierran arc, it was largely emergent in the Paleocene. This map also does not show skinny Nevada in its pre-Miocene shape. (b) Early to middle Eocene paleogeography of the western United States. The basement rocks of oceanic crust that make up western Oregon, Washington, and northern California had not arrived yet. The deep forearc basin in front of the Sierran arc filled with huge thicknesses of marine Eocene sediments. This map also does not show skinny Nevada in its pre-Miocene shape. (c) Oligocene paleogeography of the western United States. The basement rocks of oceanic crust that make up western Oregon, Washington, and northern California had finally docked, forming a deep-marine forearc basin. Meanwhile, the forearc basins in California dried up due to a global drop in sea level, and mostly non-marine deposits (Sespe, Lospe formations) were formed. This map also does not show skinny Nevada in its pre-Miocene shape.

Source: After Nilsen and McKee, 1979, by permission of the Pacific Section SEPM.

to Orange County (upper Newport Bay). These same basins continued to sink and accumulate thick marine Pliocene packages as well, from the shallow marine Etchegoin Group in the San Joaquin Basin to the diatomaceous earth of the Sisquoc Formation in Santa Barbara County, to the enormous piles of Pico and Repetto sediments in the Ventura and Los Angeles Basins, which were subsiding at phenomenal rates.

15.5.5   Pliocene–Pleistocene

Finally, the Pliocene and Pleistocene mark a significant change in California tectonics and sedimentation. Much of

California was undergoing major tectonic uplift. This was especially rapid in the Transverse Ranges region, where the blocks were rotating, crumpling, and faulting (Figure 13.26). The gentle summit peneplane on the Klamaths (once buried to the top) was exhumed, and the Klamath canyons were carved as the range rose (Figure 9.10). Likewise, the gentle summit surface of the west slope of the Sierras began to be incised even as it rose up along faults.

In addition, the appearance of global ice caps drew the sea level down, so the seas retreated from basins that had long been under deep water, such as the San Joaquin, Los Angeles, and Ventura Basins. Glaciers also affected

(c)

FIGURE 15.5 (Continued)

California directly, as major ice caps covered much of the Sierras, as well as the highest peaks of the Klamaths and Mt. Shasta. Indirectly, the cooler, wetter climates of the ice ages produced enormous Ice Age lakes (Chapter 8) in nearly all the basins of the Mojave Desert, as well as the central San Joaquin Basin (Tulare Formation).

15.6 THE LAST 30 MILLION YEARS OF COMPLEX TECTONICS

From the Jurassic–Cretaceous up until about 30 Ma, the tectonics of California was relatively simple and straightforward. The dominant feature is the Klamath–Sierra–Salinian–Peninsular Range volcanic arc (and its remnant plutons that are now uplifted and eroded), and the features that are associated with Andean-style arcs (Figure 4.14): an enormous forearc basin (the Great Valley and other marine basins) and a large accretionary wedge complex (the Franciscan and its southern equivalents). Even after the arc volcanism shut off during the Laramide orogeny (70–40

Ma), the forearc basin continued to sink and trap sediment (Figure 7.21). Subduction did not end, so additional slices were added to the accretionary wedge, especially in the western edge of the Eastern Franciscan Complex in the northern Coast Ranges. The general uplift of the region increased, thanks to the Oligocene volcanism of the Nevadaplano, plus the global drop in sea level in the early Oligocene changed sedimentation patterns in California, but the fundamental tectonics of extinct arc edifice, forearc basin, and accretionary prism remained intact.

After about 30 Ma, the tectonics of California became very complicated, with multiple effects occurring in many different regions. For generations, California geologists had puzzled about the complicated patterns of faulting, folding, and uplift that their mapping uncovered, the terranes that didn't match across faults, and the other peculiar features and struggled to understand what could be the cause. Phenomena such as the extension of the Basin and Range Province and the eruption of the Cascade arc were well documented, but no one could give a satisfactory tectonic explanation for them.

The breakthrough occurred with a landmark publication in 1970 by Tanya Atwater, now retired from University of California, Santa Barbara, but at that time still a young graduate student at Scripps Institution of Oceanography. It was one of the very first papers of the plate tectonic revolution to apply tectonic models to understanding complicated rocks on land. Rather than trying to solve the puzzle using the many twisted and deformed puzzle pieces on land, as most geologists had done for generations, she took a look at the seafloor off California instead.

By 1970, the seafloor spreading record of relatively simple oceans like the Atlantic was well understood, but the fossilized evidence of the motion of the plates beneath the Pacific (preserved in marine magnetic anomaly profiles) had still not been fully deciphered. Atwater and others recognized that the complicated pattern of magnetic anomalies showed that the Pacific Ocean had once been floored by three different plates spreading away from one another (Figure 15.6). One plate had been almost completely subducted beneath the Alaska trench, leaving only its magnetic traces on the seafloor (and a tiny slice trapped beneath the Bering Sea). Walter Pitman III first documented this and named this vanished plate the "Kula plate" (*Kula* is the Tlingit word for "all gone"). The second plate was the Pacific plate, which currently underlies nearly all the Pacific Ocean.

But the third plate is now mostly gone. At one time (Figure 15.6), it formed a large single plate, the Farallon plate, which spread away from mid-ocean ridges between it and the other two plates. As it did so, it subducted beneath western North America to melt and form the volcanoes of the Coast Range–Sierra–Salinian–Peninsular Range arc. Today, most of the Farallon plate has subducted so far that it is split into two plates (Figure 15.7a): a northern Juan de Fuca plate (subducting beneath Oregon, Washington, and British Columbia to form the Cascade arc) and a

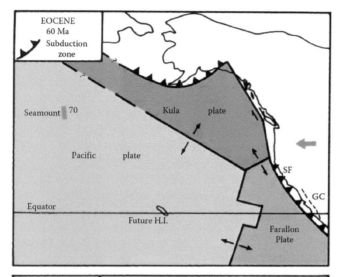

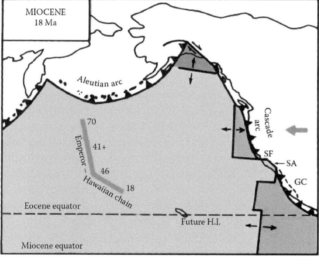

FIGURE 15.6   Early history of Pacific seafloor spreading.

Source: Modified from Atwater, T., *Geological Society of America Bulletin*, 81, 3513–3536, 1970, Figure 18. With permission from the Geological Society of America.

southern Rivera–Cocos plate (now subducting beneath Central America to form their active volcanic chain).

In between these two subducting slabs of the Farallon plate, the San Andreas fault forms a transform plate boundary, since the Pacific plate is moving to the northwest relative to the North American plate and has no convergent motion that would promote further subduction (Figure 15.7a).

Atwater (1970) studied the modern configuration of the plates carefully, then used the fossilized plate motions recorded as magnetic anomalies on the seafloor and backtracked the plates in time. She realized that about 30 Ma, the peculiar "corner" of the junction between the Pacific and Farallon plates (Figure 15.7a) would first begin to touch the old Farallon trench. Once the corner began to be subducted, it would cut the Farallon plate into its two modern remnants and form the San Andreas transform boundary between them. The subduction zone would gradually

vanish, as would the volcanic arc, and in its place would be a region of strike-slip motion.

But what took place behind the San Andreas transform once subduction ceased in the region? And how did it connect to the extension of the Basin and Range Province, or the disappearance of the Cascade arc volcanoes? In these pioneering days of plate tectonics, there were lots of speculative models trying to explain it. Some favored the idea of extending the seafloor spreading ridge beneath the Gulf of California north and underneath the Basin and Range Province. Other models related the opening of the Basin and Range to the lack of compression due to subduction and the occurrence of a shearing motion on the edge of the San Andreas plate boundary.

The most widely accepted model, however, was proposed by Bill Dickinson and Walt Snyder in 1979. Back in 1975, they had published the first paper explaining the Laramide orogeny as the product of horizontal subduction, which is now the most widely accepted model. In their 1979 paper, they focused on Atwater's discovery that as the corner of the old Farallon plate–Pacific plate boundary (Figure 15.6) impinged on the trench, it changed from a subduction zone into a transform boundary. As a result, there is no longer any plate plunging into the mantle east of the San Andreas transform. The motion on this transform is entirely northwest–southeast shear, with the Pacific plate headed northwest relative to North America. Thus, a "slab gap" or "slab window" (Figure 15.7b) appeared to the east and behind the San Andreas transform, between the down-going Farallon remnant to the north (now the Juan de Fuca plate) and the Farallon remnant to the south (now the Rivera–Cocos plate). In the slab gap, there would be no Farallon slab intervening between the North American plate and the mantle, so the hot mantle material could well up through the slab gap and force its way upward beneath the overlying North American plate.

As Atwater (1970) originally realized and Dickinson and Snyder (1979) and later Severinghaus and Atwater (1990) confirmed, this has important implications and predictions about plate motions and local geology (Figure 15.8). Plotting the geometry of the slab window predicts that the mantle bulge through the window would start out small at about 30–20 Ma and would be located in southern Arizona. This is consistent with the fact that the earliest extension of the Basin and Range does indeed occur in Arizona. In addition, the Oligocene arc volcanism in Arizona–New Mexico should also shut off at this time, which it does. Next, the subducting triangular corner of the slab window would expand north and south as more and more of it was subducted. Thus, the northern subduction boundary between the Juan de Fuca and North American plates would be changed into a lengthening San Andreas transform fault zone. As the slab window enlarged, it would expand the upwelling mantle beneath the North American plate up into Nevada and Utah, meaning, the Basin and Range would open from south to north (as it does—see Chapter 7). The change of the subduction zone into a transform zone would

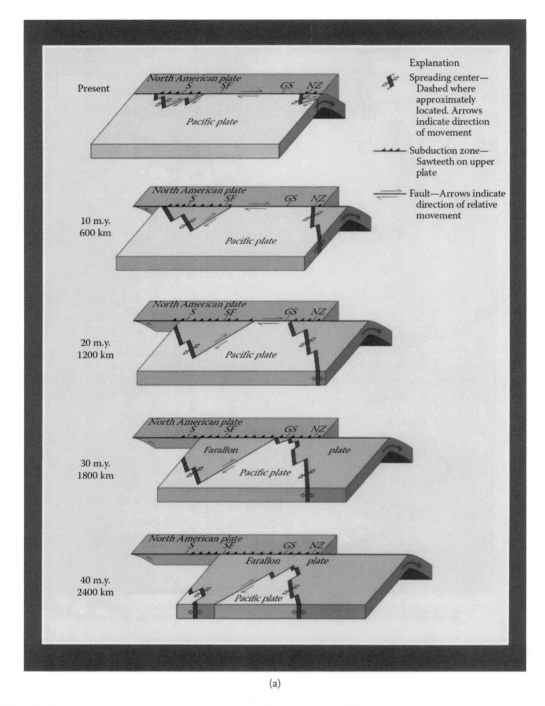

(a)

FIGURE 15.7 (a) Geometry of the interaction between the Pacific, Farallon, and North American plates, making a region of subduction turn into the San Andreas transform. (b) Slab gap model of Dickinson and Snyder.

Source: (a) From Atwater, T., *Geological Society of America Bulletin*, 81, 3513–3536, 1970. (b) Modified from Prothero, D. R., and Dott, R. H. Jr. *Evolution of the Earth* (8th ed.). McGraw-Hill: New York, 2010.

also shut off the arc volcanism from south to north. Indeed, we see very little Miocene or Pliocene arc volcanic activity in Nevada as the Basin and Range expands. More to the point, this suggests that the Cascade arc is gradually shutting down at its south end (currently, its southernmost eruptive center is Shasta and Lassen—see Chapter 6). If the Pleistocene volcanoes in the Clear Lake volcanic field (Chapter 12) and Sutter Buttes (Chapter 10) were once part

of this arc but are now extinct, this, too, is consistent with the slab gap model (Figure 15.9).

The slab gap model predicts not only the south-to-north opening of the Basin and Range and the south-to-north shutoff of the arc volcanism but also other features as well (Figure 15.8). When the mantle bulge got large enough through the slab window to expand to the east of the Basin and Range, it apparently caused the Colorado Plateau to lift

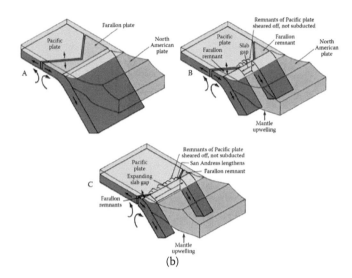

FIGURE 15.7 (Continued)

more than 3,500 m (11,000 ft) into the sky in the Pliocene, cutting the Grand Canyon and many other canyons in the region. It would explain why the Rocky Mountains renewed their Laramide uplift and incised their canyons after being nearly buried in the late Miocene (Figure 15.8c—d).

Not every feature is directly related to the slab gap model. As Atwater (1970) first predicted, the plate geometries dictate that the San Andreas transform should expand in both directions, and we can see that during the Neogene history of California (Chapter 11). The final stage was the ripping open of the Gulf of California (Chapter 14) as the San Andreas stretched south to the Mexican border and brought Baja California north over the past 10 Ma. The peculiar clockwise tectonic rotation of the Transverse Ranges (Chapter 13) is a secondary result of the shear along the San Andreas transform, where crustal blocks do not all behave in a simple shear motion.

Nevertheless, plate tectonics has managed to produce an explanation for the many peculiar features of California's amazing geology, especially its strange behavior over the past 30 million years. But not all the pieces of the jigsaw puzzle are in place. There are still some unsolved problems, like how to explain the rotation of the southern Coast Ranges (Chapter 13). What was the nature of the proto–San Andreas fault that accounts for the extreme offset of ancient Jurassic and Cretaceous piercing points that happened long before the modern San Andreas was born 30 m.y. ago? Some of it is clearly due to the additional shear within the Salinian block, but there are still mysteries to solve. The

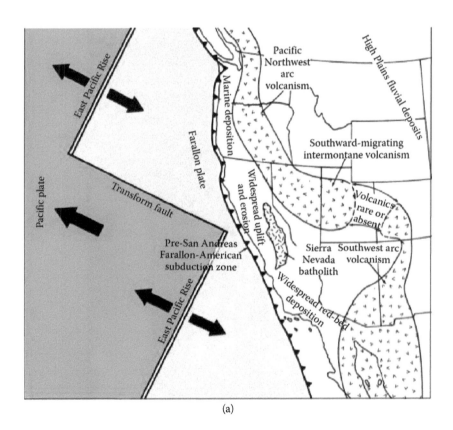

(a)

FIGURE 15.8 Successive stages of evolution of California and the western United States during the Cenozoic: (a) Oligocene (30 Ma), (b) middle Miocene (15 Ma), (c) Pliocene (5 Ma), (d) Pleistocene (1 Ma).

Source: Modified from Dickinson, W. R., in Armentrout, J. M., Cole, M. R., and TerBest, H. Jr. (eds.), *Cenozoic Paleogeography of the Western United States I,* Pacific Coast Paleogeography Symposium 3, Pacific Section SEPM, Fullerton, CA, 1979, 1–13.

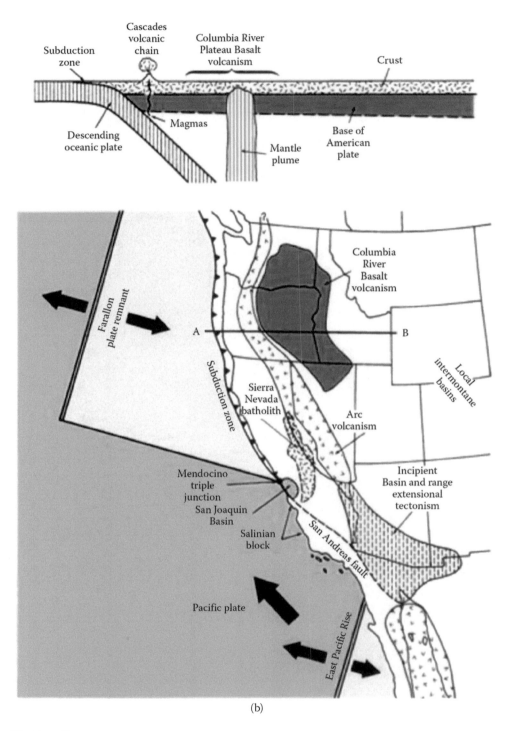

(b)

FIGURE 15.8 (Continued)

jigsaw puzzle is not yet complete, but the big picture is much clearer than it was just 45 years ago.

15.7 SUMMARY

California is a complex jigsaw puzzle of many different pieces that arrived at different times, all of which have experienced some extreme tectonic forces. The oldest rocks in California are Proterozoic (1.7–1.6 Ga) metamorphics and granitics in the Mojave Desert and Transverse Ranges and lower Paleozoic shallow marine sedimentary rocks. These were the only parts of California that existed before the mid-Paleozoic. Starting with the Devonian Antler orogeny, and again with the Permian Sonoma orogeny, many exotic terranes were accreted to the western Sierran foothills and to the Klamath Mountains. In the early Mesozoic (Triassic–Jurassic), a small Andean-style arc complex with volcanic arc rocks and accretionary wedge deposits began to form. This arc reached full size by the Cretaceous, when the Sierra–Salinia–Peninsular Ranges arc volcanoes and

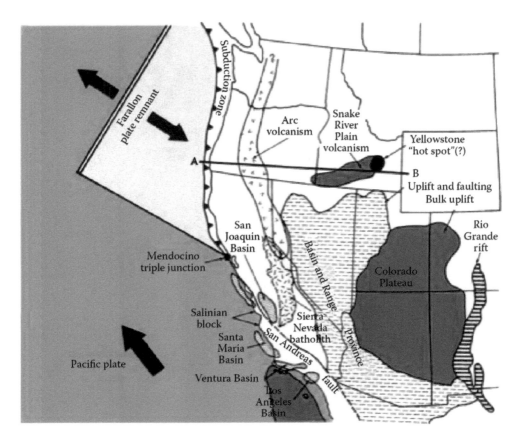

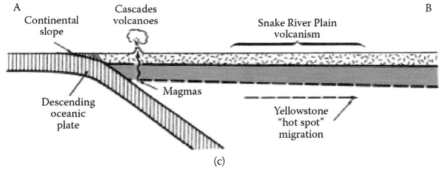

FIGURE 15.8   (Continued)

their remaining magma chambers were intruded, and the huge Great Valley forearc basin and Franciscan accretionary wedge developed.

During the Laramide orogeny (70–40 Ma), Sierran arc volcanism ceased as the subducting slab went horizontal, but extensive Paleocene and Eocene forearc basin deposits were still accumulating. By the Oligocene, California was subjected to uplift, extensive volcanic ash from the Nevadaplano, and a global drop in sea level, so most of the deposits are nonmarine red beds. During the Neogene, thick marine deposits were formed in many of the basins as the region switched from a subducting plate boundary to an expanding San Andreas transform boundary. This transform boundary had a slab gap behind it with no intervening Farallon remnant slab to prevent mantle upwelling. Consequently, the area behind the San Andreas experienced

a mantle bulge that opened the Basin and Range Province from south to north and gradually shut off the arc volcanism of the Cascades from south to north as well. Independently of these processes, the Transverse Range block swung clockwise by about 90°, while the Salinian block, and then Baja California, ripped away from Mexico to slide north along the San Andreas.

Finally, in the Pleistocene there was widespread uplift and a global drop of sea level, causing many ranges (the Klamaths and Sierras) to be uplifted and incised, while nonmarine deposits filled the formerly marine basins in the Great Valley and Los Angeles and Ventura Basins. The highest parts of California (Sierras, Mt. Shasta, and the Klamaths) were also glaciated, while the low basins of the Mojave filled with lakes during the cooler, wetter parts of the ice ages. Most of California is still undergoing rapid

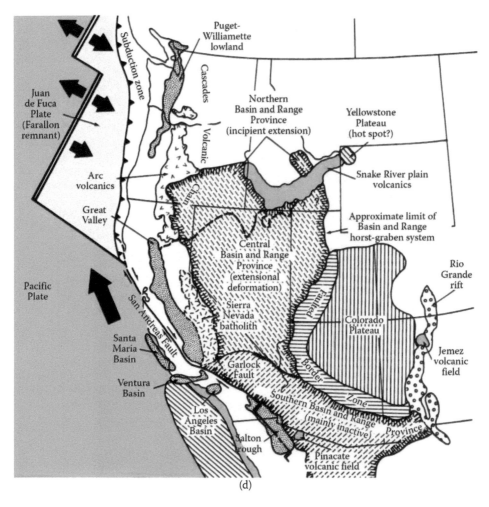

(d)

FIGURE 15.8 (Continued)

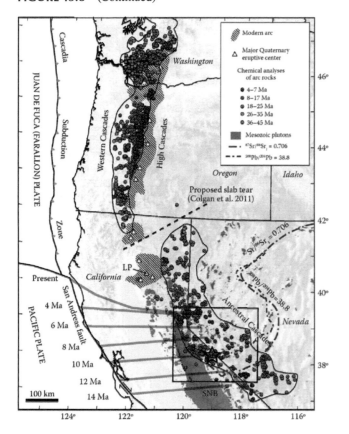

FIGURE 15.9 The gradual northward movement of the Mendocino triple junction, the connection between the Cascadia subduction zone off the coast of Northern California–Oregon–Washington, the San Andreas transform to the south, and the Mendocino Fracture Zone to the west. As the edge of the subducting slab retreated to the north, arc volcanism should have shut off from south to north as well. Currently, the southernmost active Cascade volcano is Lassen Peak (LP). But the Pleistocene Sutter Buttes eruption (see Chapter 6) and the Pliocene Clear Lakes volcanoes in Sonoma County (see Chapter 11) may have been formed when the subducting plate edge (shown by the broad gray lines) was farther south. Whether there is a south–north shutoff of the volcanoes east of the Sierras in the Ancestral Cascades is less clear.

Source: Courtesy of Geological Society of America.

uplift and tectonism, since the San Andreas is still an active plate boundary, and mountains adjacent to it are still experiencing some of the most rapid uplift rates in the world.

RESOURCES

Armentrout, J.M., Cole, M.R., and TerBest, H., Jr. (eds.). 1979. *Cenozoic Paleogeography of the Western United States I.* Pacific Coast Paleogeography Symposium 3. Pacific Section SEPM, Fullerton, CA.

Atwater, T. 1970. Implications of plate tectonics for the Cenozoic tectonic evolution of western North America. *Geological Society of America Bulletin, 81,* 3513–3536.

Cooper, J.D. (ed.). 1991. *Paleozoic Paleogeography of the Western United States II.* Pacific Section SEPM Book 67. Pacific Section SEPM, Fullerton, CA.

Dickinson, W.R. 1979. Cenozoic plate tectonic setting of the Cordilleran region of the United States, 1–13. In Armentrout, J.M., Cole, M.R., and TerBest, H., Jr. (eds.), *Cenozoic Paleogeography of the Western United States I.* Pacific Coast Paleogeography Symposium 3. Pacific Section SEPM, Fullerton, CA.

Dickinson, W.R. 1981. Plate tectonics and the continental margin of California, 1–28. In Ernst, W.G. (ed.), *The Geotectonic Development of California.* Prentice Hall, Englewood Cliffs, NJ.

Dickinson, W.R., and Snyder, W.S. 1979. Geometry of subducted slabs related to the San Andreas transform. *Journal of Geology, 87,* 609–627.

Fritsche, A.E. (ed.). 1995. *Cenozoic Paleogeography of the Western United States II.* Pacific Section SEPM Book 75. Pacific Section SEPM, Fullerton, CA.

Howell, D.G., and McDougall, K.A. (eds.). 1978. *Mesozoic Paleogeography of the Western United States.* Pacific Section SEPM Book 8. Pacific Section SEPM, Fullerton, CA.

Ingersoll, R.V., and Ernst, W.G. (eds.). 1987. *Cenozoic Basin Development of Coastal California.* Prentice Hall, Englewood Cliffs, NJ.

McPhee, J. 1981. *Basin and Range.* Farrar, Straus, and Giroux, New York.

Prothero, D.R. (ed.). 2001. *Magnetic Stratigraphy of the Pacific Coast Cenozoic.* Pacific Section SEPM Special Publication 91. Pacific Section SEPM, Fullerton, CA.

Prothero, D.R., and Dott, R.H., Jr. 2010. *Evolution of the Earth* (8th ed.). McGraw-Hill, New York.

Severinghaus, J., and Atwater, T. 1990. Cenozoic geometry and thermal state of the subducting slabs beneath western North America. *Geological Society of America Memoir, 176,* 1–22.

Stewart, J.H., and Poole, F.G. 1975. Extension of the Cordilleran miogeosynclinal belt to the San Andreas fault, southern California. *Geological Society of America Bulletin, 86,* 205–212.

Whitmeyer, S.J., and Karlstrom, K.J. 2007. Tectonic model for the Proterozoic growth of North America. *Geosphere, 3,* 220–259.

Woodburne, M.O. 1975. *Cenozoic Stratigraphy of the Transverse Ranges and Adjacent Areas, Southern California.* Geological Society of America Special Paper 162. Geological Society of America, Boulder, CO.

VIDEOS

California tectonics and geology are also animated in these videos:
www.youtube.com/watch?v=bjU2ueb1Rvg
www.youtube.com/watch?v=2TSTVCIZ-dg
www.youtube.com/watch?v=rzVU0PoZLE4

Animations

In addition, Tanya Atwater maintains a website (http://emvc.geol.ucsb.edu/1_DownloadPage/Download_Page.html) with excellent animations of the tectonics of the western United States, especially California. These are highly recommended for comprehending the three-dimensional motions of the tectonic blocks of California.

16 California's Gold

The old American Dream . . . was the dream of the Puritans, of Benjamin Franklin's "Poor Richard" . . . of men and women content to accumulate their modest fortunes a little at a time, year by year by year. The new dream was the dream of instant wealth, won in a twinkling by audacity and good luck. [This] golden dream . . . became a prominent part of the American psyche only after Sutter's Mill.

—H. W. Brands
Historian, 2003

"Hey boys, come up here!" Lee's excited shout bounced from rock to rock down the gulch. "I've got all of California right here in this pan!"

—Phyllis Flanders Dorset
The New Eldorado: The Story of Colorado's Gold and Silver Rushes

16.1 THAR'S GOLD IN THEM THAR HILLS!

Of all its geological resources, gold (Figure 16.1) is the one that transformed California and put it on the map. Before the Gold Rush of 1848–1855, California had long been a sleepy, remote little Mexican province, ignored by the government down in Mexico City and coveted by ambitious American settlers. They launched the "Bear Flag Revolt" in 1846, but this uprising was put down. By 1847, America had taken California after winning the Mexican–American War, which officially ended on February 2, 1848, with the Treaty of Guadalupe Hidalgo. Still, California was an underpopulated territory with no clear path to statehood and barely 8,000 Mexican and American residents.

Here's a trivia question: Where was the first discovery of gold in California? It was not the famous event in 1848 in the Sierra Nevada foothills but happened six years earlier near Los Angeles. In 1842, a Mexican vaquero named Jose Francisco de Garcia Lopez was driving his cattle in a valley now known as Placerita Canyon, in the San Gabriel Mountains south of the modern city of Santa Clarita. He took a nap under the "Oak of the Golden Dream" and dreamt of finding gold. After his nap, he pulled up some wild onions from the creek bed and found gold flakes clinging to the roots. The word spread, and about 2,000 Mexican prospectors rushed to the area, mostly from Sonora, Mexico. By 1848, they had recovered a total of about 57 kg (125 lb) of gold by panning through the creek gravels. At that point, the gold was almost exhausted, and the news of richer gold fields up north led to the abandonment of this prospect. Today, Placerita Canyon is a state park, but after heavy winter rains that move the big boulders in the creek,

it still yields gold that has been uncovered and dislodged during the flooding.

A few weeks before the Treaty of Guadalupe Hidalgo formally transferred California from Mexico to the United States, pioneering Swiss rancher Johann Sutter (who owned a big agricultural empire in the Sacramento Valley) sent his foreman John Marshall up to the American River at Coloma to build a sawmill for cutting lumber (Figure 16.2). On January 28, 1848, Marshall was inspecting the crew's work one evening and happened to see a gleam in the gravel of the millrace (the trough beneath the mill where the water is diverted into the waterwheel that drives the saw blades). He bent down and found several small nuggets that he believed to be gold. Marshall tried to keep his discovery a secret. He rode down to Sutter's Fort (now the site of Sacramento) to show his boss. They tested the metal by stamping it into flat sheets without shattering it, trying to burn it or dissolve it in acid (gold resists this), and weighing it to get its density. All the tests confirmed it was gold. Sutter was dismayed, because he was trying to maintain control over his huge ranch and was justifiably afraid that a stampede of prospectors would destroy his empire. In fact, as soon as gold fever spread, squatters took over his land and stole his livestock, and no one in the government had the interest or the power to enforce his claims, so Sutter lost everything. John Marshall also suffered and died sick and penniless when the Gold Rush ruined his life as well.

However, much Sutter and Marshall tried to keep the secret, it finally leaked out. In March 1848, a San Francisco newspaperman and merchant by the name of Samuel Brannan got word of the discovery. He first set up a shop for mining supplies and then published the news and walked through the streets of San Francisco shouting, "Gold! Gold! Gold from the American River!" He was the first of many merchants, bankers, and businessmen who got rich mining the miners. Most of the miners themselves made almost nothing.

One of these was dry goods merchant Levi Strauss, who took surplus denim from tents and invented a new kind of pants, the now-familiar denim jeans known as Levis, with their trademark copper-riveted stress points. They were rugged and durable and popular with the miners, whose hard work ruined most hand-made clothes.

Some of the richest merchants sold a huge variety of hardware and mining supplies and dry goods. Four men who started out in the gold fields and then found riches selling supplies to miners were Leland Stanford Jr., Hollis P. Huntington, Mark Hopkins, and Charles Crocker. To outsiders, they were the "Big Four," but they called themselves "The Associates." After hearing a presentation from an enterprising civil engineer named Theodore Judah that there was a feasible route for the railroad to cross the

DOI: 10.1201/9781003301837-16

FIGURE 16.1 A 44 lb gold nugget, one of the largest still in existence, now on display at Ironstone Winery in Murphys. It was found by the Sonora Mining Company on Christmas Day 1992.

Source: Photo by G. Hayes.

FIGURE 16.2 Modern replica of Marshall's and Sutter's sawmill on the American River at Coloma, California, just up the bank from the original gold discovery site.

Source: Photo by the author.

Sierras, they pooled their wealth and created what became the Central Pacific Railroad Company. The US government had talked for years about building a transcontinental railroad to link the gold fields of California to the East, but the vote was always stalled by the issue of slavery. But when the South seceded and a unified Republican Party and president ruled in Washington, Congress passed the Pacific Railroad Act of 1862 (in the middle of the Civil War), authorizing government financing, which the Big Four then drew upon to start construction. The railroad ran from San Francisco to Promontory Point, Utah, where it joined in 1869 with the Union Pacific Railroad coming across the Great Plains and the Rockies from Missouri. It was a miracle of engineering at that time, going up impossibly steep grades in the Sierras and then crossing the Basin and Range of Nevada. It was built by huge construction crews consisting largely of 12,000 Chinese coolies who came for the work and proved their mettle in working hard but were abused by their white bosses.

When it was completed, Stanford, Huntington, Crocker, and Hopkins became four of the richest men not only in California but also in the world. Stanford served as California governor and also as a US senator and founded Stanford University. Huntington went on to develop other railroads, but his name is best known today for the Huntington Library and Art Gallery and gardens south of Pasadena, California, built by his son Henry. Crocker was the man who actually supervised the construction of the railroad and later went into banking, running Wells

Fargo Bank before selling out at a profit and founding what became Crocker Bank (which, ironically, was bought by Wells Fargo in 1986). Mark Hopkins was the treasurer of the Central Pacific Railroad Company and is best known today for the legendary Mark Hopkins Hotel in San Francisco.

By August 1848, newspapers in New York City and other eastern cities were abuzz with news of the gold discovery. Soon, thousands of men (including my great-great-grandfather William E. Prothero, a recent immigrant to Pennsylvania from the Welsh coal mines) were risking the dangerous 33,000 km (18,000 NM) voyage down around the tip of South America, a trip that took five to eight months. Others sailed down to the Caribbean coast of Panama, crossed the dangerous jungles on mules and in canoes for a week, and then had to wait for a ship to San Francisco on the Pacific shore. Even fewer tried to cross overland via the California Trail branching out from the Oregon Trail, the most hazardous route of all. Hundreds of ships bearing miners and high-priced merchandise for this huge market converged on San Francisco, where most ships were abandoned when the sailors rushed off to the gold fields. The derelict ships formed an enormous forest of bare masts and rigging, much of which was scavenged for housing and supplies or sunk into San Francisco Bay for landfill.

During the stampede into the Sierra foothills, thousands of "forty-niners" rushed to the Sierra foothills, hoping to strike it rich—and most of them failed. There they founded hundreds of small boomtowns with colorful names like Rough and Ready, Poverty Hill, Poker Flat, Red Dog, You

Bet, Yankee Jim's (who was actually an Australian, not a Yankee), Gouge Eye, Humbug City, Chinese Camp, Fair Play, Chili Gulch, Mokelumne Hill, Placerville (formerly "Hangtown"), Enterprise, Whiskeyville, El Dorado, and Angels Camp. In this last town, a young prospector and writer by the name of Samuel Clemens (writing under the pen name "Mark Twain") first made his reputation with his short story "The Celebrated Jumping Frog of Calaveras County." These towns were indeed "rough and ready," with hundreds of hardworking, fast-shooting, hard-carousing prospectors who relied on vigilantes for justice and supported more saloons and brothels than any other kind of business. Today, you can drive down through the "gold country" on California Highway 49 through most of these historic towns (some restored to their Gold Rush appearance). California's state seal shows a miner panning for gold, and its motto is "Eureka" ("I have found it"). Even the shape of the California state highway signs resembles a miner's shovel.

Most of the miners made almost nothing for all their hard work, and they risked hazards of life in the rough gold towns. But the earliest miners to arrive in 1848 were able to pick up huge nuggets just lying on the riverbed or line up elbow to elbow on the shallows of the river with gold pans and make some money. As described by John McPhee in *Assembling California*:

In a deep remote canyon on the east branch of the north fork of the Feather River, two Germans roll a boulder aside and under it find lump gold. Another couple of arriving miners wash 400 ounces there in eight hours. A single claim yields fifteen hundred dollars. The ground is so rich that claims are limited to 48 inches square. In one week, the population grows from 2 to 500. The place is named Rich Bar. At Goodyears Bar, on the Yuba, one wheel-barrow-load of placer is worth $2000. From hard rock above Carson Creek comes a single piece of gold weighing 112 pounds. After black powder is packed in a nearby crack, the blast throws out $110,000 in gold. A miner is buried in Rough and Ready. As shovels move, gold appears in his grave. Services continue while mourners stake claims. So goes the story, dust to dust. From the auriferous gravels of Iowa Hill two men remove $30,000 in a single day. A nugget weighing only a little less than Leland Stanford comes out of hard rock in Carson Hill. Size of a shoebox and nearly pure gold, it weighs just under 200 pounds (troy)—the largest piece ever found in California.

But soon those easy-to-find deposits were gone, and eventually only gangs of miners building huge canals and dams and reservoirs were able to make a living, and the solo miner was pushed out of the richest areas. Large corporations eventually bought out these operations, and then the miners had to work for a company rather than for themselves.

In the meantime, the huge numbers of immigrants changed California radically. In the year 1849 alone, about 90,000 people arrived, and by 1850, more than 300,000 people had made the trip. Although most were from the United States, there were many from China, Australia, Europe, and nearly every continent. About 8,000 Californians of European ancestry (Spaniards, Mexicans, and Americans) lived in the territory before the Gold Rush. Afterward, the population increased by more than ten times to more than 92,000 by 1850 and quadrupled again to 380,000 by 1860. San Francisco went from a sleepy town of a few hundred in 1846 to 25,000 by 1850. The growth of California was so rapid that it prompted Congress to enact the Compromise of 1850, admitting California to the Union as a free state while rejecting Northern efforts to ban slavery in all the new Western territories.

Without the gold pouring into the US Treasury in 1861–1865, the United States might not have survived. The huge influx of wealth allowed the government to issue lots of bonds to pay for the enormous costs of the Civil War and prevented them from defaulting and causing a financial crash that might have occurred and forced the United States into bankruptcy. The historian John Bidwell wrote:

It is a question whether the United States could have stood the shock of the great rebellion of 1861 had the California gold discovery not been made. Bankers and businessmen of New York in 1864 did not hesitate to admit that but for the gold of California which monthly poured in five to six millions into that financial center, the bottom would have dropped out of everything. These timely arrivals so strengthened the nerve of trade and stimulated business as to enable the government to sell its bonds at a time when its credit was its life-blood and the main reliance by which to feed, clothe and maintain its armies. Once our bonds went down to 38 cents on the dollar, California gold averted a total collapse and enabled a preserved Union to come forth from the great conflict.

16.2 MINING METHODS

As described earlier, the earliest prospectors found gold deposits that were so rich that nuggets were just lying on the streambed for anyone to pick up. These deposits were soon exhausted by the flood of prospectors, so most switched to the old cheap and reliable method of panning for gold (Figure 16.3a). A *gold pan* is a simple large shallow dish (at first, many prospectors used ordinary plates or cooking pans) that you fill with a scoop of gravel from the stream bottom. As you slosh the gravel and water around, the lighter sand and gravel grains spill over the side, while the heavy gold particles are trapped in the bottom. It is a backbreaking, tedious job that makes for sore shoulders and arms after hours of work while crouching in ice-cold rivers, but it costs relatively little and requires only simple equipment.

Soon the prospectors found better methods. The most common was to divert the water and stream gravels through a wooden trough called a sluice box (Figure 16.3b), which had baffles on the bottom to trap the heavy gold. Another device was the "rocker," a sluice box on curved rocking cradle legs (Figure 16.3c) that tipped side to side to slosh the water and sand out of it and trap the gold; a long version of this was called a "long tom" (Figure 16.3c and d). Eventually, crews of miners would dam up entire rivers and direct the flow through sluice boxes to extract every bit of gold from it (Figure 16.2e), or excavate the exposed river bottom below the dam and reservoir. According to the US Geological Survey, by 1854, these crude **placer** techniques (separating gold from river sediments by manual techniques) retrieved about 370 t (12 million ounces) of gold, worth about $16 billion at today's prices.

As soon as 1853, crude placer methods had exhausted most of the easily exploited streams and rivers. At this point, some miners began to use **hydraulic mining** (Figure 16.4). They would set up huge hoses and nozzles and force water through them from a nearby reservoir and canal system they had built (Figure 16.4a). The water pressure would blast away at ancient Eocene river gravel that filled the paleovalleys (Chapter 8), washing it down to waiting sluice boxes to retrieve gold that had been trapped for 40 million years or more. These methods scoured away enormous areas of the hillside, stripping them bare and leaving gigantic mining scars and piles of tailings of gravel everywhere. Some of these mining scars, like Malakoff Diggins (Figure 16.4b), cover many acres of land and have still not healed or grown back more than 160 years later. It was one of the most environmentally destructive mining operations ever conducted, but few cared about such issues then. All that mattered was recovering as much gold as possible and then moving on to the next deposit—consequences be damned. Through these methods, miners extracted another 343 t (11 million ounces) of gold, worth about $6.6 billion.

The scars of the hydraulic mining were not the only environmental damage to the mined areas. The huge amounts of sediment that had been washed down the rivers clogged their channels. This made their channels very shallow and made flooding downriver even worse. A series of very bad floods in the 1870s and early 1880s destroyed numerous farms below the mining areas, and the farmers banded together to sue the miners in court. In a landmark

FIGURE 16.3 Placer mining methods: (a) Gold panning. (b) The more efficient sluice box replaced gold pans. (c) Antique diagram showing the construction of a rocker or "cradle" and a long tom. (d) A crew working on a long tom. (e) Eventually, large crews resorted to damming rivers to run their flow through sluice boxes.

Source: Courtesy of Wikimedia Commons.

FIGURE 16.4 Hydraulic mining: (a) Hoses and water jets eroding away tons of ancient river gravels and their gold. (b) Malakoff Diggins, one of the largest scars of hydraulic mining in the Sierras.

Source: Courtesy of Wikimedia Commons.

FIGURE 16.5   A dredging barge, with its big excavator arm in front.

Source: Courtesy of California Division of Mines and Geology.

decision (*Edwards Woodruff v. North Bloomfield Mining and Gravel Company*) on January 7, 1884, Federal Judge Lorenzo Sawyer ruled in favor of the farmers and declared that hydraulic mining was a "public and private nuisance." For the time being, this stopped hydraulic mining altogether, although a later act by Congress in 1893 allowed such mining to resume if they built large dams downstream to trap the sediment. However, massive floods in 1891 had destroyed most of the hydraulic mining equipment, so it never resumed on the scale it once had been practiced.

During the late 1890s, another technique arose in importance: dredging. Huge barges (Figure 16.5) of shallow draft would anchor in the river and use big scoops or steam shovels to pour enormous volumes of river sediments through sluice boxes. These became the preferred method for another decade or so, before they, too, ran out of deposits of gold economical enough to mine. By the time the dredging phase of mining ended, they had recovered 622 t of gold (20 million ounces), or about $12 billion worth.

Although mining placer gold eroded from the bedrock was profitable for a long time, other miners went to the source of the gold: the "Mother Lode," where the gold had formed in the first place. They began digging tunnels and shafts (Figure 16.6) to follow gold-bearing seams into the hillside (especially quartz veins, where most of the gold was concentrated). This operation required much hard work breaking rock and the risks of using explosives, such as dynamite, along with the dangers of mine collapse or of suffocation in the mineshaft. In fact, the dangerous tendencies of older explosives like nitroglycerin or gunpowder led to the invention of dynamite in 1867, which became essential to the mining industry and made its inventor, Alfred Nobel, immensely rich. (That fortune went on to support the Nobel Prizes after his death.) Nevertheless, huge volumes of rock were mined this way. Once the raw ore had been hauled out of the shaft, it had to be crushed into a fine powder with a stamp mill, which pulverized the ore with heavy pistons. The crushed ore could then be floated

FIGURE 16.6   Shaft mining in California.

Source: Courtesy of California Division of Mines and Geology.

in water, which would separate the gold from the lighter waste material.

More efficient, however, was to wash the ore powder over copper plates covered with mercury, an element that was mined in several places in the northern Coast Ranges (discussed in Chapter 12). Mercury binds to gold to form an amalgam that can easily be separated from other rock. Then the mercury is stripped from the amalgam by heating it in a furnace, leaving the pure gold. Unfortunately, people then did not realize that mercury causes brain damage. As you might imagine, working with a dangerous chemical like mercury not only was hazardous to the miners but also resulted in huge areas of landscape with deadly mercury contamination. Some of these mine tailings areas are still

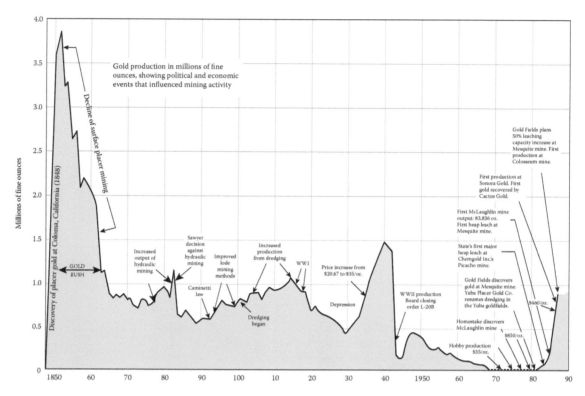

FIGURE 16.7   History of gold production in California showing the frequent boom-and-bust cycles as old mining methods lost effectiveness and new methods were developed.

Source: Courtesy of California Division of Mines and Geology.

hazardous waste sites, even 160 years later. Hard-rock mining in the Sierras continued until the 1940s, when most of the mines finally closed due to the lack of gold and the high costs of mining. Today, none of these mines could be reopened without passing stringent environmental regulations that (unfortunately) were not in place during the early days of gold mining, when men took the wealth from the land and left a devastated landscape for their descendants to clean up.

The history of gold production follows a series of boom-and-bust cycles (Figure 16.7) as each mining method peaked in production and then declined as most of the gold was found and exhausted. This was followed by another boom, when a new technology for extraction was found, followed by a bust, and then yet another boom and bust for each new technological breakthrough. In Figure 16.7, for example, the first big boom was the initial Gold Rush from 1849 to 1855, when the easily obtained surface gold was extracted by placer techniques. Then came the hydraulic mining boom in the late 1870s and 1880s, before it became illegal in 1884. Up to about 1915, most of the production came from barges dredging the bottoms of the rivers, but that method eventually produced diminishing returns. Some of the production booms and busts were driven by economic conditions, such as the downward cycle during the Great Depression, followed by a boom in the 1930s due to the increased price and demand for gold, and then another bust during World War II.

In the late twentieth century, production of gold (and many other mineral resources, like silver, copper, nickel, and uranium) was done by heap leaching. In this method, miners reprocess the low-grade ore from the mine waste in the tailings, which had too little gold in the ore to be profitably extracted by old techniques. Heap leaching takes these tailings and leaches them with chemicals like cyanide, which tend to bind to gold. The resulting chemical mixture is then extracted and refined. Just like mercury, using cyanide for extraction is also dangerous, but today there are many environmental regulations to keep it from leaking out of the leach ponds. Today, there are at least 16 gold mines still active in California, although their operations are mostly restricted to leaching gold out of old mine waste, not digging new mine shafts or placer mining.

16.3   GOLD MINERALIZATION

Since it is a very valuable substance, an enormous amount of research has gone into how gold got into the rocks of the Mother Lode. Early hard-rock miners noticed that gold most often formed in association (Figure 16.8) with veins of quartz (known as lodes), and so their digging usually followed quartz veins. These, in turn, filled in ancient fractures and fault lines. This suggests that the quartz is formed in one of the last stages of the cooling of magma chamber, when the excess silica, plus all the rare elements that cannot fit into the crystals of the silicate minerals that previously crystallized into granitic rocks, is left behind as a residue (Figure 16.9). Many of these quartz veins apparently forced their way up into cracks and fissures that formed after the rest of the magma had cooled and crystallized.

Another observation is critical as well. The richest gold veins in the Mother Lode don't occur in the main granitic batholith bodies but primarily where granitic dikes and quartz veins have intruded up through the metasedimentary and metavolcanic wall rocks and roof pendants of the western Sierras: the Foothills terrane (particularly the Mariposa Slate), or rocks of the Calaveras Complex, or especially in the Melones and Bear Mountain fault zones crossing these blocks (Figure 8.13). As we discussed in Chapter 8, the Foothills terrane is probably an ancient Jurassic accretionary wedge complex that formed to the west of the Jurassic

predecessor of the Sierra Nevada volcanic arc, while the Calaveras Complex is probably a Permian–Triassic accretionary wedge plastered onto the edge of the Shoo Fly Complex of early Paleozoic origin. This suggests to many geologists that the solutions of dissolved silica to make quartz apparently scavenged many of their rare elements from these metasedimentary and metavolcanic rocks (especially the serpentinites, which started as basalts or peridotite-rich rare elements) as they melted their way upward. The original granitic magma is not very rich in gold or most precious metals.

Finally, in some cases it appears that water is important in the process. Incorporating water into the melt lowers the melting temperature (especially of silica). Many of the quartz veins probably originated as a superheated mixture of water and dissolved silica, known as a **hydrothermal deposit**. This hot solution was very effective in scavenging rare elements out of the country rock as it intruded upward. In addition to quartz and gold, other rare precious metals (silver and copper), calcite, and ankerite (iron carbonate), along with minor amounts of iron pyrite ("fool's gold") and other metal sulfides, plus clays and micas like chlorite and sericite (fine-grained muscovite), are commonly found in the veins. Most hydrothermal deposits are thought to be associated with volcanic sources of heat, where the deep magma below heats up the groundwater and injects a mixture of water, rare elements, and dissolved silica into the surrounding rocks.

16.4 SUMMARY

The discovery of gold at Placerita Canyon in 1842, and especially in the Sierras in 1848, started the California Gold Rush, transforming California and making it into a state. During the Gold Rush, miners obtained gold from placer deposits (modern sedimentary deposits with gold, as well as Eocene sediments eroded from the bedrock), as well as from hard-rock

FIGURE 16.8 Gold in vein quartz.

Source: Courtesy of Wikimedia Commons.

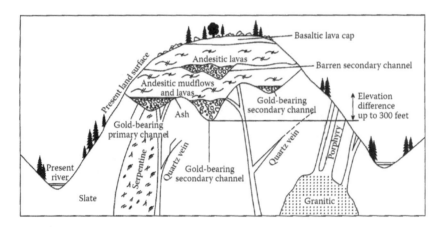

FIGURE 16.9 Mode of occurrence of gold in the Mother Lode. Not only were modern river placer deposits rich, but so, too, were ancient Eocene river gravels in the paleovalleys of the Sierras. The source for this placer gold were quartz veins and other hydrothermal veins, especially if they had leached gold and other rare elements from nongranitic country rock, such as serpentine or ophiolite.

Source: Modified from Chakarun, J. D., *California Geology*, 123–126, 1987. With permission from the California Division of Mines and Geology.

mining of the source gold veins. Most of the Mother Lode came from quartz veins that were the final melt of cooling magma chambers, which concentrated elements that could not fit into the previously crystallized minerals. Other quartz veins are formed hydrothermally as solutions of superheated water full of dissolved silica, which scavenge rare elements from the metamorphic country rock of the Calaveras and Foothill terranes, and then concentrate it in quartz veins.

RESOURCES

Ambrose, S.E. 2000. *Nothing Like It in the World: The Men Who Built the Transcontinental Railroad 1863–1869*. Simon & Schuster, New York.

Bidwell, J. 1948. *In California Before the Gold Rush*. Ward Ritchie Press, New York.

Brands, H.W. 2003. *The Age of Gold: The California Gold Rush and the New American Dream*. Doubleday, New York.

Chakarun, J.D. 1987. Tertiary gold-bearing gravels, northern Sierra Nevada. *California Geology*, 123–126.

Hill, M. 1999. *Gold: The California Story*. University of California Press, Berkeley.

McPhee, J. 1993. *Assembling California*. Farrar, Straus, and Giroux, New York.

Rawls, J.J., and Orsi, R.J. (eds.). 1999. *A Golden State: Mining and Economic Development in Gold Rush California*. California History Sesquicentennial Series 2. University of California Press, Berkeley.

Robert, F., Poulsen, K.H., and Dubé, B. 1997. Gold deposits and their geological classification. *Exploration Geochemistry*, 29, 209–220.

Yongfeng, Z., An, F., and Juanjuan, T. 2011. Geochemistry of hydrothermal gold deposits: A review. *Geoscience Frontiers*, 2, 367–374.

VIDEOS

The Gold Rush

How the Earth Was Made: Episode 13: America's Gold (History Channel); excellent documentary about the geology of how California's gold was formed, as well as how it was recovered: www.youtube.com/watch?v=sxwfRmzOnPO

Also on Gold Mining

www.youtube.com/watch?v=bwGNCwxD8No
www.youtube.com/watch?v=Dt7hTCeoXUM
www.youtube.com/watch?v=vXghYXPE32c
www.youtube.com/watch?v=MXyvrmsPZ80
www.youtube.com/watch?v=mx3OeQ_Vlj0
www.youtube.com/watch?v=NIMXhmvuiQA

Gold Panning

www.youtube.com/watch?v=ZCL6FKQZyoM
www.youtube.com/watch?v=mTn5aAYQKUE
www.youtube.com/watch?v=eLiwl9zHbW4

Sluice Boxes and Rocker Boxes

www.youtube.com/watch?v=gxe4N9fUhuU
www.youtube.com/watch?v=0SY-dIPV7vI
www.youtube.com/watch?v=IpZEPvKd9V8
www.youtube.com/watch?v=2z0-DWV0wdM
www.youtube.com/watch?v=9xUJYJXAgVo

Gold Mining and Gold Deposits

www.youtube.com/watch?v=bAKu2YuEN3w
www.youtube.com/watch?v=yCoULfk8pXE
www.youtube.com/watch?v=Qi8b_uFbNgM

17 California's Oil and Gas

Formula for success: rise early, work hard, strike oil.

—J. Paul Getty

17.1 BLACK GOLD

For more than a century, the world has run on petroleum. It powers nearly all vehicles and vessels, it creates electrical power, and it is distilled into thousands of products, from synthetic fabrics (nylon, polyester, etc.) to all the cheap plastic we throw away to heating our homes. We even eat oil, since enormous quantities of oil are used to make our food. A single acre of corn currently uses about 80 gallons of oil in the form of pesticides, fertilizers, and fuels for the machinery.

The price of oil can make or break the world economy, as well as the economies of states and countries. During the 1970s, the Organization of Petroleum Exporting Countries (OPEC) twice brought the world to its knees and created shortages, rationing, and gas lines by restricting the production of oil. The United States has gone to war twice in the Persian Gulf to protect supplies there, supporting theocratic dictator states that happen to sit on the world's largest supply of oil, while tolerating even worse regimes elsewhere in the world.

This influence on the economy is exacerbated by the fact that the overall supply is limited, and it is becoming more and more difficult and expensive to find supplies to feed our huge demand. The world consumes more than 210 million barrels a day, or about 77 billion barrels (1 barrel [bbl] = 42 US gallons, or about 159 L) of oil a year, and over the long run that demand is increasing, even if in 2014–2020 there was a short-term oil glut due to Saudi Arabia flooding the market with cheap oil to drive down the price and drive out competition with small oil companies and to punish Russia and Iran, followed by another glut in 2020 and 2021 when COVID shut down the economies of the world. The United States alone consumes more than 18 million barrels a day, 6.8 billion barrels a year, more than twice the demand of China, the next largest consumer. World oil companies must resort to more and more expensive and difficult techniques to find sources of oil to keep up with demand, although the recent boom in hydraulic fracturing (known as "fracking") has helped somewhat.

What about the future of oil resources? Back in 1956, the brilliant geologist M. King Hubbert predicted that the United States would reach its peak in oil production in the early 1970s. Peak production actually occurred in 1971 (Figure 17.1). Since then, US oil production has declined steadily, just as supplies of all non-renewable natural resources decline once they have passed their peak. The current boom in fracking has reversed the trend somewhat (Figure 17.1) but has nowhere near made up the difference compared to the peak in 1971. The last giant and supergiant fields were found in the United States in the 1950s and 1960s, and none has been found since then. This inevitably means that the US supply declined 30–35 years after the peak of discovery (Figure 17.2).

Likewise, many geologists (both inside and outside the oil industry) see evidence that the world supply of oil has already peaked and will begin a slow and steady decline through the rest of the century. We know this because no supergiant oil fields have been found in decades (Figure 17.3), and there are good geological reasons to think there are no more to be found. The overall peak of global discovery occurred in the 1960s and has been slowly declining since then, despite greater efforts and better technologies to find oil.

Of course, there are always short-term fluctuations in world economic conditions, which can mislead those who don't understand commodities. For example, oil producers often temporarily flood the market with cheap oil to drive out the high-priced competition (as the Persian Gulf nations did in 2014–2016 to depress the price of oil at a time when the oil consumption in China was dropping due to its slow economy). This gives people the false impression that oil is abundant, but geologists and commodity traders know better. Someday, within the lifetime of people reading this book, oil will truly become the scarce commodity that it really was all along. We will no longer be able to import millions of tons of cheap plastics or fabrics or use it on a large scale for agriculture, let alone afford to waste it powering our transportation or fueling our energy needs.

17.2 WHERE DOES OIL COME FROM?

Contrary to popular myth, oil is not produced from the bodies of dinosaurs. Instead, it is an organic material that is formed by the decomposition of trillions of marine phytoplankton as they become buried in sediments. Oil is actually a mixture of many different kinds of complex organic molecules, mostly long chains of carbon atoms with hydrogen atoms attached, or **hydrocarbons**. Chemically, it is about 85% hydrogen and 13% carbon, with minor amounts of sulfur, nitrogen, and oxygen. The simplest of these hydrocarbons is **methane** (CH_4), which is the major component of natural gas, along with longer-chain molecules like ethane (C_2H_6), propane (C_3H_8), and butane (C_4H_{10}).

These organic molecules start out as a complex mixture of decayed organic matter in sedimentary rocks known as **kerogen**. The most common source of rich kerogen is deep-water shales, which often trap lots of decaying matter and

DOI: 10.1201/9781003301837-17

FIGURE 17.1 History of oil production in the United States as of 2011. Back in 1956, M. King Hubbert predicted that the United States would reach its "peak oil" in the early 1970s, and it came to pass in 1971. Since then, US oil production has been steadily declining from 1971 levels, despite new discoveries in Alaska, deep offshore, and more recently, the boom in the Bakken Shale in the Williston Basin of North Dakota and Montana. As shown by this plot, however, the recent "boom" nowhere near makes up for the lost production since 1971.

Source: From Prothero, D. R., *Reality Check: How Science Deniers Threaten Our Future*. Indiana University Press, Bloomington, IN, 2013.

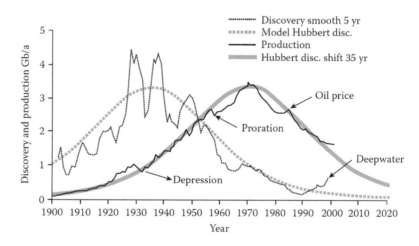

FIGURE 17.2 The Hubbert peak oil curve (curves to the right peaking in the early 1970s) can be predicted based on the curve of the rate of discovery of major oil fields (peaks to the left). Typically, the peak in oil production falls about 35 years after the peak in major oil discoveries. Since discoveries tailed off after 1935, it was expected that production would slow down 35 years later, around 1970.

Source: From Prothero, D. R., *Reality Check: How Science Deniers Threaten Our Future*. Indiana University Press, Bloomington, IN, 2013; modified from several sources.

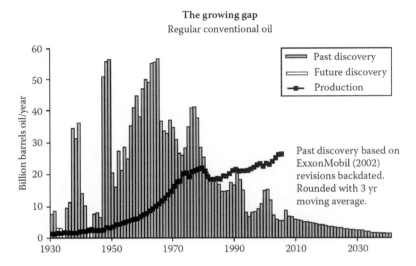

The growing gap
Regular conventional oil

Past discovery based on ExxonMobil (2002) revisions backdated. Rounded with 3 yr moving average.

FIGURE 17.3 The peak of worldwide discoveries of giant oil fields was in 1965–1975 and has been declining since. No supergiant oil discoveries have been made in decades. This would predict that the Hubbert peak of production should occur between 2000 and 2025. At the moment, the manipulation of oil production by Saudi Arabia and other oil producers makes it hard to tell whether we have passed Hubbert's peak worldwide, but by 2030, we should definitely be past the peak worldwide.

Source: From Prothero, D. R., *Reality Check: How Science Deniers Threaten Our Future*. Indiana University Press, Bloomington, IN, 2013; modified from several sources.

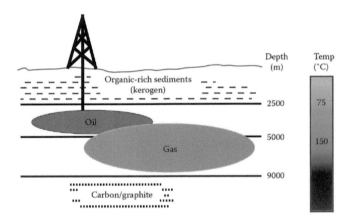

FIGURE 17.4 Oil window.

Source: Redrawn from several sources by E. T. Prothero.

are formed where there is not much oxygen to break it down. In some cases, large volumes of megascopic plant matter can also produce hydrocarbons rather than conventional coal. Rocks rich in kerogen are known as **source rocks**.

However, most source rocks are highly impermeable shales, so there is no point in drilling them directly. In addition, the kerogens in them are not yet oil when they first form. Instead, the kerogens must be cooked under pressure so they break down into simpler liquid hydrocarbons we know as petroleum. This chemical transformation is known as **maturation**. These restrictive conditions greatly limit which rocks will produce oil and which will not. Under a normal geothermal gradient, the source rock must be buried about 2,500 m (8,000 ft) in the Earth's crust, where the

temperatures are at least 65°C (150°F), so the kerogen will break down into liquid petroleum (Figure 17.4). If they are buried too deep (greater than 4,600 m, or 18,000 ft), they will be heated too much (above 150°C, or about 350°F) and the kerogen will break down into natural gas. We call this narrow range of suitable depths and pressures the **oil window**. The vast majority of the organic-rich rocks of the world have not resided in that range of ideal conditions long enough to produce oil. Time is also a critical factor. The kerogens must remain in the oil window long enough for oil to be produced, but not too long, or they will be overcooked and the oil will break down.

If all the conditions are right for the source rock to remain in the oil window, liquid oil will be produced and then migrate out of the source rock to a **reservoir rock**, which has high porosity so oil can saturate it. Contrary to popular myth, a "reservoir" is not some big underground cavern full of oil. Instead, it is a solid rock with a high volume of **porosity** between the grains (sometimes up to 40% pore space in sandstones that are well sorted) and high **permeability** (interconnectedness between the pore spaces so the fluid can flow through). Typically, these are well-sorted quartz sandstones, so there are lots of pores, with no fine clay or silt to clog them up, or cement in the pores. Other rocks, such as limestones or granitic rocks, have no natural pores, but if they are fractured, they can develop high porosity and permeability along the fractures.

Once the oil migrates from the source rock to the reservoir rock, a third condition is essential: an impermeable **seal** on top of the reservoir rock to trap the oil and prevent it from migrating further upward. The trapped oil then stays in the

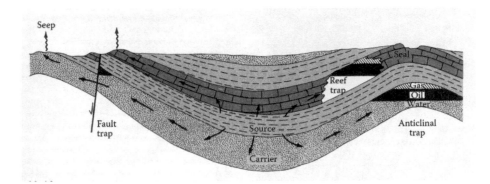

FIGURE 17.5 Different kinds of structural traps for oil deposits in the subsurface. The most common types of traps in California are anticlines and faults.

Source: From Prothero, D. R., and Dott, R. H. Jr. *Evolution of the Earth* (8th ed.). McGraw-Hill: New York, 2010.

reservoir rock until drilling releases it. This is most often the case when an impermeable shale lies on top of a sandstone reservoir, but often there can be structural seals, where a fault displaces a body of impermeable rock over the top of reservoir rocks. A very common kind of structural trap is the crest of an anticline, where oil migrates up the limb until a lid of the folded seal rock covers it, where it resides until drilling releases it (Figure 17.5). A fault on top of a reservoir often forms a good trap, since the fault gouge along the plane of movement is usually impermeable. There are also cases where the source rock shale both underlies the reservoir rock, providing the oil, and overlies it, providing the seal.

Thus, the ideal conditions to produce oil—abundant source rocks cooked for a long time in the narrow oil window, with a good reservoir above the source with a seal on top—are really rare. This is one of the many reasons that oil is actually quite scarce in most crustal rocks, and also that it is so difficult to find and obtain. There are far more rocks below our feet that would be good for producing oil but didn't meet all these conditions: source rocks that never reached the oil window or were too heated; source rocks with no reservoir rocks above them; potential reservoir rocks with no seal above them; and so on. Vast quantities of oil have been generated in the geologic past, and even trapped, only to be destroyed by tectonics and erosion.

Oil geology is a business of failure. Most promising prospects turn out not to have economical deposits of oil in them, and most holes that are drilled come up dry. But all oil geologists need is one really good producing well out of dozens of dry holes, and the payoff may be enough to keep them going, even when it costs more than $2 million to drill a typical hole.

So how do geologists find oil? Contrary to the myth from *The Beverly Hillbillies* that you can "shoot at your food" and oil will come bubbling up, oil exploration is a very sophisticated and expensive business. In the early days, geologists just dug holes or drilled where there were already natural tar seeps at the surface. But by the middle of the twentieth

century, it evolved from simple guesswork and "wildcat drilling" to a very precise business that uses many different sources of geologic information. The geologist starts with making and compiling geologic maps and reports of an existing region and studies the surface outcrops to see what can be learned. This is the type of fundamental exploration that led to our understanding of much of California's geology discussed in this book. The exploration geologist also tries to get the data on oil fields that have already produced, or all the wells that have been previously drilled in the area (if it's not proprietary data owned by another drilling company). In addition, geologists use a lot of **seismic stratigraphy** to see the structure underground. This method generates sound waves that bounce off subsurface layers and give a sort of cross section of the structure and stratigraphy beneath our feet.

Once all the available geologic information has been obtained and the exploration geologist has plotted it all in three dimensions, decisions must be made. If it looks like there is both a promising source rock at the right depth and a reservoir with a good seal, the exploration geologist makes recommendations of where to drill. Then the bosses make crucial decisions about which areas to drill, how to obtain the drilling rights on the land at a reasonable cost, and what the likely payout will be. A rule of thumb is that a typical drilled hole on land costs at least $2 million, and in the deep offshore area, about $150 million, so the decision to drill is not taken lightly.

In recent years, not only the methods for searching for oil but also drilling techniques have become more sophisticated. In the past, you could only drill straight down from your derrick to a potential reservoir, but today they can drill at angles, or make the drill bit and string of pipes turn or change direction at depth, and even drill horizontally. This allows them to reach pockets of oil that were previously inaccessible, or drill on land where the leasing rights are reasonable, and then reach pockets of oil deep below the surface more than six miles away.

17.3 CALIFORNIA OIL

California has long been a major player in the exploration and discovery of oil in the United States. A century ago, California was the biggest producer of all the states, pumping more than 100 million barrels a year, or 38% of the nation's supply. Over most of the last century, California's average production of about 500,000 barrels a day meant that it was the third largest producer, after Alaska and Texas. Recently, North Dakota has moved up to number two, displacing Alaska, and California is now in the fourth position. Still, California produces about 10% of the US supply, or about 200 million barrels a year. The petroleum industry is also one of the largest employers of geologists around the world and still has many thousands of employees in California. Oil wealth and interests powered much of the geological research in California, especially in its sedimentary basins and structures that are relevant to oil discovery.

Tar seeps and natural bitumen were coming up through the ground in many places in California. Natural tar pits, like those at Rancho La Brea in Los Angeles (see Chapter 20), Carpinteria east of Santa Barbara, and McKittrick north of Taft, have the proper conditions to preserve millions of fossils of Ice Age animals, including California's state fossil, the saber-toothed cat *Smilodon fatalis*. Pioneers used tar as a lubricant for wooden wagon axles and to seal the gaps in roofs, but otherwise there was very little demand for oil. The oldest asphalt mine in California was at McKittrick in 1864, followed by tunneling for oil at Sulphur Mountain in Ventura County in 1866. The first commercial oil field in California was discovered at Pico Canyon in Los Angeles in 1875.

Most of the early wells were dug by hand, using simple tools, and could not go very deep (Figure 17.6). The first oil derrick was a simple flimsy wooden structure, erected in Kern County in 1878 (Figure 17.7). By 1889, steam-powered drill rigs had produced oil in California's largest field, Midway-Sunset, near Taft. The first rotary drill rig was employed at Coalinga in 1903, and in 1910, the Lakeview gusher blew out near Taft, the largest gusher in US history. Meanwhile, new discoveries kept occurring in the Los Angeles and Ventura areas as well as the Kern River field near Bakersfield so that by 1903 California was the largest oil-producing state in the country. Just two fields (Midway-Sunset and Kern River) produced more than the entire state of Texas. By 1914, California produced 38% of the national supply of oil. In that same year, scientists began a decade of collecting Ice Age fossils at former oil fields at La Brea tar pits, transforming our understanding of the ice ages.

Meanwhile, other developments spurred the demand for oil. The invention of the Ford Model T made gasoline-powered cars more practical and affordable than earlier automobiles run by steam and other sources of power and created a demand in oil for transportation. The opening of the Panama Canal (which used California oil during

FIGURE 17.6 Before the age of rotary drilling, oil workers used a large rig that dropped a long steel spike into the ground to dig for oil. To the left of this photo is a long wooden rocker beam, lifted up and down by a flywheel that repeatedly slams the spike down, and then lifts it again, guided by the roughneck. This century-old antique rig is now on display at the Santa Paula Oil Museum.

Source: Photo by the author.

FIGURE 17.7 Antique oil derrick used a century ago in the Midway-Sunset field, now on display at the West Kern Oil Museum.

Source: Photo by the author.

construction) made it possible to ship California oil to the East Coast and Europe and the rest of the world.

Some of the "oil barons" in California in the early twentieth century became immensely rich. One of the most notorious was Edward L. Doheny, the first to drill in Los Angeles. His cutthroat business tactics and obscene displays of wealth made him a target for muckrakers. Among these was novelist Upton Sinclair, who based his best-seller *Oil!* on Doheny. That novel eventually became the basis for the Oscar-winning movie *There Will Be Blood*, with Daniel Day-Lewis portraying the ruthless and relentless oil baron who will stop at nothing to get at oil and get rich. The movie does an excellent job of portraying the conditions in the early days of California oil production, showing the old hand-excavated wells (Figure 17.6) and early rotary drilling technology.

The real-life Doheny was equally unscrupulous and driven to succeed at any cost. In the early 1920s, he bribed Interior Secretary Albert Fall with a $100,000 "loan" (worth about $1.32 million in today's dollars). In turn, Doheny got the federal government to lease his Elk Hills Oil Field (in the west San Joaquin Basin) as the Naval Petroleum Reserve Field (which it still is today) at very favorable rates with no competitive bidding. Another oil baron, Harry F. Sinclair of Mammoth Oil, got a similar deal on his fields at Teapot Dome in Wyoming. When the deals were exposed and Congress investigated, it became the "Teapot Dome Scandal." The entire scandal revealed the corruption in the White House and might have brought down the Harding administration in 1924 if Harding hadn't died in office first. Fall resigned and was eventually convicted of receiving bribes from Doheny, and Sinclair went to jail, but somehow Doheny was never convicted of the same charge.

Over the twentieth century, exploration and production and the price of oil went through ups and downs due to external forces, such as the Great Depression and World War II. In the 1930s, Texas and Oklahoma discoveries permanently passed California for the top oil-producing states in the country, followed by the discoveries in Alaska in the 1950s. Yet by 1953, California had the deepest oil well in the world, Richfield 67–29, which drilled down to 17,895 ft.

In the late twentieth century, production began to slow down in California as other regions could produce more oil more cheaply. The 1969 Santa Barbara oil spill deeply changed public attitudes about the risks and potential for environmental disaster due to oil and meant more environmental regulations and higher costs and delays in production. On top of this, multinational oil companies bought up all the California fields and only chose to operate those that could compete with their international supplies. Many of California's leading companies merged with other oil companies and lost their California focus. For example, pioneering companies like Union Oil of California (later Unocal 76) merged with Phillips 66 and Conoco. Standard Oil of California, once a part of John D. Rockefeller's Standard

Oil monopoly before it was broken up by the Supreme Court in 1911, was renamed Chevron and eventually merged with Texaco. Two other Standard remnants, Exxon (formerly Standard Oil of New Jersey, and then Esso) and Mobil (formerly Standard Oil of New York, or Socony), were recombined as ExxonMobil. Ironically, most of the pieces of Rockefeller's Standard Oil empire that trustbusters once broke up have recombined into a few large oil giants.

By the 1970s, the limits of production began to show in California. The last 100-million-barrel fields (Tule Elk and Yowlumne fields) in California were drilled in 1973 right at the US peak production predicted by M. King Hubbert (Figure 17.1). Kern County, the number one producing county in the state, reached its peak in 1985 at 256 million barrels per year and has been declining since. That same year, California produced 424 million barrels per year, and production has since slowly decreased statewide (Figure 17.8). Since 2010, fracking has tried to boost production in the state, but so far the results have been disappointing. For some reason, it doesn't work nearly as well in most California oil fields as it does in other places, and the environmental concerns mean that it may not be legal in the state eventually. Not the least of these concerns is the huge amount of scarce California water that is wasted in fracking, which might mean it will be outlawed in this state. In addition, California geologists have not yet found any suitable targets in the Monterey Formation that would make it suitable for fracking.

17.4   SAN JOAQUIN BASIN

As discussed in Chapter 10, the configuration of the San Joaquin Basin (Figure 10.10) makes it ideal for oil production. The trough-shaped deposit of Cenozoic sedimentary rocks includes the primary source rock, the Monterey Shale, often with sandstone reservoir rocks above the shales, and more shale sealing rocks above the reservoir rocks. The thick deposits in the center of the basin thin and pinch out rapidly on the flanks, so oil migrates up from the deep rocks in the center of the basin to margins near Taft on the west, and Bakersfield on the east, where it can be trapped. On the basin margins, there are often anticlines or faults that create structural traps. This is especially true in the fields around Taft and McKittrick, which are subject to the stresses of the nearby San Andreas fault (Figure 17.5).

For these reasons, the San Joaquin Basin is one of the largest oil-producing regions in the United States, with 68% of California's oil coming from Kern County alone. Before 1899, the San Joaquin Valley was famous primarily for its agriculture (especially cotton and fruit trees). The 1899 discovery of "black gold" in a shallow hand-dug oil well on the west bank of the Kern River changed all that. The Kern River discovery started an oil boom, and a forest of wooden derricks sprang up overnight on the floodplain just north of Bakersfield, then a sleepy farm town known at that time as "Bakers Swamp." Soon Kern River

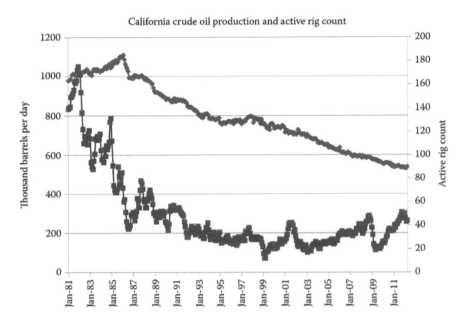

FIGURE 17.8 Declining production in California oil fields (barrels per day shown in the red curve) and decreasing active rig count (blue curve).

Source: Courtesy of California Department of Conservation.

production accounted for seven out of every ten barrels of oil that came from California, and Kern River field by 1903 had made California the top oil-producing state in the country. Inspired by the Kern River discovery, "oil prospectors" spread out over the San Joaquin Valley, and derricks began to pop up everywhere. Many discoveries followed, and a string of spectacular gushers at Coalinga, McKittrick, and Midway-Sunset fields kept the valley in the oil news. Lakeview Gusher was the greatest oil well the West, or for that matter, the country, has ever known.

More than a century later, the San Joaquin Valley still produces a lot of oil. In fact, just the Kern County part of the valley in 2015 had more than 42,000 producing oil wells that provided about 68% of the oil produced in California, 10% of the entire US production, and close to 1% of the total world oil production. In addition, there are another producing 2,000 wells in Fresno County. If the San Joaquin Valley were a state in its own right, it would rank right behind Texas, Alaska, North Dakota, and Louisiana as the fifth largest oil producer in the country.

The San Joaquin Valley is also home to 21 giant oil fields that have produced more than 100 million barrels of oil each, with four supergiants that have produced more than 1 billion barrels of oil. Among these supergiants are Midway-Sunset, the largest oil field in the lower 48 United States, and Elk Hills, the former US Naval Petroleum Reserve (Table 17.1).

There is not enough space in this book to discuss all these oil fields, but we will discuss Midway-Sunset field (Figures 17.9 and 17.10), the largest of them all, as an excellent example of their typical geology. Midway-Sunset is the largest field in California and the third largest in the United

States. It is a supergiant field, producing almost 3 billion barrels of oil so far, and its estimated reserves are thought to be about 532 million barrels still remaining, or 18% of California's total. The area of the field is enormous, running about 30 km (20 mi) from northwest to southeast and 5–6 km (3–4 mi) wide, from beneath Maricopa in the south and up beneath Taft and McKittrick. A drive along Highway 33 or any of the parallel roads like Midoli Road or Petroleum Club Road runs along the axis of this enormous field, through miles and miles of pumpjacks and storage tanks (Figure 17.10a), packed as closely as possible.

What makes Midway-Sunset so rich is that it combines all the elements needed for oil formation and lots of structural and stratigraphic traps as well. Structurally, it is on the west flank of the basin, so the saturated reservoir sandstones rise up from the basin center to the east and then either pinch out to nothing (trapping oil in their upper part) or are folded over anticlinal ridges or truncated by faults that seal their edges (Figure 17.10b). The field is full of many small and large anticlines and synclines, all chopped up by numerous faults, so the geology is complex and required much research to find all the oil pockets. Even more important, Midway-Sunset has at least 22 separate identifiable reservoir rock bodies at various depths. The shallowest is the Pleistocene Tulare Formation, which was the first to be discovered by following natural tar seeps at the surface, such as at McKittrick. In most places, the Tulare Formation provides the sealing caprock, which kept all the oil rising from below trapped until shallow drilling hit the main pools. Drilling deeper, geologists struck numerous small sandstone bodies and layers in the Monterey Formation at

TABLE 17.1

Rank of the San Joaquin Valley Oil Fields and Their Recent Production

Rank	Field	Date	Total Production through 2008
1	Midway-Sunset	1894	2,947 million barrels of oil
2	Kern River	1899	2,035 million barrels of oil
3	South Belridge	1911	1,535 million barrels of oil
4	Elk Hills	1911	1,317 million barrels of oil
	Coalinga[a]	1887	923 million barrels of oil
5	Buena Vista	1909	669 million barrels of oil
	Coalinga East Extension[a]	1928	504 million barrels of oil
6	Cymric	1909	497 million barrels of oil
	Kettleman North Dome[a]	1928	459 million barrels of oil
7	Lost Hills	1910	391 million barrels of oil
8	McKittrick	1896	309 million barrels of oil
9	Mt. Poso	1926	299 million barrels of oil
10	Kern Front	1912	212 million barrels of oil
11	North Coles Levee	1938	165 million barrels of oil
12	Edison	1928	149 million barrels of oil
13	North Belridge	1912	143 million barrels of oil
14	Fruitvale	1928	125 million barrels of oil
15	Rio Bravo	1937	118 million barrels of oil
16	Greeley	1936	116 million barrels of oil
17	Round Mountain	1927	113 million barrels of oil
18	Yowlumne	1974	111 million barrels of oil
19	Mountain View	1933	91 million barrels of oil
20	Poso Creek	1938	88 million barrels of oil
	Ten Section	1936	85 million barrels of oil
	Paloma	1934	61 million barrels of oil
	South Coles Levee	1938	59 million barrels of oil
	San Joaquin Valley		14.3 billion barrels of oil
	District 4 (Kern County)		12.2 billion barrels
	District 5 (Fresno County)		2.1 billion barrels
	California		28.3 billion barrels of oil

[a] San Joaquin Valley Fields Located in Fresno County.

Source: Modified from http://sjvgeology.org/oil/index.html.

depths of 1,500 m (4,900 ft), which are the primary producers of the field, although the PULV pool was 2,700 m (8,700 ft) down. Below these are the middle Miocene sandstones of the Temblor Formation, which are found in the deepest wells and apparently trapped oils that rose from the Eocene Kreyenhagen and other shales at greater depth.

Midway-Sunset field has produced so much because it went through multiple phases of exploration and production. First was the primary oil production from the reservoirs under natural pressure, which were then extended by water injection to flush the original oil out. Then they tried flushing steam down through the older fields, which produced far more oil than originally expected. In addition, the field theoretically should not have produced much at all, since most of the traps were breached naturally by

faults even before it was discovered. However, the oil that seeped out first solidified and provided a seal on the fault lines, preventing further escape until humans drilled down.

Midway-Sunset field continues to be productive, with only a slight decline since the 1980s, and new fields have been discovered in this complex system even in the past few decades, despite the fact that the field has been continuously operating since 1894.

So much oil is in some of the reservoirs at such pressure that gushers were not uncommon in the early days of production, before blowout preventers were routinely fitted to the drilling rigs. The most famous of these was the Lakeview Gusher, which exploded through the drilling rig on March 14, 1910. It spewed 9 million barrels of oil all over the landscape for 18 months before it was finally brought under control in September 1911 (Figure 17.10c and d). Initially, the daily flow was 18,800 barrels, which formed a river of crude oil across the landscape; crews rushed to contain it with sandbags and earthen dikes. Eventually, it was pouring out 90,000 barrels a day, but the crews managed to divert some of the flow into their pipelines that ran over the Coast Ranges to the coast at Port Avila. Unfortunately, less than half of the 9.4 million barrels that were released were recovered over the 544 days of the spill. The rest evaporated and formed crusts of congealed asphalt or seeped into the ground, contaminating the area for years.

This is why blowout protection is now required in all drilling operations. Gushers may be spectacular to watch, but they are enormously expensive in destroying equipment and wasteful in lost oil. In addition, they are often deadly not only to the crews who experience the blowout but also, ultimately, to everyone who lives in the area. Today the site has collapsed into a crater, and there is a historic monument on the site of the gusher.

17.5 LOS ANGELES BASIN

The Los Angeles Basin is the second largest area of oil production in California, after the San Joaquin Valley. It has been producing since 1892 (drilling below what is now Dodger Stadium) and produced almost half of the state's total oil production until the 1990s (Table 17.2). In 1904, there were more than 1,150 wells in just the city of Los Angeles. Most of the early wells were pumped dry pretty quickly, so production has diminished in the last century. In the late 1970s, the Los Angeles Basin was still producing a billion barrels a year, although as of 2013 it produced only 255 million barrels.

Even though it is almost completely covered by urbanization, the Los Angeles Basin still has about 40 active oil fields (Figure 17.11) and more than 4,000 producing wells. Some are right in downtown Los Angeles but carefully hidden by structures that make them invisible to the public. These rigs were part of the historic Los Angeles City Field, which

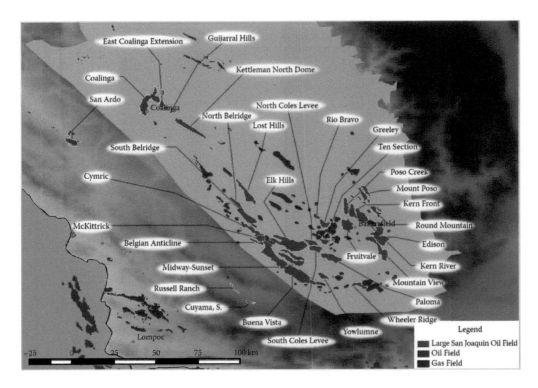

FIGURE 17.9 Map of the major fields in the San Joaquin Basin.

Source: Modified from several sources.

once went right through downtown Los Angeles and Dodger Stadium, covering an area of about 6 km (4 mi) long and about 3.2 km² (780 ac), or what is now Koreatown, Westlake, Echo Park, Chinatown, and Elysian Park. In the 1890s, this field was "Oil Queen of California," the top producer in the state. It yielded almost half of the oil in the state and made oil barons like Edward Doheny immensely rich. At its peak in 1901, there were 200 separate oil companies in the area. Forests of derricks once covered the hills north of Los Angeles, especially in Elysian Park, around what is now Dodger Stadium. Today, only one well remains, producing 3.5 barrels a day in a carefully concealed fenced enclosure on South Mountain View Avenue near Alvarado in Westlake.

The oil was relatively shallow and easy to reach with 1890s technology. It comes from the Pliocene Puente Formation in an east-to-west-trending faulted anticline, and with pinch-outs producing stratigraphic traps. There are three separate producing horizons that are about 300 m (900 ft), 350 m (1,100 ft), and 450 m (1,500 ft) below the surface. Some companies drilled all the way down to 2,300 m (7,500 ft) in the Topanga Formation, but no oil was produced. By the 1960s, most of the old oil fields were gone, covered by the huge growth of Los Angeles.

However, many of these fields still had significant oil and gas, and people just built right over them with no concern or thought of the consequences. In 1985, the "Ross Dress for Less" store in the Fairfax District was leveled by an explosion caused by the buildup of methane gas from below, and now most modern buildings in the area of the oil fields

have methane detectors and ventilation in their basements. During the 1990s, construction of the Belmont Learning Center, a proposed magnet high school for the Los Angeles Unified School District, was delayed for many years and cost $377 million because of concerns of groundwater contamination due to oil field chemicals, as well as also worries about methane leakage.

Even ritzy cities like Beverly Hills sit on top of oil fields. The Beverly Hills Oil Field was discovered in 1900 (long before the city was a haven for the movie stars and rich celebrities) and produced more than 150 million barrels of oil. There are still 97 currently active wells in the field, confined to "drilling islands," or areas of real estate with a cluster of many wellheads concealed by fake building façades. The wells all drill out diagonally or horizontally from these drilling islands so that the oil may be actually coming from beneath the mansions of millionaires or the expensive stores on Rodeo Drive. Beverly Hills is one of the longest actively producing fields in the state, with more than a million barrels in 2006, and probably 11 million barrels still in reserve.

Other fields, like Rancho La Brea, were once huge producing fields (the Salt Lake Oil Field) with many derricks, but decades ago production had slowed down. Instead, it is now famous for its tar pits full of Ice Age fossils. Today, it is Hancock Park, with the La Brea Tar Pits Museum (see Chapter 20), the Los Angeles County Art Museum, and other museums nearby on "the Miracle Mile" of Wilshire Boulevard.

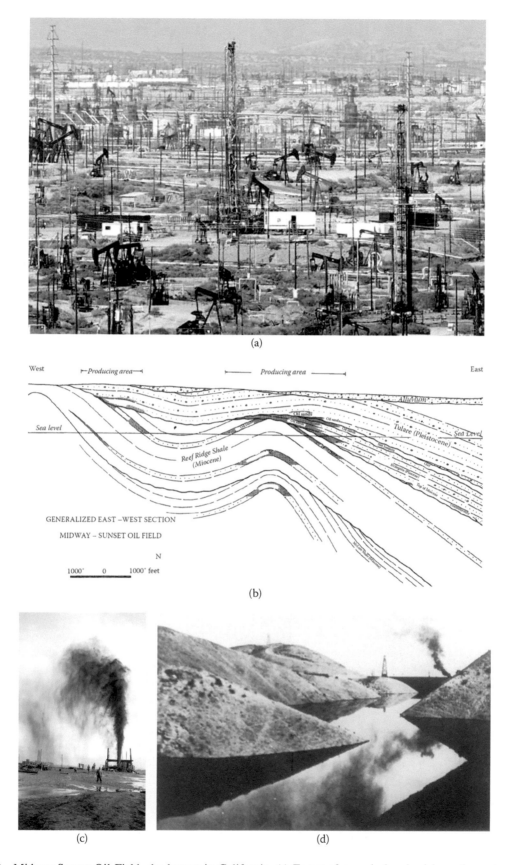

(a)

(b)

(c)   (d)

FIGURE 17.10 Midway-Sunset Oil Field, the largest in California. (a) Forest of pumpjacks, derricks, valves, storage tanks, and pipelines that cover the landscape for miles. (b) Stratigraphic cross section of Midway-Sunset field as it was understood in 1954. In the past 62 years, much more structural complexity has been documented. (c) Lakeview Gusher in 1910. (d) The river of oil draining from Lakeview Gusher in 1910.

Source: (a, c, d) Photos courtesy of Wikimedia Commons. (b) Map from *California Division of Mines and Geology Bulletin*, 170, 1954.

TABLE 17.2
Oil Production in the Los Angeles Basin

Oil Field	Year Discovered	Cumulative Production (mbbl[a])	Estimated Reserves (mbbl[a]) 12/31/08	2008 Production (mbbl[a])	Producing Wells (2008)
Wilmington[b]	1932	2,687,674	296,682	14,586	1,428
Long Beach	1921	942,727	3,469	1,498	299
Santa Fe Springs	1919	628,264	5,490	647	106
Brea-Olinda	1880	412,116	19,175	1,127	475
Inglewood	1924	396,467	33,081	3,107	449
Dominguez	1923	274,030	60	8	3
Coyote, West[c]	1909	252,960	0	0	0
Torrance	1922	225,944	5,595	387	110
Seal Beach	1924	214,289	6,649	491	129
Montebello	1917	203,935	5,839	648	106
Beverly Hills	1900	149,082	10,129	884	95

[a] Thousands of barrels.
[b] Second largest in California.
[c] Overlapped into Orange County.

FIGURE 17.11 Map of the major oil fields in the Los Angeles Basin.

Source: Courtesy of US Geological Survey.

We discussed the structure of the Los Angeles Basin in Chapter 12. As Figure 13.19 shows, the basin is fault-bounded and extremely deep and narrow, with many deep marine shales (such as the Monterey/Modelo, Repetto, Pico, and Puente Formations) acting as source rocks, and sands within the Monterey or Puente Formation acting as reservoirs. Structurally, it is chopped into numerous fault blocks (Figure 13.19b), and there are folds along many of these blocks, so there are structural traps, such as anticlines, as well as places where impermeable rocks

(a)

(b)

(c)

FIGURE 17.12 (a) Image of Long Beach Oil Field at Signal Hill in the 1930s showing the amazing density of derricks. (b) THUMS Island Grissom, with its camouflage of fake trees and buildings to make it look less like an oil platform. The "condo tower" covers the oil derrick and can move around the island as they change drilling positions. (c) THUMS Island Freeman from the air, showing the conventional offshore oil operations just inside the screen of trees.

Source: (a–b) Courtesy of Wikimedia Commons. (c) Photo by D. Clarke.

are faulted over reservoirs. Most of the biggest oil fields (La Brea tar pits, Inglewood, Dominguez Hills, Long Beach, Seal Beach, and Huntington Beach) follow the Newport–Inglewood fault (Figure 17.11), the site of the

devastating 1933 Long Beach earthquake (Figure 13.19a). This fault zone has many complicated structures, so there is much production beneath anticlines and from fault-sealed reservoirs. The structural trend of Torrance–Wilmington–Belmont fields is similar, since they are anticlinal structures faulted up against the edge of the basin and against the upthrown Palos Verdes Hills block. The Santa Fe Springs, Brea-Olinda, and Richfield fields (Figure 17.11) are on the trend of the Whittier fault zone.

Some of the oil fields in the Los Angeles Basin are huge. The immense Long Beach Oil Field was discovered in 1921, and it soon became the largest field in America, producing almost 20% of the nation's oil supply. In 1923, it produced more than 68 million barrels, the richest field by surface area in the United States, and about 20% of the total world output. By the 1930s, there were so many derricks on Signal Hill that it was known as "Porcupine Hill" (Figure 17.2a). It became the eighth largest field in the history of California, although it is mostly depleted now, with only 5 million barrels left of the original 950 million barrels once in it. There are still 294 wells in operation, producing about 1.5 million barrels of oil a year. You can still find many pumpjacks scattered around the city of Long Beach, mostly behind fences, but many sit right in the middle of public parks or even in the middle of parking lots and mini-malls. They are not disguised like those in some other fields in Los Angeles.

The main structural feature of the field is a huge anticline, and Signal Hill is just a surface expression of this buried structure. Like other large oil fields, it is composed of a series of reservoir sands surrounded by impermeable source rock shales. Drilling revealed that this area of the Los Angeles Basin is incredibly deep (Figure 13.19b), with deeper and deeper drilling reaching 4,560 m (14,950 ft, or almost 3 mi) in the 1950s, reaching Catalina Schist basement. Seven separate pools have been found, ranging from only 610 m (2,000 ft) to more than 2,300 m (7,500 ft) deep. The thick oil-bearing sands of the Miocene Puente Formation and Pliocene Repetto Formation were the reason the field produced so much early in its history.

Nearby is the gigantic Wilmington Oil Field, the third largest oil field in the United States by cumulative production (Figure 17.11). Discovered in 1932, it originally contained about 8 billion barrels of oil, making it a supergiant. Although much of that is now gone, geologists are optimistic that more can be produced using enhanced recovery methods. It is trapped by an anticlinal structure caught between the blocks bounded by the Newport–Inglewood fault and the Palos Verdes fault. Drilling revealed that the Catalina Schist basement is 2,400 m (8,000 ft) below the surface, and the oil comes from the Puente and Repetto Formations, as in many Los Angeles Basin fields.

Many of the surface operations on the Wilmington field are now closed down or hidden, but there is still significant production on artificial islands off the coast of Long Beach. They were known as the THUMS Islands (Figure 17.12b and c), an acronym for Texaco, Humble, Union, Mobil, and Shell, the original consortium of oil companies that built it, although Occidental Petroleum currently owns the

FIGURE 17.13 Natural tar seeps on Highway 150 north of Santa Paula.

Source: Photo by the author.

fields. The THUMS Islands are actually a series of offshore man-made islands covered with pumping operations. But on the outside of each island is a screen of artificial structures, trees and waterfalls, sculptures, and fake high-rise buildings that look like condo towers covering the derricks, so to the casual boater they just look like a natural island (Figure 17.12b). This "aesthetic mitigation" was overseen by theme park designer Joseph Linesch and cost $10 million at that time—but saved the owners all the grief that the public outcry over an unsightly oil platform would cause. The screens and waterfalls also help discourage swimmers and boaters from trespassing. The *Los Angeles Times* called them "part Disney, part Jetsons, part Swiss Family Robinson." Originally built in 1965 with prosaic names A, B, C, and D, in 1967 the four islands were renamed Island Grissom, Island Chafee, and Island White, after the Apollo 1 astronauts who died when their capsule exploded on the launch pad, and Island Freeman, after a pilot who died flying for the National Aeronautics and Space Administration (NASA).

THUMS production peaked at 148,000 barrels a day in 1969. There were 300 million barrels pumped by 1974, 500 million barrels by 1980, and 900 million barrels by 2002. Production declined to 44,444 barrels a day in 1992, because most of the oil is gone, and they must inject huge amounts of water to flush the remaining oil out. Only 20% of the pumped oil was water in 1965, but by 1994 it was 94% water.

17.6 VENTURA BASIN

The third largest producing region of California is the Ventura Basin. As discussed in Chapter 12, it is like the Los Angeles Basin in that it is a deep (almost 7 km, or 23,000 ft) basin full of young (upper Miocene–Pliocene) deep-marine shales and turbidites that poured into the basin as it sank rapidly along bounding faults (Figure 13.20b). Although we think of its surface as the flat Oxnard Plain on which Ventura–Oxnard and Camarillo lie, it is actually much bigger than this and includes the folded and faulted Cenozoic rocks on Transverse Ranges north to the Big Pine and Pine

Mountain faults, and south to the Santa Monica Mountains (Chapter 12). The east–west trough of the basin then continues offshore from Oxnard to form the deep marine basin between the Santa Barbara coast and the Channel Islands, where immense quantities of oil are produced on offshore platforms that are visible as you drive the coast along Highway 101. These same oil fields were responsible for the disastrous 1969 Santa Barbara oil spill, which transformed the environmental movement in this country, and the more recent pipeline leak in 2015, which also despoiled some of the most scenic and valuable beach property in the state.

The Ventura Basin was the site of the first oil field to be discovered in California. In 1861, natural tar seeps north of Santa Paula (Figure 17.13) were excavated with hand tools (Figure 17.7). Today, there is an excellent oil museum in Santa Paula commemorating the original discovery of oil in California and the long history of oil production in the Ventura–Santa Paula region. Since 1861, more than 97 oil fields have been found, 66 of which have produced more than a million barrels of oil. The three largest fields are giant fields, with more than 100 million barrels, and combined have produced at least 1,530 million barrels of oil: Ventura Avenue, Rincon, and San Miguelito. The deepest wells in the basin reached 6,550 m (21,500 ft) in the Rincon field, but even at that depth the rocks were still upper Miocene.

The sources of the oil here are different from those in other parts of California. Although much of it comes from the Monterey/Modelo, there is also some from the Santa Margarita Formation, the Pliocene diatomites of the Sisquoc Formation, and the lower Miocene Rincon Shale. Some oil may even come from the Paleocene and Eocene Santa Susana Shale and Llajas Formation. The reservoirs are a bit different, too, and include Miocene sandstones not only in the Monterey but also in the Pico Formation, the Santa Margarita Formation, and the Sisquoc Formation. Unlike other basins, there are also Paleogene reservoir rocks, such as the Oligocene Sespe red beds and the shallow marine Oligocene–Miocene Vaqueros Sandstone, and even older rocks like the Eocene Matilija, Coldwater, and Llajas Formations. Most of the oil is trapped in crests of huge anticlines that occur on the flanks of the basin (Figure 13.20a), especially where thrust faulting has pushed impermeable rocks over the top of reservoir rocks.

The largest field in the basin is Ventura Oil Field (Figure 17.14a), just northwest of the city of Ventura. Highway 33 travels right through the heart of the field. It is 13 km (8 mi) long by 3 km (2 mi) wide, oriented east–west across the valley of the Ventura River. Oil was first discovered in 1919, and this field has produced almost a billion barrels of oil, making it the seventh largest field in California. Its production peaked in the 1950s, when it ranked as the 12th most productive field in the United States. In 1954, it produced more than 31 million barrels of oil. There are still about 50 million barrels in reserve, and it has more than 423 wells still producing 11,000 barrels a day.

The oil field is at the crest of the Ventura Anticline (Figure 17.14b), a large fold that is 26 km (16 mi) long, and forms a long east–west ridge just north of Ventura. As you drive through it on Highway 33, you can see the change of the dip of the beds in the roadcuts as you go from one limb of the anticline to the other. Most of the production comes from the Pliocene Pico Formation, where the porous turbidite sands are good reservoir rocks with up to 20% porosity. Below that, oil is also trapped in the upper Miocene Santa Margarita Formation. There are actually eight different oil-bearing levels in the field, ranging from 1,120 m (3,680 m) down to more than 3,700 m (12,000 ft). The deepest wells reached 6,600 m (21,500 ft) but produced nothing at this depth.

The source rock of the oil is the Monterey Formation beneath, which probably began to produce in the late Neogene as it became buried under the Pico Formation. Late Pleistocene folding and faulting on the basin created the anticlinal trap, along with a series of faults that also serve to trap oil.

17.7 THE MONTEREY FORMATION

Since at least 38 billion barrels of California's oil (including 84% of the oil in the San Joaquin Basin) comes from a single unit, the Monterey Formation, it is worthwhile to discuss it in greater detail here. The Monterey covers at least 4,532 km² (1,750 mi²) of surface outcrops and in the subsurface underlies most of the big coastal basins from Point Arena in the

(a)

(b)

FIGURE 17.14 Ventura Oil Field: (a) Aerial shot of the oil field, with Ventura and the Pacific in the background to the south. (b) Diagram of the structure of the Ventura Avenue Anticline and related features in the Ventura Basin.

Source: (a) Courtesy of Wikimedia Commons. (b) Courtesy of Lamont-Doherty Earth Observatory.

Gualala block (Figure 12.13f) to Berkeley Hills to its type area in Monterey Bay down to Orange County (Figure 17.15). It ranges from almost 17 million years (early–middle Miocene) in some of the oldest basins to as young as 7 near the top of the unit, and it spans four of California's middle and late Miocene benthic foraminiferal stages (Saucesian, Luisian, Mohnian, and Delmontian). Most of the recent dating, however, has been done using the siliceous plankton in the formation, especially the diatoms and the radiolarians. They give a very precise zonation of different parts of the Monterey Formation, so if the fossils have been well preserved and are not too diagenetically altered, it is possible to correlate the units with great precision. A number of volcanic ashes have been dated in the Monterey as well, giving numerical age constraints to the chronostratigraphy.

The thickness of the unit varies dramatically from one basin to the next, but in some places it is more than 1,000 m (3,300 ft) thick. The local stratigraphy also varies tremendously from basin to basin, with some outcrops ranging from shallow marine to deep marine deposits, but most of the Monterey Formation is deep-water shales, siliceous shales, and cherts. In many places, the shales are finely laminated with millimeter-scale dark and light layers, indicating very still deep water with fluctuating oxygen conditions. The darker layers are richer in kerogens, and these become the hydrocarbons that mature into the oil that makes the formation so productive. In most cases, the outcrops appear as

FIGURE 17.15 Distribution of the Monterey Formation in California.

Source: Courtesy of Wikimedia Commons.

finely laminated beds of mudstone or shale alternating with siliceous shales and sometimes hard resistant chert layers with minimal clay content (Figure 17.16).

In many cases, the Monterey is so rich in siliceous plankton that it is not just a siliceous shale but a diatomite or diatomaceous earth. Above the Monterey in westernmost Santa Barbara County, just south of Lompoc and La Purisima Mission, is the 1,000 m (3,300 ft) thick Pliocene Sisquoc Formation, made almost entirely of pure diatomaceous earth (Figure 17.17a and b). The Lompoc Hills are made entirely of this diatomite, the largest such deposit in the world. Diatomaceous earth or diatomite is a tightly packed rock made of the shells (called frustules) of the planktonic algae known as diatoms (Figure 17.17c). The pore spaces between the shells, as well as the tiny pores inside the shell, make diatomite excellent for filtration purposes, since almost any size of particle or impurity is trapped as the water flows through.

Diatomite is mined on a huge scale for many purposes. Diatomite is used not only in filters of every kind and as an absorbing agent (e.g., in kitty litter) but also as a decolorizing agent, as a metal polishing grit, in matte paint formulas, in match heads, and in many other industrial purposes. Diatomaceous earths are found in many parts of the California coastal region, such as the Valmont Diatomite in the Palos Verdes (Figure 17.17d).

So what kinds of conditions are responsible for producing such immense deep-water basins that accumulated such a high volume of organic-rich shale, and also such high volumes of silica? The most popular model (Figure 17.18) uses the fault-bounded basins of the modern California Borderland off the Channel Islands and Catalina as an inspiration. Deep-water basins close to the coastline (e.g., the ancient Cuyama Basin in Figure 17.18) serve as sediment traps, so nearly all the turbidite sands and most of the mud will be trapped in them as they subside. The next basin offshore (the ancient Santa Maria Basin in Figure 17.18) will get a much smaller supply of muds from the land and will have a lower sedimentation rate, so it can be dominated by siliceous ooze from the blooms of planktonic algae at the surface. The farthest basin offshore (Santa Barbara Basin in Figure 17.18) will get almost no muds from the land, so it may be filled with pure siliceous ooze, even though it is only a few kilometers offshore from the land source of sediment. Normally, thick deposits of siliceous ooze occur only in the deepest ocean basins far from land, but the unusual tectonics of California, with its multiple fault-bounded basins, makes it possible for it to form on a basin not far from the land and eventually become uplifted into California's mountains.

But this is not the whole story. The Monterey (and units like the Sisquoc as well) is extremely silica-rich compared to most deep-water shales, especially those forming in the ocean today. Rich blooms of siliceous plankton (diatoms and radiolarians) that make siliceous ooze require something to bring a lot of dissolved silica to the surface waters, where they live, since most surface

(a)   (b)

(c)   (d)

FIGURE 17.16   Typical outcrops of the Monterey Formation: (a) Finely laminated shales and resistant chert layers at Montana del Oro beach, south of Morro Bay. (b) Resistant siliceous shales near Naples Beach, Goleta. (c) Highly contorted Monterey shales and cherts near the boathouse, Vandenburgh Air Force Base. (d) Typical outcrops of the Altamira Shale Member of the Monterey Formation in the Palos Verdes Peninsula.

Source: Photos by R. Behl.

waters are highly depleted in silica. Extensive siliceous volcanism in the middle Miocene of California has been suggested, although it does not explain the whole story. Instead, we know that most modern siliceous oozes are produced where upwelling currents from the deep ocean bring scarce nutrients like silica to the surface, where the diatoms can then bloom by the trillions once their limiting nutrient becomes available. The middle Miocene was an important time in global climate and oceanography. The last of the warm "middle Miocene climatic optimum" ended about 16 Ma, and by 14 Ma, the Antarctic ice cap grew to almost its present size. This expansion of Antarctic ice is thought to have driven oceanic circulation, especially cold deep-water currents. These currents then moved around the ocean basins and triggered upwelling events in many coastlines (especially California, which today gets the cold upwelling California Current from the Gulf of Alaska). This is confirmed by the fact that there is a rapid turnover of diatom diversity in the middle Miocene, and that they tend to be mostly modern cold-water lineages that originate at that time. Not only that, but the late Miocene diatoms are much smaller and thinner-shelled, suggesting that they are growing faster and blooming and reproducing at an earlier age to take advantage of huge new resources in dissolved silica.

The Monterey Formation is truly an amazing story, not just because it has produced so much oil, but also because of the tectonic and climatic story behind it.

17.8   THE ENVIRONMENTAL COSTS OF OIL

So far, we have only focused on how oil is formed and where it has been trapped and recovered in California. But there is a big downside to oil drilling and production as well. The immediate negative effects are things like oil spills contaminating the land or the groundwater in certain areas. The much bigger problem, however, is that fossil fuels like oil are the reason for global warming and other related environmental disasters due to climate change (see Chapter 24). As the world is now beginning to come to terms with the problem, the race is not to extract and burn more oil but how to phase out fossil fuels altogether and replace them with renewable energy (Chapter 23).

To some degree, this transition had already happened. Due to environmental regulations adding to the costs of drilling and exploration, new drilling and new exploration in California have almost completely ended, and most oil production in the state is from pre-existing wells and oil fields. Major oil companies, like Occidental Petroleum, closed their California offices, and most of the remaining

FIGURE 17.17 Diatomaceous earth (or diatomite) is common in the upper Miocene rocks of California, both in the upper Monterey Formation and in overlying units: (a) Mine exposures of the immense deposits of white diatomite of the Sisquoc Formation in the mountains south of Lompoc. (b) Outcrop of Sisquoc diatomite near the entrance to the Lompoc quarry, showing the well-developed bedding, here folded by later tectonism. (c) Microphotograph of the dense packing of diatom shells ("frustules"), which have lots of tiny pores and permeability, so they trap even the smallest impurities in the water passing through. (d) Even boulders of diatomite are surprisingly light, since most of the rock is air space. This example is from the Valmont Diatomite, Malaga Cove, on the Palos Verdes Peninsula.

Source: (a) Photo by R. Stanley. (b, d) Photos by the author. (c) Photo by A. Chang.

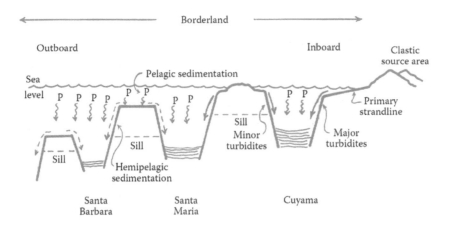

FIGURE 17.18 Depositional model for the Monterey Formation (see text for discussion).

Source: Modified from Blake, G. H., *Pacific Section SEPM, Special Publication*, 15: 1–14, 1981.

major oil company offices in the state are either focused on maintaining existing fields and production or else are doing exploration in other parts of the world from their California offices. There is also some offshore oil and gas production in California, but there is now a permanent moratorium on new offshore oil and gas leasing and new offshore platforms in both California and federal waters, although new wells can be drilled from existing platforms.

California is a state with a high degree of environmental consciousness, so the political popularity of oil in California is at an all-time low, and most voters elect politicians who are not taking money from the oil lobbyists. In 2021, Governor Gavin Newsom directed the Department of Conservation's Geologic Energy Management (CalGEM) Division to initiate regulatory action to end the issuance of new permits for hydraulic fracturing ("fracking") by January 2024. Additionally, Governor Newsom requested that the California Air Resources Board (CARB) analyze pathways to phase out oil extraction across the state by no later than 2045. These measures mean that oil production in California will be phased out in a little more than 20 years, no matter how much still remains in the ground.

The bad reputation of oil in the state started with the legendary Santa Barbara oil spill (Figure 17.19). On January 28, 1969, a blowout occurred on Union Oil (now Unocal) Platform A in the Dos Cuadras Oil field, about 6 mi (10 km) off the coast of Santa Barbara. About 80,000 to 100,000 barrels (3,400,000 to 4,200,000 gallons) of oil spilled into the ocean, fouling the beaches up and down the coast from Ventura to beyond Santa Barbara and Goleta, as well the northern shores of the four northern Channel Islands. The spill was a disaster for marine life, with at least 3,500 sea birds and hundreds of marine mammals such as dolphins, sea lions, and elephant seals killed. Images of dying sea birds with their wings fouled by oil (Figure 17.19b) and dead sea lions covered in petroleum filled the news media at that time. This oil spill was particularly prominent because it fouled one of the most beautiful stretches of coast in California, and rich communities like Santa Barbara and Ventura have a lot of political clout. The outrage reached epic proportions as people reacted to the spill with strong political pressure on politicians at both the state level and federal level. The oil spill was still not contained when President Richard Nixon visited on March 21, largely because the oil industry had little experience with big oil spills and had not yet developed all the measures they now use to prevent, contain, and clean up the mess. The official cleanup took about 45 days, although globs of tar were floating up for many months, and the beaches were not reopened until June 1.

This was one of the first offshore major oil spills in US history and was an important factor in the environmental movement that exploded in the 1970s. After the public outrage over this oil spill and other pollution problems nationwide, such as the Cuyahoga River in Cleveland catching fire, almost everyone was outraged at what was happening to the environment. The first Earth Day was set up by Senator Gaylord Nelson of Wisconsin on April 22, 1970, and the environmental movement was popular throughout the country. It was a time when the United States was already undergoing radical change from the civil rights movement, the feminist movement, and heated opposition to the Vietnam War and the draft by the younger generation. In July 1970, Congress and President Nixon established the Environmental Protection Agency (EPA) and the Clean Water Act, with almost unanimous support of both Republicans and Democrats and a Republican president. At this point in history, environmentalism was not a partisan issue, and few politicians were in the pockets of big polluters.

Locally, in 1969 the state of California banned any further offshore drilling in California inside the 3 mi limit of state control, and there was a lot of pressure to prevent new drilling even outside the state's boundaries. In 1970, the California Coastal Commission was set up to regulate not just oil drilling but any use of the public coastlines. The California Environmental Quality Act (CEQA) was passed, which not only regulated oil drilling and other sources of pollution but also set up the requirements for an Environmental Impact Report (EIR) for almost all new construction of housing projects and new industrial plants in the state as well. This law had interesting and sometimes unintended side effects, such as using it to block many kinds of construction and development. Today, almost any large construction site is required by CEQA to hire an environmental consulting firm to survey the land as it is graded and bulldozed, not only to assess the environmental effect, but also to protect artifacts or fossil sites if they are found. This is why environmental firms are a major employer of geologists, paleontologists, and archeologists in California, not matched by any other state.

The problem with oil spills has continued to plague the oil industry, as bigger and bigger spills have occurred. Even though a lot of new technology, like blowout preventers, has been developed, there are still problems which produce gigantic horrendous consequences. The 1989 Exxon Valdez oil spill in Alaska was much larger than the 1969 Santa Barbara spill but was the fault of negligence by the captain. The gigantic 2010 Deepwater Horizon spill in the Gulf of Mexico dwarfed both of those disasters and occurred even though the technology of blowout preventers was being used.

Oil spills will continue to remain a problem on the California coast, simply because there are still so many platforms and other related facilities in the region. Most recently, the October 1, 2021, Orange County oil spill was caused when two ships dragged their anchors across the sea floor and ruptured seafloor oil pipes from an offshore platform. The leaked oil covered 8,320 acres and fouled beaches from Long Beach down to Newport. This spill was the final straw for many people and their politicians, which led to the passage of laws in California to phase out oil drilling completely in the state (as mentioned earlier).

17.9 SUMMARY

California has long been one of the nation's biggest oil producers and continues to produce significant reserves, even if it is no longer the richest oil-bearing state as it was a century ago. California oil helped propel the economy of the early twentieth century as more and more uses for oil were found in gasoline-powered automobiles and aircraft. The oil fields of the San Joaquin Basin (especially Midway-Sunset) are among the nation's largest, and some have produced incredible amounts of oil for more than a century. No less impressive is the oil wealth of the Los Angeles Basin,

FIGURE 17.19 The 1969 Santa Barbara oil spill: (a) The oil slick visible around Platform A in the Santa Barbara Channel emanated from fissures in the seabed. (b) A duck covered in a thick coating of crude oil, picked up when it lighted on waters off Carpinteria State Beach in Santa Barbara County, California, after the oil spill in January 1969. (c) Cleanup crews try to scrape up and collect the oil on the beaches, while tugboats offshore drag booms and nets to scoop up the oil slick on the surface.

Source: (a) Courtesy of the USGS. (b–c) Courtesy Wikimedia Commons.

although most of that production happened in the early twentieth century. The Ventura Basin also has very large oil fields, including the offshore fields between Santa Barbara and the Channel Islands. However, the increasing cost of producing oil in California, as well as the depletion of many of the oil fields and the environmental damage produced by oil production, suggests that oil will be less important in California's future than it once was.

RESOURCES

Blake, G.H. 1981. Biostratigraphic relationship of Neogene benthic foraminifera from the California outer continental borderland to the Monterey Formation. *Pacific Section SEPM, Special Publication, 15*, 1–14.

Davis, M.L. 2001. *The Dark Side of Fortune: Triumph and Scandal in the Life of Oil Tycoon Edward L. Doheny*. University of California Press, Berkeley.

Deffeyes, K. 2001. *Hubbert's Peak: The Impending World Oil Shortage*. Princeton University Press, Princeton, NJ.

Deffeyes, K. 2006. *Beyond Oil: The View from Hubbert's Peak*. Hill and Wang, New York.

Deffeyes, K. 2010. *When Oil Peaked*. Hill and Wang, New York.

Downey, M. 2009. *Oil 101*. Wooden Table Press, New York.

Franks, K.A., and Lambert, P.A. 1985. *Early California Oil: A Photographic History, 1865–1940*. Texas A&M University Press, College Station.

Garrison, R.E., and Douglas, R.G. 1981. *The Monterey Formation and Related Siliceous Rocks of California. Pacific Section SEPM Book 15*. Pacific Section SEPM, Fullerton, CA.

Inman, M. 2016. *Oracle of Oil: A Maverick Geologist's Quest for a Sustainable Future*. W.W. Norton, New York. [Excellent new biography of legendary oil geologist and prophet M. King Hubbert].

Lazarus, D., Barron, J., Renaudie, J., Diver, P., and Turke, A. 2014. Cenozoic planktonic marine diatom diversity and correlation to climate change. *PLoS One, 9*(1), e84857.

McCartney, L. 2008. *The Teapot Dome Scandal: How Big Oil Bought the Harding White House and Tried to Steal the Country*. Random House, New York.

Prothero, D.R. 2013. *Reality Check: How Science Deniers Threaten Our Future*. Indiana University Press, Bloomington, IN.

Prothero, D.R., and Dott, R.H., Jr. 2010. *Evolution of the Earth* (8th ed.). McGraw-Hill, New York.

Rintoul, W. 1976. *Spudding In: Recollections of Pioneer Days in California Oil Fields*. California Historical Society, Sacramento.

Rintoul, W. 1990. *Drilling through Time: 75 Years with California's Division of Oil and Gas*. California Division of Oil and Gas, Sacramento.

Sabin, P. 2004. *Crude Politics: The California Oil Market, 1900–1940*. University of California Press, Berkeley.

Sampson, A. 1975. *The Seven Sisters: The Great Oil Companies and the World They Shaped*. Viking, New York.

Sinclair, U. 1927. *Oil!* (2007 ed.). Penguin, New York.

Williams, J.C. 1997. *Energy and the Making of Modern California*. University of Akron Press, Akron, OH.

Yergin, D. 1994. *The Prize: The Epic Quest for Oil, Money, and Power*. Buccaneer Books, New York.

VIDEOS

Videos About Early Oil Drilling in California

www.youtube.com/watch?v=aMbL49KBYe0

www.youtube.com/watch?v=29g8yYMgE3c

Videos about the drilling and exploration techniques:

www.youtube.com/watch?v=krkigplNUXU

www.youtube.com/watch?v=a3anE3PEPP8

www.youtube.com/watch?v=SfazJ6P_g7w

www.youtube.com/watch?v=oftgbYm9d7U

An episode of *California Gold* with Huell Howser, originally aired on PBS stations, about the THUMS islands: www.youtube.com/watch?v=Ls6WN0cnOBE

Peak Oil

www.youtube.com/watch?v=3t5jVzmaxTU

www.youtube.com/watch?v=PJfSVNowTLY

Video version of Daniel Yergin's book *The Prize*: www.youtube.com/watch?v=Qspu35JG59Q

Other

In addition, the Oscar-winning movie *There Will Be Blood* is a dramatic account of the early California oil industry, with accurate portrayals of oil production in the days of simple hand tools and early drill rigs, as well as the ruthless tactics of the oil barons.

It was loosely based on Upton Sinclair's muckraking novel, *Oil!*, which in turn was based on Los Angeles oil baron Edward Doheny.

California Division of Oil and Gas, "Oil and Gas Production History in California": http://ftp://ftp.consrv.ca.gov/pub/oil/history/History_of_Calif.pdf

18 California's Water

Water is the essence of our existence. Without it there can be no life. Civilizations grew next to water. But as populations grow, especially in arid regions, there is not enough to go around.

—**Marc Reisner**
Cadillac Desert, **1983**

18.1 THE STUFF OF LIFE

Gold and oil have been very important resources over the course of history, but water is essential to all life. Living things are largely made of water. Humans and other animals can often survive for long periods without much food, but not without water. Water is essential not only to human civilization (for drinking, bathing, cooking, and many other uses) but also to the animals and plants we eat, so all of agriculture is largely dictated by water as well. In California, water and the politics surrounding it have dictated events in the state's history, as well as affected the economic strength of the state.

18.2 PRECIPITATION

The average precipitation (both rain and snow) for California is quite low, only 58 cm (22.8 in) falling per year. By contrast, the wetter parts of eastern North America get at least 127–254 cm (50–100 in) per year, and the rain forests of the Olympic Peninsula of Washington get as much as 380 cm (150 in) per year. But the "average rainfall" is meaningless in this state. California receives water as rain or snow in very unequal amounts (Figure 18.1). Much of the northern parts of the state are typically well-watered (more than 254 cm, or 100 in, of rain per year), and the mountains (especially the western flank of the Sierras) get significant rainfall and snowfall, amounting to about 75% of the total rainfall in the state. The southern half of the state is semidesert or full desert, with only about 25 cm (10 in) of rain a year, which is inadequate water to support its people and agriculture, so it demands about 80% of California's total supply.

But average or total rainfall isn't the whole story. It's also important to know how often the rain falls. In California, the rainfall patterns are not like those of the East or Midwest, which have rain year-round. Instead, much of California gets nearly all its rain and snow in a few big winter storms out of the Pacific, and almost no rainfall in spring, summer, or fall. Those are the normal patterns, but there are occasional long periods of extreme drought as well. For example, there were horrendous droughts between 1928 and 1934, 1976 and 1977, 1987 and 1992, and as of this writing, drought has prevailed from 209 through 2021.

The native plants (particularly the coastal sage scrub and chaparral) are adapted to growing quickly during the brief rainy season, flowering and spreading their seeds, and then going dormant and dying back during months of drought. But most agricultural crops, and many ornamental and landscaping plants, need water continuously, so they must be artificially watered. In the cities, about half of the total water usage is for landscaping, and the many Southern California households use almost 200 gal a day watering outdoor plants.

The extreme example of this waste are lawns and large areas of grass for parks and golf courses, which are a landscaping habit picked up from the rainy British Isles and successful in the rainy eastern and central states. They make no sense in the drought-prone regions of California and end up being huge water wasters. Most places can be landscaped just as attractively with drought-tolerant native plants and other forms of xeriscaping (landscaping using drought-tolerant plants).

18.3 DRAINAGE AND RUNOFF

When streams and rivers flow across the Earth's surface, they create a network of trunk streams and tributaries known as a **drainage** (Figure 18.2). Between the drainages of two different rivers is a **drainage divide**, an elevated area (typically a hill, ridge, or mountain crest) that separates water going down one river from water going down another. For example, on the Continental Divide in the Rocky Mountains, a raindrop that falls west of that line will flow toward the Pacific, but if it falls to the east of that boundary, it will flow to the Atlantic or Gulf of Mexico. Scientists typically map the Earth's surface as a series of drainages and their basins (Figure 18.3a) separated by divides. Another term for a drainage basin and its divides is **watershed**.

A glance at Figure 18.3a shows the major drainages and basins in California. By far, the largest is the San Joaquin–Sacramento Basin (the Central Valley), which gets its water from the wettest parts of the state: the Sierras and the northern mountains. Their drainage basin covers almost half of the state and nearly the entire state where abundant water flows. Table 18.1 shows the annual discharge and runoff of the major California rivers. Naturally, the Sacramento is the largest river in discharge or runoff, followed by rivers of the wet north, like the Klamath, Eel, and Russian Rivers—plus the San Joaquin River, which is the largest river in the dry parts of the state. By contrast, desert rivers like the Mojave have only tiny amounts of discharge (typically during rare winter floods) and are dry most of the year, with their snowmelt sinking directly back down into the riverbed. The Los

DOI: 10.1201/9781003301837-18

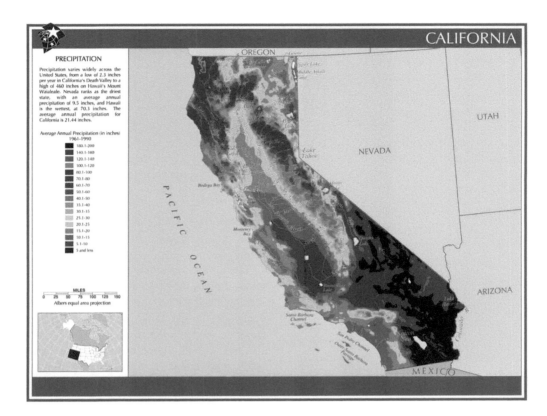

FIGURE 18.1   Rainfall map of California.

Source: Courtesy of US Geological Survey.

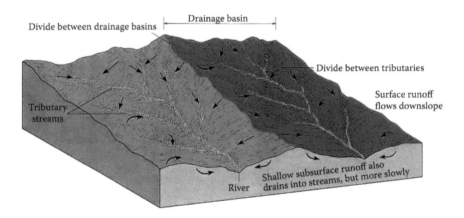

FIGURE 18.2   Drainage network and drainage divide.

Source: Courtesy of US Geological Survey.

Angeles River is mostly lined in concrete, and most of its natural flow during big winter rainstorms goes straight into the ocean. The flow you see the rest of the year is treated sewage.

The major drainages can be clustered into larger units called **hydrologic regions** (Figure 18.3b), which are convenient for water management purposes. They serve as useful subdivisions of the landscape that describe areas with similar amounts of rainfall and river discharge.

The typical precipitation and runoff of each of the hydrologic regions in California are given in Table 18.2.

As we already discussed, the annual precipitation and annual runoff are largest in the North Coast and Sacramento River hydrologic regions. The San Joaquin River region receives less than half of the values of either the North Coast or Sacramento River regions. Every other hydrologic region receives barely 10% or less of the two wettest regions in the north. Driest of all is the Colorado River region, with barely

1% of the runoff of the two wet northern regions. These numbers capture the overall inequality of the distribution of precipitation and surface runoff. Average precipitation means nothing in a state that has either very wet regions or very dry regions, with no region getting the amount of rainfall that approaches the mathematical average for the state.

18.4 GROUNDWATER

In most parts of the world, the vast majority of the rain falling on the landscape does not flow out to the ocean via rivers but sinks into the ground and infiltrates through until it saturates the sediments below. It is known as **groundwater**. The supply of water is critical in many parts of the world, especially when the surface rivers have dried up. This is particularly true of California, which has little or no surface water in the drier parts of the state. In a typical year, 30% of the state's water supply comes from groundwater, but in times of drought, it can be 60% or more. California

has about 450 known groundwater reservoirs, storing about 850,000 ac-ft (1.04 million m³) of water.

As rainfall seeps down into the ground, it flows down (infiltrates) through the pore spaces between the sedimentary grains in the subsurface (Figure 18.4). This unsaturated soil with mostly air in the pore spaces is known as the **zone of aeration**. Eventually, it accumulates deeper underground to form a **zone of saturation**. The top surface of the saturated zone is called the **water table**. The water table is a dynamic feature, moving up and down in response to the rate of inflow of water. If there are heavy rains, the water table will rise, and sometimes it will completely saturate the highest layers of soil and flood the entire landscape. If there is a long period of drought, the water table will drop.

In addition, the water table is generally not a flat surface but forms underground "hills of water," or high spots, and "valleys," or low spots. The hills of water usually correspond to the topographic hills and ridges, where the water flowing down from the tops of rainy hills must

(a)

FIGURE 18.3 (a) Major river drainages of California. (b) Hydrologic regions used in water management.

Source: Courtesy of California Department of Conservation.

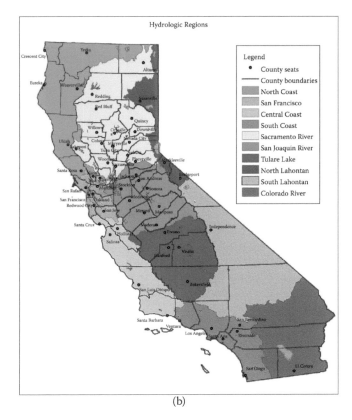

Hydrologic Regions

Legend
- County seats
--- County boundaries
North Coast
San Francisco
Central Coast
South Coast
Sacramento River
San Joaquin River
Tulare Lake
North Lahontan
South Lahontan
Colorado River

(b)

FIGURE 18.3 (Continued)

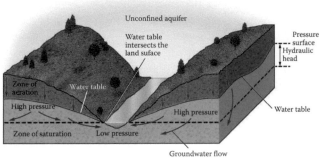

FIGURE 18.4 Schematic diagram of groundwater flow showing the water table, zones of aeration and saturation, and effects of pumping, creating a cone of depression.

Source: Courtesy of Oxford University Press.

sink down from a higher elevation to reach the water table. Consequently, in any zone of saturation, gravity drives a natural flow of water from the high spots in the water table down to the low spots (Figure 18.4). The low spots often correspond to the rivers and lakes on the landscape, which are replenished by groundwater flowing down to them even when there is no rain for a long time. Where groundwater bubbles up at the surface, it forms a spring.

Not all kinds of subsurface material transmit water equally well. A very porous and permeable material (like the sand and gravel that generally underlie most valleys, or a porous sandstone or rock shattered with lots of cracks) is an **aquifer** (Latin for "water-bearing"). Sediments or rocks that are nonporous and impermeable, so water cannot pass through them, are known as **aquicludes** (Latin for "closed to water"). Layers of mud or shale are the most common types of aquicludes, along with solid rock masses with no pores or cracks. A rock or sediment that lets water slowly pass through is called an **aquitard**.

If people dig or drill wells down to the water table and then pump it faster than it is replenished, the water table will drop as well (Figure 18.5). Lots of heavy pumping in the well will cause the place where the well bottom penetrates the water table to suck all the water around it and form a depleted area at the well called a **cone of depression**. This is a common problem where water is scarce. The first wells in an area may not be that deep and just barely reach the water table. If they pump water

TABLE 18.1

Data from Selected Rivers within the Boundaries of California, Compared with the Two Largest Rivers in the New World

River	Drainage Area, mi² (km²)	Average Discharge, cfs (m³/s)	Average Annual Runoff, Thousands of Acre-Feet
Sacramento	8,900 (23,050)	24,670 (699)	17,870
Klamath	12,000 (31,080)	17,785 (504)	12,900
Eel	3,113 (8,063)	7,410 (210)	5,379
San Joaquin	1,676 (4,341)	4,375 (124)	3,179
Pit	4,711 (12,200)	3,880 (110)	2,819
Russian	793 (2,054)	2,365 (67)	1,712
Kings	1,545 (4,001)	2,225 (63)	1,624
Los Angeles	827 (2,142)	215 (6)	156
Mojave	2,121 (5,493)	6 (0.17)	4
Mississippi	1,250,965 (3,240,000)	635,200 (18,000)	460,360
Amazon	2,374,500 (6,150,000)	7,100,000 (200,000)	5,115,000

Source: Data from *California Water Atlas*, US Geological Survey.

TABLE 18.2

Annual Precipitation and Runoff of the Major Hydrologic Regions Shown in Figure 18.3b

Hydrologic Region	Annual Precipitation, Acre-Feet (km³)	Annual Runoff, Acre-Feet (km³)
North Coast	55,900,000	28,900,000
Sacramento River	52,400,000	22,400,000
North Lahontan	6,000,000	1,900,000
San Francisco Bay	5,500,000	1,200,000
San Joaquin River	21,800,000	7,900,000
Central Coast	12,300,000	2,500,000
Tulare Lake	13,900,000	3,300,000
South Lahontan	9,300,000	1,300,000
South Coast	10,800,000	1,200,000
Colorado River	4,300,000	200,000

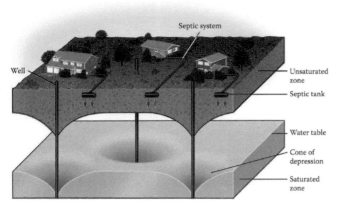

FIGURE 18.5 Effects of drilling deeper wells and then over-pumping, causing a depression of the water table, which makes the shallow wells go dry.

Source: Courtesy of Oxford University Press.

too fast compared with its rate of replenishment, or if newer, deeper wells are installed that pump even faster, the water table will drop and the shallower wells will go dry (Figure 18.5).

Water always flows downhill or toward the direction of least pressure. Thus, if you sink a very deep well and pump very hard, the water will flow away from the higher parts of the water table and the shallower wells and down to the deepest well, which depresses the water table the most. This can create problems. If, for example, the natural flow of water goes away from a source of contamination and then the water pumping changes the high spots on the water table, the flow will reverse and the contaminants will be sucked into the deeper wells (Figure 18.6).

In coastal regions, the natural groundwater flows down from its mountain headwaters through the coastal plain toward the ocean. But the sediments beneath the seafloor are saturated with salty seawater. If something depletes the

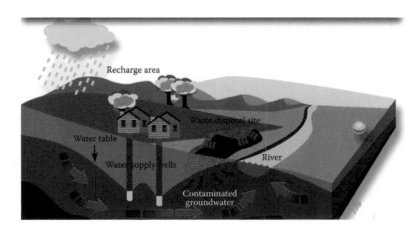

FIGURE 18.6 Problems with groundwater contamination. If a deep well sucks all the water toward it, it may bring contaminants from other areas into the water system.

Source: Courtesy of US Geological Survey.

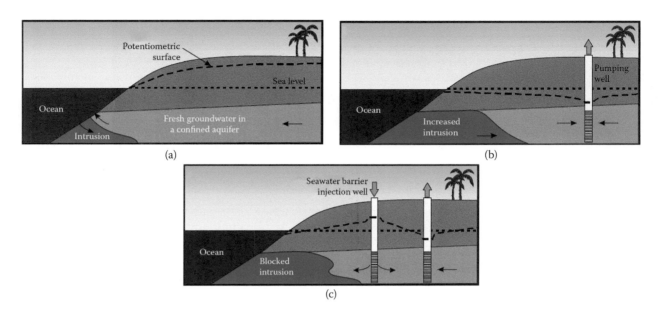

FIGURE 18.7 Problems with saltwater intrusion. An injection well near the coast is drilled, and fresh water is pumped back down into it to create a freshwater barrier to salt water intruding into the aquifer as its water pressure drops.

Source: Courtesy of US Geological Survey.

freshwater flow toward the ocean, then salt water will flow up into the coastal wells because there is not enough fresh water to keep it back (Figure 18.7). This problem happens in nearly all the coastal communities in California. To prevent this, several measures are possible. One solution would be to reduce the overpumping on the freshwater supply so that the fresh water can flow toward the sea and keep the salt water back. However, this is not feasible in areas of rapid growth, because water demand is increasing. The most widely used solution is to put a series of injection wells near the coast to act as barriers. Fresh water is pumped down into them, increasing the pressure toward the ocean and keeping the salt water back (Figure 18.7). The fresh water also flows back through the remaining aquifers, keeping them supplied with water as well. This uses more energy and is more expensive than letting the water flow naturally by gravity, but it is the simplest means of preventing saltwater intrusion when there is big demand on groundwater in coastal regions.

The Orange County Basin is a typical example of a major groundwater storage system, which supplies most of the needs of its residents despite the explosion in population to more than 3 million people over the past 30 years. Nevertheless, the groundwater basin can hold 10 million–40 million acre-feet of water, of which about 1.5 million acre-feet is usable. More than 200 production wells draw groundwater that provides for 70% of the water supply for the county. In cross section, the basin (Figure 18.8a) has recharge areas in the mountains to the east, which provide runoff that flows into the porous gravels and sands that cover the floodplain. The rivers themselves (especially the Santa Ana River, which comes from the San Bernardino Mountains and the eastern San Gabriel Mountains) are broad sandy washes with porous gravel bottoms, so most of the rains during storms sink down into the ground before reaching the ocean. In addition, the river channels are surrounded by broad **spreading ponds**, gravel-lined areas of the former floodplain of the river that are left undeveloped, so that during heavy rains the water will be trapped and infiltrate into the groundwater supply (Figure 18.8b). This is important because much of the land surface of Orange County (and most of the developed areas of California) is covered mostly by huge areas of impermeable concrete and asphalt and buildings. In heavy storms, most of the rain runs off into concrete flood control channels and is lost to the ocean, rather than sinking in and replenishing the groundwater supply.

The Orange County Basin is fault-bounded and sinking in the middle (Figure 18.8a), with the bottom more than 700 m (2,200 ft) below sea level, so there are many deep aquifer layers alternating with aquicludes. At the coast, the bedrock comes up near the surface along the Newport–Inglewood fault, confining the western edge of the basin into a narrow shallow layer of aquifers. Like most coastal groundwater operations, there are large injection wells near the ocean to prevent saltwater intrusion into the rest of the basin (Figure 18.8a).

A special type of groundwater configuration is an **artesian spring** or well (Figure 18.9a). The word *artesian* is often touted in commercials for beer or water, but it has no bearing on how good the water or beer tastes. All it means is that the water comes from a **confined aquifer**, or an aquifer sandwiched between two aquicludes. The aquifer is typically recharged up in the mountain foothills, so as it seeps down into the subsurface, it cannot flow up because it

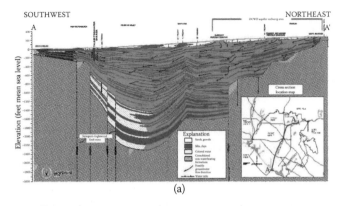

(b)

FIGURE 18.8 (a) Orange County water basin in cross section, showing the aquifers and aquicludes, the source water from the mountains to the east (right), and the injection wells near the coast to keep the salt water (purple) from invading. (b) Aerial view of some of the spreading ponds along the rivers in Orange County. These trap rainwater adjacent to the riverbeds and allow the excess water to seep into the ground.

Source: Courtesy of the Orange County Water District.

is capped by an aquiclude, and thus it is under pressure. All it takes is a well or fault to break through the confining layer and the water will rise naturally under pressure. Artesian springs were once common in the foothills of the Sierras (Figure 18.9b) and other California ranges, where the sedimentary beds dip down from the mountains into the basin, so that the aquicludes can form a confined aquifer.

The depletion of groundwater has another serious side effect. If there is too much pumping, or if there is a long drought, the soils and sands will sink because there is no water between the pore spaces to hold the grains apart. During the middle part of the twentieth century, the explosion in agriculture in the San Joaquin Valley caused too much overpumping of the groundwater. By the 1940s and 1950s, the ground was sinking dramatically (Figure 18.10a). In many places, the pavement buckled and houses began to crack apart as the ground sank and foundered. Trees whose roots could no longer reach the water table died off. In some places, the

ground sank so dramatically that it began to crack and buckle, forming huge fractured areas and craters (Figure 18.10b).

Since then, scientists and state agencies have been monitoring the groundwater withdrawal so farms in the Central Valley cannot overpump the groundwater to the degree they once did. However, natural causes are causing similar groundwater depletion and subsidence thanks to the recent droughts from 2005 to 2021. In many places in the Central Valley, the ground has sunk dramatically (Figure 18.10c). If the long-term drought since 2005 persists much longer, it will not only have drastic effects on the pavement and buildings in the regions where sinking is the greatest (such as around Tulare and Corcoran, or between Merced and Modesto) but also eventually cause a crisis in the water supply for both the farmers and residents of these areas, which have recently exploded in population due to their cheap cost of living and housing.

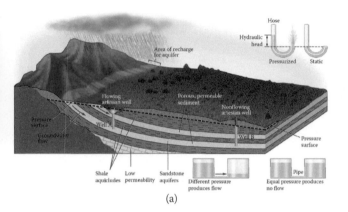

(a)

(b)

FIGURE 18.9 Artesian wells: (a) Diagram of typical configuration of an artesian water system showing the confined aquifer capped by an aquiclude so that water recharged from high elevation is under pressure as it sinks down into the aquifer. (b) Natural artesian spring in Kern County, about 1890.

Source: (a) Courtesy of Oxford University Press. (b) Courtesy of US Geological Survey.

(c)

FIGURE 18.9   (Continued)

18.5   WATERWORKS

The supply imbalance between north and south long held back growth in the drier areas until huge dams and waterworks could trap the rain from the north and from the Sierras and divert it down to the arid central and southern part of the state. Most vulnerable are the huge urban populations of Southern California. The places that are now inhabited by tens of millions of people originally could only support small populations of indigenous peoples and Spanish and Mexican colonials on the natural water supply of the region. None of this population growth could have come without bringing water in from long distances.

The Central Valley is the richest agricultural region in the history of the world, yet it is naturally drier than North Africa. Like much of Southern California, it goes through months of drought in the spring, summer, and fall and then may be flooded by a few storms in the winter. When there was no imported water, the natural lakes, like Tulare Lake and Lake Buena Vista, would dry up each summer, and the natural landscape across the plains would be covered in sagebrush. To compensate, California has built the world's largest and most expensive waterworks (Figure 18.11). It is the most completely transformed landscape of its size anywhere on Earth. Tulare Lake was completely transformed into fields and irrigation ditches, although the record rains of 2023 flooded it and destroyed most of the farms that had been built on its bed. Today the Central Valley churns out about 25% of America's food. It would not have been possible without huge irrigation systems and dams.

California's water systems serve more than 38 million people, which use more than 266 billion gallons per year, or 192 gallons per day per person, or 70,080 gallons per year per person. Farmers in the state also irrigate 2.3 million ha (5.68 million ac) of farmland. California has the world's largest and most complicated, yet also most productive, water system, bigger than that seen in any other country, let alone any other state. Yet the battles over water in the state have been long and contentious, especially when it comes to the struggle between two sides. One side is a handful of large farming corporations who use about 80% of the state's water to grow crops (but have disproportionate power in Sacramento and Washington, DC). On the other side are the huge populations of the cities, who use only 20% of the water but often resent having to make sacrifices in their water usage that amount to only a tiny percentage of the total California water budget. The story of how water is collected, distributed, and used is one of the most important aspects of California history in the twentieth century and beyond.

As soon as California became a state in 1850, water issues were central to its political and economic development. In the Great Flood of 1861–1862, the state experienced 45 days of record-breaking rainfall that flooded 300 square miles of the Central Valley and caused mass devastation and loss of life. Sacramento was flooded by more than a meter of water, and people had to get around in boats. The Los Angeles and Orange County Basins were also flooded in a meter of water, and 200,000 cattle drowned. The following winter was one of the driest ever recorded, and the drought ruined many of the farmers who were still recovering from the previous year's floods. These events showed California's vulnerability, not only to problems of getting water to dry areas, but also to extremes of seasonal fluctuations.

18.5.1   WATER TO LOS ANGELES

The first serious efforts at major water relocation were focused on the booming metropolis of Los Angeles, which had been outgrowing its natural water supply since its first

(a)

(b)

FIGURE 18.10 Effects of excessive groundwater depletion on the Central Valley: (a) Iconic image of how much the ground had sunk in the San Joaquin Valley between 1925 and 1955 as a result of overpumping. (b) Areas of subsidence led to ground sinking and breaking into cracks and crevices. (c) Satellite image of the current subsidence of the Central Valley due to ground subsidence from the drought of 2012–2015.

Source: (a–b) Courtesy of Wikimedia Commons. (c) Courtesy of National Aeronautics and Space Administration.

wave of expansion in the 1880s. The most powerful men in the city had long been looking for a source of water that could be brought to Los Angeles to keep up with the growth. One of these was Fred Eaton, the mayor of Los Angeles in 1898, who was casting around for a water source that could be brought to the thirsty growing city. Another was the self-trained engineer William Mulholland (Figure 18.12a), who did the hard work of figuring out whether building such an aqueduct was feasible.

By 1902, Eaton and Mulholland realized that the Sierra snowmelt flowing into the Owens Valley was the best candidate for a source. Eaton and his agents secretly bought up all the water rights to the Sierra drainages and bribed the regional engineer of the Bureau of Reclamation, Joseph

Lippincott, to give them critical information and assistance. Between 1902 and 1907, this deceptive purchase of the water rights to the Owens Valley continued. Meanwhile, Eaton's powerful friends among the city fathers, using their insiders' information, secretly bought up the worthless sagebrush land of the San Fernando Valley. These people included the publishers of the *Los Angeles Times*, Harrison Gray Otis, and his son-in-law Harry Chandler, who promoted the need for a water supply in their paper, as well as Henry Huntington, the railroad baron. When the aqueduct was announced in 1908, they instantly made millions because the land would soon be suitable for agriculture. This story was fictionalized in the Roman Polanski movie *Chinatown* (with Jack Nicholson as the detective

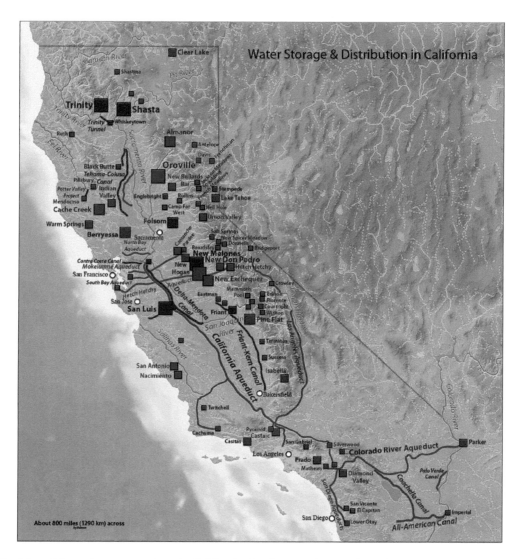

FIGURE 18.11 Map of the major water projects in California showing aqueducts and canals and the dams and reservoirs (indicated by squares, with relative size indicated by the size of the square). Blue indicates the California State Water Project; red, the Central Valley Project; purple, shared infrastructure; green, other federal structures; and gray, state and private structures.

Source: Courtesy of Wikimedia Commons.

who learns of the conspiracy from the millionaire played by John Huston).

Using $1.5 million in bond funds voted on by Los Angeles residents in 1905, construction of the Los Angeles Aqueduct (Figure 18.11) began in 1908 and was finished in less than five years in 1913, ahead of schedule and under budget. It was a miraculous feat of engineering for its time, comparable to the building of the Panama Canal. It carried water over a distance of 359 km (223 mi) down the flank of the Sierras, along the side of the Tehachapi Mountains, and then over the Transverse Ranges, all by gravity flow; no uphill pumping was required. It transports 0.49 km³ (400,000 ac-ft) of water every year. Yet it was built almost entirely by crews of men and horses with wooden wagons and dynamite, with only a few gasoline-powered tractors and primitive construction equipment (Figure 18.12b and c). It used more than 2,000 workers, who dug 164 tunnels, plus many miles of both open canals

and closed pipelines. On the downhill slopes, it also generated hydroelectric power, making it cost-efficient. When it was finally officially dedicated on November 5, 1913, several politicians got up and made long-winded speeches. When Mulholland's turn came, he simply said, "There it is. Take it." Then they released the water into the cascades near Sylmar (Figure 18.12d).

At first, the farmers and ranchers of the Owens Valley were unhappy with the way their water had been stolen, but they still seemed to have enough water to survive. By 1924, however, as the city took more and more water, the Owens Valley farmers rebelled. They began using sabotage and started dynamiting the aqueduct at night, leading to brutal repression by Mulholland's army of private security forces, along with state and federal troops. The climax of the "Owens Valley War" occurred when a large group took over the Alabama Gates, the largest controlling structure in the valley, and held it closed for several days.

FIGURE 18.12 Los Angeles Aqueduct: (a) William Mulholland surveying the work. (b) Work crews hauling huge sections of pipe with mule teams. (c) Huge pipes crossing Jawbone Canyon in the Tehachapis. The water flow is powered entirely by gravity without pumps. (d) Cascades at the terminus of the aqueduct above Sylmar.

Source: Courtesy of Wikimedia Commons.

The war would have continued except that in August 1927, the Inyo County Bank collapsed. Most Owens Valley farmers lost everything and had to sell off their farms. The Owens Valley, which had once been a fertile valley famous for its fruit orchards, went dry, and Owens Lake is now dry most of the year. In later years, thirsty Los Angeles added a second aqueduct parallel to the first. It even tried to take the water from Mono Lake, until federal courts stopped the city in 1994 because it was destroying the fragile ecosystem.

But once the Los Angeles Aqueduct water began flowing in 1913, the city of Los Angeles began using it more and more, especially as the city grew dramatically during the boom years of the 1920s. Soon, it needed to look for even more water. In 1923, Mulholland began plans for an aqueduct to draw water from the Colorado River and bring it across the Mojave Desert but did not live to see this project built. In 1931, the Metropolitan Water District began construction on the Colorado River Aqueduct (Figure 18.11). First, it built Parker Dam on the Colorado to create a reservoir downstream from Hoover Dam. By 1939, construction had been completed on the Colorado River Aqueduct. It carries 1.5 km^3 (1.2 million acre-feet) of water over 389 km (242 mi) across the desert, using 101 km (63 mi) of canals, 148 km (98 mi) of tunnels, and 135 km (84 mi) of buried conduits and siphons.

After the Colorado River Aqueduct was built, it was assumed that California could have all the water it needed. The original pact between the states that draw on the Colorado River (Colorado, Utah, Arizona, Nevada, and California) was written with an overly generous estimate of the annual flow rate based on a few really wet years, so each state's allotment is too large and their total is greater than the Colorado can actually deliver. As Reisner (1986) put it, "the Colorado River is unable to satisfy all the demands on it, so it is referred to as a 'deficit' river, as if the river were somehow at fault for its overuse." Since then, the authorities have reduced each state's allocation, so California can draw only 0.512 km^3 (415,000 ac-ft), rather than the 1.5 km^3 (1.2 million acre-feet) of water that it once drew.

Thanks to the postwar boom of the 1940s and 1950s, by the 1960s Los Angeles had outgrown its water supply again. This time it was forced to turn to Northern California.

18.5.2 CENTRAL VALLEY WATER

Private ranchers and farmers were building small dams and canals from early in California's history, but in the twentieth century, it took on greater urgency. During the Great Depression, Congress and President Franklin Roosevelt's New Deal were eager to authorize water projects for California. Not only was this good for the economy of the nation, but also the construction provided jobs for many men who were unemployed during the Depression. In addition, California agriculture provided jobs for the unemployed, especially the impoverished farmers from the "dust bowl" states in the High Plains who had lost their farms due to the drought and bad farming practices. These people were often called "Okies," although only some of them were from Oklahoma; there were many from Kansas, Nebraska, Texas, Colorado, and other states hit by the extreme drought. They were made famous by John Steinbeck's classic novel *The Grapes of Wrath*, which described their miserably poor lives as transients living in filthy camps, desperate for work and slaving in the fields for almost nothing just to earn a meager living.

In the first years of the New Deal, the Central Valley Project (Figure 18.11) had poured millions of dollars in construction funds into a huge project of dams and canals. These include the huge Shasta Dam, the Folsom Dam, and big dams on other major Sierra rivers, along with more than a dozen other reservoirs. It is still one of the largest water systems in the world, with 7 million acre-feet (8.6 km^3) of water stored, providing water for 3 million acres (12,000 km^2).

Within a decade or so, however, even the huge Central Valley Project was still not big enough. In the early 1960s, the energetic governor of California, Edmund G. "Pat" Brown (father of Governor Edmund "Jerry" Brown, who served four terms in his own right), managed to get funding for a large amount of infrastructure, including many freeways and public works. The biggest, however, was the immense California State Water Project (Figure 18.11), the largest state-built water project in the United States. It taps the Feather River in the northern Sierras, carrying about 2.8 km^3 (2.3 million acre-feet) of water to about 29 different water agencies in the state. Begun in 1960, it has 29 dams, 18 pumping plants, 5 hydroelectric plants, and 970 km (600 mi) of canals and pipelines. It pumps water from the Central Valley over the Tehachapi Mountains, using more energy than it makes, so it is the state's largest energy consumer.

Its main feature is a series of canals and reservoirs in the Central Valley that supply the delta region and link with the older Central Valley Project to irrigate areas that were previously without water. It also includes the enormous California Aqueduct (Figure 18.11), the largest aqueduct in North America. This aqueduct is more than 1,129 km (702 mi) long and carries 370 m^3/s (13,000 cu ft/sec) of water. Its main job is to carry water from the wet north to the dry regions in the south, especially Los Angeles (West Branch of the aqueduct); the Inland Empire of San Bernardino and Riverside (East Branch); Santa Barbara (Coastal Branch); and much of the south San Joaquin Basin and Mojave Desert. It can be seen in aerial view (Figure 18.13) snaking across the Central Valley close to the route of Interstate 5,

FIGURE 18.13 Aerial view of the California Aqueduct crossing the Central Valley near Tracy, California.

Source: Courtesy of Wikimedia Commons.

which crosses it several times. Like all the aqueducts that must cross the San Andreas fault to reach Los Angeles, it crosses the fault line around Palmdale. However, each segment of the canal has seismic gates that will close in an earthquake, so only one or two sections will be ruptured and leak, and the rest of the water will be held back until repairs can be made.

These waterworks have made life and agriculture possible in California. However, continued growth and agricultural demand always threaten the state's economy with a water shortage. This is especially true when the supply of water is below normal, such as the major drought of 2012–2021.

18.6 DROUGHTS

Along with hotter temperatures and longer and more frequent heat waves, rainfall patterns have also changed. The changing climate is likely to increase the need for water but reduce the supply. Rising temperatures increase the rate at which water evaporates into the air from soils and surface waters. Rising temperatures also increase the rate at which plants transpire water into the air to keep cool, so irrigated farmland would need more water. But less water is likely to be available because precipitation is unlikely to increase as much as evaporation. Soils are likely to be drier, and periods without rain are likely to become longer, making droughts more severe. Increasing temperatures and declining rainfall in nearby states have reduced the flow of water

in the Colorado River, a key source of irrigation water in Southern California.

By looking at climate indicators like tree rings, it is possible to study the pattern of dry and wet years over many centuries. Southern California and the Southwest, in particular, have a persistent pattern of long-term drought, broken by one or two rainy years, then another pattern of drought (Figure 18.14). This pattern is roughly cyclical but irregular in its spacing, so there's no consistent duration of periods of drought or rainy years. But the study of centuries of tree rings has shown that during certain periods, there were megadroughts that lasted many years, with some that lasted decades, with few or no rainy years to break the spell. For example, a drought that began in 1276 CE and lasted decades is thought to have destroyed the civilizations of the Anasazi people (also known as the Ancestral Puebloans) of the Southwest. They were forced to abandon their elaborate cliff villages and flee to the lowlands for water and food. Their descendants are thought to be the modern Hopi and Zuni people of New Mexico and Arizona.

Similarly, the southwestern part of the United States (including Southern California) has been in a pattern of drought since the year 2000, in what is now called the "southwestern North America megadrought." It is the driest period in this region since 800 CE, following the period from 1980 to 1998, the wettest period in the region in at least 1,200 years. The causes are complex, including many drier La Niña years and fewer wet El Niño years than normal. But there is no doubt in the scientific community that

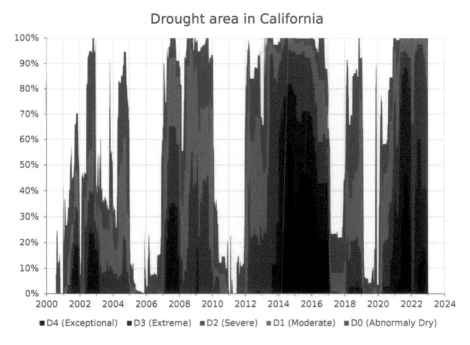

FIGURE 18.14 The pattern of drought since the year 2000. Most of the interval from 2001 to 2023 has been characterized by five to seven drought years in a row, punctuated by a rainy year or two, so the cumulative effect is a long-term water imbalance and dropping of reservoir levels and groundwater levels that a single wet year cannot compensate for.

Source: Courtesy of Wikimedia Commons)

FIGURE 18.15 The severity of California's recent drought is illustrated in these images of Folsom Lake, a reservoir in Northern California located 25 mi (40 km) northeast of Sacramento. The lake is formed by Folsom Dam, in the foreground, which is part of the US Bureau of Reclamation's Central Valley Project. In the July 20, 2011, view, the lake was at 97% of total capacity and 130% of its historical average for that date. In the January 16, 2014, shot, the lake was at 17% of capacity and 35% of its historical average. It reached its lowest levels in 2022, although the heavy rains of spring 2023 temporarily refilled it. However, another stretch of dry years could easily reduce its water levels again.

Source: Courtesy California Department of Water Resources.

the underlying driver of this hot dry weather is climate change, and the warming of the planet. Only a handful of rainy years, like 2010, 2019, and the record winter rains of 2023 broke this pattern, and the rains of 2023–2024 may or may not end the drought completely. Before the rains of 2023 came, many of the reservoirs in the western United States were so low that they had enormous "bathtub rings" of deposits along their banks where minerals precipitated on the banks of the lake or reservoir that were normally underwater. Some were reduced to tiny pools in the center of the reservoir with huge areas of dry exposed banks (Figure 18.15). Some had almost reached "dead pool" stage, where the water was too low to go through the intake channels near the base of their dams, so the lake would become stagnant and no longer produce any hydroelectric power. The record rains of early 2023 refilled almost all the reservoirs in California but did not fully end the drought situation. The 23 years of drought since 2000, plus the heavy pumping of groundwater for agriculture in the San Joaquin Basin, meant that the groundwater levels of the region are still dangerously low. One or two wet years is not enough to replenish it. Climate models looking at the centuries of tree ring data, and the meteorological data for the last century, predict that the Mediterranean climates of California are in for a continued drying trend as global climate gets hotter (Polade et al., 2017).

With the continual drought in the Southwest, another important factor is also affected: snowpack in the mountains. Dry years mean a lot less snowfall each winter, so the snowpack gets smaller and smaller. Since the 1950s, the snowpack in California (primarily the Sierra Nevada Mountains) has been getting less each winter. This is not just an inconvenience to skiers and snowboarders but has other important implications as well. If there is less snow, the high mountains get drier and drier, and the vegetation belts may shift. The tree line may even shift upward as the

alpine ecosystems get smaller and smaller. Animals which require specific vegetation and habitat are also affected, most of them forced to reduce their range or even migrate away if the cold winters and snowy conditions that produce their food supply vanish.

For humans, a smaller snowpack is crucial to their water supplies for the rest of the year. Most of the reservoirs in California are there to trap snowmelt in the spring and summer and store the winter's snows as water for human use. As their volume shrank due to the overall drought, a big part of their water loss came from the poor snowfalls of the previous winters. By the end of the 2021–2022 winter snow season, the snowpack was alarmingly small and the reservoirs were distressingly low (Figure 18.15). The heavy snows of the winter of 2022–2023 seem to have refilled most of California's reservoirs to capacity again, but another dry winter or two and that water surplus will be gone and we are back in the persistent drought pattern since 2000. As mentioned already, the single rainy spring of 2023 is not enough to compensate for 20 years of continuous drought. In particular, the groundwater levels in many places, such as the San Joaquin Basin, are perilously low and will take many wet, snowy years in a row to be replenished.

18.7 SUMMARY

Water is essential to almost everything in California, yet the northern counties and Sierras get most of the water, while the southern half of the state is semidesert or desert in average rainfall amounts. Water drains to the ocean in a series of drainage basins separated by divides, and the largest of these are in the northern parts of the state. Water also moves as groundwater by infiltration through the soil. Many areas have large groundwater reservoirs, although they have to be managed to keep down saltwater intrusion, contamination by waste sites, and overpumping by individuals. In the arid southern half of the state, groundwater is the major source of natural water, since there is so little precipitation.

To compensate for the imbalance of supply versus demand for water in California, the state has some of the largest waterworks in the world. These include three large aqueducts that supply the dry region of Southern California: the Los Angeles Aqueduct from the Owens Valley (built 1908–1913); the Colorado River Aqueduct, from Parker Dam on the Colorado (built 1931–1939); and the California Aqueduct, built in the 1960s. The agriculture and cities of the Central Valley are served by the Central Valley Project (built in the 1930s), and then in the 1960s by the California Water Project from the Feather River in the northern Sierras (which also supplies the California Aqueduct). However, the persistence of long periods of drought threatens whether these water supplies are enough in the face of continued development and population growth.

RESOURCES

Carle, D. 2009. *Introduction to Water in California*. University of California Press, Berkeley.

Davis, M.L. 1993. *Rivers in the Desert: William Mulholland and the Inventing of Los Angeles*. HarperCollins, New York.

Hundley, N. 2001. *The Great Thirst: Californians and Water—A History*. University of California Press, Berkeley.

Ingram, B.L., and Malamud-Roam, F. 2013. *The West without Water: What Past Floods, Droughts, and Other Climatic Clues Tell Us about Tomorrow*. University of California Press, Berkeley.

Kahri, W.L. 1983. *Water and Power: The Conflict over the Los Angeles Water Supply in the Owens Valley*. University of California Press, Berkeley.

Kelley, R. 1989. *Battling the Inland Sea: Floods, Public Policy, and the Sacramento Valley*. University of California Press, Berkeley.

Mulholland, C. 2000. *William Mulholland and the Rise of Los Angeles*. University of California Press, Berkeley.

Pisani, D. 1984. *From the Family Farm to Agribusiness: The Irrigation Crusade in California and the West, 1850–1931*. University of California Press, Berkeley.

Polade, S.J., Gershunov, A., Cayan, D.R., Dettinger, M.D., and Pierce, D.W. 2017, Precipitation in a warming world: Assessing project hydro-climatic changes in California and other Mediterranean climate regions. *Scientific Reports*, 7(10783). https://doi.org/10.1038/s41598-017-11285-y

Reisner, M. 1986. *Cadillac Desert: The American West and Its Disappearing Water*. Penguin, New York.

Standiford, L. 2015. *Water to the Angels: William Mulholland, His Monumental Aqueduct, and the Rise of Los Angeles*. Ecco, New York.

VIDEOS

Marc Reisner's classic book *Cadillac Desert* (1983) has been made into an excellent four-part PBS documentary series. It can be seen on YouTube.

Episode 1, on Mulholland and the Los Angeles water supply, "Mulholland's Dream": www.youtube.com/watch?v=hkbebOhnCjA

Episode 2, about the damming of the Colorado River and its effect on California, "An American Nile": www.youtube.com/watch?v=Mis-CU9oZO0

Episode 3, about the Central Valley water projects and agriculture, "The Mercy of Nature": www.youtube.com/watch?v=_yeXlMN6Fo0

Episode 4, "The Last Oasis": www.youtube.com/watch?v=OiSdjs4BHUU

19 California's Coasts

You'd catch 'em surfin' at Del Mar, Ventura County line,
Santa Cruz and Trestle, Australia's Narrabeen
All over Manhattan, And down Doheny Way,
Everybody's gone surfin', Surfin' U.S.A.

—The Beach Boys
Surfin' U.S.A.

19.1 SURF'S UP!

California is legendary for its famous surfing, sunny beaches, and scenic rocky coasts. The surfing craze of the early 1960s was imported from Hawaii to California, where it became a big part of American culture for a while. Surfing is still a major part of the California lifestyle. Not just surfing, but also sailing, beachcombing, beach volleyball, and other beach and water sports make California distinctive.

In addition to its sandy beaches and great surf, California is also famous for its rocky coasts, particularly in places like Big Sur and around the Monterey Peninsula and along the northern coast. The rugged rocks and crashing waves are iconic images of California.

Surfing, sailing, water sports, and especially tourism along the coast are a major part of the California economy, so in many ways, California's coasts are an important resource. This is especially apparent whenever a major oil or sewage spill closes part of the coast, and there are immediate economic losses. In addition, much of California's mystique and properties values are tied to the coast. The richest communities nearly all lie along the ocean, not only for the beautiful scenery, but also for the mild coastal weather. Finally, coastal erosion and destruction of coastal developments are an important environmental problem that can cost millions of dollars. California's coasts are an important economic resource in the same way its gold, oil, and water power its economy.

19.2 LIFE'S A BEACH!

Every year, millions of people visit California's coasts, especially its sandy beaches. Yet very few of them pay attention to how the beach system works or understand what is going on beneath the crashing waves and shifting sand. Most of what people think about beaches and waves is wrong. For example, except where they are breaking, the water in the waves is not actually moving forward but cycling back in a circular motion to a starting point. The sand on the beach in the surf zone is not permanent but flowing down the coast. If you stand in the surf and then return to the same spot a few hours later, the sand is completely different from what was there hours ago. Beaches are not permanent static piles of sand but dynamic features with the water and sand in constant motion.

Often, this misunderstanding can have serious consequences. For example, dozens of people every year drown from rip currents, but if they had known to swim parallel to the beach, they would have lived. On a larger scale, developers and coastal towns often build structures on the beach that end up destroying their supply of sand or ruining the harbor they intended to protect. In other cases, natural processes erode beaches and cause cliff collapse, costing millions of dollars. Some coastal commissions and other large-scale organizations that are charged with making expensive decisions about coastal development don't know the fundamentals of coastal geology and end up wasting millions of dollars of taxpayers' money.

The motion of water waves in the beach is not as simple as it appears to be. If you just focus on the surf zone, it appears that the water is constantly moving toward the land. But if you put a floating object out in the waves, you discover something surprising: the water particles in the waves are *not* moving forward very far but are traveling in circular orbits and return to their starting positions with each wave cycle (Figure 19.1). Only the *energy pulse* of the wave moves forward. The water particle at the top of its orbit is on the crest of the wave, and then it drops to the bottom of the orbit when the trough of the energy pulse moves through.

A good analogy is the wavy motion of a field of tall grain. It appears that the grain is flowing forward, but we know that this is an illusion because each stalk is actually rooted to the ground. Or think about the "wave" illusion when the spectators in a stadium or arena stand up and sit down in a coordinated fashion. We know that each "particle" (person) did not move sideways but only stood up and sat down—but it appears as if a wave is traveling around the seats.

If you place floating objects or tracers in the wave, you will notice something else: the orbits of the crests and troughs are largest at the surface, but they get smaller and smaller with depth (Figure 19.1). At a certain depth (called **wave base**), the circular orbital motion dies out altogether, and the wave energy does not reach any deeper. The wave base is about half the **wavelength** (the distance between two crests, or two troughs), so if you are looking down on the waves and have a way to measure their wavelength, you can estimate the depth of the wave base.

Below the wave base, the water is relatively calm and rarely experiences any strong currents or water motion. This is the quiet water, where most of the sea creatures live, undisturbed by normal surface activity. It is also where submarines can lurk, unaffected by surface waves. But when

DOI: 10.1201/9781003301837-19

storm waves (with their longer wavelengths) cause the wave base to go deeper and disturb the deeper waters, then these quiet areas are also disturbed. This is why so many seashells wash onshore after storms.

The orbital motions of the waves are driven by the winds and retain their energy across many miles of ocean without falling apart. But as the waves move onto the shore, they "feel" the bottom as the wave base intersects the seafloor. At this point, the orderly circular orbits of the water particles are disrupted, and the crests get higher and bunch up as the shallow bottom pushes them up (Figure 19.1). They also crowd together as they slow down due to friction with the bottom, but faster-moving waves behind them come up quickly. Eventually, the bottom gets so shallow that the orbital motion falls completely apart and you end up with the "curl" of a breaking wave that surfers require. The wave crashes on the shoreline into foam and surges into the **swash zone** and then retreats back to the sea (known as **backwash**), repeating this cycle hundreds of times an hour.

This is the basic operation of waves, and yet most people don't understand this. Sadly, this ignorance extends to novelists and Hollywood screenwriters. For example, *The Poseidon Adventure* is a 1969 novel about an ocean liner that is capsized by a huge breaking wave in the ocean. The survivors must find a way to climb up through the upside-down ship to get rescued. It has been made into a movie three times—and yet the basic premise of the plot is completely impossible. Waves break into surf only where the bottom is so shallow that they cannot continue their orbital motion—and that water would be too shallow for any boat like the *Poseidon*. Out in the open ocean, there may be wind blowing small whitecaps and foam off wave crests, but they do *not* break into surf in deep water. Open-ocean waves (even tsunamis, as in the original plot of the novel, or "rogue waves," in one of the movie versions) do not form a breaking wave, just an unusually high swell. This is typical of most Hollywood movies that concern science—the writers are only interested in a good story and building the story arc and characters, not scientific accuracy.

Take another cult Hollywood movie, the 1966 surfer classic *The Endless Summer*. It follows three surfer dudes traveling the world on the search for the "perfect wave," following the advice of surfers they encounter. The movie even launched a "surf-and-travel" culture. But the surfer dudes could have saved themselves the trouble if they had just looked at the map of the sea bottom. Good surfing only occurs on coastlines where the continental shelf is narrow and rises up steeply out of the ocean. Beaches on the Atlantic or Gulf Coasts of North America have broad, shallow continental shelves that slope gently offshore for many kilometers, so any wave energy coming from the ocean "feels bottom" and breaks long before it reaches their beaches. That's why big-wave surfing never became popular in New Jersey or Texas, since there are rarely big waves on those coasts (except during storms, when surfing is too

dangerous). Surfing originated in the Hawaiian Islands, which are volcanoes that rise out of the ocean steeply. They have no real continental shelf and are pounded by waves with maximum energy. Likewise, California beaches have

(a)

FIGURE 19.1   (a) Orbital motion of water particles (and floating objects, like birds) in an ocean wave. The particles travel in small circles and only move forward and backward a small amount, returning to their starting point. Only the energy moves forward through the water. (b) Orbital motion of water particles in an oceanic wave dies down with depth, until there is no motion at the wave base. As long as the bottom is deeper than the wave base, the orbital motions are stable. (c) As the bottom slopes up and becomes shallower than the wave base, the orbital motions are disrupted, and the water at the bottom tends to be held back by friction compared to the water at the top of the wave. In addition, the waves become steeper and higher as they bunch up, like cars in a traffic jam. Eventually, the water is so shallow that the orbital motions cannot be completed and they break into the curl of surf.

Source: (a, c) Courtesy Oxford University Press.

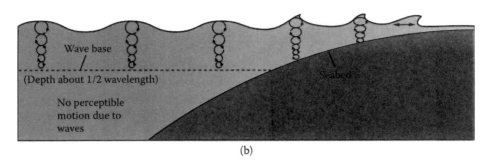

(b)

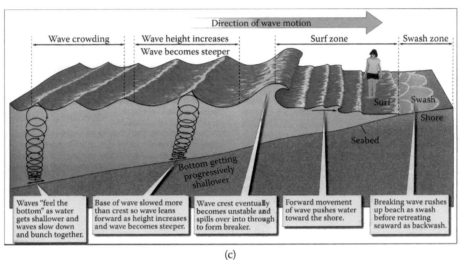

(c)

FIGURE 19.1   (Continued)

a relatively narrow offshore continental shelf, so most of the wave energy hits the shoreline without being dissipated. That's why California adopted the surf culture from Hawaii and spread it across the country in the 1960s. A simple map of the seafloor could have saved the three surfers in the movie a lot of fruitless searching. (Nowadays, it's even easier to look online for websites that list the world's great surfing beaches. There are even apps for that!)

When the water piles up on the shoreline, especially during heavy surf, sometimes it does not flow back to the ocean evenly. Instead, it flows back in a narrow swift current called a **rip current** (improperly called a "riptide," but it has nothing to do with tides). Rip currents can be spotted from above (Figure 19.2a), because they form narrow lanes of smooth water perpendicular to the shoreline, where their offshore current disrupts the incoming wave crests. They are very dangerous, because they move faster than a person can swim, so if you try to swim shoreward against them, they will wash you farther out to sea until you get tired and drown. Many people drown each year trying to swim against them. Here's a bit of advice that may save your life someday: if you are in a rip current, swim parallel to the shoreline (Figure 19.2b). The rip is very narrow, so if you swim just a few meters up or down the beach, you'll get back into water that is heading shoreward and it will carry you to safety.

Also, the rip current ends a few hundred meters from the shore, so even if you do nothing at all, you will be

eventually freed from the rip current. Every year, a number of dog owners die when they panic and enter the water to rescue dogs being pulled out to sea by a rip current. Often, the dogs return to the shore safely under their own power, but the owners are not so lucky. The lesson here is to wait and walk to one side of the rip current, encouraging the dog to swim back via a safer route.

In cross section, most beaches have a standard profile (Figure 19.3). The **foreshore** region slopes from the spot where the wave base intersects the bottom up to the top of the surf zone or shoreline. At the top of the foreshore is a ridge or crest of sand, known as a **berm**. This is where sunbathers like to place their towels, umbrellas, and folding chairs. Behind the berm, the sand slopes gently landward in the **backshore** area. The backshore is rarely immersed in seawater except during very big storms, especially during high tides. At those times, the storm waves wash out the berm temporarily and deposit driftwood and trash in the backshore region, which may remain there until the next storm. Behind the backshore may be an undercut sea cliff (typical in California) or a sand dune field (typical in the Gulf and Atlantic Coasts).

Most people do not visit the beach frequently enough to notice, but if you come to the same beach during the summer and after winter storms, you will see a remarkable difference. Driven by the gentle waves that push sand up onto the shore, summer beaches are wide and very sandy

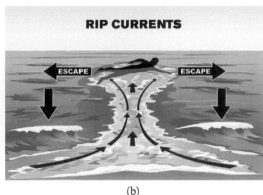

(a)   (b)

FIGURE 19.2 (a) A rip current can be spotted by the smooth lane of water that interrupts the breakers. (b) The strategy for surviving in a rip current is to swim parallel to the beach until you are out of it, and then you can swim to shore.

Source: Courtesy of the National Oceanic and Atmospheric Administration.

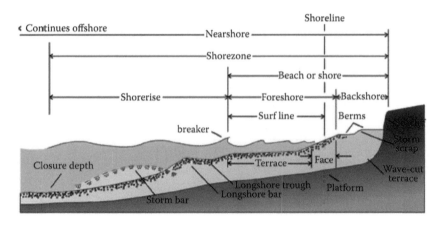

FIGURE 19.3 Idealized beach profile.

Source: Redrawn by E. T. Prothero.

and slope gently up to the berm (Figure 19.4a and b). By contrast, the huge storms of winter have high-energy waves that erode the sand and deposit it in an offshore sandbar below the wave base. The winter beach (Figure 19.4a and c) is typically much narrower and coarser-grained (often covered by more gravel than sand), with more exposures of bedrock below the beach sand.

So far, we have looked at the beach in cross-sectional profile. If you look down on the surf from a cliff top or get an aerial view from Google Earth or an air photo, you will notice something called **wave refraction** (Figure 19.5). Ocean waves are generated by storms out at sea, so unless the storm is directly offshore at right angles to the beach, it will be somewhere up- or downcoast. In California, the majority of Pacific storms occur in the North Pacific, so on most days the waves come from the north. As the waves approach land, one side of the wave feels bottom first and begins to slow down, while the other end is still in deeper water and keeps moving fast. This makes the angle of the wave to the shoreline bend, or **refract**, so that the wave

crests begin to pivot until they are almost, but not quite, parallel to the shoreline as they break (Figure 19.5b). That is why waves always break into surf from one end to the other, something that all surfers learn in order to get a good long ride.

The fact that waves are coming in at an angle means that the water approaches the surf zone at an angle as well. During each onrushing surf wave, the water runs up the beach at an angle to the shoreline. It then flows straight back during the backwash phase, tracing a zigzag path of the water (Figure 19.6). As this happens hundreds of times each hour, with every wave the water moves slightly down the shore in the direction in which the waves are driving it. Not only does the water steadily move downshore with a zigzag motion but so also do sand, driftwood, beach toys, and any other object caught in the waves.

This downshore flow of the water, sand, and other material is called the **longshore current** or **beach drift** (Figure 19.6). Thus, if you stand in the surf zone in one spot and then come back to the same spot a few hours later, all the sand you

FIGURE 19.4 (a) Contrast between winter and summer beaches. During summer, the beach is broad and sandy with a gentle slope. During winter, the sand is dragged seaward by the storm waves into offshore sandbars below the wave base, leaving a narrow, more gravelly beach. (b) Summer at Cove Beach, Año Nuevo State Beach, San Mateo County. (c) The same beach in winter, with very little sand and a narrow strip of gravel remaining.

Source: Courtesy of Wikimedia Commons.

once stood on has moved downshore. It is comparable to the famous paradox posed by the Greek philosopher Heraclitus: "No man ever steps in the same river twice." Likewise, if you leave a floating beach toy in the surf and ignore it for a few minutes, it will float parallel to the shoreline soon after you leave it. Contrary to our impressions, the beach is not a static pile of sand but a dynamic "river of sand" that is moving in the longshore current direction.

So where does this river of sand start and end? In most cases, the sand supply comes from the rivers and streams that wash sediment to the ocean. Some of the sand supply also comes from the erosion of sea cliffs. Once the sand reaches the surf zone, it is caught up in the longshore current and moves downshore, sometimes for many tens of kilometers (Figure 19.7). In California, however, this

beach drift doesn't continue forever. Eventually, the beach encounters a submarine canyon incised into the continental shelf. These canyons used to be ancient river valleys carved into the shelf when the sea level was much lower during the last ice age. The beach sand then flows down into the submarine canyon and eventually ends in the deep sea as a submarine fan.

This continuous one-way river of sand from the mountains to the deep ocean is called a **beach compartment**. Southern California has many excellent examples (Figure 19.7). The longshore current direction on this coast is nearly always to the south, since the major storms that create the waves come from the North Pacific. The sand that flows down from the mountains above Santa Barbara moves east along the shoreline past Ventura and

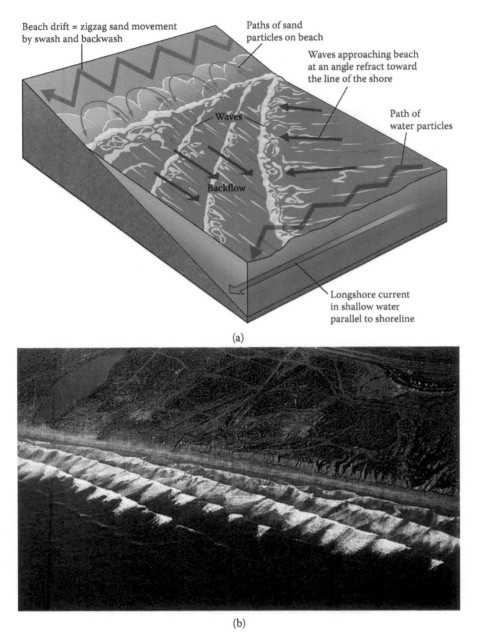

FIGURE 19.5 Wave refraction. (a) Diagram showing how the wave fronts approach the shoreline at an angle but then bend or refract in their pathway until they strike the shoreline nearly (but not quite) parallel. (b) Aerial photo of strong wave refraction.

Source: (a) Courtesy of Oxford University Press. (b) Courtesy of US Geological Survey.

Oxnard, until it comes to a complex of submarine canyons off Port Hueneme and Point Mugu, where it vanishes into the deep ocean. A new supply of sand eroded from the Santa Monica Mountains flows east through Santa Monica Bay down to Redondo Beach, where it disappears at Malaga Creek, the head of the Redondo submarine canyon. Very little of this sand flows around the rocky Palos Verdes Peninsula. Further down the coast, the creeks on the Palos Verdes supply new sand that moves southeast and adds the sand supply from the Los Angeles River in San Pedro. This sand continues all the way past Long Beach, Huntington Beach, and down into Orange County, where eventually it heads down into the Newport submarine canyon. Sand from creeks at Laguna Beach, Dana Point, Capistrano, San Clemente, Oceanside, and the coast in San Diego County flows along the beach until it reaches the cliffs of La Jolla, where the beach vanishes down the La Jolla submarine canyon. Similar maps of beach compartments could be drawn for the entire California coast, but this example is typical.

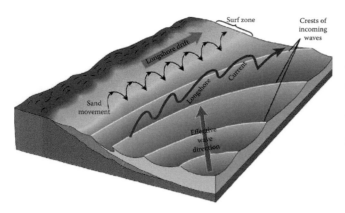

FIGURE 19.6 Dynamics of longshore drift.

Source: Courtesy of Wikimedia Commons.

19.3 BEACH EROSION AND HARD STABILIZATION

The river of sand is very sensitive to anything that might block its flow or supply of new sand. Most people don't realize the sand is flowing and cannot be obstructed without consequences, so they make costly mistakes. There is a tendency for humans to build hard structures of concrete and rocks (**hard stabilization**) with the intent of protecting "their" stretch of beach, as if it were just a permanent sand pile. But they don't realize the sand is in motion, so this effort at "stabilizing" the beach usually ruins it. As the old TV commercial once said, "It's not nice to fool mother nature." Once these poorly thought-out structures are built, they disrupt the entire system and do more damage than good. In many cases, there are expensive lawsuits brought by neighbors negatively affected by someone else's construction that ruined their stretch of beach.

Dams built up in the headwaters of the river cut the downstream sand supply and often end up starving the beaches of sand along the coastline. This can be mitigated by treating the sand in the river as a resource that costs money, just like the water or any other natural resource. If a dam shuts off the downstream flow of sand, then we are forced to pay to extract sand from the dry upper part of the reservoir and put it back in the river or truck it directly to the beach.

On many beaches, the local landowners construct straight rocky walls out into the surf perpendicular to the shore called **groins**. Their intent is to protect the beach next to their property, but they become a barrier to the river of sand (Figure 19.8). After a while, the sand flowing down the beach will pile up on the upstream side of the groin. At the same time, it will be depleted from the downstream side,

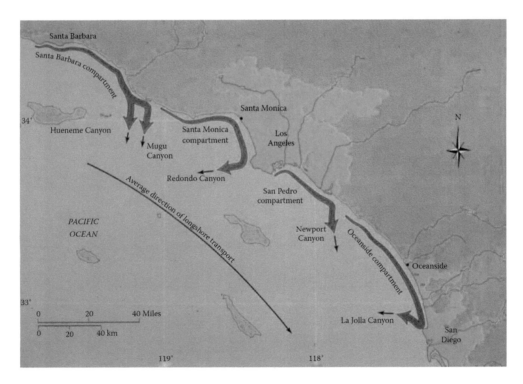

FIGURE 19.7 The beach is a river of sand flowing down from the mountains and rivers, along the beach front, and eventually down into submarine canyons. Each separate pathway from river to deep ocean is called a compartment. For example, sand from western Santa Barbara County travels southeast and flows down Hueneme and Mugu submarine canyons. Sand from the Santa Monica Mountains and Ballona Creek in Marina del Rey flows down to Redondo submarine canyon and does not continue around the Palos Verdes points. Sand from the Palos Verdes and the Los Angeles River, San Pedro, and Long Beach continues down to Newport submarine canyon. Sand from southern Orange County and San Diego County continues down to La Jolla submarine canyon.

Source: Courtesy of the California Division of Boating and Waterways; available at www.dbw.ca.gov/csmw/PDF/GriggslSed BudgetTechProposal061704.pdf.

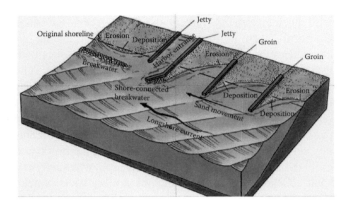

FIGURE 19.8 Sand buildups and depletion due to beach obstructions.

Source: Redrawn from several sources.

usually resulting in lawsuits from the owner of the depleted beach. Several groins in a row create a mutilated beach that looks like a bunch of saw teeth (Figure 19.8).

Another obstruction is a pair of parallel rock walls at the mouth of a harbor called a **jetty** (Figure 19.8). The intent of the jetty is to provide shelter at the harbor mouth so that boats can sail through without experiencing the breaking surf. But just like groins, the jetty blocks the sand flow past the mouth of the harbor, so the upstream side of the jetty is overwhelmed with sand, while the downstream side is depleted (Figure 19.8). Eventually, the sand-clogged upstream side will cause the surf to break right at the tip of the jetty, completely defeating the purpose of the jetty as a shelter for boats moving in and out of the harbor.

Two examples of California harbors illustrate the difficulties. When Santa Cruz Harbor was developed (Figure 19.9a), a jetty was built out into the ocean. The longshore drift coming from the north has since completely filled in the northwest side of the jetty, while the sand on the southeast side is now depleted. In Santa Barbara Harbor, a long curved breakwater was built to enclose the boat basin. Sand coming down from the north was first trapped against the north side of the breakwater, completely engulfing it. After the obstruction could block no more sand, the beach drift wrapped around the southern tip of the breakwater and built a sand spit into the mouth of the harbor (Figure 19.9b). If left alone, this sand would eventually close off the mouth of the harbor and render it useless. Santa Barbara must spend a lot of its budget every year for a big dredge that pulls the sand from the spit and pumps it past the harbor and down the coast, where the longshore drift can pick it up again. If you mess with the natural system, you must pay to accomplish what nature used to do for free.

Another kind of obstacle is a breakwater wall set up parallel to the shoreline (Figure 19.8). In principle, this wall provides shelter for a boat anchorage behind it. But it also causes the waves to refract around both ends, and they wrap around the tips and force the sand to pile up in the center of the bay behind the breakwater. If the sand continues to build

up, it will form a salient or bulge that will grow outward until it turns the breakwater into a peninsula (Figure 19.8).

This happened to Santa Monica harbor in 1933 when a large breakwater wall was built at the cost of almost 11 million of today's dollars. By 1940 (Figure 19.9c), a sandy peninsula had grown so far out from the original beach that the legendary Santa Monica Pier was about to turn into a boardwalk over sand, with no fishing off its end. Eventually, the expensive project had to be torn down (costing even more money), because it was damaging more of the coastline than it was protecting. Its main function, the boat anchorage, was going to be filled with sand anyway. Once again, we see how people make costly mistakes that would not happen if they knew the basic mechanics about how waves and beaches work. Unfortunately, they don't—or they ignore these principles.

19.4 HEADLANDS, SEA CLIFFS, AND COASTAL UPLIFT

The shorelines of the world look very different from one another because they are dominated by different geologic processes. In the Mississippi Delta, the coastline is growing out into the Gulf of Mexico because it is dominated by the huge sediment supply of the Mississippi River and the erosion by longshore currents is much weaker. On the many beaches of the Atlantic states (especially Long Island, New Jersey, and the Outer Banks of the Carolinas) or Texas, the longshore currents are very strong, and they

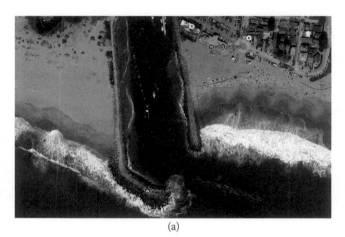

(a)

FIGURE 19.9 Satellite images showing how harbor modifications have altered California beaches: (a) Santa Cruz, where a pair of jetties has caused sand buildup on the upstream side (left) and depletion on the downcoast side (right). (b) Santa Barbara Harbor, where sand flowed from the northwest (left) and built up to the west of the breakwater. It now flows southeast along the outside of the breakwater and is building a sand spit that would close off the harbor if not for a permanent dredge removing the sand and pumping it downcoast. (c) Aerial view of Santa Monica coastline in 1940 showing how the breakwater built in 1933 caused the sand to build out and swamp the pier.

Source: Courtesy of US Geological Survey.

(b)

(c)

FIGURE 19.9   (Continued)

with much tectonic movement and uplift. Nearly the entire California coastline is rising out of the sea, producing our characteristic sea cliffs and often irregular coastlines with rocky headlands and promontories (Figure 19.10). The waves are refracted around the points and promontories as they hit the coastline, so they not only erode the points but also pound the sides of the points. In the process, they carve sea caves into the sides of the cliffs, which eventually cut right through and make an **arch** (Figure 19.10b and c). If the waves continue to erode the arch, eventually the span collapses and all that remains are big rocky **stacks** (Figure 19.10d) standing in the surf, doomed to be eroded away someday. Rugged uplifted coasts, such as those of Big Sur or Northern California, often have many arches and stacks as they are chopped down by the waves.

Over the course of thousands of years in geologic time, the forces of erosion in uplifted headlands are working to chop off the points and promontories and flatten them out. Meanwhile, the waves refracted at the headlands mean that they are bent away from the embayments between them, so the wave energy is weak and sand fills in the bays. If this process continues long enough without interruption, eventually the cliffs will be straightened and the bays filled out to produce a flat coastline.

Look at the diagram of the coastal sea cliffs (Figure 19.10a) again. Waves constantly undercut the cliffs at their base (especially during storms), so the cliffs are always being undermined and destabilized. After enough time, big blocks break off and collapse onto the beach. Sadly, lots of people like to build or live on top of a beach cliff for the amazing views and fresh sea breezes. They spend millions of dollars for such prime real estate. Then they become very upset when the natural process of erosion undercuts "their property" and causes their patio or backyard or even house

spread out the abundant river sand supply in a series of narrow sandy islands parallel to the coast called **barrier islands**. In other coastlines, like in Chesapeake Bay, the irregular shape of the bays and estuaries of the Delaware, Chesapeake, Potomac, and Rappahannock Rivers is due to the fact that these areas were once branching river and stream valleys during the low sea levels of the last ice age, and the rise of sea level since then has drowned them. In other coastlines, like in British Columbia or Scandinavia, glaciers cut vertical-walled valleys down to the ocean during the last ice age. Since the glaciers melted, the rise of sea level has drowned these glacial valleys, making **fjords**.

Barrier islands, drowned river valleys, and large deltas building out occur on coastlines that are sinking or just barely above sea level, usually on passive tectonic margins above passive margin wedges (Figure 4.10). But the California coastline (along with the entire Pacific Coast of North America) is an active margin or a transform margin,

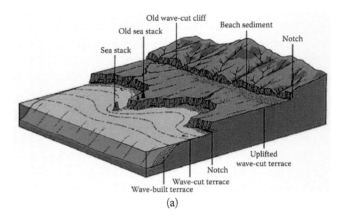

(a)

FIGURE 19.10   (a) Diagram showing the major features of a rising coastline. (b) Sea stack with an arch carved into it at Point Arena. Eventually, this arch will collapse and form two separate stacks. (c) Rugged sea arch at Big Sur. (d) Series of sea stacks on the Pacific Coast near Bodega Bay. (e) Collapsing cliff and vulnerable buildings on Solana Beach, California.

Source: (a, e) Image and photo courtesy of Wikimedia Commons. (b–d) Photos by G. Hayes.

(b)

(c)

(d)

(e)

FIGURE 19.10 (Continued)

to collapse into the surf (Figure 19.10e). Millions of dollars are spent armoring the cliff with concrete and other substances to stabilize it.

But this ignores the natural processes at work here. A seawall or cliff-face "stabilization" feature will slow erosion for a while, but at a cost. First, it depletes the beach of some of its sand supply, since the high-energy waves are reflected backward, which may ruin the beach below. Years of experience with the disastrous construction of seawalls has shown that in the long run, they do more harm than good. If one landowner "protects" his cliff but his neighbor doesn't, the waves will cut back the neighbor's property—and he can sue the landowner whose armored cliff diverted all the wave energy onto adjacent property. Even if you build a complete seawall along a shoreline, the wave energy must go somewhere. Usually, it ends up eroding and undermining the cliff and beach *below* the seawall, causing the entire expensive structure to collapse—and with it the property above the cliff. Humans are arrogant and constantly think that they can "fool mother nature." But in the end, everyone ends up paying a lot more and doing more damage when they build structures on the edge of sea cliffs. Sooner or later, nature will win and that sea cliff will collapse.

In addition to steep sea cliffs, wave erosion produces another feature of rising coastlines (Figure 19.10a). At the base of the cliff and down beneath the beach sand is a gently sloping surface carved into the bedrock called a **wave-cut platform** or wave-cut terrace. This feature is hard to see on beaches with lots of sand, but they are apparent on rocky beaches, especially at very low tides. The **inner edge point** of this platform at the base of the cliff was right around mean sea level at the time it formed.

Not only do many areas of California's coast have a platform or terrace caused by the waves currently pounding the shoreline, but also above the cliff is another terrace, and in some cases there may be as many as a half dozen or more discrete stair-step **marine terraces** above the ocean (Figure 19.11). These uplifted marine terraces occur on rising shorelines in many parts of the world (such as New Guinea), although the California examples have been studied the longest. Dating the first terrace above sea level has shown that it is about 125,000 years old (so it is called the "125 k" terrace by geologists). The next one up is often 230,000 years old, then another terrace at 330,000 years old, and another at about 100,000 years older, and so on.

The terraces can be seen in many places up and down the California coast. In San Diego, the "mesas" on which Balboa Park and San Diego Zoo are built, along with most of the city above the coastal cliffs, are on the 125 k terrace. In northern San Diego County near San Onofre, Interstate 5 runs on top of the 125 k terrace (Figure 19.11b), and where the natural cliffs are visible on the Camp Pendleton Marine Base, you can make out several more terraces above. Almost all of San Clemente and Dana Point is on the 125 k terrace, as are the parts of Newport Beach above the cliffs (Figure 19.11c). The 125 k terrace makes up the first row of houses above the cliffs on the Palos Verdes Peninsula (Figure 19.11d), and higher terraces are present, although hard to see with all the housing and trees built over them. The upper flat parts of Malibu above the cliffs (such as the location of Pepperdine University) are on the 125 k and higher terraces. Many of the coastal towns in Santa Barbara County and up the coast are built on the 125 k terrace, and several terraces can be seen on the Big Sur Coast. The terraces are particularly striking near Santa Cruz (Figure 19.11e), on uninhabited islands like San Clemente Island off Oceanside (Figure 19.11f), and in many other places (like Sea Ranch near Gualala) all the way to the Oregon border and beyond.

This pattern of terrace dates spaced about 100,000 years apart was first noticed in the 1970s, when geologists showed that the 100,000- to 110,000-year Milankovitch cycle of the shape of the Earth's orbit (known as the **eccentricity cycle**) is the major controller of glacial–interglacial cycles. This cycle was first discovered in deep-sea cores of Ice Age sediments. All the terrace dates correspond to times when the Earth was at peak interglacial and ice caps had melted back and the sea level rose to its highest level.

So what explains this correlation? Remember, the coastline is constantly rising out of the surf zone, so it is like a conveyor belt, or an old-fashioned tape recorder, moving along as new things happen to it and carrying older things along. Meanwhile, sea level fluctuates up and down like a yo-yo with the changes in glacial ice volume (Figure 19.12). Each time the sea level reaches its highest point during an interglacial melting episode, it cuts a new terrace. The sea level then drops and fluctuates at a lower level for the next 100,000 years of glacial fluctuations. Meanwhile, the cliffs keep rising, putting the old marine terrace high in the air and out of reach of the next rise in sea level 100,000 years later, so it cuts a new terrace below the last one. If the coast keeps rising without interruption, each new terrace will be cut below the last one, and the old ones keep rising out of reach of the ocean. However, they also erode from rainfall and streams and landslides as they get older. The oldest terraces (500,000 years or older) tend to be indistinct and hard to spot, while the most recent one (125 k) is usually clearly visible.

Another analogy might help with understanding this dynamic process. In basketball, the rim is supposed to be 10 ft above the ground. Most NBA players and college players can jump high enough and are tall enough that they can

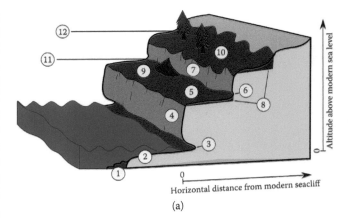

(a)

(b)

(c)

FIGURE 19.11 Dynamics of a rising set of marine terraces. (a) Diagram showing the major features: (1) low-tide cliff/ramp with deposition; (2) modern shore (wave-cut/abrasion) platform; (3) notch/inner edge, modern shoreline angle; (4) modern sea cliff; (5) old shore (wave-cut/abrasion) platform; (6) paleo-shoreline angle; (7) paleo--sea cliff; (8) terrace cover deposits/marine deposits, colluvium; (9) alluvial fan; (10) decayed and covered sea cliff and shore platform; (11) paleo-sea level I; (12) paleo-sea level II. (b) Marine terraces on Camp Pendleton Marine Base, as seen from San Onofre State Beach just west of Interstate 5. The photo is taken from the 125 k terrace, and several stair-step terraces can be seen on the cliff profile. (c) Dana Point sits on top of the 125 k terrace, while the harbor below is at modern sea level. In the distance, the hills of San Clemente are mostly on the 125 k terrace. (d) Old aerial photo of the south side of Palos Verdes Peninsula showing the distinct terraces. Today most of them are obscured by houses, trees, and other development or destroyed by the Portuguese Bend landslide. (e) Modern example from near Santa Cruz. (f) Stair-step terraces on the south shore of San Clemente Island.

Source: (a, e) Image and photo courtesy of Wikimedia Commons. (d) Photo by J Shelton. (f) Photo by permission of the Geological Society of America. Remaining photos by the author.

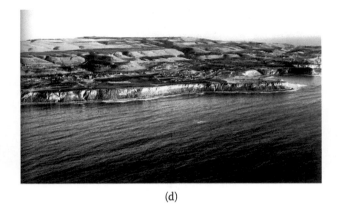

(d)

(e)

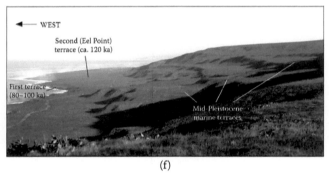

← WEST

Second (Eel Point)
terrace (ca. 120 ka)

First terrace
(80–100 ka)

Mid-Pleistocene
marine terraces

(f)

FIGURE 19.11 (Continued)

slam-dunk on a 10 ft rim. But suppose the rim kept rising so that the next time the player tries to slam, the rim is now 11 ft above the court level. Most basketball players cannot dunk on a rim that high, and they would miss the rim by a foot. There are some really tall players, like former NBA star Dwight Howard, who can slam on a higher rim. Howard actually did complete a slam on a 12 ft rim in the 2009 NBA All-Star Game Slam Dunk Contest. But if the maximum rise of sea level is always to the same height (just like the 10 ft most NBA players can jump for a slam dunk), then raising the rim (or the cliffs and coastline) will pull it out of reach.

19.5 CALIFORNIA ESTUARIES AND BAYS

Most of California's coastline is rising because it is being uplifted along fault lines tied to the transform tectonics of the San Andreas fault, or the active margin of the Cascadia

subduction zone north of Cape Mendocino. But there are places where the faulting has down-dropped the coastline, especially in the tectonic basins discussed in Chapter 12. For example, the sinking Los Angeles Basin has large rivers flowing down through it, such as the Los Angeles River and Santa Ana River. The Ventura Basin creates the broad Oxnard Plain, with the Santa Clara River and Calleguas Creek draining across it. There are estuaries in most of the large rivers that drain San Diego County, which can be seen as you drive along Interstate 5. Morro Bay is the mouth of a big estuary, as are Castroville and Marina, where the Salinas River drains into Monterey Bay. Nearly the entire San Francisco Bay has a large complex of shallow tidal mudflats along its shores, with several rivers (besides the largest of all, the Sacramento–San Joaquin River) draining into it. There are large bays like Humboldt Bay, which receives the water from the Humboldt River at one end and the Trinity River on the north side of the lagoon.

These settings create estuaries where the river water mixes with seawater at the mouth of the river and, in many cases, coastal bays and lagoons. These regions are incredibly diverse biological habitats, with a wide array of marine life adapted to the sandy surf zone, the quiet tidal mudflats, the estuarine waters where the salinity fluctuates from fresh water to full seawater, and so on. They are particularly valued as habitat for water birds and migratory birds, which rely on them for food during their strenuous migrations up and down the flyways.

Sadly, most of these once-pristine wetlands have been modified or even destroyed by human activity. In most cases, people have turned them into harbors or marinas or boat basins for navigation and recreational boating and covered the natural sand and mud with concrete and piers. This is true of Marina del Rey in Los Angeles, the natural estuary of Ballona Creek that is now a marina, Newport Bay in Orange County (the natural mouth of the Santa Ana River), and the San Pedro–Long Beach harbor area, one of the largest shipping ports in the world. In many cases, they have been filled in with sediment and refuse as waste dumps, or turned into landfills to expand room for development, as has happened to much of the natural shoreline of San Francisco Bay.

In some places, however, the classic configuration of the barrier island complex (so typical of most of the East and Gulf Coasts) can develop. Barrier islands (Figure 19.13) are long sandy islands that run parallel to shoreline and enclose a lagoon behind them. They tend to occur mostly on coastlines that are stable or subsiding and where there is not only an abundant sand supply but also a very strong longshore current to spread the sand along the shoreline.

A typical barrier island has several characteristic features. The foreshore region is a beach much like those we have already discussed. The backshore region and the highest part of the island is a coastal sand dune complex, sometimes stabilized by salt-tolerant plants. Behind the barrier complex is a coastal lagoon, which is typically brackish but can fluctuate from completely normal seawater salinities

FIGURE 19.12 Mechanism by which marine terraces are carved.

Source: Redrawn from several sources.

(3.5% salt per volume of water) during high tides to almost fresh water when there is a large flood coming down from the rivers. These brackish lagoons can be small wetlands or huge inland seas like Pamlico Sound behind the Outer Banks of North Carolina. The lagoon has its own distinct assemblage of plants and animals that can tolerate the wide range in salinity, from salt-tolerant marsh grasses (such as *Spartina*) to animals like oysters, certain kinds of crabs, and a few species of snails, plus a wealth of seabirds that feed there.

The barrier island complex appears static, but it is actually very dynamic. In the short term, the water flows in and out through a gap in the barrier called the tidal inlet (Figure 19.13) during the twice-daily fluctuation of the tides. But this tidal inlet is constantly shifting up and down the coast. Longshore currents tend to make it built down the

coast in the direction of the beach drift until it has migrated a long way down the barrier. But a large surge of fresh water coming down the river during a rare flood will breach the barrier island and form a new inlet up the coast from the old inlet. This new inlet will begin to migrate downshore in the years that follow until yet another storm resets it and starts the process all over again.

Another consideration is that many barrier islands (especially on the East and Gulf Coasts) experience huge hurricanes every decade or so. Geologists have dug trenches through barrier islands in many parts of the world to examine their sedimentary record and consistently find that nearly all the older sediments in the island are hurricane deposits from a single large storm. The quiet daily activity of tides and waves leaves almost no sedimentary record. Instead, a huge hurricane will wash the entire barrier island

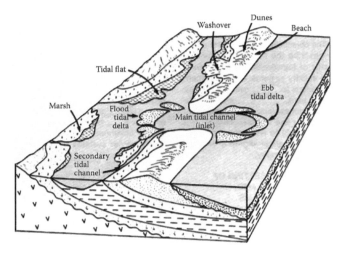

FIGURE 19.13 Diagram of the typical features of a barrier island complex.

Source: Redrawn from several sources.

(typically only a few meters above sea level at its highest point) completely away and then redeposit the sediment as the storm wanes and normal longshore sedimentation resumes.

This is why it is extremely foolhardy to build houses or other structures on barrier islands, because sooner or later a hurricane is guaranteed to wipe them out. The barrier islands of the East and Gulf Coasts are heavily developed with houses and towns that are doomed to destruction someday—yet almost no one warns people buying these death traps that barrier islands are just temporary features formed between hurricanes. When the inevitable storm comes along, people lose their lives and property, and we all pay for this foolish development through our higher taxes and higher insurance premiums.

Not only do barrier islands change due to the migration of the tidal inlet, and hurricanes washing them away and redepositing their sediment, but they also change over the longer term. Their shape is closely tied to the sea level, so as the sea level changes, the entire complex migrates landward or seaward. If sea level rises, for example, the beach and dune sands will shift landward and migrate across the old organic-rich lagoonal muds. If these become buried in

the rock record, the dune and beach sands form a reservoir sand above an organic-rich lagoonal layer that could be a source rock for oil. Indeed, a number of major oil fields have been found in ancient migrating dune deposits. In other cases, the barrier islands shift seaward with falling sea level (Figure 19.14).

At one time, there were a handful of barrier island complexes on the California coast, especially in the mouth of creeks in down-dropping tectonic basins. As mentioned already, however, nearly all of them have been so altered by humans that they no longer show most of the features of barrier islands or estuarine lagoons. The barrier complex and back-barrier bay of Humboldt Bay (Figure 19.15) have many of the features of a typical barrier island, but in other ways, it is completely artificial. For example, the natural tidal inlet no longer migrates because it has been stabilized by a set of jetties. Most of the natural back-barrier marsh has been dredged with ship channels to allow deeper-draft ships to travel through the normally shallow bay.

One of the few barrier complexes that is still in a nearly natural state is Mugu Lagoon in Ventura County (Figure 19.16). It has a small but pristine barrier island, with a natural lagoon behind it filled with marsh grass and a diverse community of wildlife, including an incredible number of shorebirds that migrate through here. The only reason it has not been destroyed by development is that it became a military base in World War II and is still under military control as US Naval Air Station Mugu. The Navy still uses it to test long-range radar systems and other classified projects, and naval planes come and go from its airstrip, especially on their way to bombing ranges in the Channel Islands.

Thanks to the US Navy, no private developer has been allowed to turn it into a marina or a dense cluster of beach condominiums. Its beaches are still unspoiled since it is not overrun by hordes of people; only authorized naval personnel are allowed on base. The Navy has not left it completely natural, of course. During World War II, they tried to dig a deeper basin in the central lagoon to create room to turn ships and submarines. That effort failed, as sediment filled it in almost as quickly as it was dug. In addition, much of the dry land adjacent to the marsh is covered by naval buildings and airstrips and the small community for naval personnel. In the past, there was careless dumping of toxic chemicals from

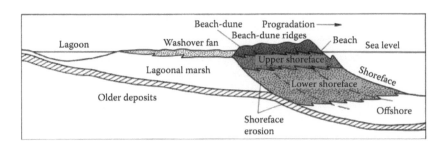

FIGURE 19.14 Diagram of seaward migration of a barrier island complex due to falling sea level\ or progradation from high sediment supply.

Source: Courtesy of SEPM.

FIGURE 19.15 Satellite view of Humboldt Bay and barrier island complex.

Source: Courtesy of US Geological Survey.

FIGURE 19.16 Satellite view of Mugu Lagoon.

Source: Courtesy of US Geological Survey.

Behind the beach and dune complex is a long back-barrier lagoon. It has been divided into a larger western lagoon just south of the base complex and west of Calleguas Creek and a smaller eastern lagoon east of Calleguas Creek. It experiences a dramatic change with the twice-daily tidal cycle, so while visiting the naval base, you can see the tides come in and out over the course of a few hours. The lagoon is full of marsh grasses, oysters, crabs, and many other creatures tolerant of the variation of salinity and the low-tide exposure. It is particularly rich in aquatic birds and shorebirds, who use it as a feeding and resting place during their annual migrations.

Detailed studies have been done using geographic information systems (computer mapping software known as GIS) to overlay a series of maps and aerial photographs over decades, showing how dynamic the lagoon really is (Sadd and Karpilo, 2001). Each map shows the tidal inlet in a different position, since it migrates down the coast for years until a major flood in Calleguas Creek cuts a new inlet opposite its mouth. The west lagoon changed shape over time, especially after the Navy abandoned its efforts to dredge a deep turning basin in the lagoon, and sediment quickly filled in the hole that they dug. The size and shape of the barrier island have changed, largely due to the shift from heavy agricultural use of the land upstream (which supplied a lot of sediment to the bay). Today, there is a more urbanized landscape upstream, with dams and flood control structures cutting down the supply of sand coming down Calleguas Creek. This has depleted the supply of sediment, so the beach is being eroded back as sediment supply decreases. Much of it is disappearing down the offshore submarine canyon and is not being replaced. During the 1960s, a lot of federal money was spent bringing in sand from dredging the coastal marinas upshore and dumping it to replenish the beaches at Port Hueneme and Mugu Lagoon, but this is no longer being done, and the area is eroding again.

Finally, barrier complexes like Mugu Lagoon are also subject to another effect: the global rise of sea level due to climate change. This will eventually wipe out and flood the entire barrier complex of the East and Gulf Coasts of the United States, so those barrier islands and their huge populations will vanish beneath the waves. More likely, they will be destroyed by the higher frequency of huge hurricanes that are occurring due to global warming, so that they will probably

aviation fuels into the lagoon. The Navy now has an extensive mitigation effort to monitor these old wastes, and a strict protocol for dumping any other fuels or toxic chemicals. But those problems are offset by the excellent condition of the barrier island and lagoon, as proven by the high diversity of marine life and seabirds (many endangered or threatened) that depend on it. Even if the naval base were to be closed (which has been threatened several times), the sheer diversity of endangered wildlife will keep the land in public hands and protected from private developers for the long term.

The barrier complex (Figures 19.16 and 19.17) is affected by a number of processes. Longshore current comes down out of the northwest, so the barrier island sand (and the tidal inlet) tends to migrate to the southeast. Much of the sand also flows down into submarine canyons just upcoast at Port Hueneme and at Mugu submarine canyon. Countering this effect is the fresh water coming down from Calleguas Creek, which floods the estuary and lagoons with fresh water during major storms. The tidal inlet has a habit of migrating east with the longshore current until another major flooding event on Calleguas Creek cuts a new inlet opposite the creek mouth, and then the process of inlet migration starts all over again.

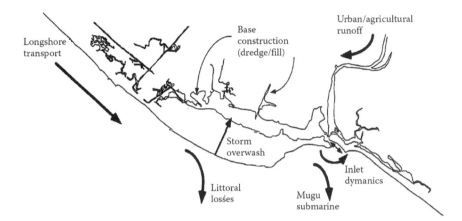

FIGURE 19.17   Diagram of the processes affecting geomorphology of Mugu Lagoon.

Source: From Sadd, J. L., and Karpilo, R., in Prothero, D. R. (ed.), *Modern and Ancient Barrier, Lagoonal, and Marine Environments, Ventura County, California*, Pacific Section SEPM Book 90, Pacific Section SEPM, Fullerton, CA, 2001, 29–36; courtesy of Pacific Section SEPM.

be damaged beyond rebuilding even before sea level rises and drowns these coastlines. The same processes will also eventually affect all the long sandy barrier island complexes in California. This is true from the Silver Strand on San Diego Bay to Balboa Island and Newport on Newport Bay, to Long Beach and San Pedro, to the Marina del Rey area, to Mugu Lagoon, to Morro Bay, to most of the low-elevation regions around San Francisco Bay, all the way to Bodega Bay, Point Reyes, and Humboldt Bay. People love to buy property and live near the beach, but over the long term, the region is vulnerable to severe storms, rising sea levels, and (on the Pacific Coast) tsunamis, which can utterly destroy human efforts to live on the glorious beaches and coast.

RESOURCES

Bascom, W. 1980. *Waves and Beaches. Anchor Books*, Garden City, NY.

Emery, K.O. 1960. *The Sea Off Southern California*. Wiley, New York.

Griggs, G. 2010. *Introduction to California's Beaches and Coasts*. University of California Press, Berkeley.

Griggs, G., and Ross, D.S. 2014. *California Coast from the Air: Images of a Changing Landscape*. Mountain Press, Missoula, MT.

Griggs, G., and Savoy, L. 1985. *Living with the California Coast*. Duke University Press, Durham.

Hayes, M.O., Michel, J., and Holmes, J.M. 2010. *Coast to Explore: The Coastal Geology and Ecology of Central California*. Pandion Books, New York.

Hobbs, C. 2012. *The Beach Book: Science of the Shore*. Columbia University Press, New York.

Kaufman, W., and Pilkey, O. 1979. *The Beaches Are Moving*. Anchor Press, Garden City, NY.

Meldahl, K.H. 2015. *Surf, Sand, and Stone: How Waves, Earthquakes, and Other Forces Shape the Southern California Coast*. University of California Press, Berkeley, CA.

Pilkey, O. 2011. *The World's Beaches: A Global Guide to the Science of the Shoreline*. University of California Press, Berkeley.

Sadd, J.L., and Karpilo, R. 2001. Holocene shoreline change and sediment supply in Mugu Lagoon/Barrier System, California, 29–36. In Prothero, D.R. (ed.), *Modern and Ancient Barrier, Lagoonal, and Marine Environments, Ventura County, California*. Pacific Section SEPM Book 90. Pacific Section SEPM, Fullerton, CA.

Sharp, R.P. 1978. *Coastal Southern California*. Kendall-Hunt Publishing, Dubuque, IA.

VIDEOS

A classic old video, "The Beach: A River of Sand," with excellent time-lapse sequences: www.youtube.com/watch?v=FqT1g2riQ30

Beach Processes

www.youtube.com/watch?annotation_id=annotation_4120056893
&feature=iv&src_vid=sTDsQNXwn04&v=kJNB0dScA3o
www.youtube.com/watch?v=U9EhVa4MmEs
www.youtube.com/watch?v=G1FIBuybN78

Marine Terraces and Uplifted Coastlines

www.youtube.com/watch?v=p_AUkB9EGBo
www.youtube.com/watch?v=Pgiy34t3t9I
www.youtube.com/watch?v=z1swjSvgx6A
www.youtube.com/watch?v=D0902_iYBrQ
www.youtube.com/watch?v=UWIO90HxhEQ
www.youtube.com/watch?v=ITv6gSUmTjc
www.youtube.com/watch?v=fztT507KivI
www.youtube.com/watch?v=WEqq4kIXzOM

Barrier Islands

www.youtube.com/watch?v=u-MFerc-pis
www.youtube.com/watch?v=jFqE9FYSe4A
www.youtube.com/watch?v=IvCL2TLSDUQ
www.youtube.com/watch?v=Rn7x0f4msYg
www.youtube.com/watch?v=utDERYua404
www.youtube.com/watch?v=8WDaP_HuT50

Coastal Erosion

www.youtube.com/watch?v=5NMF1sqfR3Q
www.youtube.com/watch?v=nujYG_b8lI8
www.youtube.com/watch?v=ucNdCgfjjDc

Animations

Excellent animations by Tanya Atwater, especially the one showing the process of carving uplifted marine terraces: http://emvc.geol.ucsb.edu/1_DownloadPage/Download_Page.html#IceAge

20 California's Fossil Resources

I grew up in Los Angeles, and I was always fascinated by the La Brea Tar Pits. Right in the middle of the city, in an area called the Miracle Mile, for crying out loud, we have these eldritch ponds of dark, bubbling goo. And down in the muck, there're all these amazing fossils: mammoth and saber tooth cat and dire wolf.

—**Greg Van Eekhout**

Through the study of fossils I had already been initiated into the mysteries of prehistoric creations.

—**Pierre Loti**

20.1 CALIFORNIA'S AMAZING FOSSIL RECORD

It is easy to think of gold and oil as natural resources, but fossils are a kind of natural resource as well. Not only do we crush fossiliferous limestone into powder to make cement and use blocks of limestone full of fossils for building stone, but California's fossils also have other implications as well. They tell us about the history of life on this planet and about the strange and amazing ways animals and plants have adapted and changed. They tell us how life has evolved through time, and give us clues about ancient environments. As we saw in previous chapters, certain fossils tell us how far exotic terranes have traveled across the Pacific to reach California. In many cases, fossils have been important for museums to create a sense of wonder and excitement about the world and the prehistoric past. Hollywood has made millions with movies about prehistoric life, especially the *Jurassic Park* and *Jurassic World* series. Last but not the least, fossils are the key to establishing the age of rocks and correlating them over long distances. Nearly all the oil wealth of California was found by paleontologists doing biostratigraphy of microfossils to correlate their wells and cores and determine where the oil might be found. Indeed, the organic material that produced the oil in the first place comes from the bodies of trillions of fossil plankton, or from the undecayed remains of plants to make coal, hence the name *fossil fuels*.

In contrast to some other states, California is not as rich in the total volume or area of fossils. For example, across great stretches of the Midwest, such as the Cincinnati Arch in Ohio–Indiana–Kentucky, or large parts of Michigan, Iowa, Kansas, and Missouri, there are immense volumes of limestone made up of trillions of fossils (especially crinoids or "sea lilies"), or deep-water shales with their own incredible creatures. In some places, you cannot take a step without walking on dozens of brachiopods and corals and crinoids. By contrast, most of California was not even here in the Paleozoic or much of the Mesozoic, so there are not as many square miles of rich fossiliferous rocks of those eras.

But what California lacks in volume, it makes up for in quality, from the incredible Paleozoic fossils (especially trilobites) of the Mojave Desert to the amazing marine reptiles and dinosaurs of San Diego and the Central Valley, to a long record of Cenozoic mammals and marine fossils. Then there are extraordinary deposits like the La Brea tar pits, which preserve millions of bones of everything from mammals like saber-toothed cats down to insects, delicate birds, and lizards, and even plant fossils and pollen.

The fossil deposits discussed in this chapter are not an exhaustive list. In fact, there are websites online that provide lists of nearly all the fossil localities known in California. Instead, the examples listed in what follow are the highlights of some of the more important fossil localities in the state, giving a sampling from most geologic periods (Figure 20.1).

20.2 PALEOZOIC FOSSILS

20.2.1 THE PRECAMBRIAN–CAMBRIAN OF THE DEATH VALLEY REGION

The earliest known evidence of life are structures known as stromatolites. Composed of many finely laminated layers of sediment forming a cabbagelike or domed structure, they are formed when mats of blue-green bacteria trap layer after layer of sediment in their sticky filaments and then grow through the old sediment layer and trap more sediment. The oldest stromatolites occur in rocks 3.4–3.5 billion years in age in both Australia and South Africa. They are the only megascopic evidence of life for its first 3 billion years, when nearly all life was single-celled and microscopic. Then, at the end of the Proterozoic and into the Early Cambrian, stromatolites declined rapidly, so they nearly vanished from the fossil record. The reasons for this are not clear, although most paleontologists think it was due to the evolution of the first grazing mollusks and other creatures that could crop these dense mats for the first time. In the Upper Proterozoic Deep Springs Formation in the White Mountains, stromatolites are quite common (Figure 20.2), but these are among the last of the stromatolites in California or anywhere.

The oldest evidence of multicellular life is known as the Ediacara (ee-dee-AK-ar-a) biota. Originally described from the Ediacara Hills of the Flinders Ranges in Australia, it is now known from localities all over the world, including England, Russia, China, Newfoundland, and Namibia. They also occur in the uppermost Precambrian and Lower Cambrian rocks of California (Figure 20.2c).

DOI: 10.1201/9781003301837-20

The Ediacara biota is typically preserved as impressions of soft-bodied creatures on seafloor sands or muds (now turned to sandstones or shales), with no parts of the original creature left. Although most Ediacarans were small, many were quite large, such as the worm-like fossil *Dickinsonia*, which was more than a meter across. Some of these impressions resemble sea jellies and sea pens, and others are worm-like, so the conventional interpretation has interpreted them as these familiar animals.

Other scientists point out that the Ediacarans did not have the same symmetry as sea jellies and sea pens and worms but were constructed in a totally novel fashion not seen in any living animal group. Instead, they think that the Ediacarans were the earliest experiment in multicellular evolution, which vanished when modern groups evolved. Apparently, Ediacarans had a huge surface area to allow them to exchange food, gases, and waste products, so they did not evolve a true digestive and respiratory system, as found in higher animals. Some paleontologists compare the Ediacarans to a quilted air mattress filled with fluid.

Yet another idea, the "Garden of Ediacara" model, suggests that they were large flat creatures that lay on the shallow seafloor and used symbiotic algae in their tissues (as modern reef corals, giant clams, and several other types of animals do) to get their oxygen and get rid of carbon dioxide. There are even scientists who argue that they were not animals at all but instead were fungi or even soil structures.

Whatever the Ediacarans were, they were the first large multicellular organisms on the planet and spread worldwide from about 600 million years ago (Ma) until the Early Cambrian. A number of these are found in the late Proterozoic rocks (especially the Stirling Quartzite and the Wood Canyon Formation) of the Mojave Desert, Death Valley, and White and Inyo Mountains of California (Figure 20.2), as described by Hagadorn and Waggoner (2000) and Hagadorn et al. (2000).

In the earliest Cambrian, most of these large soft-bodied Ediacarans vanished, and we find the earliest fossils with hard shells. Nicknamed the "little shellies" or the "small shelly fossils" (SSFs), they are tiny shells only a few millimeters in size. Most are shaped like simple tubes or cones, such as the cone-in-cone structure of *Cloudina* (Figure 20.3a), while others look like simple cap-shaped limpet shells and may have belonged to the first mollusks. Still others looked like tiny spiky jacks (from the old kids' game ball and jacks). These were probably pieces of the external armor in the skins of larger creatures, or spicules that were woven together to form a sponge. They are all so tiny that for a long time they were missed by paleontologists, until they learned to find layers full of these tiny shells (Figure 20.3b) and take samples that were then etched in acid, releasing the tiny shells inside. These fossils are particularly common in the lower part (Figure 20.2a) of the Upper Proterozoic–Lower Cambrian Wood Canyon Formation, found over much of the Death Valley–Mojave Desert region.

After about 15 m.y. dominated by the small shellies (from 545 to 520 Ma, the first two stages of the Cambrian), finally, during the third stage of the Cambrian (Atdabanian state, 520–515 Ma), we see larger fossils with shells. These include not only trilobites but also a lot of other very primitive and experimental groups of jointed-legged arthropods, archaic limpet-like mollusks, and weird experiments in other body plans. These included the archaeocyathans, the first reef builders on the planet, which were built

FIGURE 20.1 Index map showing the locations of major California fossil localities mentioned in the chapter: (1) Precambrian and Paleozoic fossils, Death Valley area; (2) Precambrian–Cambrian fossils, White Mountains; (3) Cambrian fossils, Marble Mountains; (4) Inyo Mountains: Paleozoic fossils, Mazourka Canyon, and Triassic ammonoids, Union Wash; (5) Cretaceous ammonites, Great Valley Group, near Chico and Redding and the northern rim of the Sacramento Basin; (6) Cretaceous marine reptiles and dinosaurs, Panoche Hills; (7) Cretaceous and Cenozoic fossils, San Diego area; (8) late Cretaceous and Cenozoic fossils, Silverado Canyon, Orange County; (9) Eocene plant localities at LaPorte and Chalk Bluffs, Sierra Nevadas; (10) Eocene–Oligocene plant localities, Warner Range; (11) Oligocene fossils, Otay Mesa, San Diego County; (12) localities in Los Angeles area, including the middle Miocene Topanga Canyon Formation and Rancho La Brea tar pits; (13) middle Miocene localities at Sharktooth Hill and Pyramid Hill, northeast of Bakersfield; (14) middle Miocene fossils, Barstow Formation; (15) middle–late Miocene fossils, Red Rock Canyon State Park; (16) late Miocene fossils, Blackhawk Ranch, Mt. Diablo; (17) Pliocene–Pleistocene fossils, Anza-Borrego Badlands; (18) Calistoga Petrified Forest.

Source: Base map courtesy of California Division of Mines and Geology.

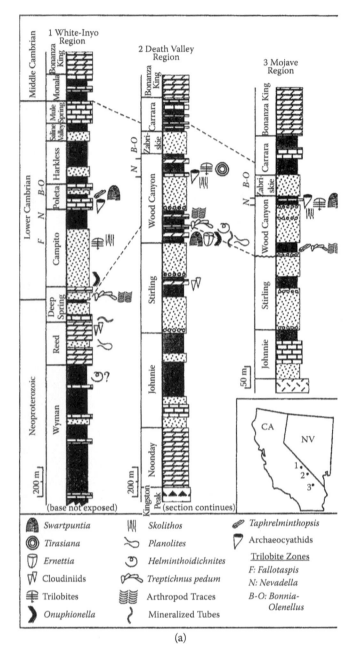

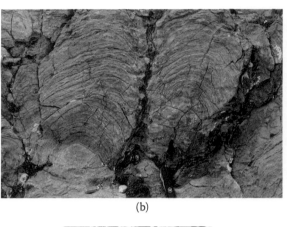

(b)

(c)

FIGURE 20.2   (Continued)

FIGURE 20.2 Ediacaran fossils of eastern California: (a) Stratigraphic sequence of Ediacaran fossils, as well as *Cloudina* and other small shelly fossils and other trace and body fossils of the late Proterozoic–Cambrian of California. (b) Layered domed structures known as stromatolites, formed by sediment trapped by sticky mats of cyanobacteria or algae, Upper Proterozoic Deep Springs Formation, White Mountains. (c) Ediacaran fossil impression *Swartpuntia* from the lower Wood Canyon Formation.

Source: (a) After Hagadorn, J. W., et al., *Journal of Paleontology*, 74: 731–740, 2000. (b–c) Photos by J. W. Hagadorn.

somewhat like sponges with a simple conical body structure. Archaeocyathans are found in many Lower Cambrian rocks in California (Figure 20.2a), and huge reefs of them are preserved in mounds in the White Mountains just east of California in Lida, Nevada.

Another experimental group are the helicoplacoids (Figure 20.4), odd-looking relatives of sea stars and sea urchins that are built as a corkscrew spiral of plates around a spindle-like body form. How these creatures lived and fed is still mysterious, but they are known primarily from the Lower Cambrian beds of the White Mountains of California and almost nowhere else. Just a few complete specimens have ever been found, although broken pieces of their dumbbell-shaped plates are common in the middle part of the Poleta Formation near Westgard Pass. These odd experimental creatures apparently lived with their points stuck in the muddy bottom (nicknamed "mudstickers"). They vanished by the end of the Early Cambrian, possibly because the seafloor began to be churned up by worms for the first time in the late Early Cambrian, which would destabilize the bottom for mudstickers like the helicoplacoids.

20.2.2   EARLY CAMBRIAN TRILOBITES OF THE MOJAVE DESERT

The most common and typical fossils of the Cambrian are the trilobites. When trilobites first appear in the late Early Cambrian (Atdabanian stage), they are the primitive group known as the olenellids (Figure 20.5a and b). Olenellids are so archaic they still have not fused all their tail segments into a single plate, as more advanced trilobites did. They are also distinctive in having large crescent-shaped eyes on

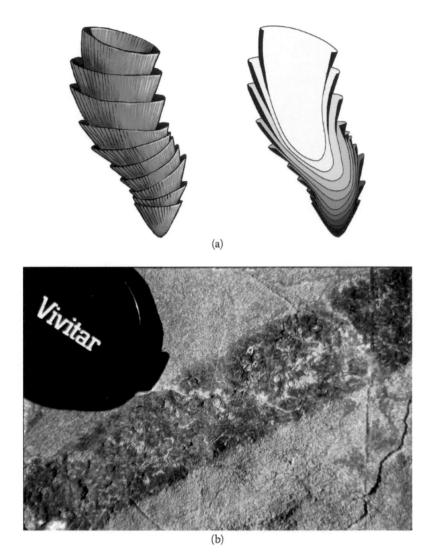

(a)

(b)

FIGURE 20.3 The latest Proterozoic–earliest Cambrian small shelly fossils were the first tiny-shelled animals to evolve, long before trilobites and larger-shelled organisms appeared: (a) Reconstruction of the fossil *Cloudina*, built of a series of conical shells inside one another, common in latest Proterozoic rocks. (b) Close-up of a limestone layer with lots of tiny conical shells inside, easily missed by less-observant fossil collectors. Rocks like this can be carefully dissolved in acid to release the tiny shells. Wood Canyon Formation in the White Mountains.

Source: (a) Drawing by M. P. Williams. (b) Photo by the author.

their head shield (cephalon), spines on the back corner of the head shield, and many spiny segments in their thorax, including large backward-pointing spines near the front of the thorax (Figure 20.5a). They are such common and easily collected fossils in California that they were considered one of the two finalists for California's official state fossil. (The saber-toothed cat won out, even though no one can collect their fossils legally as they can trilobites.)

There are several places where olenellids can be collected. One of the most famous collecting areas is in the southern Marble Mountains in the central Mojave Desert, where they occur in a distinctive layer known as the Latham Shale (Figure 20.5c). There students and amateur collectors have mined these fossils for decades. Take Interstate 40 to the Kelbaker Road exit, and then get off the interstate and drive 1 mi south to the T junction. Turn left (east) and drive

to the now-extinct town of Chambless along the old Route 66 (now the National Trails Highway). Turn southeast on the road to Cadiz. After the paved road curves due east, but just before the paved road goes due south to cross the railroad tracks, turn on a dirt road on the left that goes due north. Go about 1 mi north on the dirt road until you get to a junction with a well-traveled east–west dirt road, and then turn east. About half a mile along this, you'll see the dirt road heading northeast toward the old quarry. Follow this as far as the road is passable, and then hike up to the Latham Shale (the brown shale unit below the gray cliff-forming Chambless Limestone). Look for old "glory holes" of serious collectors, and sift through and turn over the larger shale pieces. You will see many good cephala (head shields) of every size, although complete trilobites are extremely rare. A good website about it is http://inyo.coffeecup.com/site/latham/

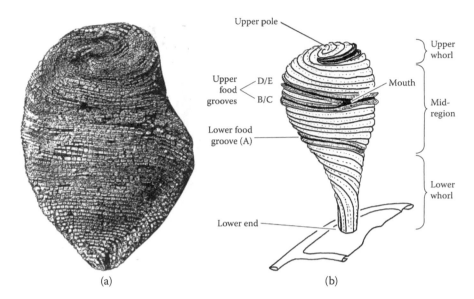

(a)   (b)

FIGURE 20.4 The bizarre spindle-shaped echinoderms called helicoplacoids were common in the Lower Cambrian beds (Poleta Formation) in the White Mountains and found almost nowhere else: (a) One of the complete and articulated (but crushed flat) specimens. (b) Reconstruction of what they probably looked like in life.

Source: From Prothero, D. R., *Bringing Fossils to Life: An Introduction to Paleobiology* (2nd ed.). Columbia University Press, New York, 2013.

latham.html. Another is www.trilobites.info/CA.htm. In addition, the Providence Mountains to the northeast of the Marble Mountains are rich in olenellids, although they are not as accessible as those in the Marble Mountains.

Another famous locality is Emigrant Pass in the Nopah Range near Tecopa. Take Interstate 15 to Baker, California, and exit at Baker and proceed north on CA127 (Death Valley Road) to the junction with Old Spanish Trail. At 77 km (48 mi), turn right on Old Spanish Trail and proceed through Tecopa to Emigrant Pass. The exposure is just to the west of the summit of the pass on the south side of the road, 23 km (14.3 mi) from Tecopa. The GPS coordinates are 35.8856°N, −116.0603°W. A good website is http://inyo.coffeecup.com/site/cf/carfieldtrip.html. Another is http://donaldkenney.x10.mx/SITES/CANOPAH/CANOPAH.HTM.

20.2.3 ORDOVICIAN–DEVONIAN FOSSILS OF MAZOURKA CANYON, INYO MOUNTAINS

Although Paleozoic rocks and fossils are relatively rare in California, there are a number of fossiliferous limestones in the Mojave Desert and White-Inyo Mountains and in Death Valley. Perhaps the best and most accessible sequence is in Mazourka Canyon (Figure 20.6a), in the Inyo Mountains, just southeast of the town of Independence in the Owens Valley. After you pass through the ghost town of Kearsarge on the valley floor, you follow a rough four-wheel-drive track up the main branch of Mazourka Canyon itself. The left (north) branch of the road goes up to Badger Flat, where you can find outcrops of the Lower Ordovician Al Rose Formation (limestone and shale), full of graptolites. The main (east) branch of the road goes up Mazourka Canyon

itself. After you pass the ruins of several old mining operations, where you can park your vehicle, you hike up the narrow canyon until you find numerous resistant limestone ledges full of fossils (Figure 20.6b–d). The main units in the Inyo Mountains are:

- Rest Spring Shale (Mississippian): 800 m of metamorphosed red shale, sandstone, and siltstone
- Perdido Formation (Mississippian): 200 m of reddish sandstone, quartzite, and chert
- Squares Tunnel (Upper Devonian): Argillites and sandstones with no macrofossils
- Sunday Canyon (Silurian–Devonian?): 230 m of graptolitic shales with some coral-rich layers
- Vaughn Gulch (Silurian): 500 m of bioclastic limestones with fairly abundant invertebrates
- Ely Springs Dolostone (Upper Ordovician): 100 m of dolostone and chert
- Johnson Spring Formation (middle Ordovician): 70 m of sparsely fossiliferous limestones interbedded with dolomites and quartzites
- Barrel Springs Formation (middle Ordovician): 50 m of fossiliferous limestones, shales, and quartzites
- Badger Flat Formation (middle Ordovician): 200 m of abundantly fossiliferous gray silty limestone with some quartzite and chert

The first outcrops you encounter as you go up the canyon are the steeply dipping Badger Flat and Barrel Springs Formations, where most of the limestones yield abundant middle Ordovician brachiopods, corals, sponges,

FIGURE 20.5 The earliest known trilobites are the primitive forms known as olenellids, from the third stage (Atdabanian) of the Cambrian: (a) Typical olenellid, showing the distinctive D-shaped head shield with spines on the corner, crescent-shaped eyes, and furrows ending in a round knob down the center axis (glabella). The thorax had many segments (some with spines on their tips), and instead of an advanced fused tail plate, they had separate tail segments and a tail spike. (b) Collection of typical olenellids from the Latham Shale in the Marble Mountains. The earliest known trilobites are the primitive forms known as olenellids, from the third stage (Atdabanian) of the Cambrian. (c) Collecting in the Latham Shale in the southern Marble Mountains. The quarry is shown by the big holes with piles of shale chips beneath them and collectors working away.

Source: (a) Courtesy of Wikimedia Commons. (b–c) Photos by the author.

crinoids, and bryozoans, with a few trilobites and other typical Ordovician fossils. Further information can be found at the website http://donaldkenney.x10.mx/SITES/CAMAZOURKA/CAMAZOURKA.HTM#toc8.

20.3 MESOZOIC FOSSILS

20.3.1 Triassic Ammonoids of the Inyo Mountains

Triassic rocks are generally rare in California, but there is one locality in the west flank of the central Inyo Mountains, east of Lone Pine in the Owens Valley. It is the famous Union Wash section, which exposes a thick sequence of

Upper Permian and Lower Triassic rocks (Figure 20.7a). As early as 1896, the legendary paleontologist Charles Doolittle Walcott discovered and collected these beds. By 1914, scientists were publishing on the limestone beds (Smith, 1905, 1914, 1932). They were full of the typical ceratitic ammonoids of the Early Triassic (about 240 Ma), which have distinctive U-shaped suture patterns (Figure 20.7b and c). The most famous of these was *Meekoceras*, but there are thick limestone beds (now standing as nearly vertical ridges on the south wall of Union Wash) that are full of the smaller ceratitic ammonoid *Parapopenoceras*. Almost a dozen other genera and species have since been named from these rich deposits (Smith, 1905, 1932; see Stone et al., 1991).

FIGURE 20.6 Mazourka Canyon rocks and fossils: (a) Outcrops of middle Ordovician Johnson Spring Formation in upper Mazourka Canyon. (b) Large macluritid snail in a fossil hash of typical Ordovician animals. (c) Orthid brachiopod. (d) Etched horn corals (rugosids) and brachiopods.

Source: Photos by D. Sloan and D. Strauss, available at www.DSComposition.com.

According to Smith (1905, 1932), these included *Ophiceras* (four species), *Owenites* (four species), *Xenodiscus* (four species), *Anasibirites* (three species), *Sturia* (two species), *Lanceolites* (two species), *Clypeoceras* (two species), *Lecanites* (two species), *Inyoites*, *Proptychites*, *Aspenites*, *Flemingites*, *Pseudosageceras*, *Prophingites*, *Danubites*, *Juvenites*, and six additional species of *Meekoceras*.

The site is still accessible by hiking up the canyon after you drive to the end of the dirt road. It is on Bureau of Land Management (BLM) land, so the only collecting allowed now are loose specimens that have fallen from the limestone bed (unless you have a BLM research permit). Further information can be found at http://inyo.coffeecup.com/site/uw/uwfieldtrip.html.

20.3.2 Great Valley Group of the Central Valley

Most of California during the Mesozoic consisted of the Sierra Nevada volcanic arc complex, which is not suitable for fossil preservation. Nor are the rocks of the accretionary prism, which are highly sheared and metamorphosed, good for fossils. But the forearc basin of the Central Valley was an important marine basin, with many places that trapped not only abundant microfossils but also megafossils like ammonites and other mollusks, and in a few places dinosaurs and marine reptiles.

In addition to excellent exposures of deep marine sediments, the Great Valley group is legendary for its fossils.

The ammonites, nautiloids, and belemnites are abundant and well preserved, and their sequence has been well studied and worked out by generations of paleontologists (Figure 20.8). The sandstone beds and shallower deposits are also rich in a variety of mollusks, especially scallops and other clams, such as the huge flat clams known as inoceramids. There are also a number of different kinds of distinctive Cretaceous snails. In addition to ammonites, these shales are rich in both planktonic and benthic foraminiferans, which are also used to create biostratigraphic zonations and help date the strata.

In addition to invertebrates, some localities also produced vertebrates that swam in these Cretaceous seas. The best-known of these are the Moreno Shale and the Panoche Formation in the Panoche Hills and Tumey Hills, in the western San Joaquin Basin margin east of Fresno (Figure 20.9a). In the 1930s, Dr. Samuel P. Welles of the University of California at Berkeley began exploring these deposits, along with Dr. Chester Stock and crews from Caltech. They recovered a rich fauna of Cretaceous marine reptiles, including the giant marine Komodo dragons known as mosasaurs (Figure 20.9b), giant sea turtles, dolphin-like ichthyosaurs, flying pterosaurs, and a spectacular complete skeleton of a long-necked plesiosaur, about 13 m (40 ft) long, known as *Morenosaurus* (Figure 20.9c). It is now on display at the Natural History Museum of Los Angeles. In addition, the best dinosaurs known from California were found in these beds,

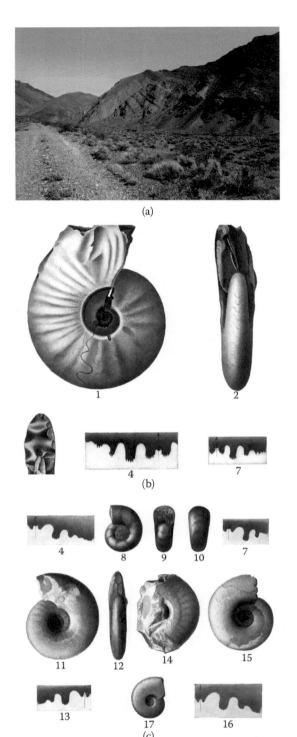

(a)

1 2

(b)

4 7

4 8 9 10 7

11 12 14 15

13 17 16

(c)

FIGURE 20.7 Triassic fossils of Union Wash, Inyo Mountains: (a) Main exposures in Union Wash. The resistant ridges on the right are typically full of ceratitic ammonoids. (b) Large ammonoid *Meekoceras*, showing the corrugated septum (lower left) that produces U-shaped sutures (images 4 and 7). (c) Ammonoid *Ophiceras*, with diagrams showing their diagnostic U-shaped ceratitic suture patterns on the outer shell.

Source: (a) Photo by G. Retallack. (b) From Smith, J. P., *US Geological Survey Professional Paper*, 40: 1–394; 1905; courtesy of US Geological Survey. (c) From Smith, J. P., *US Geological Survey Professional Paper*, 40: 1–394, 1905, courtesy of US Geological Survey.

including partial skeletons of several duck-billed dinosaurs (Figure 20.9d).

20.3.3 CRETACEOUS FOSSILS OF SAN DIEGO COUNTY AND BAJA CALIFORNIA

The Upper Cretaceous Point Loma Formation (72–76 Ma) of the La Jolla Group crops out in several places in San Diego County, including La Jolla and Sunset Cliffs, but also up near Carlsbad and Palomar Airport. It yields a rich fauna of Late Cretaceous ammonites (Figure 20.10a), including not only huge planispiral forms reaching 1 m across but also ammonites that no longer spiral in a plane but have weird uncoiled shell shapes, known as heteromorphs. In certain places, the La Jolla Group also yields terrestrial fossils, especially dinosaurs. In 1987, an 80% complete skeleton of a nodosaur ankylosaur, approximately 4 m (13 ft) in length, was found near Carlsbad (Figure 20.10b and c). In addition, there are jaws of lambeosaurine duck-billed dinosaurs (much like those from the Moreno Hills) found in La Jolla and Sunset Cliffs, discovered during excavation for a housing project near Palomar Airport in 1983 (Figure 20.10d). All these fossils are on display at the San Diego Natural History Museum.

South of the border in Baja California, a huge area of Cretaceous badlands of the Rosario Formation near El Rosario has long produced not only petrified wood but also several kinds of duck-billed dinosaurs (Figure 20.10d) and predatory dinosaurs like albertosaurs, as well as the only Mesozoic mammal fossils west of the Rocky Mountains.

20.3.4 CRETACEOUS FOSSILS OF SILVERADO CANYON, ORANGE COUNTY

The Holz Shale Member of the Ladd Formation (Figure 20.11a) in Silverado Canyon area (Chapter 14) is particularly famous for its rich fossil fauna of deep marine bivalves and gastropods. The collections (Figure 20.11b) are especially rich in the heavily ribbed shells of *Pterotrigonia*, large inoceramid clams, and the high-spired gastropod *Anchura* with its broad pointed flange on its aperture, as well as rare ammonites (Figure 20.11c). More information about the locality, the geology and stratigraphy, the fossils, and their paleoecology can be found in Bottjer et al. (1982).

20.4 CENOZOIC

20.4.1 EOCENE FOSSILS OF THE SAN DIEGO AREA

Paleocene marine fossils (mainly mollusks of the "Martinez stage") have been collected from a handful of localities around the state, such as the type section of the Martinez Stage near Martinez, California, as well as Paleocene units like the San Francisquito Formation, the Pattiway Formation, and the Silverado Formation. However, they are rare and typically poorly preserved and often difficult to extract from hard sandy matrix that is well cemented.

FIGURE 20.8 Selection of Cretaceous cephalopod fossils from the Great Valley Group collected mostly by Peter Rodda, now in the California Academy of Sciences collections. Ammonites include (a) *Acanthoplites gardneri*, (b) *Acanthoplites spathi*, (c) *Ancyloceras ajax*, (d) *Annuloceras summersi*, (e) *Arleticeras lupheri*, (f) *Asthenoceras delicatum*, (g) *Desmoceras colusae*, (h) straight-shelled baculitid *Baculites ovatoides*, (i) Nautiloid *Aturoidea olssoni*, and (j) Belemnite *Acroteuthis watsonensis*.

Source: Courtesy of P. Roopnarine and the California Academy of Sciences.

(i)   (j)

FIGURE 20.8   (Continued)

(a)   (b)

(c)   (d)

FIGURE 20.9   (a) Upper Cretaceous beds of the Moreno Shale and Panoche Formation in the Panoche Hills. (b) Skeleton of the mosa-saur *Plotosaurus* from the Panoche Hills. (c) The plesiosaur *Morenosaurus* from the Moreno Hills. (d) Partial skull of the duck-billed dinosaur *Saurolophus* from the Panoche Hills.

Source: Specimens on display at the Natural History Museum of Los Angeles County. (a) Photo by S. McLeod. (b–d) Photos by the author.

Much more abundant and well preserved are Eocene fossils, especially mollusks from the Capay and Domengine stages up and down the Coast Ranges, from San Diego to Simi Valley to Coalinga (type area of the Domengine and Yokut faunas) to the area near Winters (type area of the Capay fauna). Perhaps the best place to see Eocene fossils is in western San Diego County, where there is a thick Eocene sequence of both marine and nonmarine beds

(Figure 14.11). Not only are there abundant fossils (mainly mollusks) in the marine formations, but the nonmarine units also yield abundant and well-preserved mammal fossils (Walsh, 1996; Walsh et al., 1996). In the past 50 years, the explosion of development of new housing projects and other buildings in San Diego County has created numerous new exposures, which yielded many important fossils to salvage paleontologists, mostly stored in the San Diego

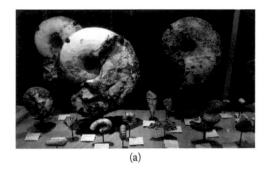

(a)

(c)

(b)

(d)

FIGURE 20.10 (Continued)

FIGURE 20.10 Late Cretaceous fossils of the La Jolla Group in San Diego County, on display at the San Diego Natural History Museum: (a) Spectrum of different kinds of ammonites (top and lower left), as well as rudistid and inoceramid clams (lower right). (b) Reconstruction of the San Diego ankylosaur at life size. (c) Crushed skeleton and body armor of the nearly complete fossil ankylosaur. (d) Lambeosaurine duck-billed dinosaurs like this one have been found in the La Jolla Group, in the Panoche Hills, and El Rosario in Baja California.

Source: Photos by the author.

Natural History Museum. Sadly, the rapid pace of development has meant that most of those exposures have since been covered up and built over. Fossil mammals occur in the middle to upper Eocene formations, the Friars, Mission Valley, and Pomerado Formations (Figure 14.11) in the southern part of the county, and in the Santiago Formation in the north part of San Diego County.

The Eocene mammals of San Diego County (Figure 20.12a) include a wide spectrum of families, very similar to those from the middle Eocene of Utah, Wyoming, and other localities. They range from the big brontotheres (Figure 20.12b) *Metarhinus pater, Duchesneodus uintensis*, and *Parvicornis occidentalis* (rhino-like hoofed mammals with paired bony nasal horns, distantly related to horses) to primitive three-toed horses (*Epihippus*), the primitive rhinocerotoid *Hyrachyus*, hippo-like rhinos *Amynodon* and *Amynodontopsis*, and the first fossil tapir that had a proboscis (*Hesperaletes*) (Figure 20.12c). Among even-toed hoofed mammals (Artiodactyla), they include a number of primitive groups, as well as the very earliest known camel, *Poebrodon*, the protoceratid *Leptoreodon*, and abundant even-toed hoofed mammals known as *Protoreodon* (Figure 20.12a and d). There are rare predators, such as the primitive weasel-like "miacids" *Tapocyon, Procyonodictis*, and *Miocyon*, the creodonts *Limnocyon* and *Apataelurus*, and *Harpagolestes*, one of the wolf-sized hoofed predators known as mesonychids

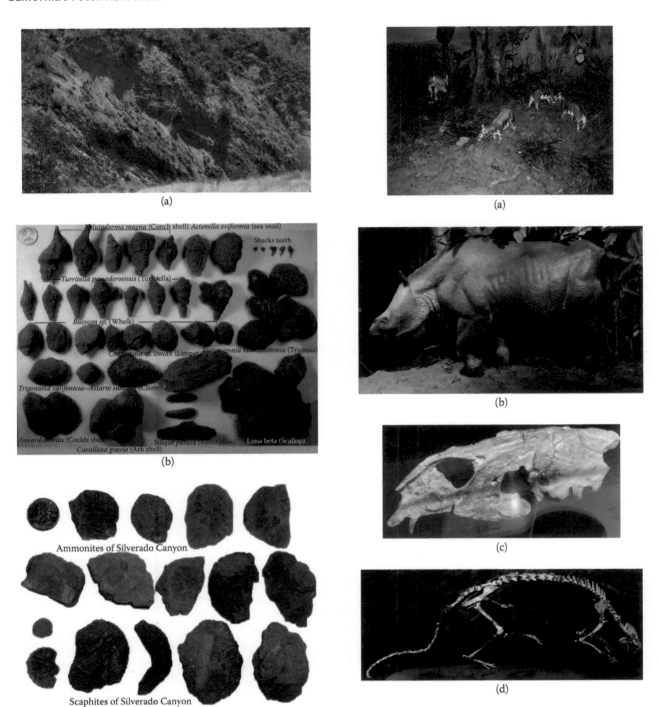

FIGURE 20.11 Upper Cretaceous fossil beds of the Holz Shale Member, Ladd Formation, in Silverado Canyon: (a) Fossiliferous shale exposures in the ravine just west of the narrows. Upper Cretaceous fossil beds of the Holz Shale Member, Ladd Formation, in Silverado Canyon. (b) Selection of typical snails and clams from the Holz Shale. (c) Ammonoids from the Holz Shale.

Source: (a) Photo by the author.

FIGURE 20.12 Eocene fossils of San Diego County: (a) Reconstruction of the middle Eocene forests of San Diego, with the primitive oreodont *Protoreodon* being hunted by the hoofed predator *Harpagolestes*. (b) The largest mammals were the extinct brontotheres, distant relatives of horses and rhinos. (c) The nearly complete skull of *Hesperaletes*, the oldest-known tapir with a proboscis. (d) The skeleton of the most common late middle Eocene mammal, an even-toed hoofed mammal known as *Protoreodon*. (e) Reconstruction of one of the lemur-like primates common in the middle Eocene jungles of San Diego.

Source: Photos of displays at the San Diego Natural History Museum, taken by the author.

(e)

FIGURE 20.12 (Continued)

(Figure 20.12d). Far more abundant, however, are the small mammals, which are known from hundreds of specimens. These include primitive opossums, a variety of archaic insectivorous mammals, lemur-like and tarsier-like primates, the first rabbit relatives in North America, and a huge diversity of primitive rodents.

20.4.2 EOCENE PLANT FOSSILS OF THE SIERRA NEVADA MOUNTAINS AND WARNER RANGE

As discussed in Chapter 8, the Sierra Nevada mountains were blanketed by thick river deposits and some volcanic ashes that flowed down from the Nevadaplano to the east. These deposits were mined extensively for the gold trapped in the gravels of their riverbeds (Chapter 8), but in places they also had fine-grained deposits (especially volcanic ash layers) that preserved amazing fossil leaves. Two of the main localities in the northern Sierras are LaPorte and Chalk Bluffs (Figure 20.13a). Both were extensively cut back by hydraulic mining during the Gold Rush (Chapter 16), but the remaining exposures are still rich in leaf fossils.

At LaPorte, a volcanic ash dated at 34.3 Ma caps the gold-bearing river sediments (Prothero et al., 2011), and this fine ash bed yields extraordinary late Eocene leaf fossils. Potbury (1935) described more than 41 species of plants, including cycads, ferns, conifers, palms, oaks, elms, laurels, *Liquidambar*, holly, legumes, euphorbs, and many other kinds of trees and shrubs (Figure 20.13b). Based on the analysis of the leaves, the mean annual temperature of

LaPorte was about 21°C (70°F), and the annual precipitation was 1,650 mm, which is temperate, even though it was high in the Sierras at that time (see Chapter 8).

At Chalk Bluffs, there is an abundant middle Eocene flora (Figure 20.13c), including palms, pines, firs,

(a)

(b)

(c)

FIGURE 20.13 Eocene plant fossils of the Sierra Nevada mountains: (a) The abandoned hydraulic mining exposures of gold-bearing river sediments at LaPorte capped the white volcanic ash at the top (dated around 34.3 Ma), which yields the latest Oligocene fossil leaves. (b) *Sterculia* leaf from LaPorte. (c) *Macginitea* leaf from Chalk Bluffs.

Source: (a) Photo by the author. (b–c) Photos by D. Erwin and the University of California Museum of Paleontology.

magnolias, laurels, breadfruit, moonseed vines, willows, poplars, legumes, and a wide diversity of other types of plants (MacGinitie, 1941). Many of these plants are characteristic of a subtropical rainforest, with a mean annual temperature of 22°C (72°F) and more than 1,625 mm of annual precipitation. Thus, the middle Eocene Chalk Bluffs flora is representative of the last of the warm tropical conditions of the "greenhouse world" of the Eocene, while LaPorte shows the beginning of the cooling trend of the late Eocene.

Finally, the Warner Range (Figure 20.14a) in the northeast corner of California, just east of Cedarville and the Nevada border, is important for California paleobotany. As discussed in Chapters 6 and 7, the Warner Range is the westernmost part of the Basin and Range Province in northeast California and northwest Nevada, and to the east are the vast volcanic plateaus of the Modoc Plateau. The Warner Range produces a rich flora (Figure 20.14b—d) that also spans the middle–late Eocene global cooling event. It yields several sequential magnetostratigraphically dated floras (Myers, 2003, 2006; Prothero and Upton, 2015), from the 35 Ma Steamboat flora (mean annual temperature = 17.1°C, or 63°F; 2,340 mm of annual precipitation) to the 34 Ma Badger's Nose flora (mean annual temperature = 12.5°C, or 54.5°F; 1,780 mm of precipitation) to the early Oligocene (31.5 Ma) Granger Canyon flora (mean annual temperature = 9.6°C, or 49°F; 2,600 mm of annual precipitation).

20.4.3 OLIGOCENE FOSSILS

Oligocene fossils are rare in California, since most of the Oligocene section is missing due to a major unconformity caused by a global drop in sea level. There are some Oligocene mollusks and foraminifera in the San Joaquin Basin, but they are mostly located on private ranch property, such as Twisselman Ranch in the Temblor Range. The upper part of the Sespe Formation in Ventura and Santa Barbara Counties yields a few late Oligocene mammal fossils, mostly rodents. Perhaps the best Oligocene exposures come from the Otay Mesa area south of San Diego and just north of the Mexican border. In the past 30 years, excavations for housing projects have produced a huge collection of late Oligocene (early Arikareean) mammals, now stored and on display at the San Diego Natural History Museum. They include dozens of specimens of the tiny rabbit-like oreodont *Sespia* (also known from the Sespe Formation and from Nebraska) (Figure 20.15a and b), the sheep-sized oreodont *Mesoreodon* (Figure 20.15a), the primitive camel *Miotylopus* (Figure 20.15a), the primitive dog *Mesocyon*, and a large assemblage of other mammals typical of the late Oligocene in North America.

20.4.4 MIOCENE MOLLUSKS OF TOPANGA CANYON

Many budding young paleontologists first got experience collecting fossils at the legendary Topanga Canyon locality on Old Topanga Canyon Road (Figure 20.16a). That locality

(a)

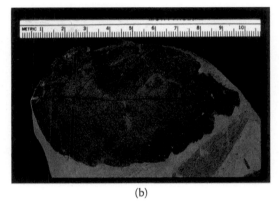

(b)

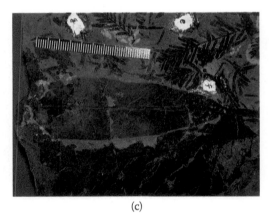

(c)

FIGURE 20.14   (a) Exposures of Eocene–Oligocene rocks in the Warner Range just west of Cedarville. (b–f) Photos of plant fossils from the latest Eocene Badger's Nose flora: (b) alder *Alnus*, (c) laurel and some pine litter, (d) magnolia leaf, (e) cone of the "dawn redwood" *Metasequoia*, and (f) leaf of *Mahonia*, the "Oregon grape."

Source: Source: (a) Photo by the author. (b–f) Photos by J. Myers.

has been collected by so many fossil hunters that the outcrop has now been mined more than 7 m away from the edge of the road into a broad amphitheater, giving it the nickname "Amphitheater Roadcut." Unfortunately, that site is now closed, but there are other Topanga Canyon exposures up and down the road and elsewhere in the Santa Monica Mountains (Chapter 13).

The middle Miocene Topanga Canyon Formation (Cold Creek Member, 16–17 Ma) yields a huge diversity of shallow marine mollusks (Figure 20.16b). By far, the

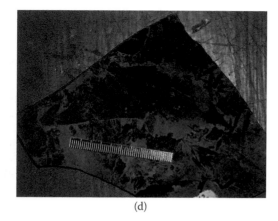

(d)

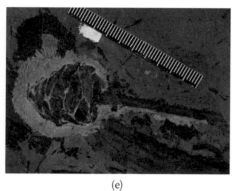

(e)

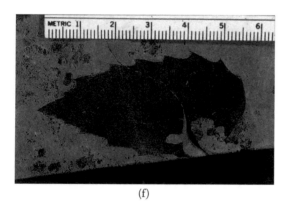

(f)

FIGURE 20.14   (Continued)

most common is the high-spired snail *Turritella ocoyana*, which is represented by thousands of specimens. In certain places, there are beds full of thick oyster (*Crassostrea gigantea*) shells. After these two taxa, the most common fossils are big scallops (*Lyropecten crassicardo*) and clams, such as *Dosinia mathewsoni*, *Spisula abscissa*, and *Tellina arctata*. There are also rare moon snails (*Neverita recluziana*), whelks, cowries, top shells, razor clams, and dozens of other molluscan species. In addition, the Topanga Canyon Formation has yielded rare barnacles, sand dollars (*Scutella morrisi*), fish scales, whale bones (including the oldest fossil of the pygmy baleen whale), shark teeth, and sea lion bones. Downhill around the next bend from the Amphitheater, and higher in the steeply dipping stratigraphic section, are outcrops that

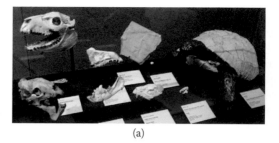

(a)

(b)

FIGURE 20.15   Oligocene fossils of the Otay Formation in Otay Mesa, near the Mexican border, now on display at the San Diego Natural History Museum: (a) Typical fossils from the Otay fauna, including the large oreodont *Mesoreodon* (upper left), the saber-toothed "false cat" *Eusmilus* (lower left corner), the snout of the camel *Miotylopus* (top center), the skull of the primitive dog *Mesocyon* (lower right), a skull of the rodent *Allomys* (lower right corner), and the shell of a tortoise (extreme right). (b) Reconstruction of the most common mammal, the rabbit-like oreodont *Sespia*.

Source: (a) Photo by the author. (b) Illustration by M. P. Williams.

seem to be predominantly oyster beds, suggesting that the normal subtidal marine conditions of the Amphitheater have shallowed into brackish lagoonal deposits. Finally, the highest outcrops of the section are beach and near-shore sands yielding sand dollar fragments, razor clams, and scallops, completing the shallowing-upward trend in these outcrops.

20.4.5   Miocene Fossils of Sharktooth Hill, Kern County

Sharktooth Hill bone bed (Figure 20.17a and b) is a famous locality in the Round Mountain Silt northeast of Bakersfield. It yields about 30 different species of sharks, including the famous gigantic great white shark *Otodus* (formerly *Carcharocles*) *megalodon*, which was the size of a whale (Figure 20.17c and d). *O. megalodon* might have grown so huge because the middle Miocene was a period of diversification of marine mammals—many different dolphins, toothed whales, and baleen whales (more than a dozen species known from Sharktooth Hill bone bed);

(a)

(b)

FIGURE 20.16 Middle Miocene fossils of the Topanga Canyon Formation: (a) "Amphitheater" locality on Old Topanga Canyon Road. (b) Collection of high-spired *Turritella* snails, *Dosinia* clams, and *Crassostrea* oysters (top) from the Topanga Canyon Formation.

Source: Photos by the author.

manatees; and the extinct marine mammals known as desmostylians—and the giant shark preyed on all of them. The bone bed also yields the earliest known fossil seals and sea lions (Figure 20.17e). They evolved from bears in the early Miocene; the oldest known seal is *Enaliarctos mealsi*, which comes from Pyramid Hill nearby. There are even bones from land mammals whose carcasses must have floated out to sea and then sank to the bottom. These include mastodonts, rhinos, horses, camels, tapirs, bear dogs, dogs, weasels, and several other surprising finds. The high concentration of bones has long been a puzzle, but recent research shows that the deep marine basin near Bakersfield was extremely starved for sediment, so that the bone bed represents a lag accumulation in a setting where very little mud accumulates or buries the fossils over decades to centuries.

20.4.6 Barstow Formation, Mojave Desert

To the north of the city of Barstow is the Mud Hills. In these mountains are the classic fossil-bearing beds of the Barstow Formation (Figure 20.18a), famous for the fossil mammals dated between 14 and 17 Ma old that gave the Barstovian Land Mammal Age its name. These rocks have been collected for almost a century and have produced a classic middle Miocene assemblage ranging from mastodonts, horses, camels, pronghorns, peccaries, and rhinos to predators such as primitive dogs and bears, huge bear dogs (Figure 20.18b), and a wide range of rodents. If you drive north out of town to the Rainbow Basin, you can see excellent exposures of the Barstow Formation, all visible from the dirt road (complete with the dated volcanic ashes), and now folded into a gentle trough-like syncline (Figure 20.18a).

20.4.7 Red Rock Canyon State Park, Mojave Desert

Driving on Highway 395 between the Owens Valley and Southern California, you pass through Red Rock Canyon State Park (Figure 20.19a). The highway passes through ledges of sandstones and siltstones of the Dove Spring Formation of the Ricardo Group that form the main outcrops of Red Rock Canyon. These lowest beds yield volcanic ash dates of 11.8 ± 0.9 Ma at the base of Member 3 of the Dove Spring Formation, and the paleomagnetic signature of the rocks (Whistler and Burbank, 1992) shows that they date as far back as 14.0 Ma at the base of the Dove Spring Formation (Figure 20.19b). To the north, the Dove Spring Formation overlies the even older (early Miocene) Cudahy Camp Formation. Partway up the cliffs, there is a distinctive black layer that forms a resistant ridge. This is a basalt lava flow that marks the middle of the Dove Spring Formation (boundary between Members 4 and 5), and it has been radiometrically dated at 10.05 ± 0.25 Ma old, and it is at the base of a long zone of normal polarity correlated to magnetic Chron C5n (9.0–10.5 Ma). Above the basalt ridge are Members 5 and 6 of the Dove Spring Formation, which yield volcanic ash dates of 8.4 ± 1.8 Ma and 8.5 ± 0.13 Ma on rocks correlated with Chron C9 (8.0–8.5 Ma) in the top of Member 5 (Figure 20.19b). The highest rocks in the Dove Spring Formation correlate with Chron C7 (7.0–7.5 Ma). Thus, the combined volcanic ash dates and magnetics tell us that the entire unit spans between 7 and 14 Ma, one of the longest and most complete records of the middle and early late Miocene anywhere in the world.

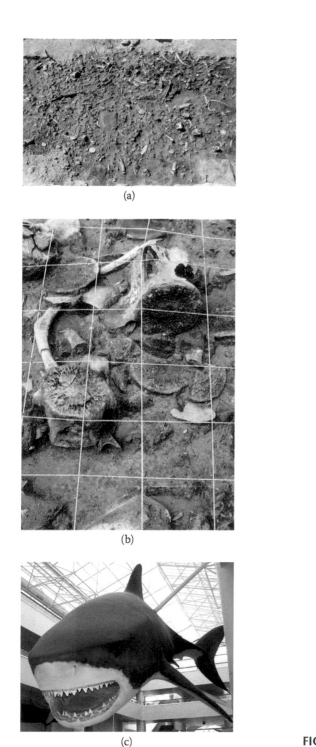

(a)

(b)

(c)

(d)

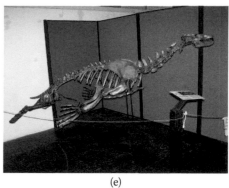

(e)

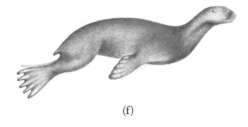

(f)

FIGURE 20.17   (Continued)

FIGURE 20.17 Middle Miocene Sharktooth Hill locality, northeast of Bakersfield: (a, b) Outcrop shots showing the density of teeth and bones in the bone bed, especially of whale vertebrate. (c) Life-sized model of the supershark *Carcharocles megalodon*, the largest shark, at Sharktooth Hill, on display at the San Diego Natural History Museum. (d) Assemblage of typical shark teeth, with *C. megalodon* in the middle and other sharks (especially the common mako shark, *Isurus*) surrounding it. (e) Primitive sea lion relative *Allodesmus*, in the Buena Vista Museum. (f) Restoration of the earliest known pinniped *Enaliarctos*, from the early Miocene of Pyramid Hill, northwest of Sharktooth Hill.

Source: (a–b) Photos by L. Barnes. (d) Photo by R. Irmis. (e) Photo by the author. (f) Drawing by H. Galiano.

Driving into Red Rock Canyon gives you a sense of déjà vu, for indeed you have seen this place before no matter whether you have gone there (Figure 20.19a). Red Rock Canyon is a relatively short distance from the movie studios, so it is also one of the most popular film locations whenever they want to shoot rugged sandstone outcrops. According to www.redrockcanyonmovies.com, it has been in more than 100 films (mostly Westerns and sci-fi) since the 1920s, starting with the 1925 silent film *Wild Horse Canyon* and including many classics, such as *The Mummy*, *Stagecoach*, *Hopalong Cassidy*, *Petrified Forest*, *Zorro Ridges Again*, *Gunsmoke Ranch*, *The Painted Desert*, *The Lone Ranger*

(a)

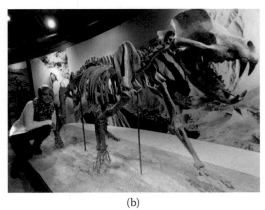

(b)

FIGURE 20.18 Classic middle Miocene Barstow fossil beds, north and west of Barstow, California: (a) Exposures of the Barstow Syncline in Rainbow Basin. (b) Composite skeleton of the huge Barstovian bear dog *Amphicyon ingens*, from the Raymond Alf Museum in Claremont. This specimen is based on fossils from Colorado and Nebraska, but also found at Barstow. Paleontologist Ashley Fragomeni Hall for scale.

Source: (a) Photo by the author. (b) Photo by the author.

Rides Again, *Buck Rogers* (1939), *Flash Gordon Conquers the Universe*, and *Badlands of Dakota*, as well as more recent films, such as *Silverado*, *Poltergeist 2*, *Homer and Eddie*, *Highlander 2*, *Holes*, *I'll Be Home for Christmas*, and *Joshua Tree*. In 1996, it was in the early sequences in the first *Jurassic Park* film, where it is called "Snakewater, Montana" (despite the fact that no place in Montana looks even remotely like Red Rock Canyon).

The Ricardo Group at Red Rock Canyon was first collected over a century ago by crews from the University of California Museum of Paleontology (UCMP) in Berkeley (Merriam, 1919). These rocks and fossils have been studied and collected and mapped by many since, especially Dr. David Whistler, now retired from the Los Angeles County Museum. Red Rock Canyon State Park yields eight species of fossil plants known from beautifully preserved petrified woods, some opalized. It is a typical plant assemblage from the late Miocene, roughly 10 million years old. Its closest modern representatives now live in the upper Sonoran zone of the San Jacinto Range between San Jacinto Peak and Santa Rosa Mountains, Southern California. The

plants include black locust, Mexican pinyon pine, cypress, California live oak, red-root (New Jersey tea), acacia, desert thorn, and palm like the ubiquitous *Washingtonia* palm trees introduced all over Southern California since the 1880s.

A wide variety of vertebrate fossils have been collected. The lake beds yield a small fish (sucker), frogs, toads, three kinds of salamanders, a pond turtle, an extinct goose, an otter, and a beaver. The mammals that once roamed this plains-like habitat include ten species of horse, four kinds of camels, two varieties of rhinos, three pronghorns, a vulture, a pika, two species of ground squirrels, rabbits, deer mice, and two kinds of extinct gomphothere mastodonts. There are also species that might be found in more brushy habitats, such as a peccary, two extinct sheep-like animals called oreodonts, a species of extinct three-toed browsing horse, one ring-tailed cat, a small skunk, a species of short-legged camel, a wolverine, two kinds of weasel-like animals, two varieties of foxes, four different kinds of spiny lizards, a night lizard, a rosy boa, racer snakes, a chipmunk, a hedgehog, two species of gopher-like rodents, two kinds of pocket mice, a bat, and three species of small perching birds. The wetter habitats probably were the homes for two types of alligator lizards, one species of mole, one kind of small rear-fanged snake, and four different types of shrews. In addition, there are large bear dogs (Figure 20.18b), true bears, six different species of dog, and three large cats, including a saber-toothed one.

20.4.8 BLACKHAWK RANCH QUARRY, MT. DIABLO

Blackhawk Ranch Quarry is a legendary locality in the Sycamore Formation on the south slope of Mt. Diablo (see Chapter 12). First discovered in the 1920s, it was extensively excavated by the UC Berkeley during the 1930s, using workers from the Works Progress Administration (WPA) program (Figure 20.20a and b). Ownership of the quarry passed from one rich landowner to the next over the years, but today it is part of the Mt. Diablo State Park. Through all this time, the UCMP and UC Berkeley students and volunteer groups have continued to collect hundreds of fossils from it (all stored at the UCMP).

Based on the large late middle Miocene fauna of mammals, it became the basis for the Montediablan land mammal age. The Blackhawk Ranch Quarry yielded thousands of fossils, including several large mastodonts (*Gomphotherium simpsoni*) (Figure 20.20c and d), three different species of horses, three genera of camels, the recently described peccary *Woodburnehyus grenaderae*, primitive pronghorns, saber-toothed cats, coatimundis, ringtails, bone-crushing borophagine dogs, foxes, ground squirrels, beavers, and rabbits (Figure 20.20d). The plant fossils from the same formation include leaves of poplars, willows, elms, sycamores, oaks, sumac, and mountain mahogany, indicating a much wetter climate than today. Many of these fossils used to be on display in a museum in San Ramon, but that exhibit has since closed and the fossils

(a)

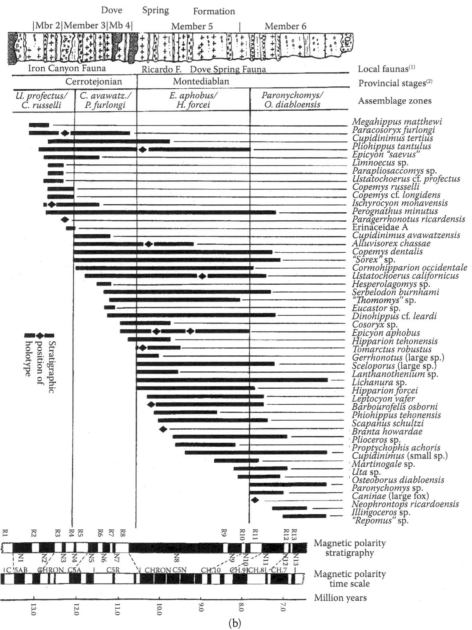

(b)

FIGURE 20.19 Middle–upper Miocene beds at Red Rock Canyon. (a) Classic exposure of the Dove Spring Formation, Ricardo Group, showing river channel sandstones capping floodplain mudstones. (b) Biostratigraphic zonation of the key mammal fossils from Red Rock Canyon.

Source: (a) Photo by the author. (b) Modified from Whistler, D. W., and Burbank. D. P. *Geological Society of America Bulletin*, 104: 644–658, 1992.

FIGURE 20.20 Upper Miocene Blackhawk Ranch Quarry, on the south flank of Mt. Diablo: (a) Archival image of the quarry in 1940. (b) The quarry as it appears today, completely grown over and nearly invisible. (c) The mastodont *Gomphotherium simpsoni*, the largest mammal in the Blackhawk Ranch fauna. (d) Reconstruction of the middle–late Miocene savanna mammals of the western United States, including gomphothere mastodonts (left center), deer-like dromomerycines (left foreground), camels (center distance), and three-toed horses (right).

Source: Images courtesy of P. Holroyd and the University of California Museum of Paleontology.

are back in the collections of the UCMP. Magnetic stratigraphy (Prothero and Tedford, 2000) established the age of this part of the Sycamore Formation at 8–9 Ma.

20.5 PLIOCENE

20.5.1 SAN DIEGO FORMATION

The Pliocene San Diego Formation crops out widely in southern San Diego County and northern Mexico. Almost every time a new exposure is created during construction projects, a rich assemblage of fossil mollusks is found, including many different kinds of clams, cockles, oysters, scallops, marine snails, sea stars, and sea urchins (Figure 20.21a). Even more impressive is the wide range of mammal fossils that have been found there, including many types of baleen whales (Figure 20.21b), such as extinct gray whales, humpback whales, and rorquals like the blue whale. Toothed whales include many types of dolphins, such as the extinct river dolphin *Parapontophoria* with a long narrow beak (Figure 20.21c), and half-beaked whales with the lower jaw sticking out farther than the upper one. There are also pinnipeds, including primitive walruses (Figure 20.21d)

with short tusks (*Valenictis* and *Dusignathus*), and relatives of seals and sea lions. Finally, the San Diego Formation yields nearly complete skeletons of *Hydrodamalis cuestae*, an extinct relative of the whale-sized Steller's sea cow (Figure 20.21e), a species living in the Gulf of Alaska that was discovered by Vitus Bering's expedition in 1741 and wiped out by Russian fur traders only 27 years after it was discovered.

20.5.2 ANZA-BORREGO STATE PARK

The badlands sequence (Figure 20.22a) in the Anza-Borrego has been well dated by volcanic ash dates and magnetic stratigraphy, as discussed in Chapter 14. It is an important section for calibrating when different kinds of mammals first appeared in North America. This includes the first appearance of South American mammals, such as ground sloths and porcupines, which migrated up the Panama land bridge about 3.5 Ma; the last appearance of the archaic horse *Pliohippus* and the first appearance of true modern horses (*Equus*); the first appearance of true deer from the Old World; and the first appearance of Eurasian mammals, like the musk ox *Euceratherium*; *Mammuthus imperator*,

(a)

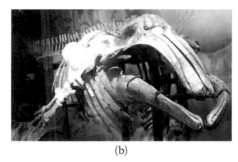

(b)

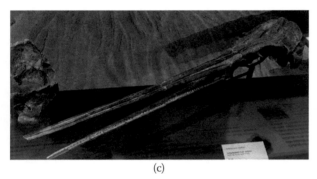

(c)

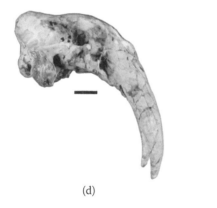

(d)

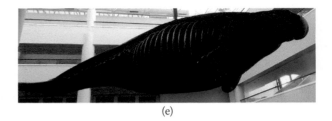

(e)

FIGURE 20.21 (Continued)

FIGURE 20.21 Typical fossils of the marine Pliocene San Diego Formation, as displayed in the San Diego Museum of Natural History: (a) Marine shells, including clams (left); sand dollars, high-spired *Turritella* snails, cockle shells, and moon snails (center); mussels (upper right); and scallops, sea stars, and abalones (right front). (b) Complete fossil skeleton of an ancestral gray whale *Eschrichtius*. (c) Skull of the extinct long-beaked river dolphin, *Pontophoria*. (d) Skull of the primitive short-tusked walrus *Valenictis*. (e) Skeleton and restoration of the extinct relative of Steller's sea cow, *Hydrodamalis*.

Source: Photos by the author.

the largest known mammoth; *Tapirus*, an extinct tapir; *Equus enormis* and *Equus scotti*, two species of extinct Pleistocene horse; *Gigantocamelus*, a giant camel; and *Capromeryx*, the dwarf pronghorn.

The bestiary for this savannah landscape reads like a "who's who" for some of the most bizarre creatures to inhabit North America. These include animals such as *Geochelone*, a giant bathtub-sized tortoise; *Aiolornis incredibilis*, the largest flying bird of the Northern Hemisphere, with a 5.2 m (17 ft) wingspan (a giant relative of the vulture-like *Teratornis* from La Brea tar pits); *Paramylodon*, *Megalonyx*, and *Nothrotheriops*, giant ground sloths, some with bony armor within their skin; *Pewelagus*, a peewee-sized rabbit (paleontologists can name with a sense of humor); *Borophagus*, a hyena-like dog with bone-crushing teeth; *Arctodus*, a giant short-faced bear; *Smilodon*, a saber-toothed cat; and *Miracinonyx*, the North American cheetah.

Beneath the mammal-bearing Palm Springs Group in the Anza-Borrego is the marine Imperial Group (Latrania and DeGuynos Formations). It formed in the early Pliocene during the early opening of the Gulf of California, when marine seawater rushed into the rift and flooded it with a shallow sand marine seaway. The Imperial Group produces a rich fauna of marine fossils, especially scallops, oysters, and many other kinds of clams; many kinds of marine snails, such as conchs and *Turritella*; sea biscuits and sand dollars; and other typical Pliocene marine organisms (Figure 20.22d).

20.5.3 Calistoga Petrified Forest

Pliocene and Pleistocene plant fossils are found in many parts of the state, but the most spectacular is the Calistoga Petrified Forest in the Napa Valley of Sonoma County, near Santa Rosa and Calistoga Hot Springs. The Petrified Forest is a privately owned tourist facility, but it is famous for one of the best Pliocene floras in the state, and one of the most completely preserved petrified forests anywhere, with the largest fossil trees ever found.

First discovered by the pioneering Yale paleontologist O. C. Marsh in 1870, it became famous shortly thereafter. Robert Louis Stevenson (author of *Treasure Island* and *Kidnapped*) mentioned it in his 1883 book *The Silverado Squatters*, written during his years of living in California. It was long owned by the Calistoga family, who also

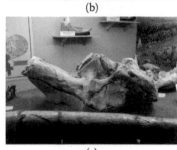

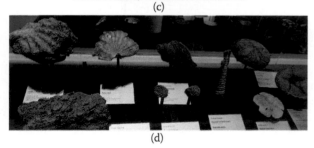

FIGURE 20.22 Fossils of the Anza-Borrego area: (a) Panorama of the Anza-Borrego badlands. (b) Giant tortoise shell from the Anza-Borrego Visitors' Center. (c) Upside-down mammoth skull (center), tusk (foreground), and extinct horse skull and jaws (background) on display at Anza-Borrego Visitors' Center. (d) Selection of Imperial Valley Formation marine fossils on display at the San Diego Natural History Museum, including oysters (left), scallops (left center), marine snails including high-spired *Turritella* (right center), and sea biscuits and sand dollars (right).

Source: (a) Photo by D. Patton. (b–c) Photos by C. Croulet. (d) Photo by the author.

developed the nearby Calistoga Hot Springs (still a popular resort today). The property was sold in 1912 to Ollie Orre Bockee, who made it into a popular tourist attraction, and it remains so under its current owners, members of the Bockee family.

The Calistoga Petrified Forest is legendary for the huge fallen logs and stumps of enormous *Sequoia langsdorfii* trees, similar to the modern coast redwoods (Figure 20.23). There is also one petrified pine tree and excellent pollen fossils. They are evidence that the flora dominated by coast redwoods once lived much farther inland than it does now, possibly because cooler and wetter conditions occurred in Sonoma County during the Pliocene. Many of the trees have nicknames, such as the "Giant" (Figure 20.23a), which is 9.7 m (60 ft) long and 2 m (6 ft) wide; the "Queen," which is 10 m (65 ft) long and 2.2 m (8 ft) wide; and the "Pit Tree" (the only pine in the flora), as well as trees named for Ollie Bockee and Robert Louis Stevenson. The trees were fossilized because they were buried in volcanic ash from an old eruption of Mt. Saint Helens about 3.4 Ma. The silica in the ash permeated through the fallen tree trunks, preserving them in great detail, including even their tree rings. The Queen, for example, has 2,000 years of tree rings visible in its trunk.

20.6 PLEISTOCENE

California has many places with Ice Age mammals excavated from shallow gravel pits and dry lake beds, from the mammoth in Tecopa dry lake to the occasional finding of a fossil mammoth or bison or horse that occurs nearly every year somewhere in the state. Many of the local natural history museums often have Ice Age mammals on display. Some, like the "Eagle Rock mammoth," was found during the excavations for the Sparkletts water plant in Eagle Rock, proof that this natural spring was operating during the Ice Age.

Some of the most spectacular discoveries include the Fairmead Landfill site in Madera County, which yielded thousands of early Pleistocene (Irvingtonian) fossils from 67 different species, including Columbian mammoths, horses, camels, ground sloths, saber-toothed cats, dire wolves, and many species of reptiles, amphibians, fish, and even diatoms. These are now on display at the Fossil Discovery Center of Madera County.

Excavations for the Diamond Valley Reservoir (nicknamed "Valley of the Mastodonts") near Hemet produced a large Ice Age fauna (from 13,000 years old to at least 60,000 years) that is now on display at the Western Science Center near Hemet. These include mammoths, mastodonts, horses, camels, long-horned bison, peccaries, pronghorns, several kinds of ground sloths, saber-toothed cats, dire wolves, American lions, short-faced bears, coyotes, rodents, rabbits, and turtles.

California is also unusual in having natural tar seeps that trapped Ice Age mammals in abundance. The tar pits at McKittrick in the western San Joaquin Valley north

(a)

(b)

FIGURE 20.23 Fossil *Sequoia* logs of Calistoga Petrified Forest: (a) The tallest log in the park, known as the "Queen." (b) The log known as the "Giant."

Source: Photos courtesy of Wikimedia Commons.

of Taft (Chapter 10) yielded many Ice Age fossils to UC Berkeley paleontologists in the early twentieth century, which ended up at the UCMP in Berkeley. Smaller tar seeps near Carpinteria just east of Santa Barbara also yielded a significant Ice Age fauna. Unfortunately, neither the fossils nor the tar pits at McKittrick or Carpinteria are on display and open for public viewing.

20.6.1 Rancho La Brea

By far, the most famous of the tar pits are those at Rancho La Brea in Los Angeles, on the "Miracle Mile" in the Wilshire District. The tar was first discovered and used by Chumash and Tongva peoples to seal the cracks in their log canoes

and mentioned by the Portola Expedition in 1769. It became a Spanish land grant known as Rancho La Brea. The tar was used by the Spaniards to lubricate the wooden wheels and axles of their wagons and to seal their roofs, as did the later Mexicans and, finally, the Americans who lived in the region after the Spaniards lost control of California. In the late nineteenth century, the area was extensively drilled for oil because it lies along the 6th Street Fault and the Salt Lake Oil District (see Chapter 17), which underlies much of the Fairfax District of Los Angeles (Figure 20.24a). But in 1901, Union Oil geologist W. W. Orcutt realized that the bones coming out of the tar pits were not simply modern cattle, horses, and pronghorns that had gotten trapped but extinct Ice Age mammals.

During the period from 1913 to 1915, crews from the UC Berkeley, Occidental College, the California Academy of Sciences, and especially the Los Angeles County Museum conducted extensive excavations in the tar pits, opening more than 100 named and numbered pits altogether. The early collections were done in great haste, recovering only the best fossils, with no effort to record the details of how the fossils were found, or at what level in each pit. This was typical in the early days of paleontology. The millions of fossils they recovered took years to clean (by dipping the fossils in solvents like turpentine), identify, curate, and store and were not fully studied for decades. Some of them were finally put on display at the Los Angeles County Museum in the 1940s and 1950s. Meanwhile, the old oil fields in the area were eventually shut down, and the land donated to the city by the landowner, G. Allan Hancock. Today, Hancock Park hosts not only the tar pits and the museum containing their fossils but also the Los Angeles County Museum of Art and several other museums along the Miracle Mile.

In 1969–1974, a newer generation of paleontologists decided to conduct a modern excavation of one of the pits not previously excavated (Pit 91), using careful methods to record the exact location and position of every fossil as it was found and recovering all the matrix around the bones to screen-wash and sort for microfossils (something not even dreamed of in 1913). Pit 91 remains open for visitors to look at the excavations (Figure 20.24b), and every summer a small crew continues to dig there.

By the late 1960s, it was becoming apparent that there was not enough room in the old museum downtown to store and display the old fossils and all the new ones from Pit 91, so millionaire George C. Page (who made his fortune by selling mail-order Christmas fruit baskets under the name "Mission Pak") endowed a modern museum at the site (now part of Hancock Park) to store the La Brea collections. There they are on display with state-of-the-art exhibits, including holographic saber-toothed cats and a demonstration of how hard it is to pull yourself out of tar. Excavations for the La Brea Tar Pits Museum (Figure 20.24c) started in 1975 and immediately hit another bone deposit, which had to be salvaged. Originally named the Page Museum of La Brea Discoveries, it is now called the La Brea Tar Pits

Museum, and it has been the main facility for La Brea fossils ever since it opened in 1977.

In 2006, another excavation to the west of Hancock Park, for a new parking structure for the Los Angeles County Museum of Art, hit 16 new bone deposits. Known as Project 23, the fossils were salvaged by encasing each bone deposit in 23 huge boxes normally used to enclose the roots of a large tree. In 2008, a crane and a flatbed truck lifted them and carried them to a fenced-in yard just north of Pit 91. There scientists can excavate each tree box slowly, carefully, and properly and keep detailed records without hurrying under the pressure of construction crews

waiting to do their jobs. The bulldozer that discovered the first bones initially scraped away the top of a nearly complete mammoth skeleton, nicknamed "Zed." It is the only articulated skeleton in the tar pits, since nearly all the other bones were disassociated by the churning tar. After 18 years, most of the work on the Project 23 boxes has been completed, and that effort is winding down. However, there are so many new fossils from Project 23 that there is not enough room in the La Brea Tar Pits Museum to store them all.

New bone deposits are found nearly every time someone digs in the area, including the 2014 discovery of fossils

(a) (b)

(c) (d)

FIGURE 20.24 The world-famous tar pits at Rancho La Brea, and their incredible Ice Age fossils: (a) Archival image of the area in the early twentieth century, when the area was mostly used for excavating asphalt and drilling for oil. (b) Pit 91, re-excavated starting in 1969 and carefully gridded off so that every bit of information about each fossil is collected. (c) The "Lake Pit" on the south side of Hancock Park. It was originally just a pit for excavating asphalt but has since filled in with water. Life-sized models of Imperial mammoths struggle for traction in the tar. Behind them is the La Brea Tar Pits Museum, shaped like a Mesoamerican pyramid. (d) Display of more than 400 dire wolf skulls, showing the incredibly large sample sizes of almost all the mammals and birds found at Rancho La Brea. (e) Imperial mammoth, largest mammal found in the tar pits. (f) Diorama of life in Rancho La Brea during the ice ages. Large ground sloths (*Paramylodon*) wander down to the tar pits for a drink, stalked by saber-toothed cats and vultures. (g) The fishbowl room, where museum staff and volunteers clean and repair fossils (left), sort the washed matrix for microfossils (right foreground), and identify and catalogue the collections, all in view of the public. The large mammoth bones in the background are from Zed, the complete articulated mammoth found during the initial excavation of Project 23. (h) The collections storage area for the La Brea Tar Pits Museum runs in pairs of corridors concealed beneath the sloping bank on all four sides of the museum. It is out of view of the public and only open to qualified researchers. Each pair of corridors is lined with metal racks holding thousands of plastic trays, storing hundreds of thousands of carefully curated fossils, all organized by species, which bony element is preserved, and what pit they came from. This is paleontologist Kristina Raymond Smith measuring sloth bones for a research study. (i) Diorama showing an animated robotic saber-toothed cat attacking a ground sloth, with Imperial mammoths, horses, camels, and bison shown in the background. (j) Juvenile fossils are often poorly preserved because their bones are thinner and more delicate. Here, an adult and calf American mastodont are on display, and juveniles of many other species are known as well. (k) Exhibit of just a few of the many diseased and pathological specimens found in the tar pits, which tended to be fossilized more often at Rancho La Brea than at any other locality.

Source: All photos by the author, except (a) and (f), which are courtesy of Wikimedia Commons.

FIGURE 20.24 (Continued)

during the excavation of the Metro Purple Line tunnel. With all the new specimens, the La Brea Tar Pits Museum no longer has the space to store all the La Brea fossils, and planning for additional storage is still underway.

Over the years, more than 3 million fossils representing more than 660 species have been recovered from the Rancho La Brea area. They range in age from 46,000 years old (late Pleistocene) to 8,000 years old (early Holocene), spanning the entire last glacial–interglacial cycle. More than 100 radiocarbon dates have been done to establish the ages of each important pit and their fossils. Based on more than 3 million fossils, Rancho La Brea yields a huge diversity of life of the late Pleistocene in the region, including 55 species of mammals, 133 species of birds, abundant remains of reptiles, amphibians, fish, beautifully preserved insects (the beetle wing covers are still iridescent), mollusks, and plants ranging from large tree stumps and logs down to dozens of species of pollen.

Of the mammal fossils, the most abundant are predators and scavengers, which make up 90% of the mammals. These include the dire wolf (*Aenocyon dirus*, formerly *Canis dirus*), with more than 4,000 fossils (Figure 20.24d); the saber-toothed cat (*Smilodon fatalis*), with more than 2,000 fossils; the immense short-faced bear (*Arctodus simus*); and the Ice Age lion or jaguar (*Panthera atrox*)—plus smaller carnivorans (coyotes, foxes, weasels, skunks, badgers, raccoons, ringtails, cougars, bobcats, lynxes, and black bears). The prey species were less abundant, but they included the imperial mammoth (Figure 20.24e) and American mastodon, three species of ground sloth (*Paramylodon harlani*, *Megalonyx jeffersonii*, and *Nothrotheriops shastensis*), horses and tapirs, bison, camels and llamas, peccaries, pronghorns, deer, and a number of smaller mammals, including many species of rodents, rabbits, shrews, and bats. Last but not the least, a Paleoindian skeleton nicknamed "La Brea Woman" was found in a Holocene pit dated to around 9,000 years ago, apparently murdered and buried in the tar pit.

The overrepresentation of predators and scavengers is unusual, because in normal ecological settings, prey must be about ten times more abundant than their predators to support their populations. This suggests that the tar pits were a predator death trap. A prey species would walk down to the edge of the tar seep (Figure 20.24f), which might be covered with a film of water or dry leaves disguising the tar beneath. Once it had stepped into the sticky tar deep enough, it could not escape, and its panicked cries would attract predators and scavengers looking for an easy meal. After leaping on the dying prey animal, they, too, would get entrapped, and soon their carcasses would all disappear beneath the surface of the tar pit, resetting the trap for another victim.

The bird fauna of Rancho La Brea is extraordinary, since it yields more than 133 species, nearly all based on complete and unbroken bones (although disarticulated and jumbled together). Normally, fossil bird bones are delicate and rarely preserved completely, so Rancho La Brea gives us some of the best records of birds in the fossil record. Once again, large predators and scavengers dominate. The most abundant bird by far is the golden eagle, followed by the bald eagle, the caracara, the huge vulture-like *Teratornis*, condors, turkey vultures, black vultures, owls, and many other birds of prey and scavengers. The second most abundant bird is the La Brea turkey, and there are almost a hundred species of ducks and geese, goatsuckers, storks, plovers, doves and pigeons, cuckoos, quail, cranes, and perching birds (Passeriformes).

Research at Rancho La Brea is ongoing. Out in Hancock Park, visitors can watch the summer work at Pit 91, or the continuous work in the tree boxes of Project 23. At the La Brea Tar Pits Museum, the large "fishbowl room" (Figure 20.24g) allows visitors to watch the staff prepare, sort, and identify the fossils through glass windows. Not visible to the public is the immense fossil storage area (Figure 20.24h), which wraps completely around the entire perimeter of the La Brea Tar Pits Museum, concealed beneath the sloping sides of the "Mayan pyramid" design of the building. There, the remaining millions of specimens are carefully laid out in large plastic sliding drawers on metal racks, available for study by qualified researchers.

All sorts of research is constantly being conducted with these collections, since the size and the quality and quantity of the collections allow research that is impossible in any other collection. For example, it is possible to examine the evolutionary response of the birds and mammals through the climatic changes of the last ice age, and surprisingly, there is almost no change in the size or robustness of the common mammals or birds (Prothero et al., 2012). Pollen studies have been conducted to determine the changes in vegetation, showing that during the peak glacial about 20,000 years ago, Hollywood was much cooler and wetter and covered by pine trees and snow, much like the tops of the San Gabriel Mountains to the north. Many researchers look at the functional morphology of the impressive specimens, especially trying to study how predators like saber-toothed cats and dire wolves once fed and killed their prey (Figure 20.24i). Rancho La Brea is a particularly good place to study diseased and injured specimens, since many sick animals were trapped there when they became desperate for food (Figure 20.24k). The La Brea Tar Pits Museum has one of the best collections of pathological fossils in the world. It also preserves delicate bones that are usually missing in most fossil skeletons, so the opportunities are endless.

RESOURCES

Bottjer, D.J., Colburn, I.P., and Cooper, J.D. (eds.). 1982. Late cretaceous depositional environments and paleogeography, Santa Ana Mountains, Southern California. *Pacific Section, SEPM*, 129.

Hagadorn, J.W., Fedo, C.W., and Waggoner, B.M. 2000. Lower Cambrian Ediacaran fossils from the Great Basin, U.S.A. *Journal of Paleontology*, 74, 731–740.

Hagadorn, J.W., and Waggoner, B. 2000. Ediacaran fossils from the southwestern Great Basin, United States. *Journal of Paleontology, 74*, 349–359.

MacGinitie, H. 1941. A middle Eocene flora from the central Sierra Nevada. *Carnegie Institution of Washington, Publication, 534*, 1–178.

Merriam, J.C. 1919. Tertiary mammalian faunas of the Mohave Desert. *University of California Publications, Department of Geology, Bulletin, 11*, 437–585.

Myers, J.A. 2003. Terrestrial Eocene-Oligocene vegetation and climate in the Pacific Northwest, 171–188. In Prothero, D.R., Ivany, L.C., and Nesbitt, E.A. (eds.), *From Greenhouse to Icehouse: The Marine Eocene-Oligocene Transition.* Columbia University Press, New York.

Myers, J.A. 2006. The latest Eocene Badger's Nose flora of the Warner Mountains, northeast California: The "in between" flora. *PaleoBios, 26*(1), 11–29.

Potbury, S. 1935. The La Porte flora of Plumas County, California. *Carnegie Institute of Washington, Publication, 465*, 29–81.

Prothero, D.R. 2013. *Bringing Fossils to Life: An Introduction to Paleobiology* (2nd ed.). Columbia University Press, New York.

Prothero, D.R., Syverson, V., Raymond, K.R., Madan, M.A., Fragomeni, A., Molina, S., Sutyagina, A., DeSantis, S., and Gage, G.L. 2012. Stasis in the face of climatic change in late Pleistocene mammals and birds from Rancho La Brea, California. *Quaternary Science Reviews, 56*, 1–10.

Prothero, D.R., and Tedford, R.H. 2000. Magnetic stratigraphy of the type Montediablan Stage (Late Miocene), Black Hawk Ranch, Contra Costa County, California: Implications for regional correlations. *Paleobios, 20*(3), 1–10.

Prothero, D.R., Thompson, A., and DeSantis, S. 2011. Magnetic stratigraphy of the late Eocene La Porte flora, northern Sierras, California. *New Mexico Museum of Natural History Bulletin, 53*, 629–635.

Prothero, D.R., and Upton, E. 2015. Magnetic stratigraphy of Eocene-Oligocene fossil plant localities, Warner Range, northeastern California. *New Mexico Museum of Natural History Bulletin, 65*, 265–272.

Smith, J.P. 1905. The Triassic cephalopod genera of North America. *U.S. Geological Survey Professional Paper, 40*, 1–394.

Smith, J.P. 1914. The middle Triassic marine invertebrate fauna of North America. *U.S. Geological Survey Professional Papers, 83*, 1–254.

Smith, J.P. 1932. Lower Triassic ammonoids of North America. *U.S. Geological Survey Professional Paper, 176*, 199.

Stone, P., Stevens, C.H., and Orchard, M.J. 1991, Stratigraphy of the lower and middle(?) Triassic Union wash formation, east-central California. *U.S. Geologic Survey Bulletin, 1928*, 1–26.

Walsh, S.A. 1996. Middle Eocene mammalian faunas of San Diego County, California, 75–119. In Prothero, D.R., and Emry, R.J. (eds.), *The Terrestrial Eocene-Oligocene Transition in North America.* Cambridge University Press, Cambridge.

Walsh, S.A., Prothero, D.R., and Lundquist, D. 1996. Stratigraphy and paleomagnetic correlation of middle Eocene Friars Formation and Poway Group in southwestern San Diego County, California, 120–154. In Prothero, D.R., and Emry, R.J. (eds.), *The Terrestrial Eocene-Oligocene Transition in North America.* Cambridge University Press, Cambridge.

Whistler, D.W., and Burbank, D.P. 1992. Miocene biostratigraphy and biochronology of the Dove Spring Formation, Mojave Desert, California, and the characterization of the Clarendonian land mammal age (late Miocene) in California. *Geological Society of America Bulletin, 104*, 644–658.

21 California's Slippery Slopes

It was just one big black thing coming at us, rolling, with a lot of water in front of it, pushing the water, this big black thing. It was just one big black hill coming toward us.

—Jackie Genofile

California tumbles into the sea.

—Steely Dan, "My Old School"

21.1 LANDSLIDES: GRAVITY ALWAYS WINS

The winter of 2004–2005 was a wet one in Southern California. Many places in the steep mountains behind the urban belt had flooded and experienced landslides. Huge amounts of rain had fallen in the last weeks of December and first weeks of January. In the sleepy coastal town of La Conchita, there was no reason to think that this winter rainy season was unlike any other. La Conchita consisted of just a handful of houses with about 300 residents, located right on the coast on Highway 101 between the rich communities of Santa Barbara and Ventura. However, it was much more laid-back and inexpensive, with small beach cottages inhabited mostly by a lot of retired surfers, artists, beachcombers, and hippies who savored their pleasant beachfront life without the high prices and congestion of cities like Santa Barbara.

It was also sited just north of the freeway and railroad tracks, at the base of a steep cliff made of loose sandstones and mudstones once deposited in ancient seas (Figure 21.1a–d). The ancient sedimentary rocks were then uplifted by faults to heights over 150 m (500 ft) above sea level. These cliffs had proven to be prone to landslides all up and down the coast, from the rich movie-star colony of Malibu to most of the sea cliffs west of Ventura. They were formed of softer sedimentary rocks, with lots of clays that soaked up water, expanded, and became slippery when they were saturated. The bluffs above the town had many landslide scars, showing a long history of instability. In March 1995, part of hillside had given way, covering up the houses on the street right up against the base of the cliff (Figure 21.1d). The 1995 slide measured 120 m (400 ft) wide, 330 m (1,100 ft) long, and spread across 4 ha (10 ac). It was greater than 30 m (100 ft) deep, with a volume estimated at 1.3 million m³ (1.7 million yd³). The year 1995 had also been a wet year, with 390 mm

(15 in) of rain in the two weeks prior to the slide. No one was killed in the event, because it moved relatively slowly and people were warned to get out of the way, but a number of houses in the street at the base of the hillside were destroyed. After it was over, they built large retaining walls at the base of the slide debris in hopes of stopping further movement (Figure 21.1e).

Still, the residents of La Conchita were pretty mellow during the rains of 2004–2005 and did not expect this year to be different from any other rainy winter. Early on the morning of January 10, 2005, small mudflows started moving down the nearby canyons, so they closed Highway 101, and emergency officials and TV crews were there to monitor them. Then, at 12:30 in the afternoon, the cliff suddenly gave way, as the TV news cameras were rolling. (Video footage of this event is online and amazing to watch.) A mass of earth 350 m (1,150 ft) long and 80–100 m (260–330 ft) wide remobilized from the unstable 1995 landslide, quickly moved downslope at 10 m/sec (33 ft/sec), faster than anyone could outrun it. Before people in the houses below could react, it had overrun almost half the town, burying and destroying 13 houses and damaging 23 others, and burying dozens of people (Figure 21.1a–c). Many of the houses were pushed forward as if by a mighty bulldozer, then torn apart before being buried. The retaining wall was overrun and pushed aside (Figure 21.1e). Emergency workers and townspeople rushed to the aid of people trapped under their smashed houses and rescued quite a few (Figure 21.1f). Still, some houses were so deeply buried and crushed that there was no way to dig down without the rescuers endangering their own lives, and there were no sounds or other signs of life. The rescuers eventually gave up, and these ten bodies remain buried in the slide mass, with memorials attached to the chain-link fence enclosing the site where they vanished (Figure 21.1g).

A number of descriptions of the event were given by many witnesses. This story from the *Los Angeles Times* is typical:

As the rain kept coming and mud started to flow, Greg Ray and Tony Alvis decided to lend a hand, helping move a friend out of a Quonset hut at the base of the hillside in La Conchita. Together, they had only been at it about 15 minutes when neighbor John Morgan sounded the first warning. "The mountain is coming down!" he shrieked. Standing directly underneath the tumbling hillside, hearing its terrible crackle and roar and watching a plume of earth spew toward the sky, the men broke and ran for their

DOI: 10.1201/9781003301837-21

FIGURE 21.1 The La Conchita landslide of 2005: (a) View of the slide and town from the south, along Highway 101. (b) Aerial view of the landslide. (c) View from the cliff at the top of the slide, looking down the slide. (d) Diagram showing the geology of La Conchita and the footprint of the 1995 slide and 2005 slide. (e) The retaining wall built at the foot of the 1995 slide, which was overwhelmed by the 2005 slide. (f) View standing on the slide, looking at houses at the toe of the slide destroyed by the 2005 movement. (g) Today, the slide debris is surrounded by a chain-link fence, and numerous memorials of flowers, teddy bears, and messages are often attached to the fence for the dead that were buried in the slide.

Source: Courtesy Wikimedia Commons.

(f)

(g)

FIGURE 21.1 (Continued)

lives. With the hillside exploding, shooting dirt and boulders 100 feet high, Ray sprinted down Vista del Rincon Drive, then wheeled right, onto Santa Barbara Avenue, making a beeline for the ocean. He lost sight of the others as the hillside bore down. Out of the corner of one eye, he could see a house and a trailer in hot pursuit. Instinctively, he ducked for cover, throwing himself between two cars and wedging into the smallest of lifesaving cracks, just as the mud and debris washed over him. "I just turned and dove underneath these cars, that's what saved my life," said Ray from the county hospital in Ventura, with scratches and bruises and a gaping wound where a splintered 2-by-4 had speared his right leg.

His friends weren't so lucky. Alvis and Morgan were unable to outrun the slide, succumbing to the wall of mud and debris. Ray said he learned Tuesday that the men were found dead on each side of him. "It was so horrific," he said. "It all happened so fast."

It took three hours to rescue Ray, an artist who has lived in La Conchita on and off for a decade. Badly hurt and thrown into pitch-black darkness, he could hear people crying and screaming all around him. He reached down to his injured leg, thinking a piece of wood was laying across it, only to find the stake protruding through the limb. He could turn his head a little and move his hands. While trapped, he took a picture of himself with a disposable camera he was carrying in his shirt pocket to record the moment. The weight of the cars, now under the trailer that had given chase, pressed down on him and made it hard to breathe. Two friends from the community found him within minutes and alerted rescue crews, who cut him out of the wreckage. "It was scary. I didn't really know whether I was going to make it," Ray said. "I told myself, 'I just have to bear the pain. I have to make it. I have to survive.'"

He learned after his surgery Monday that Alvis, a fellow artist and quasi-business partner, didn't make it. And the sorrow deepened when told Tuesday afternoon that Morgan also had perished. "He was the one who saved my life," Ray said of Morgan, a groundskeeper for Naval Base Ventura County in Port Hueneme. "He was a real good guy." A fourth person, named Kyle, the man Ray and Alvis had been helping, made it out alive. Set to be released from the hospital in a few days, Ray said he's eager to return to La Conchita despite the devastation. His home was not damaged. And the painter and woodcarver figures he's got plenty of work to do, especially on behalf of his good friend Alvis. Ray said Alvis would have wanted the community to push forward.

"I don't really have any fear of La Conchita. It's done everything it can to me," he said. "It's a really nice place with a lot of good people."

—Los Angeles Times, **January 12, 2005**

In the aftermath of these events, people began to rebuild their lives and mourn their dead. Naturally, they tried to blame it on someone else, rather than Mother Nature and their own foolishness for living there. Some filed lawsuits blaming it on irrigation of the croplands above the cliff, but their suits were thrown out of court since the geologic evidence clearly showed that the extraordinary rains, and the long-unstable nature of the cliff, were the primary culprits. Many people urged the remaining residents to sell out and move somewhere else so that the town could be bulldozed and made into a park or something that doesn't endanger people. But with the big drop in the value of the remaining homes, no one wanted to sell at a loss, and housing is so expensive elsewhere that they had few options to move someplace affordable. La Conchita was one of the last relatively inexpensive communities left on that stretch of coast. Today, the town is still in this state of denial, with memorials left all along the fence at the edge of the slide mass. Neither the state nor anyone else has the money or willpower to buy them out and remove them from harm's way, so there they sit, waiting for the next landslide to wipe out even more of the town and claim more lives.

21.2 BLACK MONSTERS: DEBRIS FLOWS OF THE SAN GABRIEL MOUNTAINS

The San Gabriel Mountains rise to the north of Los Angeles with summits over 3,100 m (10,000 ft) above the plain below, one of the steepest mountain fronts in the world. They are also one of the fastest-rising ranges in the world, with average rates of uplift of several millimeters per year, although they tend to move much more than that during the frequent earthquakes, when the mountain front may jump upward and southward a meter or more. Their steep relief and high elevation are due to the fact that the range is riddled with active faults, from the San Andreas fault on the northern edge to the San Gabriel fault down the middle to the Sierra Madre–Cucamonga thrust fault, which plunges beneath the range from the southern foothills (Fig. 13.10).

This steepness, combined with the fact that the rocks are heavily shattered, makes them very prone to landslides. In addition, most of Southern California is a virtual desert, with just over 25 cm (10 in) of rainfall in a normal year, and often much less than that in drought years. It is bone-dry nearly all year round, with only a handful of very large wet Pacific storms to provide rains in the winter months. When these storms arrive, they slam up against the mountains and then drop amazing amounts of rainfall in a short time. In the winter storms of 2004–2005, over 1.42 m (56 in) of rain fell in Pasadena. One of the most intense rainstorms ever recorded comes from my own backyard, the Rossmoyne station in north Glendale (Daingerfield, 1938). In a storm that ran from February 27 to March 4, 1938, over 330 mm (3 ft) of rain fell in just a week, with a maximum intensity over one 5 min period of 122 mm/hour (roughly 5 in an hour). This is still one of the most intense rainstorms ever recorded. These intense cloudbursts quickly overwhelm the natural array of channels and canyons and pour down as a huge flood choked with mud and boulders known as a debris flow. It has the viscosity of wet concrete, and it is so dense and fast-moving that it can carry huge boulders, as well as lighter objects like fire engines, coffins, cars, and houses. Debris flows are particularly common in wet winter months right after a recent brushfire has burned the slopes. Not only are the slopes free of vegetation to hold the water and soil back, but the burned chaparral brush also produces creosote oils, which permeate the topsoil and make it waterproof, so more of the water flows off the surface rather than sinking in.

These debris flows happen quite frequently in the canyons along the San Gabriel front. A particular terrifying example from February 1978 is recounted by Pulitzer Prize–winning author John McPhee in his essay "L.A. Against the Mountains":

> In Los Angeles versus the San Gabriel Mountains, it is not always clear which side is losing. For example, the Genofiles, Bob and Jackie, can claim to have lost and won.

They live on an acre of ground so high that they look across their pool and past the trunks of big pines at an aerial view over Glendale and across Los Angeles to the Pacific bays. The setting, in cool dry air, is serene and Mediterranean. It has not been everlastingly serene. On a February night some years ago [1978], the Genofiles were awakened by a crash of thunder and lightning striking the mountain front. Ordinarily, in their quiet neighborhood, only the creek beside them was likely to make much sound, dropping steeply out of Shields Canyon on its way to the Los Angeles River. The creek, like every component of all the river systems across the city from mountains to ocean, had not been left to nature. Its banks were concrete. Its bed was concrete. When boulders were running there, they sounded like a rolling freight train. On a night like this, the boulders should have been running. The creek should have been a torrent. Its unnatural sound was unnaturally absent. There was, and had been, a lot of rain. The Genofiles had two teen-age children, whose rooms were on the uphill side of the one-story house. The window in Scott's room looked straight up Pine Cone Road, a cul-de-sac, which, with hundreds like it, defined the northern limit of the city, the confrontation of the urban and the wild. Los Angeles is overmatched on one side by the Pacific Ocean and on the other by very high mountains. With respect to these principal boundaries, Los Angeles is done sprawling. The San Gabriels, in their state of tectonic youth, are rising as rapidly as any range on earth. Their loose inimical slopes flout the tolerance of the angle of repose. Rising straight up out of the megalopolis, they stand ten thousand feet above the nearby sea, and they are not kidding with this city. Shedding, spalling, self-destructing, they are disintegrating at a rate that is also among the fastest in the world. The phalanxed communities of Los Angeles have pushed themselves hard against these mountains, an aggression that requires a deep defense budget to contend with the results. Kimberlee Genofile called to her mother, who joined her in Scott's room as they looked up the street. From its high turnaround, Pine Cone Road plunges downhill like a ski run, bending left and then right and then left and then right in steep christiania turns for half a mile above a three-hundred-foot straightaway that aims directly at the Genofiles' house. Not far below the turnaround, Shields Creek passes under the street, and there a kink in its concrete profile had been plugged by a six-foot boulder. Hence the silence of the creek. The water was now spreading over the street. It descended in heavy sheets. As the young Genofiles and their mother glimpsed it in the all but total darkness, the scene was suddenly illuminated by a blue electrical flash. In the blue light they saw a massive blackness, moving. It was not a landslide, not a mudslide, not a rock avalanche; nor by any means was it the front of a conventional flood. In Jackie's words, "It was just one big black thing coming at us, rolling, with a lot of water in front of it, pushing the water, this big black thing. It was just one big black hill coming toward us."

In geology, it would be known as a debris flow. Debris flows amass in stream valleys, and more or less resemble fresh concrete. They consist of water mixed with a good deal of solid material, most of which is above sand size. Some of it is Chevrolet size. Boulders bigger than cars ride long distances in debris flows. Boulders grouped like fish eggs pour downhill in debris flows. The dark material coming toward the Genofiles was not only full of boulders; it was so full of automobiles it was like bread dough mixed with raisins. On its way down Pine Cone Road, it plucked up cars from driveways and the street. When it crashed into the Genofiles' house, the shattering of safety glass made terrific explosive sounds. A door burst open. Mud and boulders poured into the hall. We're going to go, Jackie thought. Oh, my God, what a hell of a way for the four of us to die together.

The parents' bedroom was on the far side of the house. Bob Genofile was in there kicking through the white satin draperies at the panelled glass, smashing it to provide an outlet for water, when the three others ran in to join him. The walls of the house neither moved nor shook. As a general contractor, Bob had built dams, department stores, hospitals, six schools, seven churches, and this house. It was made of concrete block with steel reinforcement, sixteen inches on center. His wife has said it was stronger than any dam in California. His crew had called it "the fort." In those days, twenty years before, the Genofiles' acre was close by the edge of the mountain brush, but a developer had come along since then and knocked down thousands of trees and put Pine Cone Road up the slope. Now, Bob Genofile was thinking, I hope the roof holds. I hope the roof is strong enough to hold. Debris was flowing over it. He told Scott to shut the bedroom door. No sooner was the door closed than it was battered down and fell into the room. Mud, rock, water poured in. It pushed everybody against the far wall. "Jump on the bed," Bob said. The bed began to rise. Kneeling on it—on a gold velvet spread—they could soon press their palms against the ceiling. The bed also moved toward the glass wall. The two teen-agers got off, to try to control the motion and were pinned between the bed's brass railing and the wall. Boulders went up against the railing, pressed it into their legs and held them fast. Bob dived into the muck to try to move the boulders, but he failed. The debris flow, entering through windows as well as doors, continued to rise. Escape was still possible for the parents but not for the children. The parents looked at each other and did not stir. Each reached for and held one of the children. Their mother felt suddenly resigned sure that her son and daughter would die and she and husband would quickly follow. The house became buried the eaves. Boulders sat on the roof [Figure 21.2a]. Thirteen automobiles were packed around the building, including five in the pool. A din of rocks kept banging against them. The stuck horn of a buried car was blaring. The family in the darkness in their fixed tableau watched one another by the light of a directional sign endlessly blinking. The house had filled up in six minutes and the mud stopped rising near the children's chins.

(from McPhee, 1989, *The Control of Nature*, Farrar Straus Giroux, New York, pp. 184–186)

Since then, the events of the 1978 debris flows were nearly repeated in the same neighborhood. In late August and early September 2009, the Station Fire burned almost 200,000 ac of the Angeles National Forest, including all the hillsides just above the neighborhood where the Genofiles lived. Even though there were no huge Santa Ana winds to push the fire, it burned for weeks because the area had

accumulated many decades' worth of dry brush and could not be stopped. Hard work by hundreds of firefighters managed to save nearly all the homes that were up against the mountain front during the fires, but then came the warnings that the barren slopes would be sliding downhill when the rains came.

In mid-January of 2010, there were four days of heavy rains (over 12 in in the foothills) that saturated the hillsides and caused some sliding, but most of it was contained in the huge debris basins in the canyons, whose purpose is to catch rocks and mud and let the water flow downhill. As soon as the rains stopped, there were dump trucks running up and down the steep mountainside streets, trying to empty the debris basins before the next rains came.

On the night of February 5–6, 2010, a rainstorm (predicted to be relatively small) reached the foothill area and became an intense thunderstorm with howling winds and pounding rain. The rain gauge measured 6 in that had dropped in just a few hours. The already-saturated hillsides soon started to move, and huge debris flows came roaring down Ocean View Boulevard in La Cañada, tumbling cars around and turning them into twisted wrecks or lifting giant 20 ft, 8000 lb pieces of concrete barriers known as K-rails, and tossing and hurling them into other objects. These huge barricades had been put up since the Station Fire to divert and slow the river of mud, but they were no match for the power of debris flows. Most of the residents of the neighborhood had already evacuated at the first sign of rain, but the few who stayed behind despite the warnings were hunkered down in their homes, since escape through the roiling river of mud and rocks and cars was impossible. Emergency services chose not to send them a reverse-911 evacuation call, since they were safer indoors than if they tried to evacuate through the dangerous storm.

Nevertheless, the danger to the homes was very real. Even though it had been cleared since the January rains, the 23 ft deep Mullally Canyon debris basin overflowed in just a few hours during the night. The houses at the top of Ocean View Boulevard were overwhelmed when the monstrous black walls of mud and rocks burst through the K-rails, fences, and sandbags and then through their windows, doors, and even walls, filling their homes with mud that even rose to the ceiling. In some places, the bent and twisted cars acted like battering rams as the debris flow smashed through the walls. Some homes were utterly destroyed (Figure 21.2b), and at least 15 were so damaged that they were red-tagged by authorities and the residents were not allowed inside for fear they could collapse.

As reported by Thomas Curwen in the *Los Angeles Times* (February 12, 2010):

Henry Laguna had never heard a sound so terrifying. Like a train, screeching and crashing as it flew off the tracks. "Get up!" he screamed. He had just left the bedroom to check on the puppy but frantically raced back to wake up his wife and son. "Get up!" he screamed again. "We have to get out!" He

(a)   (b)

FIGURE 21.2   Landslides in the hills above La Crescenta, California. (a) Cars and boulders trapped in mud, burying houses to the eaves, during the 1978 debris flows. (b) Houses caved in by debris flows pushing through their uphill walls during the 2010 slide; they are getting ready for demolition.

Source: (a) Courtesy Wikimedia Commons. (b) Photo by the author.

burst into his son's room. It was a little after 4 a.m. "Dad, what is it?" Brian sat up. "We have to get out!" They ran into the master bedroom. Henry's wife, Damaris, was already out of bed. "Let's go! Let's go!" he yelled, throwing open the slider to the patio, not thinking to turn on a light.

Just hours before they had been together, warm and safe, talking and playing with Lindy, the new golden retriever, a gift to Damaris. Then the rain came, then the downpour.

Outside in the darkness, they slipped on the wet concrete. They dashed around the pool to the end of the property, as far from the house as possible. Damaris fell once, then again. Surging water rose around their bare feet and ankles. Within seconds, a torrent hit them, bringing with it rocks and branches. Holding on to one another, they reached the cypress tree and the fence, anything to put between them and the raging water, anything to hang on to. They were drenched. . . .

Henry and his family clung to the fence and the cypress, struggling to keep their footing as the speed of the water and mud picked up. The rain beat down upon them. Debris battered the house. Henry heard glass breaking, wood cracking and snapping. Damaris saw a dresser and the couch from the television room slide into their swimming pool.

The fence began to bow under their weight. Henry thought they should jump into their neighbor's back yard, but what if the flood got worse and what if that house started to break apart as well? He decided it was best to stay. He encouraged his wife and son to climb higher. He helped them as best he could, trying to escape the torrent himself. He felt neither the cold nor the discomfort. He thought only about how to save his family.

The report describes how they eventually went back to the house to get out of the chilling downpour and find the dog, but it could not be found. They eventually sought refuge at the house of a neighbor across the street. By daybreak, they found their house (Figure 21.2b) utterly destroyed, but

they wanted to stay and salvage things. However, the house was too dangerous to walk into, and the hillsides could still slip at any moment, so eventually they were convinced to leave and let the crews finish the demolition of their house and all their possessions.

Stories such as these are typical of debris flows around the world. They are enormously powerful and capable of lifting and carrying huge amounts of material of very large size over significant distance. Some have been clocked at 100 kph (60 mph), although most are much slower. Anytime you see houses built up against the mountain fronts with huge rocks and boulders all around the neighborhood, it is a sure sign that the entire area was built by catastrophic debris flows and is likely to experience more debris flows in the future. Yet people love living in the cool mountain canyons and foothills, so they build right in the path of disaster, unaware of the story that the boulders tell. After major events like the debris flows of 2010, people demanded that the authorities put in bigger barricades to stop the walls of mud. But as geologists have often said, there is no physical barrier that will stop them if they can move 8,000 lb K-rails and carry cars and even fire engines and burst through and rip apart houses. The only real protection is to not live in such dangerous areas in the first place. It's like voluntarily standing in front of a loaded cannon and hoping that it won't fire while you're there.

21.3   EARTH ON THE MOVE

The landslides at La Conchita and the debris flows of the San Gabriel Mountains are all examples of downslope movement of material, also known as mass wasting. Anywhere the land surface is elevated (especially in areas of high relief), gravity is constantly working to move material downhill. A grain of sand on a slope is constantly feeling the pull of gravity, so the only force that prevents it from moving is the friction that holds it in place. Once something exceeds that frictional threshold, it will start to move.

California is famous for its landslides of every size and scale. Landslides cause more than $1 billion in damage each year in this state. A map of the landslide-prone areas (Figure 21.3) makes a simple prediction—the steeper the slope, the more likely there is sliding. But as the landslide map shows, the slopes in the coastal part of California are much more prone to sliding than those of the Sierra Nevada. The answer is simple: in addition to slope steepness, the other main factor that triggers slope movement is water, so the wetter coastal regions have the most areas that are slide-prone. Also, the bedrock in the coastal region is often shales and other rocks which are soft and weather deeply, while the bedrock in the Sierras is mostly granitic rocks, which crumble into sand but do not form muds as easily.

A variety of different factors determine whether a slope will stay stable or begin to move. The most important is the steepness or relief of the slope—steeper slopes are obviously less stable. If you take dry sand in a sandbox or on the beach and pour it out of the bucket, it will form a nice little conical hill, and no matter how much more sand you pour, the angle of the slope will reach what is called the angle of repose (Figure 21.4) and get no steeper. Any material that makes it steeper than the angle of repose will slide right off. The angle of repose depends on grain size, so it is very

shallow in finer sand (about 30° from the horizontal) but can be quite steep, or even vertical, in a pile of pebbles or boulders.

Another important factor is the fluid content of the material. Dry sand will never form a slope steeper than the angle of repose for a given grain size. If you have ever built sand castles, you know that if you get the sand just moist enough to stick together, you can build sand castles with vertical or even overhanging walls. In this case, there is just enough water trapped between the air spaces in the sand to form small films like the meniscus of water in a test tube. This film of water has enormous surface tension, so it is an effective binding agent. But as any sand castle builder knows, if the sand is too wet, it will flow rather than pack into a vertical wall. If there is too much fluid, there is so much water in the pore space that there are no air pockets. Instead of surface tension holding grains together, the water is now acting as a lubricant and putting pore pressure on the surrounding grains, moving them apart and causing them to flow.

You can apply the sand castle analogy to slopes. Normally, the soil and rocks are bound together by the forces that compacted them into rock in the first place, plus clay minerals and cements between the grains, and maybe groundwater to enhance surface tension. But when these same materials are oversaturated with water, the pore pressure of the fluids pushes the grains apart, and they are ready to flow. In many cases, specific layers act as lubricating surfaces along which the overlying rocks can flow. This is especially true with layer of shale or mudstone, which are typically made of smectite clays known as montmorillonite. These clay minerals expand dramatically when they absorb water into their crystal lattices. When an entire shale layer becomes saturated, it becomes a layer of slippery mud that can lubricate the motion of huge masses of rock. Then all that is needed is some sort of shock (like an earthquake) to destabilize it and trigger the slide. Sometimes it just starts when the oversaturation reaches a critical threshold, and the material spontaneously fails and breaks apart.

Such movement comes in a full range of speeds and different behaviors, from so slow that humans can't perceive it to so fast that it is free-falling with the acceleration

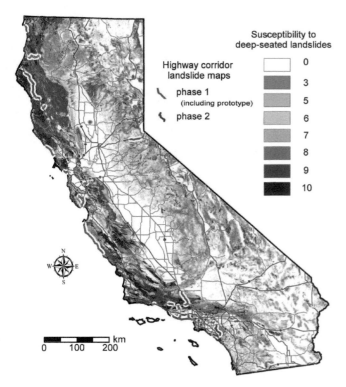

FIGURE 21.3 Map showing landslide-prone areas in California.

Source: Courtesy California Department of Conservation.

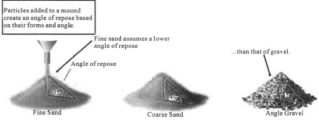

FIGURE 21.4 The angle of repose is the steepest angle that dry sediments of a given grain size will form a stable slope. If more sediment is added, the material just slides down the hill and the angle does not get any steeper.

Source: Courtesy Wikimedia Commons.

of gravity (Figure 21.5). The slowest type of movement is known as **creep**, where the top layer of soil moves slowly downhill just a few millimeters each year. It is typical of steep slopes with a lot of loose soil and moves fastest when there is frequent freezing and thawing to lift and move the soil particles upward and then downhill. Because it moves so slowly, we typically cannot see it move in real time but must detect it by how it displaces stationary objects embedded in the soil. Fence posts and telephone poles will tend to lean downhill, and tree trunks will tilt downhill at their bases, then curve upward to continue to grow vertically in compensation for the movement (Figure 21.6a). The rocks and soil itself may show visible signs of flowing downhill (Figure 21.6b). Creep may not be life-threatening, but it is found on virtually every slope in the world and sometimes moves relatively quickly, destabilizing and tearing apart houses and other structures in a matter of just a few years. Consequently, it is responsible for more damage than all other types of downslope movement combined.

Slightly faster movements are known as **slumps**, where a coherent block of earth breaks free from a scarp and slips down the curved face (Figure 21.5). If the block of earth breaks up and becomes jumbled or incoherent, it becomes an **earthflow**. This category includes both mudflows and debris flows. These can move as slowly as 1 mm/day or as fast as 10 kph (8 mph). If the flow is on very steep slopes, especially if it traps a lubricating layer of fluids or gases at the base, it can become a **slide** and can travel up to 160 kph (100 mph). Finally, the fastest movement of all is when an overhanging cliff breaks and rocks drop through the air in **free fall**. In this case, they travel without friction at the

(a)

(b)

FIGURE 21.6 *Creep* is the slow, steady movement of surface soils down a slope: (a) A creeping slope will tilt the bases of trees downhill so they can grow upward in a curve to compensate. (b) In some cases, a roadcut might cut a cross section through a creeping slope, showing the way the rocks are bent in response.

Source: Courtesy Wikimedia Commons.

FIGURE 21.5 Diagram showing the geometry of the different types of downslope movement, from slow creep to slumps and flows, to free-falling rocks.

acceleration of gravity (9.8 m/sec²) until they reach bottom and stop abruptly.

21.4 PORTUGUESE BEND LANDSLIDE

For an example of a slow-moving slump that turned into a slow earthflow, there is the famous Portuguese Bend landslide, the largest in California (Figure 21.7). It is located on the southern slopes of the beautiful Palos Verdes Peninsula (Figure 21.7a), one of the most affluent and desirable neighborhoods in Los Angeles County. Most of the Palos Verdes is covered with expensive homes with great views of the ocean, each worth several million dollars or more. The coastal cliffs that surround the peninsula are largely covered by houses or spectacular parklands, as well as several golf courses. In many places, there are paths down to the beach, where you can explore secluded coves and wonderful tide pools, or just bask on a private beach with no neighbors bothering you.

As you travel around the south side of the Peninsula along Palos Verdes Drive South, however, you reach a stretch of the road which is really bumpy and irregular (Figure 21.7b). Uphill and downhill from the road, there are no rich houses or landscaped yards at all, just open coastal sage scrublands that once mantled these hills before they were developed. In other places, you can see that all the utilities are aboveground, designed to flex under stress and be repaired quickly if a landslide moves and ruptures them (Figure 21.7b, c). The power poles became tilted at weird angles, so they eventually put them in conduits on the ground (Figure 21.7d). The landscape is bumpy and hummocky, but unless you know what to look for, you would not guess that you are on one of the world's most famous landslides.

The slide itself is 2,700 ft (820 m) wide and 3,900 ft (1,188 m) long. There is geological evidence that the slide area has been active for at least 250,000 years. The slide spans 260 ac (1.1 km²), with an average thickness of 135 ft (41 m). The ground failure occurs on an overall smooth surface approximately 100 ft (30 m) below the surface. The ground failure over the years has been due to seaward-dipping strata, rock weakness, and continual coastal erosion. Prehistoric landslides are believed to be so extensive that they destroyed the formation of higher wave-cut benches. The active slide consists of landslide rubble such as bedded blocks, which are rare among most landslides. The bedded blocks measure 10 ft (3 m) in diameter, and they appear in landslide rubble. This shows ground disturbance that could eventually cause the land to slide.

The onset of sliding has been attributed to the housing development that started in the 1950s. Homeowners tried to say it was the county's fault for constructing a road near the head scarp, but they didn't win this case. Another important factor of the sliding is the construction of Palos Verdes Drive South. Geologists associate the onset of slip to irrigation, installation of pools and septic tanks that increased groundwater levels. Houses had been built prior to the slide. Some homes were damaged, but others remain intact (Figure 21.7e).

Homes that remain occupied now have water and sewage lines available to them. The new lines were constructed aboveground, so the slide can move freely below them.

The first movement indication occurred on Friday, August 17, 1956. Cracking occurred in the foundation of a recently built structure. The cracks were repaired, but new cracking occurred just days later. Ten days after the first indication of movement, cracking propagated to Palos Verdes Drive South. By September 4, 10 cm of offset was observed on the Palos Verdes Drive South. By mid-September, more pronounced cracking was observed. Distortion became noticeable on the Portuguese Bend pier on October 4 of that year. A survey station was installed to monitor ground movement. The station recorded movement of 7–10 cm per day. By the end of October, movement slowed to 2 cm per day. Drilling projects over the next several months determined the location of the slip surface and the location of standing groundwater.

Today, the slide is still slowly moving downhill, but there is no way to stop it, since the sea cliffs at the base are constantly being cut back and eroded by the waves, removing the rocks that hold back the rest of the slope. The base of these cliffs has been armored with cages full of boulders, called gavions, which are supposed to absorb the pounding of the storm waves and slow down cliff erosion. There are several pumping operations to remove the groundwater that helps lubricate the slide, but they have not stopped the motion completely. Most people drive right through without even noticing that the landscape is bumpy, the road is always breaking and bent, and that there are no houses on some of the most valuable real estate in California. Near the edges of the slide, there are still a few houses that are occupied, at least until the slide tears them apart. Some people still live in them but have no place to go, since their house is condemned, and they can't sell the property, because no buyer would be able to build on it. In a few places, the desperate homeowners are using last-ditch efforts to keep their homes. Some of them are on stilts and supported by a huge steel trusswork beneath the floor, which can be raised or lowered if the ground starts to sink beneath them (Figure 21.8f). In one case, the house is on a strong frame for a base and then set on top of a triangle of three huge shipping containers. These act as a tripod, so the homeowner can jack up or lower any one corner of the tripod and restore it to level. These extraordinary measures seem desperate, but owning a house in the area is so expensive, and there are few places to move that are not also expensive, that these homeowners really have no choice.

21.5 CALIFORNIA TUMBLES INTO THE SEA

With hundreds of kilometers of spectacular coastline, one of the biggest problems with landslide erosion in California is the collapse of sea cliffs and all the expensive property and structures, like houses, pools, patios, and whole neighborhoods, that are built on their edge. About 72% of California's 1,760 km of coastline consists

FIGURE 21.7 The Portuguese Bend landslide: (a) Aerial view of the southern side of the Palos Verdes Peninsula showing the Portuguese Bend landslide. The slide area is the bare ground between the remaining houses. (b) Palos Verdes Drive South is all buckled and distorted as it crosses the active area of the slide. (c) Pipes and other utilities must be aboveground and built to adjust to movement, since they cannot be buried and have to be accessible to be repaired quickly when the ground moves beneath them. (d) When the slide was moving rapidly, telephone poles were tilting in all directions and power lines were stretched and snapped. (e) One of the houses torn apart by the slide. (f) The remaining houses in the area are supported by a steel-frame trusswork so they can be supported by jacks that allow them to be adjusted to level whenever they are tilted by ground movement below.

Source: Courtesy Wikimedia Commons.

FIGURE 21.8 A concrete-and-rebar seawall that has been damaged by huge storm waves. Some of the boulders that battered the seawall still remain.

of sea cliffs and bluffs. The major forces that control these cliffs are the rapid rate of tectonic uplift of much of the California coast, plus the rapid rise in sea level of about 120 m over the last 10,000 years. Other local factors contribute as well: whether the bedrock is hard and durable or soft and easily eroded, the nature of the wave energy and tidal forces that affect some coasts more strongly than other, and the frequency of storm events. The stair-step marine terraces (Chapter 19) are often made of softer sediments from the marine environment deposited during previous interglacial events and rapid rises of sea level. Thus, the entire coastline is rising faster than the forces of erosion and forms rugged cliffs that are vulnerable to landslides and collapse.

The spectacular views from sea cliff properties make them among the most valuable real estate in the country, but also one of the most dangerous places to build as well. The rich and powerful want to own these spectacular vista homes and are not happy when Mother Nature ruins their plans—and their property. Yet sea cliffs are probably the most vulnerable areas for sliding of all, since they are constantly being undercut by wave erosion (see Chapter 19). The homeowners can spend all the money they want to put seawalls on the face of their cliff to slow down erosion, but sooner or later, a big storm will undermine the seawall and make it collapse. In some cases, storms batter it with huge rocks and break through the seawall completely

(Figure 21.8). This process can go on for decades, but eventually Mother Nature will win and the sea cliff will crumble down into the sea. No amount of money or power or huge concrete seawalls can stop this inevitable process.

One of the classic examples of this process is the small beach town of Pacifica, just down the coast from San Francisco. The town was built right to the edge of cliffs of soft sandstones and shales and is undergoing some of the most rapid erosion in the state. The erosion was particularly rapid during the huge storms of major El Niño years, such as 1982–1983, 1997–1998, and 2016 (Figure 21.9). There are piles of boulders at the base, known as revetments, designed to absorb the wave energy before they erode the base of the cliff, but when a big El Niño storm hits at a really high tide, the waves rise completely above the revetments, break against the soft cliff-face sediments, and erosion is very rapid—sometimes 16 ft of cliff lost in a single storm. Consequently, the buildings the edge of the cliffs are often left hanging out over the water, or collapse completely, and have to be red-tagged, abandoned, and then demolished before they plunge into the surf.

Just south of Pacifica, and just north of Montara, is an even more dramatic example of coast erosion, ominously known as the "Devil's Slide" (Figure 21.10). A steep rocky promontory of sheer cliffs made of soft sandstones and shales is undergoing rapid erosion through coastal cliff collapse. For a long time, it formed an impassable barrier along the California coastal route, and the steep Pedro Mountain Road hugged the hills inland as the main highway through. In 1936, Caltrans completed Highway 1 along the face of the slide, making the drive shorter—but more hazardous. The first major landslide destroyed much of the road in 1940, but it was rebuilt. Another large slide in 1995 forced the road's closure for almost two years. In April 2006, the road began to develop large longitudinal cracks in the roadbed, indicating an imminent slide and forcing the highway's closure for five months, while Caltrans worked to stabilize the slide. On March 25, 2013, Caltrans gave up and shut down the landslide-prone coastal road, replacing it with the Tom Lantos Tunnels (Figure 21.11), which take the highway more inland through the promontory behind the precarious cliffs. The remnants of the old Coast Highway are still there and open to pedestrians and cyclists, but vehicles no longer try to make it across the hazardous road, with numerous sections missing.

Indeed, the problems of coast cliff collapse and landslides plague the entire Coast Highway (Highway 1), especially along the scenic Big Sur coast from Monterey to Morro Bay. The entire route is marked by one landslide after another (Figure 21.12), undermining the road, or covering it with debris from above. At any given time, some portion of Highway 1 is closed to traffic, and you can only reach most of the region from either the south or the north—but cannot drive through. The small towns along the route, as well as businesses and state parks, like Pfeiffer Redwoods State Park, are often cut off from the outside world and cannot receive any tourist traffic. Sometimes there are more than

FIGURE 21.9   Sea cliff collapse in Pacifica, California. (a–b) Collapse of properties along Palmetto Avenue after the storms of winter 1983. (c–d) Further collapse of the cliffs and loss of property after the 1998 El Niño storms. (e–f) During the 2009–2010 storms, this huge cliffside apartment complex was undermined and had to be abandoned and demolished. Only a vacant lot now stands where the remains of the cliff have not collapsed.

Source: Photos courtesy G. Griggs.

FIGURE 21.10   Aerial view of Devil's Slide, just south of Pacifica, showing the now-abandoned road around the point and the tunnels that now conduct traffic on Highway 1.

Source: Courtesy Wikimedia Commons.

FIGURE 21.11   Landslides along Highway 1 in the Big Sur area: (a) Aerial view of a huge slide with over a million tons of rocks and soil in 2002. (b) Aerial view of another landslide and collapse of the highway in 2021. (c) Ground-level view of the 2021 slide.

Source: Courtesy Wikimedia Commons.

FIGURE 21.12 Landslides in southern Orange County. (a) Collapse of Casa Romantica in San Clemente after the major storms in winter 2023. (b) Collapse of the sea cliffs in San Clemente, damaging homes and patios in 2023. (c) In many places, the Amtrak rail line between San Clemente and San Diego runs just above the beach, and the erosion has undercut the ground beneath it and closed the tracks.

Source: Courtesy Wikimedia Commons.

one closure, and the section of road cut off at both ends can only be reached by helicopter.

Coastal landslides are found up and down the entire California coast, not just in the Pacifica area, or in the Big Sur. But perhaps the most dramatic examples of cliff erosion occurred in southern Orange County, such as in San Clemente (Figure 21.12). Almost every rainy year, there are more examples of collapsing cliffs leaving patios and swimming pools hanging out into space before they finally break and collapse. A long stretch of vulnerable cliffs with multi-million-dollar homes runs from Dana Point all the way to the Camp Pendleton Marine Base in northern San Diego County. Such earth movements are expensive and hard to prevent but typically affect only a few individuals. However, there has been an ongoing problem with the rail lines that run just above the beach, particularly in southern San Clemente (Figure 21.12c). All the Amtrak train service between Los Angeles and San Diego runs along this coastal trackway, and in many places, wave erosion has undercut the tracks and threatened to completely wash them away. There has been an ongoing battle with nature. Various agencies have been trying to stabilize the tracks

and prevent further erosion, but after major storms, train service shuts down through San Clemente, and passengers have to disembark, take a bus from Oceanside or San Juan Capistrano, then reboard the train at the other end. This is hugely slow and inconvenient, but it is the price of running a rail line through the area, because there are no other suitable routes available.

21.6   WHAT GOES UP MUST COME DOWN

Even though the catastrophic, deadly events that are the focus of this chapter are dramatic, they are rare. Most of the world's landslides and downslope movement happen gradually and slowly, with little or no loss of life, but often with a tremendous loss of property and money. They happen constantly around the world on just about any surface that has elevation and relief, and they are unavoidable. If people are wise, they will avoid building in places that are prone to mass wasting. But people love mountains with their great views and cool, forested canyons, so they will insist on living in the path of danger in spite of the risk. Sometimes people are simply unaware of the hazards. Sometimes they simply do not seem to care, or think that with enough money, they can have anything they want. Mountaintop and clifftop homes are often the most expensive and popular because of their views, and their owners seem to think that as property owners, it is their right that their cliff edge be stable forever and not erode back or collapse. Yet like King Canute ordering the tide not to come in, they are trying to stop natural processes that can only be slowed down temporarily with the infusion of lots of money. Eventually, the slopes and cliffs will come down.

In many cases, people make the situation worse and then pay a heavy price. It is common practice for developers to carve housing tracts like little shelves jutting out of steep hillsides without regard to the stability of the slope they have cut. They often compound the problem by widening and extending out the width of each lot in the housing tract with loose fill dirt from the cut, further destabilizing the slope. Roads and construction often make deep cuts into slopes and mountains in order to shorten and straighten out their routes—and eventually, nature will win with rockslides and collapsing slopes. We can live *with* the landscape and try to build where the ground is most stable and make roads that follow natural river valleys (as the pioneers did)—or we can try to dominate the landscape and make it fit our convenience. Ultimately, nature will have the last laugh.

The late great political columnist and social commentator Art Buchwald said it best in a column written after one of Southern California's rainy winters and frequent slides:

Los Angeles, a Mobile Society

I came to Los Angeles last week for rest and recreation, only to discover that it had become a rain forest. I didn't realize how bad it was until I went to dinner at a friend's house. I had the right address, but when I arrived, there was nothing there. I went to a neighboring house where I found a man bailing out his swimming pool. "I beg your pardon," I said. "Could you tell me where the Cables live?"

"They used to live above us on the hill. Then, about two years ago, their house slid down in the mud, and they lived next door to us. I think it was last Monday, during the storm, that their house slid again, and now live two streets below us, down there. We were sorry to see them go-they were really nice neighbors."

I thanked him and slid straight down the hill to the new location of the Cables' house. Cable was clearing mud from his car. He apologized for not giving me the new address and explained, "Frankly, I didn't know until this morning whether the house would stay here or continue sliding down a few more blocks."

"Cable," I said, "you and your wife are intelligent people, why do you build your house on the top of a canyon, when you know that during a rainstorm it has chance of sliding away?"

"We did it for the view. It really was fantastic on a clear night up there. We could sit in our Jacuzzi and see all of Los Angeles, except of course when there were brush fires.

Even when our house slid down two years ago, we still had a great sight of the airport. Now I'm not too sure what kind of view we'll have because of the house in front of us, which slid down with ours at the same time."

"But why don't you move to safe ground so that you don't have to worry about rainstorms?"

"We've thought about it. But once you live high in a canyon, it's hard to move to the plains. Besides, this house is built solid and has about three more good mudslides in it."

"Still, it must be kind of hairy to sit in your home during a deluge and wonder where you'll wind up next. Don't you ever have the desire to just settle down in one place?"

"It's hard for people who don't live in California to understand how we people out here think. Sure we have floods, and fire and drought, but that's the price you have to pay for living the good life. When Esther and I saw this house, we knew it was a dream come true. It was located right on the tippy top of the hill, way up there. We would wake up in the morning and listen to the birds, and eat breakfast out on the patio and look down on all the smog. Then, after the first mudslide, we found ourselves living next to people. It was an entirely different experience. But by that time we were ready for a change. Now we've slid again and we're in a whole new neighborhood. You can't do that if you live on solid ground. Once you move into a house below Sunset Boulevard, you're stuck there for the rest of your life. When you live on the side of a hill in Los Angeles, you at least know it's not going to last forever."

"Then, in spite of what's happened, you don't plan to move out?"

"Are you crazy? You couldn't replace a house like this in L.A. for $500,000 [many millions of dollars in today's prices]."

"What happens if it keeps raining and you slide down the hill again?

"It's no problem. Esther and I figure if we slide down too far, we'll just pick up and go back to the top of the hill, and start all over again; that is, if the hill is still there after the earthquake."

RESOURCES

Benumof, B.T., and Griggs, G.B. 1999. The relationship between seacliff erosion rates, cliff material properties, and physical processes. *Shore and Beach*, 67(4), 29–41.

Brabb, E.E., and Harrod, B.L. (eds.). 1989. *Landslides: Extent and Economic Significance*. Balkema, Brookfield, VA.

Cornforth, D. 2005. *Landslides in Practice: Investigation, Analysis, and Preventative Mediation in Soils*. John Wiley, New York.

Costa, J.E., and Wierczorek, G.F. 1987. *Reviews in Engineering Geology. Vol. 7. Debris Flows, Avalanches: Progress, Recognition, and Mitigation*. Geological Society of America, Boulder, CO.

Dikau, R., Brunsden, D., and Schrott, L. (eds.). 1996. *Landslide Recognition: Identification, Movement, and Causes*. John Wiley, New York.

Emery, K.O., and Kuhn, G.G. 1980. Erosion of rock shores at La Jolla, California. *Marine Geology*, 37, 197–208.

Evans, S.G., and Degraff, J.V. (eds.). 2003. *Catastrophic Landslides: Effects, Occurrence, and Mechanisms*. Geological Society of America, Boulder, CO.

Glade, T., and Crozier, M.J. 2005. *Landslide Hazard and Risk.* John Wiley, New York.

Griggs, G.B., and Johnson, R.E. 1979. Coastline erosion, Santa Cruz County. *California Geology, 32*(4), 67–76.

Griggs, G.B., Pepper, J.E., and Jordan, M.E. 1992. *California's Coastal Hazards: A Critical Assessment of Existing Land-use Policies and Practices.* Berkeley, CA. Special Publication of California Policy Seminar Program. 224p.

Habel, J.S., and Armstrong, G.A. 1978. *Assessment and Atlas of Shoreline Erosion along the California Coast.* State of California, Department of Navigation and Ocean Development, Sacramento, CA.

Highland, L.M. 2009. The landslide handbook—A guide to understanding landslides. *U.S. Geological Survey Circular*, 1325. http://pubs.usgs.gov/circ/1325/

Holzer, T. 2009. *Living with Unstable Ground.* American Geological Institute, Alexandria, VA.

Jibson, R.W. 2005. *Landslide Hazards at La Conchita, California.* U.S. Geological Survey Open File Report 2005–1067. http://pubs.usgs.gov/of/2005/1067/508of05-1067.html

McPhee, J. 1989. L.A. against the mountains. In *The Control of Nature.* Farrar Straus Giroux, New York.

Norris, R.M. 1990. Sea cliff erosion: A major dilemma. *Geotimes, 11,* 16–17.

Plant, N.G., and Griggs, G.B. 1990a. Coastal landslides caused by the October 17, 1989 earthquake, Santa Cruz County, California. *California Geology, 43,* 75–84.

Plant, N.G., and Griggs, G.B. 1990b. Coastal landslides and the Loma Prieta earthquake. *Earth Science, 43,* 12–17.

Runyan, K.B., and Griggs, G.B. 2003. The effects of armoring sea-cliffs on the natural sand supply to the beaches of California. *Journal of Coastal Research, 19*(2), 336–347.

Sassa, K., Fukoka, H., Wang, F., and Wang, G. (eds.). 2007. *Progress in Landslide Science.* Springer, New York.

Slosson, J.E., Keene, A.G., and Johnson, J.A. (eds.). 1993. *Reviews in Engineering Geology. Vol. 9, Landslides/Landslide Mitigation.* Geological Society of America, Boulder, CO.

Turner, A.K. 1996. *Landslides: Investigation and Mitigation.* Special Report of the National Research Council 247: Transportation Research Board, Washington, DC.

U.S. Geological Survey. 2009. Landslides hazards program website. http://landslides.usgs.gov/

Zaruba, Q., and Mencl, V. 1969. *Landslides and Their Control.* Elsevier, New York.

22 California's Air and Water Pollution

The smog was heavy, my eyes are weeping from it, the sun was hot, the air stank, a regular hell is LA.

—Jack Kerouac

Once-ler! You're making such a smogulous smoke, my poor swomee swans, why they can't sing a note! No one can sing who has smog in his throat.

—Dr. Seuss, "The Lorax"

22.1 CALIFORNIA'S ENVIRONMENTAL CHALLENGES: POLLUTION

Although the study of the atmosphere and hydrosphere is technically not about rocks or the Earth's crust, in recent years atmospheric science, environmental geology, and hydrogeology have come to be included in a broader approach to geology, often called "earth system science." When we consider environmental issues in the Golden State, the topic of pollution has long been one of the most important issues, thanks to the legendary smog of some California cities, as well as issues of water contamination.

Pollution is a serious issue in California, as it is in most heavily developed parts of the world. Severe air pollution damages lungs and can be a leading cause of death in some parts of the world with severely polluted air, like Beijing, China, or many cities in India. *Pollution* is defined as the addition of any substance (solid, liquid, or gas) or any form of energy (such as heat, sound, or radioactivity) to the environment at a faster rate than it can be dispersed, diluted, decomposed, recycled, or stored in some harmless form. Pollution occurs all over the world, and there are many cities in the world more polluted than any in California now (such as Beijing, New Delhi, and Mexico City), but it still remains a serious issue in this state.

Over the past few decades, California has taken strong measures to deal with the issue of pollution. California is now one of the leading states in the United States not only in its strict regulations to prevent air pollution and water pollution but also in combating pollution of greenhouse gases that are causing climate change (see Chapters 23, 24).

22.2 AIR POLLUTION

22.2.1 THE PROBLEM OF SMOG

Since the Spanish colonial days in Los Angeles, there are records of smoky air from fires causing irritated eyes, burning lungs, and nausea. On July 26, 1943, there was a smog that was so sudden and severe that Los Angeles residents thought the Japanese were attacking with chemical warfare.

People would walk out on smoggy days and need to wear gas masks. But smog was not originally made famous in Los Angeles. That honor goes to London, which by the mid-1800s had such a toxic mix of industrial pollutants, smoke from coal and wood fires in the city's many hearths, mixed with their legendary foggy weather, that people got sick on a regular basis when walking around in it. In the "Great Smog of 1952," London was paralyzed by darkness and daytime and air so toxic that people were dying in large numbers.

Traditionally, it was thought that Dr. Henry Antoine Des Voeux first formally coined the term "smog" (combining "smoke" and "fog") in a paper, "Fog and Smoke," presented at a 1905 meeting of the Public Health Congress of London. On July 26, 1905, the London newspaper *Daily Graphic* summarized his paper with the words:

> He said it required no science to see that there was something produced in great cities which was not found in the country, and that was smoky fog, or what was known as "smog." . . . Dr. Des Voeux did a public service in coining a new word for the London fog.

In fact, the problem with air pollution in London goes back to 1306, when King Edward I banned coal fires in London because of the nasty air pollution.

Actually, the term "smog" was coined in California! Detailed research has shown that the term appeared twenty-five years earlier than Dr. Voeux's paper, in the *Santa Cruz & Monterey Illustrated Handbook* published in 1880, and also appears in print in a column quoting from the book in the July 3, 1880, *Santa Cruz Weekly Sentinel*. The author, Henry Meyrick, wrote:

> It is really not fog at all, but cloud of pure white mist, warmer and much less wetting than a "Scotch mist," and differing entirely from the true British fog, facetiously spelled "smog," because always colored and strongly impregnated with smoke, a mixture as unwholesome as it is unpleasant.

This reference is the first recorded instance of the term "smog" in print, but clearly, there must have been some usage of the term in Britain and the United States before 1880, because Meyrick mentioned that people were using the word.

However, neither London nor Los Angeles is as badly polluted now, thanks to environmental regulations which have greatly reduced the problem with smog. Instead, the most polluted cities in the world (Table 22.1) are found mostly in developing countries, especially China and India. New Delhi, India, has in recent years often

DOI: 10.1201/9781003301837-22

TABLE 22.1

List of the 20 Most Polluted Cities in the World

Rank	City, Country
1	Delhi, India
2	Kolkata, India
3	Kano, Nigeria
4	Lima, Peru
5	Dhaka, Bangladesh
6	Jakarta, Indonesia
7	Lagos, Nigeria
8	Karachi, Pakistan
9	Beijing, China
10	Accra, Ghana
11	Chengdu, China
12	Singapore, Singapore
13	Abidjan, Côte d'Ivoire
14	Mumbai, India
15	Bamako, Mali
16	Shanghai, China
17	Dushanbe, Tajikistan
18	Tashkent, Uzbekistan
19	Kinshasa, Democratic Republic of the Congo
20	Cairo, Egypt

Source: From www.indiatoday.in/education-today/gk-current-affairs/story/list-of-20-most-polluted-cities-in-the-world-1990041-2022-08-19).

been ranked as the most polluted city in the world, with air pollution contributing to about 10,500 deaths per year. In some rankings, India has most of the top 50 most polluted cities, along with neighboring Pakistan and Bangladesh, thanks to dense populations of these cities and their vehicles and industry, as well as the practice of burning crop stubble at the end of the winter dry season. Many of these cities have lots of small cooking fires, as well as cremations that the Hindus practice. During the winter, the cold air hugs the ground and prevents pollutants from escaping, and the Himalayas prevent the pollution from flowing off to the north, northeast, and northwest. In another ranking of major cities, Jakarta, Indonesia, comes out as most polluted, followed by Lahore in Pakistan, Mexico City in Mexico, and Shanghai, Chengdu, and Beijing in China.

22.2.2 Air Pollution and Its Effects

So what causes smog? Up until the middle part of the twentieth century, it was usually smoke from coal-burning hearths, stoves, and especially industrial plants which combined with moisture in the air to form the "pea-soup" fog and smog so famous from London, especially in mysteries like the stories of Sherlock Holmes. In many parts of the world, the emissions from industrial plants are still the major source of air pollution. In more

recent years, coal and wood are burned much less (especially in developed countries), so the pollution is mainly photochemical smog, caused by reactions of pollutants to sunlight in the atmosphere. This comes mostly from the breakdown of oil and gas and other fossil fuels used in vehicles, as well as chemicals from industrial smoke. The primary pollutants (Figure 22.1) are nitric oxide (NO) and nitrogen dioxide (NO_2), along with volatile organic compounds (VOCs). These interact with sunlight to form secondary pollutants, such as ozone (O_3), peroxylacyl nitrates (PAN), plus the major components of acid rain, nitric acid (HNO_3) and sulfuric acid (H_2SO_4). But the composition and chemical reactions involved in photochemical smog were not understood until the 1950s. In 1948, flavor chemist Arie Haagen-Smit from Caltech adapted some of his equipment to collect chemicals from polluted air. He identified ozone as a component of Los Angeles smog. He also discovered that nitrogen oxides from automotive exhausts and gaseous hydrocarbons from cars and oil refineries, exposed to sunlight, were key ingredients in the formation of ozone and photochemical smog.

So what is the danger of smog and other forms of air pollution? Sure, it's irritating and annoying to breathe the dirty air, which can sting your eyes and make it hard to breathe. But it turns out that it's much more serious than that. It's especially harmful to young children, who can grow up with impaired and scarred lungs, comparable to the effects caused by long-term smoking. It also strikes other vulnerable populations, such as the elderly and those with asthma and others with pre-existing heart and lung conditions like emphysema and bronchitis. It can inflame breathing passages, decrease the lung's working capacity, cause shortness of breath, as well as cause pain when inhaling deeply, wheezing, and coughing. It irritates the eyes and nose, and it dries out the protective membranes of the nose and throat, which interfere with the body's ability to fight infection and increase susceptibility to illness.

The effects of smog are well documented. Hospital admissions for respiratory illnesses, and also respiratory deaths, increase during periods of high ozone levels. Many other studies consistently confirm the links between air pollution and health problems. A typical study published in *Nature* magazine in 2017 showed that the smog in the city of Jinan, China, was associated with an almost 6% increase in mortality. In 2016, the Ontario Medical Association announced that smog caused an estimated 9,500 premature deaths in the province each year. A 2012 study in the San Joaquin Valley of California showed that smog is linked to several kinds of birth defects, while a 2017 paper from China showed that air pollution increased the risk of bad pregnancy outcomes. In 2013, a study published in the medical journal *The Lancet* showed that even a very small amount of exposure to particulate matter in smog resulted in a much higher percentage of low-birth-weight babies. The studies go on and on—but there is no question that air pollution is a serious health hazard, on par with smoking and drug use.

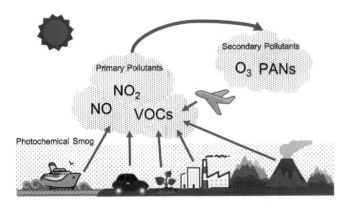

FIGURE 22.1 Cartoon showing the major polluting gases in our air that cause smog.

Source: Courtesy Wikimedia Commons.

22.2.3 AIR POLLUTION IN CALIFORNIA

Despite almost 40 years of efforts to clean up our air, California still has more cities with severe pollution than any other region of the United States, especially in levels of ozone smog. The Los Angeles–Long Beach–Riverside area continues to be the most polluted in the state and in the country, as well as Bakersfield, and Fresno-Madera, switching order as to which is worst, depending upon which pollutant is measured (Table 22.2). This is largely due to geographic features that tend to concentrate the air pollution, no matter how much our laws have done to reduce the chemicals we put out. In all these cases, these cities are at the bottom of a broad basin, with a ring of mountains around them, so air tends to get trapped and cannot flow freely. This problem is enhanced by what is known as an inversion layer (Figure 22.2).

Under normal conditions in the atmosphere, the energy from the sun passes through the atmosphere, then hits the ground and is absorbed and begins to radiate that energy as heat. The warming ground heats the air immediately above it through conduction, and that warm air begins to rise. Meanwhile, cool air above it begins to sink to replace the warm rising air, and the entire atmosphere goes through a constant circulation known as convection. As long as the cooler air is at the top and tries to sink while warm air forms at the bottom and tries to rise, there will be convection and the air will circulate, and pollution will not build up.

When inversion occurs, however (Figure 22.2), a layer of cold air forms on the ground. Especially if it is full of moisture and produces fog, the sunlight will not penetrate to the ground, and the ground will not warm up as quickly. The cold air is already at the bottom and the warm air is on top, which means air masses will neither rise nor sink. They are *inverted* from the normal configuration. As long as the cold air mass on the bottom persists, the air remains stagnant, refusing to circulate vertically, and pollutants from the ground can build up. That is how an inversion layer forms and traps pollutants and causes smog to build up easily in places that frequently have inversions.

In the case of the LA basin, there are the giant San Gabriel Mountains to the north, some of whose peaks are over 11,000 ft (3,400 m) high. On the other side is the Pacific Ocean, a source of cool, moist air (Figure 22.2). Many times in the year, a cool air mass known as the "marine layer" flows in from the ocean and slides beneath the warmer air inland. As long as this cool layer doesn't warm up, it will stagnate on the basin floor, and pollutants will be trapped within it, unable to escape by mixing with the warm air above. This polluted cool layer is dramatically visible from the high mountains around Los Angeles (Figure 22.3), where you can see the noticeable boundary at the top of the polluted ground-level air, sharply distinct from the clear, warm air above. It is a sobering fact that as you look down on it, that is the air you breathe once you are on the ground level in the LA basin. With a population of over 10 million driving millions of vehicles on the roads, plus the pollution from the ports of Los Angeles and Long Beach and many industries across the region, there are lots of sources of pollution whose emissions get trapped in that layer. We can take all sorts of measures to reduce our output of pollution in LA, but the geographic and atmospheric conditions are always working against us.

A similar dynamic occurs in the San Joaquin Basin, especially in the Bakersfield and Fresno areas (Figure 22.4). As those cities have exploded in size due to the boom in population of people seeking a cheaper place to live, they are beginning to release comparable amounts of pollutants in the process.[1] Throughout the San Joaquin Basin, the ring of mountains from the mighty Sierras to the Transverse Ranges, to the Coast Ranges, traps the air in the basin. If conditions are right to form a cool layer right on the ground (especially if it generates a ground fog called the "Tule fog"), then an inversion layer is created, and as long as that cool air hugs the ground, it can't go up (because it's colder than the air above it) or sideways (because the mountains block it). The major sources of pollution are not only cars and trucks but also all the huge oil operations that dominate the San Joaquin Basin, and also the chemicals used in pesticides, which degrade into VOCs when exposed to the sun.

In the Bay Area, the situation is different. Much of the year it gets fog from cool, moist air flowing off the ocean (known as an advection fog), and the temperatures remain cool. Although not in a confined, landlocked basin like the Los Angeles Basin or San Joaquin Valley, the Bay Area does have a ring of mountains around it, and a huge population creating the same problem with producing pollutants from vehicles and factories. Under certain circumstances, the cool air from the ocean gets trapped in an inversion, confined by those mountains, and the Bay Area experiences serious air pollution. Three of the Bay Area counties (Alameda, Contra Costa, and Santa Clara) received failing grades in a State of the Air report from the American Lung Association. Collectively, San Francisco–San Jose–Oakland metro areas rank 6th worst in the nation on pollution due to particulates, and 13th worst in ozone pollution (Table 22.2).

Inversion layers trapped up against mountains can produce smog almost anywhere that they occur. Formerly smog-free cities like Denver and Salt Lake City now get regular smoggy days because their adjacent mountains block the inversion layer from dispersing.

22.2.4 REDUCING AIR POLLUTION IN CALIFORNIA

The outrage over the horrible LA smog in the 1940s and 1950s led to more and more public support for measures which would reduce air pollution in the state. One of the biggest sources of pollution worldwide are cars and trucks. On average, each vehicle emits about 250 lb of carbon monoxides, 18.32 lb of nitrogen oxides, 29 lb of hydrocarbons, and 9,737 lb of carbon dioxide every year. In 1967, the California Air Resources Board was formed and began to address the problems of air pollution, especially from car and truck engines. In the late 1960s, the California state legislature passed a law establishing a program where every car made after 1975 must undergo a smog check on a regular basis. If the car fails the test, it cannot be registered as street legal by the Department of Motor Vehicles until it passes the test. This quickly put lots of older cars with dirty engines off the road, and the automakers had to find a way to make cars that ran cleaner, because California is the single biggest car market in the United States. They resisted and fought it in every way they could, but the state laws eventually won in the courts, and automakers and oil companies had to comply.

They reduced pollution by attaching a catalytic converter to the tailpipe of every car sold in California that has a gas or diesel engine. This device heats up the exhaust from the engine and has solid metal catalysts inside it, which help break down and neutralize the pollutants from the engine before they go out the tailpipe. Today, nearly every car in California has a catalytic converter if it has an internal combustion engine.

A significant component of the higher gas prices in California are the high taxes we pay per gallon, not only for environmental measures, but also to pay for the repairs on the roads that the drivers should be responsible for. According to one source:[2]

> The average gas tax elsewhere in the country was about 29 cents per gallon in December [2022]. California is a whole different story: there is the gasoline excise tax, 54 cents; the sales tax, averaging about 14 cents in December; and the environmental fees—cap and trade, about 21 cents; low-carbon fuel standard, about 8 cents; and the fee for abatement of leaking underground storage tanks, 2 cents. All that adds up to 99 cents, or 70 cents more than the average elsewhere in the US. That said, much of those funds go for things drivers like, such as roads and bridges, things that society needs, such as R&D on low carbon technologies and support for low-income households during the energy transition, and a small share for things that are more controversial, such as high-speed rail.

TABLE 22.2

Ranking of the Most Polluted Cities in the United States

Most Polluted Cities by Ozone Pollution

1. Los Angeles–Long Beach, CA
2. Bakersfield, CA
3. Fresno–Madera, CA
4. Visalia–Porterville–Hanford, CA
5. Phoenix–Mesa–Scottsdale, AZ
6. Modesto–Merced, CA
7. San Diego–Carlsbad, CA
8. Sacramento–Roseville, CA
9. New York–Newark, NY-NJ-CT-PA
10. Las Vegas–Henderson, NV-AZ

Most Polluted Metropolitan Regions by Average Year-Round Concentration of Particulate Matter (PM2.5)

1. Visalia–Porterville–Hanford, CA
2. Bakersfield, CA
3. Fresno–Madera, CA
4. San Jose–San Francisco–Oakland, CA
5. Los Angeles–Long Beach, CA
6. Modesto–Merced, CA
7. El Centro, CA
8. Pittsburgh–New Castle–Weirton, PA-OH-WV
9. Cleveland–Akron–Canton, OH
10. San Luis Obispo–Paso Robles–Arroyo Grande, CA

Most Polluted Metropolitan Areas by Dangerous "Spikes" in Particulate Matter (PM2.5)

1. Bakersfield, CA
2. Visalia–Porterville–Hanford, CA
3. Fresno–Madera, CA
4. Modesto–Merced, CA
5. Fairbanks, AK
6. San Jose–San Francisco–Oakland, CA
7. Salt Lake City–Provo–Orem, UT
8. Logan, UT-ID
9. Los Angeles–Long Beach, CA
10. Reno–Carson City–Fernley, NV

Source: From https://qz.com/963089/california-is-home-to-eight-of-the-10-cities-in-america-where-air-pollution-is-worst.

Another anti-pollution measure unique to California is our blend of gasoline. Since 1996, gasoline sold in California must meet strict standards of its component chemicals, so it is much less polluting than gas sold in other states. This adds to the cost of the gas, because the major oil companies must refine a special blend for California that they cannot sell anywhere else. The oil companies have no incentive to keep the price of California gas low, and they make the issue worse by keeping the refining capacity for the state at just the bare minimum. Anytime a refinery goes down for repairs, or is damaged, or is closed for the changeover

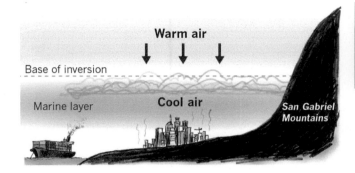

FIGURE 22.2 An atmospheric inversion occurs when a cool layer of air gets trapped beneath a warm layer and against the mountains. Since the warm air cannot sink and the cool air cannot rise, they will not circulate, and the air becomes stagnant and collects pollutants.

Source: Courtesy Wikimedia Commons.

from winter blend to summer blend and then back to winter blend, the gas supply is reduced, and the price shoots up. This is to the advantage of oil companies, who make huge profits every time this happens.[3] As the *Los Angeles Times* reported in 2012:[4]

> Memos from West Coast oil refiners from the 1990s and released years ago by Sen. Ron Wyden (D-Ore.) suggest that this is a deliberate business strategy. An internal Chevron memo, for example, stated: "A senior energy analyst at the recent API [American Petroleum Institute] convention warned that if the U.S. petroleum industry doesn't reduce its refining capacity, it will never see any substantial increase in refinery margins." It then discussed how major refiners were closing down refineries. Oil company profit reports show each dramatic gasoline price spike over the last decade has been mirrored by a corresponding corporate profit spike. This situation is well known to policymakers in California. About a decade ago, after some sharp, unexpected price hikes, then-Atty. Gen. Bill Lockyer formed a gas pricing task force that included industry experts. We viewed industry documents and cross-examined industry representatives. Among the conclusions: "Supply disruptions that contributed to major price spikes of 1999 are likely to continue . . . because (1) California refiners have little spare capacity to cover outages; (2) California refiners maintain relatively low inventory levels." The report also noted: "Refiners have significant market control." The task force recommended a series of measures, including building a strategic gasoline reserve that could flood the market when supply is most scarce. But the Legislature didn't listen. And now we are near 5 bucks a gallon.

It is a touchy political issue in the state, since every California driver hates our high gas prices, especially when there are price spikes due to refinery shutdowns. But the state does not have the power to compel big multinational oil companies to build more refineries and create excess production to cushion the blow when refining capacity is

(a)

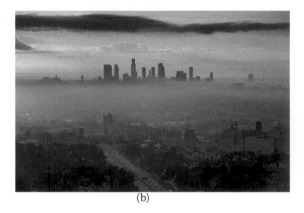

(b)

FIGURE 22.3 At one time, the LA smog was highly visible, especially when viewed from the tops of our mountains and looking down on the inversion layer full of pollutants: (a) View of downtown LA on a smoggy day from near the Griffith Park Planetarium and the Santa Monica Mountains. (b) View above the LA skyline showing the distinct smoggy layer.

Source: Courtesy Wikimedia Commons.

FIGURE 22.4 Smog hangs low in this view south toward the Tehachapi Mountains from the 18th Street parking garage in Bakersfield.

Source: Courtesy Wikimedia Commons.

cut by shutdowns. The governors and the state legislature have tried to act on the anger of Californians against their high gas prices, but unfortunately, there are no simple ways to force oil companies to work in the public interest and prevent them from price-gouging.

Yet our high gas prices are originally a result of the widespread public support of efforts to reduce our air pollution, so there are not many who would give up the efforts to clean our air just to reduce the gas price at the pump by a few cents per gallon. Sadly, most people tend to blame the wrong people: the federal government, the California governor and legislature, or even the hapless gas station owners. They do not realize that ultimately it is Big Oil that is behind our gas prices—and there is not much we can do about it.

Eventually, gasoline- and diesel-powered cars will become more and more rare. The California legislature passed a law mandating that the state will have 5 million electric vehicles by 2030, and already in 2023, California leads the nation with over 1.2 million electric vehicles sold (Florida is a distant second with only 200,000 EVs on the road). In addition, the California legislature has passed laws which will phase out gas and diesel engines altogether in the state by 2050.

The problem of air pollution is not an easy one to solve. The total carbon dioxide emissions in the state has not gone down much, despite all the incentives to move to hybrid or electric vehicles.[5] This is because the population and development continue to grow and sprawl, and as it does so, there are more and more cars on the roads producing greenhouse gases. This is an additional problem—the one regarding urban sprawl. Since the end of World War II, California has opted for more roads, more freeways, and more cars instead of building fuel-efficient public transit and is woefully underserved by less-polluting buses and trains (especially compared to many large cities in the East and Midwest with excellent subways and light rail systems). At one time, Los Angeles had an efficient system of trolleys called the Red Cars and Yellow Cars, which served much of the city. But in the 1950s, the oil companies and automobile companies managed to get the trolley system closed and taken down and their tracks ripped up and paved over.[6] (This was the basis for the plot of the movie "Who Framed Roger Rabbit?") Only now is LA slowly trying to catch up with its Metro rail system and the beleaguered city bus system. But the city is so sprawling, and most parts grew up far from any rail or bus lines, so the vast majority of commuters still have no choice but to drive cars alone to and from work and most other destinations.

Still, the news is not all discouraging. In the 1950s and 1960s, severe smog alerts were common in California (Figure 22.5). In 1977, there were 121 of them in a single year.[7] But by 1996, there were just 7, and most years now there are none. Air quality has improved remarkably over the last 40 years because of the 1990 amendments to the Clean Air Act passed by Congress in 1970. Most recently, year-over-year trends have resulted in reductions in LA air pollution of 10.6% from 2017 to 2018, and another 11.8% from 2018 to 2019.[8]

FIGURE 22.5 Comparison of a bad smog alert in Los Angeles in 1960, compared to the air quality in the same area in 2021.

Source: Courtesy Wikimedia Commons.

22.3 WATER POLLUTION

As with any region with a large population and lots of industry, California often suffers from problems of water pollution, especially groundwater contamination.

22.3.1 THE GOLD RUSH AND MERCURY CONTAMINATION

The problem with this kind of pollution goes all the way back to the days of the Gold Rush in the 1850s and 1860s (Chapter 16), when it was a free-for-all with no regulation as miners raped the landscape in their search for gold. They washed away entire hillsides with giant fire hoses to release the gold from ancient riverbed gravels and produced ugly scarred hillsides that have not recovered in over 160 years (Figure 16.4).

But this was not the only bad legacy the miners left behind. In addition to placer mining, miners also switched to shaft mining the gold in the bedrock, typically along quartz veins intruded into older metamorphic rocks. These shaft mines left huge spoil piles of waste rock dumped near their entrances. The mines were eventually exhausted and abandoned, but there was no regulation on the dangerous metals that they left behind. Rainwater would percolate down through the mine waste and wash away all sorts of dangerous heavy metals into the rivers and groundwater.

There was another method which helped extract more gold from the worthless waste rock, or from tiny flakes of gold in a hydraulic mining operation. The fine amounts of gold at the bottom of a sluice could be concentrated by adding mercury (also known as "quicksilver") to it. Hundreds of pounds of liquid mercury were added to the sluice, in 76 lb flasks, and the gold–mercury amalgam would sink to the bottom. But the fast-flowing water also washed away a high percentage of the mercury before it could be trapped and settle into the riffles. As Alpers and others (2005) described it:

Gravel and cobbles entered the sluice at high velocity caused a mercury to flour, or break into tiny particles. Flouring was aggravated by agitation, exposure of mercury to air, and other chemical reactions. Eventually, the entire bottom of the sleeves became coated with mercury. Summer mercury was lost from the sluice, either by leaking into underlying soils and bedrock or being transported downstream with the placer tailings. Minute particles of quicksilver could be found floating on surface water as far as 20 miles downstream of mining operations. Some remobilized placer sediments, especially the coarser material, remained close to their source in ravines that drained the hydraulic mines. . . . The total amount of mercury lost to the environment and placer mining operations throughout California has been estimated at 10,000,000 pounds, of which probably 80 to 90% was in the Sierra Nevada. Historical records indicated about 3,000,000 pounds of mercury was lost at hard rock or gold or silver was crushed by stamp mills.

Mercury has a unique property in that it binds well will precious metals, especially gold and silver. Once you have obtained an amalgam of gold and mercury, it can be heated and the mercury vaporizes away, leaving pure gold. This process is still important in small-scale gold mining even today.[9] In fact, even though it is banned in California, it is the most common form of mining in much of the rest of the world, producing about 15% of world gold production—and horrendous pollution problems.[10]

But there is a big problem with this method. Mercury is a highly toxic metal, even in small amounts. As one source describes it:[11]

The process begins when miners pump a mixture of water and sediment from a riverbed into a trough, where the sediment can be suspended into a slurry—a technique known as hydraulic mining. Next they add mercury, which binds to the gold particles, forming an amalgam. Mercury is heavier than pure gold, so the balls of amalgam sink to the bottom of buckets or holding ponds where they can be collected. Finally, workers burn off the mercury—often with a hand torch or in a crude stove—leaving gold metal behind. This process releases mercury to the environment in two forms. First, tailings, or waste material, can contaminate nearby land and aquatic ecosystems. Second, mercury vapor enters the atmosphere and can travel long distances before being deposited to land and water via rainfall or small dust particles. In the environment, microbes can transform mercury into a more potent form known as methylmercury. Methylmercury can be taken up by bacteria, plankton and other microorganisms that are then consumed by fish and build up to dangerous concentrations in animals higher on the food chain.

Even in the 1850s, a lot of mercury was used, and the loss of mercury was estimated at 10–30% per season (Bowie, 1905). Since over 1.5 billion cubic yards of placer sands and gravels were processed at this time, this released an enormous amount of mercury into the water. As the preceding sources indicate, about 13 million tons of mercury were released into the waters of California, and most of it is not accounted for.

Today, there is a significant industry focusing on tracing this mercury contamination in the surface waters and groundwater, and especially sampling fish to make sure they are not concentrating the mercury. There are often advisories posted to prevent fishing in areas of high contamination, because if you eat a significant amount of fish contaminated with mercury, it gets concentrated in your own tissues and can cause many illnesses or even death. Even as recently as 2005, the state of California has issued advisories for about 20 watersheds downstream from the gold mines, especially the Bear River and Yuba River watersheds, the lower American River including Lake Natoma, and the Trinity Lake area.

22.3.2 LEAD POLLUTION

Another component of pollution was the lead in gasoline, which was put in the mixture to reduce engine pinging. In the 1960s, Claire Patterson of Caltech was trying to measure the lead–lead ratios in ancient rocks to date them. He created one of the first "clean labs" for lead analysis ever developed and soon found that no matter how hard he tried, he could not eliminate the external lead contamination that was coming into his "clean" lab and ruining his measurements. Eventually, he made measurements of environmental samples and soon proved that the entire planet was being contaminated by low amounts of lead, primarily from gasoline. He published his findings in 1965 and fought hard against the smear campaign to discredit him by the combined auto, oil, and lead industries, fighting the science and the public good to maintain their profit margins.

Studies done in the late 1960s showed that there was significant lead in the drinking water in many parts of California. Not only was it coming directly from engine exhaust into the air, and then into the water, but it was also incorporated into the soils (especially urban soils) and can get resuspended in the air when the soil is disturbed. The fact that Patterson could find it everywhere—biological tissues, soil samples, water and air samples, even in our clothes and most everyday objects—showed how ubiquitous this poisoning by the lead in gasoline was.

Lead is so toxic that most environmental agencies consider *no* level of lead in air or water to be safe. As the California Air Resources Board (CARB) puts it:[12]

> Once taken into the body, the blood carries lead throughout the body and it is deposited in the bones where it accumulates. Because lead is only slowly excreted, exposures to small amounts of lead from a variety of sources can accumulate to harmful levels. Lead can adversely affect multiple organ systems of the body and people of every age group. Young children are particularly at risk of lead poisoning. They are usually exposed to lead through the normal hand-to mouth behavior that occurs through crawling or playing on the floor, and putting their hands, toys, and other items in their mouths. In children, adverse health effects of lead exposure are often irreversible and include brain damage and mental retardation. Lead poisoning is often unrecognized in children, and if undetected, it may result in behavioral problems, reduced intelligence, anemia, and liver or kidney damage.

By the 1970s, the popular support for the environmental movement pushed politicians to pass laws banning lead in gasoline. Japan was the first nation to do so in 1975, and state by state (California being among the earliest to do so), leaded gas was banned in the United States, until it was finally phased out nationwide in 1986. Lead levels within the blood of Americans are reported to have dropped by up to 80% by the late 1990s. Today, the biggest source of lead poisoning is lead-based paint (although this kind of paint has been largely banned in California) and, most often, lead pipes in older buildings and water systems. There are still reports of lead poisoning whenever one of these problematic pipe systems is found to have contaminated the drinking water. But the problem with lead in every part of the environment due to leaded gasoline is now a thing of the past.

22.3.3 REGULATION OF WATER POLLUTION

Since the passage of California's Porter-Cologne Water Quality Act and the federal Clean Water and Safe Drinking Water Acts of the 1960s and 1970s, the problem with water contamination in California has been greatly reduced. This is a welcome fact, considering how little progress we seem to make on reducing air pollution. State and federal regulatory agencies are constantly monitoring water all over the region and looking for signs of pollution that might endanger us.

Nevertheless, there are significant problems beyond mercury or lead in the water. During our long-term drought, the rivers and lakes of California have been tested and found to have poor water quality due to the heat and low flow of water leading to rapid evaporation and stagnation, producing waters that are too salty, too warm, and too low in oxygen for many aquatic organisms. This leads to toxic blooms of algae that thrive in these conditions, or death of fish and other sensitive aquatic life.

Some places, like the San Joaquin Valley, have their own peculiar issues with water pollution in their water supply. The region draws most of its drinking water from groundwater, but decades of using fertilizers and pesticides in the region has contaminated the drinking water supply. About 63% of the water in the San Joaquin Valley is contaminated by nitrates from fertilizers. Arsenic, uranium, and excessive fluoride have also been detected, mostly from pesticides, as well as from leakage from power plants and raw sewage. The famous case of the PG&E plant in Hinkley, California, leaking toxic hexavalent chromium into the drinking water supply was the subject of the Oscar-winning movie *Erin Brockovich*, and the problem was found in other PG&E facilities as well.

The problems with groundwater contamination will never go away completely, but thankfully, the regulatory agencies and powerful political forces (and investigative journalists) have been diligent in digging up the rare cases and making the water safe in most of the state.

NOTES

1 https://time.com/3399134/air-pollution-climate-change-bakersfield-caifornia/
2 https://energyathaas.wordpress.com/2023/01/09/whats-the-matter-with-californias-gasoline-prices/
3 https://energyathaas.wordpress.com/2023/01/09/whats-the-matter-with-californias-gasoline-prices/
4 www.latimes.com/opinion/la-xpm-2012-oct-12-la-oe-court-california-gas-prices-20121012-story.html
5 www.reuters.com/article/us-usa-climatechange-california-insight/a-climate-problem-even-california-cant-fix-tail-pipe-pollution-idUSKCN1PQ4MJ
6 https://inhabitat.com/what-happened-to-los-angeles-streetcars/#:~:text=In%201963%2C%20the%20Los%20Angeles,service%20in%20the%20LA%20region.

7 www.aqmd.gov/home/research/publications/50-years-of-progress#Cleaning%20Up%20Cars
8 www.iqair.com/us/usa/california/los-angeles
9 www.epa.gov/international-cooperation/artisanal-and-small-scale-gold-mining-without-mercury#:~:text=Mercury%20is%20mixed%20with%20gold,mercury%20exposure%20and%20health%20risks.
10 https://theconversation.com/gold-rush-mercury-legacy-small-scale-mining-for-gold-has-produced-long-lasting-toxic-pollution-from-1860s-california-to-modern-peru-133324
11 https://theconversation.com/gold-rush-mercury-legacy-small-scale-mining-for-gold-has-produced-long-lasting-toxic-pollution-from-1860s-california-to-modern-peru-133324
12 https://ww2.arb.ca.gov/resources/lead-and-health

RESOURCES

Almaraz, M., Bai, E., Wang, C., Trousdell, J., Conley, S., Faloona, I., and Houlton, B.Z. 2018. Agriculture is a major source of NO x pollution in California. *Science Advances, 4*(1), eaao3477.

Alpers, C.N., Hunerlach, M.P, May, J.T., and Hothern, R.L. 2005. Mercury contamination from historic gold mining in California. *U.S. Geological Survey Fact Sheet,* 2005–3014.

Bowie, A.J. 1905. *A Practical Treatise on Hydraulic Mining in California.* Van Nostrand, New York, 313 p.

Bradley, E.M. 1918. Quicksilver resources of the state of California. *California State Mining Bureau Bulletin, 78,* 1–389.

Cisneros, R., Brown, P., Cameron, L., Gaab, E., Gonzalez, M., Ramondt, S., Veloz, D., Song, A., and Schweizer, D. 2017. Understanding public views about air quality and air pollution sources in the San Joaquin Valley, California. *Journal of Environmental and Public Health, 2017,* 1–7.

Nelson, T., Chou, H., Zikalala, P., Lund, J., Hui, R., and Medellín-Azuara, J. 2016, March 23. Economic and water supply effects of ending groundwater overdraft in California's Central Valley. *San Francisco Estuary and Watershed Science, 14*(1).

Pollack, I., and Ryerson, T. 2013. Trends in ozone, its precursors, and related secondary oxidation products in Los Angeles, California: A synthesis of measurements from 1960 to 2010. *Journal of Geophysical Research: Atmospheres, 118*(11), 5893–5911.

Ritz, B. 2002. Ambient air pollution and risk of birth defects in southern California. *American Journal of Epidemiology, 155*(1), 17–25.

Schell, L. 2006. Effects of pollution on human growth and development: An introduction. *Journal of Physiological Anthropology, 25*(1), 103–112.

Van Meter, R.O. 1948. Investigating water pollution in California. *Journal (American Water Works Association, 40*(7)), 784–786.

23 California's Renewable Energy Resources

There is an urgent need to stop subsidizing the fossil fuel industry, dramatically reduce wasted energy, and significantly shift our power supplies from oil, coal, natural gas, to wind, solar, geothermal, and other renewable energy sources.

—**Bill McKibben**

23.1 RENEWABLE ENERGY IN CALIFORNIA

As discussed in Chapter 17, California has long been a major oil-producing state, the largest in the United States only a century ago. But much less appreciated is how much renewable green energy California makes and how much more it could produce if the right incentives were in place. This state creates more renewable energy than any other state except Texas. It ranks #1 in the nation as a producer of electricity from solar, geothermal, and biomass resources. As of 2017, over half (52.7%) of its energy came from renewables, and that percentage will only increase.

California has the largest geothermal complex in the world, producing about 20% of its total renewable energy. This state ranks first in the nation in total solar generation, and that lead will only expand. California is fourth in the nation in wind power generation (after Texas, Oklahoma, and Iowa) and is making a large and growing investment in wind power. Numerous dams on the major reservoirs in California produce hydroelectric power. Finally, although much reduced from its heyday in the 1960s, there are still nuclear power stations in California as well.

California is also unusual in lots of other ways. Even though its total energy consumption is the second highest in the nation (because it has the largest population of any state), its per capita energy consumption is the fourth lowest, because its mild climate means a much less demand for heating and cooling, and also because it has long had many energy-efficiency programs in place. The percentage of renewable energy by population is also remarkable, because states with similar or higher percentages of renewable energy use have much smaller populations. In 2009, California produced 8.43% of the nation's renewable energy, second only to Washington. By 2017, California was first, with 10.05% of the nation's renewable production.

23.2 IN HOT WATER

Geothermal energy is produced wherever naturally heated water or steam can be conducted through pipes and into a turbine, driving the turbine blades and creating electricity.

It uses existing systems of natural heat, usually near volcanoes or underground magma chambers, so it is completely renewable and produces no carbon dioxide or other pollution. The Earth has something like 10^{31} joules (3×10^{15} terawatt-hours) of energy, about 100 billion times the total for worldwide energy consumption, so geothermal is potentially an enormous source of heat. It is a true green and renewable energy source, since it will never run out, and it uses no fossil fuels. However, there are only a limited number of places in the planet where that heat is close enough to the surface and the water percolating down to it can be economically tapped and turned into energy in a power plant. The major limitations of geothermal are the costs of drilling down deep to tap the boiling water and the capital costs of building the plant and maintaining it. Unlike many other forms of energy production, it is highly scalable, so a small plant can power a rural village, where a huge one can supply an entire region.

23.2.1 THE GEYSERS, NORTHERN CALIFORNIA

The largest geothermal operation in the world is the Geysers (Figure 23.1), in the Mayacamas Mountains in Sonoma and Lake Counties, about 72 mi north of San Francisco. It is a complex of over 18 power plants, drawing on more than 350 wells spread over about 30 square miles. This one region produces about 20% of California's renewable energy and meets 60% of the power demands of the region between the Golden Gate and Oregon.

It produces mainly dry steam which directly powers the turbines, rather than pumping up superheated water from below. Overall, it has about 1,517 MW of installed capacity. It is located on the northern limb of the Mayacamas anticline, bounded by the Collayomi fault on the northeast and the Mercuryville fault on the southwest. The bedrock is part of the Franciscan terrane, consisting mostly of fractured and sheared graywackes. There is a large gravity anomaly and lots of seismicity at the site. These all point to a shallow magma body from the Clear Lake volcanic field beneath the geothermal field, which is the ultimate source of the heat. The intrusion is about 4 mi (6.4 km) beneath the surface and is greater than 8 mi (13 km) in diameter.

The hot springs in the area were extensively used by the local Native Americans, and in 1847, William Bell Elliott, a member of explorer John C. Fremont's party, reported on them during his survey of California and called them "the gates of hell." They were eventually called "The Geysers," even though there are no actual Yellowstone-style geysers there, only fumaroles and hot springs. Between 1848 and

DOI: 10.1201/9781003301837-23

(a)

(b)

FIGURE 23.1 The Geysers Geothermal Field. (a) The Sonoma Calpine 3 power plant is one of 18 power plants at the Geysers. (b) Drilling a geothermal well in 1977.

Source: Courtesy Wikimedia Commons.

1854, Archibald Godwin developed the hot springs of the region into a resort named the Geysers Resort Hotel, which once hosted Ulysses Grant, Theodore Roosevelt, and Mark Twain. The resort declined after the 1880s and finally closed in 1979 after landslides and fires had long ago destroyed most of the property. In 1960, the first geothermal plants were built in the area, and now there are 18 plants in active operation.

By 1999, the natural steam to power extraction had begun to deplete the field, so production began to drop. Since 1997, the Geysers field is now recharged by about 11 million gallons per day of treated sewage effluent, which is pumped down the wells to increase the volume of steam and get rid of wastewater at the same time. The water pipes for sewage effluent run about 50 km (80 mi) from the Lake County Sanitation treatment plant, which treats the wastewater of Santa Rosa.

23.2.2 Salton Sea Geothermal Field, Imperial County

As mentioned in Chapter 15, the Salton Trough is also famous for its geothermal activity and hot springs. Magma is just a few kilometers below the surface, so the groundwater that sinks down boils and rises to the surface again. Native peoples had used some of these hot springs for centuries. In the early 1900s, the hot springs were developed commercially with therapeutic spas in the foothills of the Chocolate Mountains near Bombay Beach on the east shore of the Salton Sea. This spring is still in use and is unusual for its high water temperature, ranging from 57°C to 82°C (135°F to 180°F). With few exceptions, the hot springs are

concentrated in a linear pattern along the eastern side of the valley. The line of springs extends from Desert Hot Springs into Mexico, and the arrangement strongly suggests that the warm waters are reaching the surface using fractures of the San Andreas fault system as conduits.

There are several major geothermal power plants in the Imperial Valley, extending from the south shore of the Salton Sea into Mexico. The Salton Sea geothermal field is the largest and hottest of the several fields in the Salton Valley and has the longest history of development. Over in Mexico is the Cerro Prieto geothermal field near Cerro Prieto volcano. It is a very large, productive energy source. The geothermal waters are the result of a complex subsurface heat transfer system. Convection within the mantle provides a continuously renewable source of heat. The heat is conducted through the thin continental crust. Surface waters migrating downward are heated and then dissolve chemical compounds from the rocks undergoing metamorphism and rise by convection through the water-saturated sediments to the surface.

The potential for the development of geothermal energy resources was first recognized in the mid-1920s. However, it was not until 1961 that the first commercial well in the Imperial Valley was drilled. It reached a depth of more than 1,433 m (4,700 ft). The energy crises of the 1970s spurred renewed interest in commercial development, and several wells were drilled to depths of 1,534–2,440 m (5,000–8,000 ft). In the Salton Sea geothermal field, typical brines are produced at wellhead temperatures up to 315°C (600°F). During extraction, the high temperatures and reduced pressure in the drill holes cause the superheated water to flash into steam, sending a mixture of steam and hot water at the wellhead. The steam and water solution is highly charged with chemical salts, principally sodium chloride, calcium chloride, and several metallic compounds. A unique characteristic of the brine is the high concentration of potassium in addition to sodium, and the presence of dissolved rare elements, such as lithium. The brine is slightly caustic, and severe corrosion and scaling problems complicate the production of clean turbine steam. Although the resource is large, technical problems and cost factors associated with processing the hot brines are a continuing constraint to large-scale commercial development.

There are numerous large geothermal plants (Figure 23.2a) on the south shore of the Salton Sea, and at the corner of Davis and Schrimpf Roads in Calipatria, there are striking mud volcanoes (Figure 23.2b). The area was a hotbed of geothermal research for the past 50 years, and especially so after recent efforts by the California state government to encourage renewable energy. The area also has huge solar plants because of the year-round sunshine and miles of undeveloped desert land.

Naturally, if the region has hot springs and geothermal power plants, magma must be close to the surface. Indeed, there are many volcanoes all around the area, some of them very young. Careful geophysical studies have identified the source, or magma chamber, that supplies the volcanics. Geologists think there is a hot magma chamber that runs parallel to the axis of the Salton Trough. It is about 32 km (20 mi) long by 6 km (4 mi) wide, and at least 3 km (2 mi) thick. It is at least 3,000 m (10,000 ft) and possibly as shallow as 1,200 m (4,000 ft) from the surface. The magma chamber is mostly beneath the town of Niland on the southeast shore of the Salton Sea. It is causing low-grade metamorphism of the sedimentary fill of the basin. The brines that come to the surface from deep thermal wells show that new ores are being produced by this hydrothermal solution around the pluton, especially sulfides of iron, lead, zinc, and copper.

23.2.3 OTHER GEOTHERMAL FIELDS

As mentioned earlier, the Geysers is the largest geothermal field in the world, producing over 1,590 MW. Four different fields in the Salton Sea in Imperial County produce the next most power, with the Imperial Valley plant yielding 432 MW, the Heber plant with 161.5 MW, the Ormesa plant producing 101.6 MW, and the North Brawley field yielding 64 MW. Next largest is the Coso geothermal plant in Inyo County, not far from Death Valley, which produces 272.2 MW. Finally, Mammoth Mountain ski resort is well-known for its natural hot springs, and there is one geothermal operation, the Mammoth plant, which produces 40 MW of energy.

23.3 WATER OVER THE DAM

When people think of hydroelectric power, they focus on the giant dams on the Colorado River, such as Hoover Dam and Glen Canyon Dam, or the biggest of all, Grand Coulee

Dam in Washington. In Canada, provinces like Quebec use almost 100% renewable energy, because they have so many hydroelectric facilities on the rivers in the province. But California generates a surprising amount of hydroelectric energy as well. It is a form of renewable energy because all it needs is water flowing down through a turbine to generate electricity, and only the energy of gravity pulling water downhill is required—no fossil fuels or pollution.

California has more than 50 hydroelectric plants that produce at least 50 megawatts of power each, and many smaller ones. The biggest is mighty Shasta Dam, holding back Lake Shasta, which produces 714 MW. Second biggest is Oroville Dam in the Sierra foothills, producing 644 MW (Figure 23.3). The next largest are also dams in the Sierra foothills, holding back reservoirs of water, especially snowmelt, from the Sierra Nevada Mountains. Many of these were built in the 1960s and 1970s, as part of the California State Water Project, which also built the California Aqueduct and many other structures. These include the third largest power plant in California, New Bullards Bar Dam in Yuba County (315 MW), plus New Melones Dam in Calaveras County (300 MW), White Rock Powerhouse in El Dorado County (266 MW), Folsom Dam near Folsom (198.6 MW), Mammoth Pool Powerhouse in Madera County (190 MW), and numerous others. There are a few exceptions to the dominance of Sierra foothill

(a)

(b)

FIGURE 23.2 The Salton Sea geothermal field: (a) The 38 MW Leathers Geothermal Plant in Calipatria, CA. The Salton Sea area has an unusually high level of geothermal activity, as evidenced by fields of volcano-like mud pots. This plant is one of ten geothermal generating plants owned by Calenergy in the Imperial Valley of California. (b) Salton Sea mud volcanoes.

Source: Courtesy of Wikimedia Commons.

FIGURE 23.3 The dam and hydroelectric plant for Oroville reservoir.

Source: Courtesy Wikimedia Commons.

dams, such as Parker Dam on the Colorado River near Lake Havasu (120 MW), Devil Canyon Dam in San Bernardino County (276.2 MW), and the William E. Warne dam on the Pyramid Lake reservoir in the Ridge Basin along the Grapevine on Interstate 5 (part of the California Aqueduct), which produces 74 MW of power.

More dams have been proposed for additional sites, especially in the rivers in the Klamath region, which discharge into the Pacific. But most of those proposed dams met with strong opposition and were cancelled by the 1980s. In fact, the golden days of dam building in the United States are long over. During the early 1960s, Congress was enthusiastic about building dams and paid for dams in just about every suitable canyon in the western United States. But by the 1970s, the backlash from environmentalists began to focus on the negative effects of dams and reservoirs, especially how the Glen Canyon Dam destroyed the beautiful landscapes and rock formations of Glen Canyon in Utah. Dams also create problems for fish migrations and require elaborate fish ladders to keep these species from going extinct.

In California, the drought and the reduction of flow from the San Joaquin and Sacramento rivers meant problems for the Sacramento Delta region, which becomes too salty from intrusion of seawater contamination out of San Francisco Bay if there is not enough fresh water coming down to push it back. Thanks to 23 years of drought, many people (especially farmers in the Central Valley, whose crops are by far the biggest users of water in the state) resent the dams releasing *any* water to the ocean. There is a small but vocal political movement to hold back most of the water in California's rivers and save it for agriculture. However, this fails to realize that the flow of fresh water from rivers and into the ocean is essential for these ecosystems, especially their wildlife. If the fresh water is all held back, many rivers in California will become too salty or brackish in their estuaries, and that would wipe out the wildlife in those regions. Even many ecological systems offshore depend on a steady supply of fresh water to decrease salinity and allow key marine communities to grow. Humans are very selfish about water use but don't realize that fresh water belongs not just to humans but to the entire ecosystem, and when we deplete it too much, we destroy the planet.

Other problems are also created by dams, such as excess evaporation from the surfaces of big reservoirs. Dams trap the sediment that would normally flow downstream and replenish the beaches with their supply of sand. At the moment, no more new dams are being built in California, nor are any approved for building. In fact, some dams, such as the controversial Hetch Hetchy Dam near Yosemite, which flooded a beautiful canyon once celebrated by John Muir, are now dismantled. This may change if the pattern of drought persists and more water is needed. In addition, hydroelectric power is a form of renewable energy that costs almost nothing to produce once the facilities are in place. Hydroelectric power and dam building is a complex problem with many conflicting interests at stake, so there will always be big political battles over building more dams.

23.4 THE WARMTH OF THE SUN

The most successful and rapidly growing form of renewable energy in the world is solar power. Ultimately, almost all our energy comes from the sun, whether it is absorbed directly in solar installations or retrieved indirectly from the motion of wind driven by heating and cooling of the atmosphere by the sun, or it represents solar energy long ago trapped by plants in the form of fossil fuels. (The exceptions are geothermal, which comes from the Earth's radioactive decay that warms the planet, and hydroelectric, which is powered by gravity). In many parts of the world, solar energy has grown so quickly that it is now cheaper and more efficient to use than any other power source, other than natural gas. The coal industry in most parts of the world is nearly extinct because coal cannot compete with the prices of solar and natural gas.

California's government has long focused on solar power, both under Democratic and Republican political leaders, because solar is almost universally accepted as the best solution as a low-cost source of energy that is the best in reducing greenhouse gas emissions—and California has the sunshine to make it feasible nearly everywhere. This enthusiasm for solar energy goes back as far as 1978, when Congress passed the Energy Tax Act after the Arab oil embargo, which generated the second energy crisis in the United States. A 40% tax credit was given to homes that installed solar devices after April 20, 1977, although the Reagan Administration rolled back that policy.

State laws now mandate that 60% of California's energy must come from renewables by 2030, and 100% by 2045. To meet that goal, most of that increase is going to come from solar, which has the greatest potential for expansion. At the end of 2022, California had 38,145 MW to solar capacity, enough to power 10.5 million homes, with solar power providing 27% of that. By 2027, we are expected to increase our solar generation by 27,000 MW, second only to Texas in solar capacity. Lots of different incentives and tax breaks are available to install solar facilities, and in 2020, the legislature passed a law that requires all new homes built in California to have solar panels on their roofs.

As early as 1979, California had the world's large photovoltaic cell manufacturing facility near Camarillo (owned by ARCO), and four years later, ARCO built a 6 MW solar plant on the Carrizo Plain. In 1986, the largest solar energy facility in the state at that time was built in the Mojave Desert. In 1993, PG&E built a 500 KW facility near Kernan, California. In 1996, Republican Governor Pete Wilson and a bipartisan majority of the legislature passed bills which boosted incentives for solar energy production across the state.

There are actually three major methods for generating solar power.

1. The most common form of solar energy is produced by some sort of photovoltaic cell (PV for short), where the sunlight is absorbed and converted to electricity using semi-conductors in the PV cell. These kinds of facilities can be operated at any scale, and they are the main types of solar plants that are being developed for the future. Their main limitation is that PV cells use a number of scarce metallic elements, which in turn can hamper the large-scale construction of lots of facilities. There are also environmental issues concerning the mining of these scarce elements, as well as strategic problems getting certain materials (such as rare earth elements, which are controlled by China). Thus, although solar energy itself is limitless, making solar cells has a supply and construction bottleneck that has potential to limit the widespread expansion of PV solar plants.

2. Concentrating solar power (CSP) facilities have a large array of mirrors that focus solar energy on a water-driven turbine so that when the water boils, it drives the blades of the turbine and produces electricity. One of the largest of these is the Ivanpah Solar Power Facility north of Interstate 15, just across the state line from Primm, Nevada—a familiar sight to anyone who has driven on Interstate 15 between Las Vegas and Southern California (Figure 23.4). Built in 2014, it can generate 392 MW of power using an array of 173,000 heliostat mirrors on the desert floor, which focus their energy on three towers standing 459 ft (140 m) above the surface and produce heat to power huge boilers within them. This project has been successful ever since it was first installed, although there was some controversy because the desert floor is also habitat for the protected species of desert tortoise, and some birds have been killed when passing through the intense heat of the mirror beams. Even though there are a number of other CSP facilities in Nevada and other Western states, most of the research and funding is now going toward PV systems, which are cheaper to operate and becoming even more economical as their technology improves.

3. Solar heating and cooling (SHC) systems are the kind most often installed on the roofs of private homes. These use an array of water pipes inside the solar panels, which heat up water flowing through them as the sun hits them. The hot water then goes down to an insulated storage tank, so no electricity is used to heat the water. The system also works to cool water as well. These types of systems are rapidly becoming the dominant system for domestic use in the United States, and more than 50% of all domestic solar energy devices in Austria, Switzerland, Denmark, Norway, Sweden, and Germany are of this type.

Solar is clearly the wave of the future, but there are challenges. The biggest limitation is that it only works in the daytime, so there must be battery storage, or else, some other form of power generation must kick on during the night to supply the power demands after sunset. Thus, battery research is an equally important part of the process, along with improved solar cells. In addition, during the summer months, there can be too much energy generated, more than the grid can handle, leading California to send a lot of its excess summer electricity to Arizona.[1]

As we already discussed, there are logistical and infrastructure issues as well. PV solar cells require lots of scarce metals, so their supply can experience a bottleneck and slow down the development of solar installations. The other limitation is that solar installations often are in remote areas, especially California's deserts, where they are not impinging on communities. That requires lots of power transmission lines to carry the energy long distances, which is not a trivial problem when it comes to cost and infrastructure.

Another problem hindering the transition to solar has to do with capitalism. Simply put, the profit margins for solar are not large enough (normally requiring 8–10% return on investment, and solar is less than 5%) for most major energy companies, and the big investment firms, to justify spending their money on.[2] Solar energy is unlimited and can be obtained almost anywhere, and the cheaper it gets, the less profit the moneyed interests can make off it. In addition, it is almost impossible to monopolize solar energy or restrict or control its supply to drive up the price, like it is for oil (which has *huge* profit margins). The oil supply of the world can be manipulated by just changing the supply released by the handful of powerful interests that control the world's oil supply—and they do not have to worry about competition. By contrast, once it is possible for many different organizations to set up solar energy power plants, it is almost impossible to restrict supply and drive out the competition. (In the oil business, the Saudis have the world's largest oil supply, and in the years between 2013 and 2016, they deliberately flooded the market with their cheap oil to bankrupt the smaller oil companies and wildcatters and punish Russia and Iran.)

The problem is analogous to another energy transition at the beginning of the 1800s, when the Industrial Revolution took place.[3] Water power was the first major source of energy to drive the production in factories, and factory cities were set up where there were waterfalls and other natural sources of water energy. But part of the reason that coal took over to drive energy production was that its supply could be restricted and manipulated, plus it made it possible to ship the source of energy to big cities, where labor was

cheap and abundant, allowing industrialists to exploit workers and monopolize the market. But just like today, when there is no penalty for capitalists to harm the environment with greenhouse gases from oil, back in the 1800s, the skies over Britain, Germany, and the eastern United States were heavily polluted with coal smoke, and no one could make the polluters pay for it. Thus, many economists argue that the only way solar power will finally drive out oil and gas from the energy market will be to have governments subsidize it or treat it as a public utility and remove the profit motive from energy production altogether. That is already true for many hydroelectric plants, which were built by the government when dams were built and run in the public interest and not by private firms trying to get rich.

Although some areas have difficulty finding places to put in solar plants without disturbing fragile ecosystems, most people are not as "NIMBY" ("not in my backyard") as one might expect when it comes to solar. More and more houses have solar panels on their roofs, and parking lots which have solar panels that act as shade roofs for vehicles are getting more and more common. A 2012 survey of 1,000 people in Southern California showed that 80% supported development of solar power in their communities, and many were also concerned about climate change. Two-thirds said that renewable energy is important to California's future and support further development of solar power. Solar is the future of renewable energy in California, and it will only become more and more dominant as we phase out fossil fuels.

23.5 THE ANSWER IS BLOWING IN THE WIND

Next to solar energy, the most rapidly growing area of renewable energy resources is wind. California's wind power capacity has grown 350% since 2001, when it was less than 1,700 MW; now it is over 6,000 MW. By 2016, wind supplied over 7% of California's total electricity needs, enough to power 1.3 million households. As already mentioned, California ranks fourth in the nation in wind power, after Texas, Oklahoma, and Iowa. But California was the first state with large wind farms in the 1980s, and in 1995, it produced 30% of the entire world's wind-generated electricity.

Wind farms could be sited in most places, but to be economically viable, they are best situated where the winds are very strong and steady (especially when funneled through a mountain pass). They also need to be far enough from major cities that the huge wind turbines can operate unimpeded. The major wind farms in California include two in the Bay Area (one in Altamont Pass in Alameda County, with capacity of 576 MW, and Shiloh plant in Solano County with 505 MW capacity), the Alta Energy Center (1,548 MW) and the Tehachapi Pass wind farm (705 MW) in the Tehachapi Mountains between the San Joaquin Valley and the Mojave Desert, the giant San Gorgonio Pass wind farm (Figure 23.5) along Interstate 10 between Banning and Palm Springs (capacity of 615 MW), and the Ocotillo Wind Energy Project in Imperial County, with 315 MW capacity.

(a)

(b)

FIGURE 23.4 The Ivanpah Solar Electric Generating Facility, near Primm, Nevada. (a) Looking north from Interstate 15 at the eastern boiler of the Ivanpah facility. (b). The solar energy reflected from the mirrors on the desert floor is illuminating the top of the tower and heating a boiler inside, generating steam that drives a turbine and creates electricity.

Source: Courtesy Wikimedia Commons.

(a)

FIGURE 23.5 The San Gorgonio Pass wind farm, near Interstate 10 and Whitewater, California: (a) Panoramic view shot looking north from Mt. San Jacinto at the rows and rows of giant wind turbines on the desert floor. Interstate 10 runs across the middle of the wind farm. (b) Another view of the rows of wind turbines on the windy slopes of San Gorgonio Pass.

Source: Courtesy Wikimedia Commons.

Several more wind farms are in the works, including the Tule Wind Project in San Diego County, which already has 57 wind turbines and 131 MW of production. The next stage will be offshore wind farms, which are already common throughout Europe and many other parts of the world. Although wind has not grown quite as fast as solar has, it will probably become the second largest renewable source of energy in the future in California.

23.6 BIOMASS POWER

Another renewable energy source is biomass power, which extracts the energy stored in plants and animals. California is first the nation for use of biomass in power generation. It produced 5,767 gigawatt-hours of electricity in 2017, making up about 2.8% of the state's total energy usage.

Biomass energy comes mostly from the processing of plant and animal waste, from plant composting, to capturing the gases released by decay in landfills, to processing municipal solid waste and recovering the gases produced by decay. These are the most common forms of biomass power, but there are others. One electricity plant uses chipped-up forest residue and captures the energy from its decay.

23.7 NO NUKES?

During the 1960s and 1970s, California was one of the leading states pushing for the development of nuclear power. Even though nuclear is not technically classified as a "renewable" energy source (since it's based on decay of radioactive elements in minerals, creating heat), it uses just a small quantity of radioactive materials to produce enormous amounts of energy and produces no greenhouse gases, so it does not contribute to climate change. It does, however, produce nuclear waste, which requires careful handling and storage, but almost all forms of energy production have some sort of waste product that must be disposed of carefully.

Then in 1979, the Three Mile Island accident occurred in Pennsylvania, followed by the Chernobyl nuclear reactor meltdown in the Soviet Union in 1986, and the simmering resistance to nuclear power plants reached a fever pitch. Anti-nuclear protests had been happening since the

early 1970s, but these two accidents frightened the governments of the world, so many of them either stopped building nuclear plants entirely or added so many regulations and delays that they were too costly to build. The disaster when the Tohoku quake and tsunami hit the Fukushima Daiichi plant on March 11, 2011, and spilled lots of radioactive materials into the ocean further galvanized the worldwide resistance to nuclear power. Still, France, Japan, and several other countries generate almost all their electricity through nuclear power, and most of these countries have never had an accident.

Like most states, California was heavily into building nuclear power plants in the 1960s. These included the Rancho Seco Nuclear Generating Station near Sacramento, the San Onofre Nuclear Generating Station between Camp Pendleton and San Clemente, the Vallecitos Nuclear Center 30 mi east of San Francisco in Alameda County, the Humboldt Bay Nuclear Power Plant just south of Eureka, California, and several others were planned but never built. Over the past 30 years, each of these has been permanently shut down because of political pressures and fear of a disaster. One big concern is that nuclear reactors usually need to be built near an abundant source of cooling water, so they are often located near a river, lake, or the ocean, and their hot water waste can damage the ecosystem downstream from the reactor. Another problem almost exclusive to California—reactors are almost always located near known faults, because faults crisscross nearly all of California (especially coastal California), so it's had to find places to build that are *not* near a fault. The San Onofre site, for example, is just a mile away from the Cristianitos fault, but along the beach cliffs, you can see that the fault is inactive because it is covered by a thick mantle of Pleistocene terrace deposits that are not faulted, so it has not moved since at least the late Pleistocene. Other nuclear plants were built before the nearby faults were discovered, so it is challenging to decide whether the faults are young and active and pose a real risk or not.

The only remaining active nuclear power plant in California is Diablo Canyon, just north of Avila Beach in San Luis Obispo County (Figure 23.6). It was due to be shut down in 2025, but then the strain on California's power grid during extreme heat waves made the political leaders

(b)

FIGURE 23.5 (Continued)

rethink the shutdown. Not only is it already in place to provide backup power when the grid is overwhelmed, but it is also not producing any greenhouse gases and has relatively few environmental issues (unless something goes wrong). It produces about 18 terawatt-hours of electricity a year, about 9% of the state's power already, so some consider it a valuable resource that should not be dismissed. In 2022, the California Legislature passed a bill extending its life through 2030. By then, there should be so much solar and other renewable energy in the state that it may finally be no longer necessary.

23.8 CALIFORNIA'S RENEWABLE FUTURE

California has long been a leader in focusing on environmental issues, especially since its politics is quite progressive, and big corporate and fossil fuel interests have much less influence in Sacramento than they do in many other states. The state has set many challenging environmental goals over the years. In 2006, the legislature passed the Global Warming Solutions Act, which set a goal of 33% of energy consumption coming from renewables by 2020, but that goal was achieved and exceeded, so over 50% of our energy is now renewable. In 2015, the Senate passed a bill mandating that utilities purchase at least 50% of their power from renewables. In 2018, another Senate bill shifted those targets, so that by 2030, 60% must come from renewables, and by 2045, 100% of California's energy must come from renewables.

In 2021 (Figure 23.7), the energy profile of California was still dominated by cheap natural gas (41% of electricity generated). But solar was already 15% and expected to be even more dominant as it gets cheaper. Wind was third in percentage usage, with 12%, and probably going to increase

FIGURE 23.6 Aerial view of the Diablo Canyon nuclear power plant, on a rocky peninsula just north of Avila Beach, California. The two large cooling towers are visible for miles along the coast, and the supporting buildings and large network of power transmission lines are clearly visible. It is the only active operating nuclear power station left in California, originally slated to be shut down in 2025 but now continuing until 2030.

Source: Courtesy Wikimedia Commons.

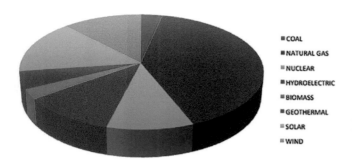

FIGURE 23.7 Pie chart showing the source of electricity generation in California as of 2021.

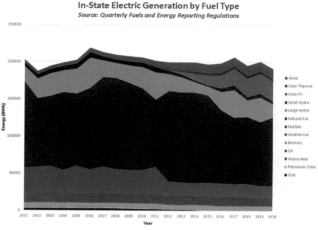

FIGURE 23.8 Trend in the relative importance in different sources of electrical power in California since 2021.

Source: Courtesy of Wikimedia Commons.

further. Hydroelectric was fourth most important with 11% (probably not going to increase much in the near future), and nuclear was fifth with 9.9% (probably not going to increase). Geothermal was sixth with 5.1% of California's electricity, and it could increase significantly. Biomass, coal, and regular petroleum were all 3% or less.

The history of power generation in California (Figure 23.8) shows some interesting trends. Back in 2001, it was almost entirely natural gas, hydroelectric, nuclear, and geothermal energy. Since about 2011, wind and solar have expanded dramatically, while energy usage from hydroelectric, nuclear, and natural gas have been reduced, so now in 2023, solar is second only to natural gas, and wind and hydroelectric are the third and fourth most important sources of energy in California.

NOTES

1 www.latimes.com/projects/la-fi-electricity-solar/
2 www.youtube.com/watch?v=3gSzzuY1Yw0
3 www.youtube.com/watch?v=3gSzzuY1Yw0

RESOURCES

BOOKS

Armstrong, J. 2021. *The Future of Energy: The 2021 Guide to the Energy Transition.* Energy Technology Publishing, New York.

Friedemann, A.J. 2021. *Life After Fossil Fuels: A Reality Check on Alternative Energy.* Springer, New York.

Hossain, E., and Petrovic, S. 2021. *Renewable Energy Crash Course.* Springer, New York.

Jelley, N. 2020. *Renewable Energy: A Very Short Introduction.* Oxford University Press, Oxford, UK.

Kanoglu, M., Cingal, Y., and Cimbala, J.M. 2019. *Fundamentals and Applications of Renewable Energy.* McGraw-Hill, New York.

Mackay, D.J.C. 2009. *Sustainable Energy—Without the Hot Air.* UIT Cambridge Ltd., Cambridge, UK.

Peake, S. 2018. *Renewable Energy: Power for the Future.* Oxford University Press, Oxford, UK.

Usher, B. 2019. *Renewable Energy: A Primer for the Twenty-First Century.* Columbia University Press, New York.

WEBSITES

California Air Resources Board. 2016. California's climate plan. https://web.archive.org/web/20160327114504/www.arb. ca.gov/cc/cleanenergy/clean_fs2.htm

Energy Information Administration. 2019a. Hydropower—Energy explained, your guide to understanding energy. www.eia. gov/energyexplained/hydropower/

Energy Information Administration. 2019b. Solar—Energy explained, your guide to understanding energy—Energy information administration. www.eia.gov/energyexplained/solar/

Energy Information Administration. 2019c. Geothermal—Energy explained, your guide to understanding energy—Energy information administration. www.eia.gov/energyexplained/geothermal/

Energy Information Administration. 2019d. Biomass—Energy explained, your guide to understanding energy—Energy information administration. www.eia.gov/energyexplained/biomass/

Energypedia. A wiki platform for collaborative knowledge exchange on renewable energy in developing countries. https://energypedia.info/wiki/Energypedia

U.S. Department of Energy. 2023. Clean energy. www.energy.gov/clean-energy

24 Climate Change and California's Future Environment

There's no debate about the greenhouse effect, just like there's no debate about gravity. If someone throws a piano off the roof, I don't care what Sarah Palin tells you, get out of the way because it's coming down on your head.

—**Jimmy Kimmel**

24.1 CLIMATE CHANGE AND SHIFTING BASELINES

We have looked at a few of the current environmental challenges faced by California, as well as some which will continue into the future. But behind all the environmental issues in the news today, the specter of a much bigger problem lurks: climate change. Many of the environmental conditions that were stable and manageable only a few years ago are now out of balance due to global warming and its side effects. And weather and other environmental conditions that were once predictable have abruptly shifted, so there are new problems to confront. Many ecologists call this phenomenon "shifting baselines." For example, many of my colleagues work on coral reefs and have been diving on them for the last 40 years have witnessed their rapid degradation and destruction in a single generation. The vibrant, healthy, diverse coral reefs they first saw as young scientists no longer exist, and the "baseline" for what is normal in the coral reef ecosystem has completely shifted. The global ecosystem has been changed irreparably in less than a century, and we are all forced to adapt and study the "new norm" in the environment. And there is no longer any doubt that it is due to humans burning greenhouse gases.

So what is the "greenhouse effect"? What are "greenhouse gases"? In a greenhouse (Figure 24.1), the temperatures inside remain warm, because energy from the sun (almost entirely in the visible part of the electromagnetic spectrum) easily passes through the glass walls, where it hits surfaces and converts the energy to heat (infrared wavelengths). But glass prevents the heat from escaping, just like greenhouse gases block the heat energy from leaving Earth. Another analogy is the inside of a parked car. Just as in a greenhouse, the interior heats up as the visible spectrum rays of sunlight pass through the windows and the solar energy is absorbed by the interior of the car, then is re-radiated back as heat. Thus, it will be hotter inside the car than outside because the glass windows let light pass through but block heat from escaping.

Because more heat arrives at the surface of the planet than can leave it, the Earth's atmosphere warms up. Back in 1896, Svante Arrhenius calculated that doubling the level of atmospheric carbon dioxide would cause global temperatures to rise by 5–6°C. This is remarkably close to the estimates of scientists in a report done in 2007 and consistent with current estimates.

What are the effects of global warming? And how do we know it is real and man-made? We will look at these big

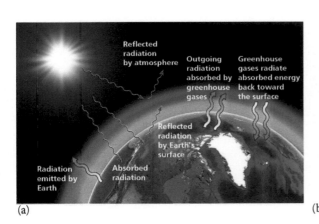
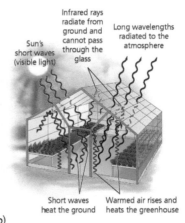

FIGURE 24.1 (a) The greenhouse effect occurs when the atmosphere lets sunlight (which is mostly in the short wavelengths, especially in the visible part of the light spectrum) pass through, but when that solar energy is absorbed by the Earth's surface, it is converted to a longer wavelength, infrared (head), and radiates back out into space, where greenhouse gases (like the glass panes in a real greenhouse) prevent it from escaping.

Source: Redrawn from several sources.

DOI: 10.1201/9781003301837-24

issues from the global perspective first, before focusing on how climate change is affecting California, and will do so even more in the future.

24.2 THE CHANGING PLANET

24.2.1 ATMOSPHERIC CHANGES

After the discovery of how the greenhouse effect works, a major breakthrough occurred when Charles David Keeling of Caltech developed one of the first devices for measuring atmospheric carbon dioxide. He tested it out in several locations, including the Big Sur in California and the Olympic rain forest in Washington, to see how reliably it measured carbon dioxide and what the influence of local plants might be on its measurements. In 1958, he began to take

measurements in two places isolated from major cities (thus minimizing effects of local pollution from cities or industry) and also isolated from the effects of plants absorbing carbon dioxide nearby: the top of Mauna Loa volcano on Hawaii, and in Antarctica. After a few years of data showing the rapid increase in carbon dioxide, the NSF cut off his Antarctic funding when they foolishly decided that he had proved his point, thus ending the potential for collecting a long-term dataset. But the Mauna Loa observatory has been running continuously for the more than 65 years, one of the longest sets of atmospheric data ever collected. By the 1960s, Keeling and his colleague, the oceanographer Roger Revelle, could see the dramatic increase in carbon dioxide in a steady upward trend (Figure 24.2). Superimposed on the overall upward trend are the annual cycles of decreasing carbon dioxide when plant growth in the Northern

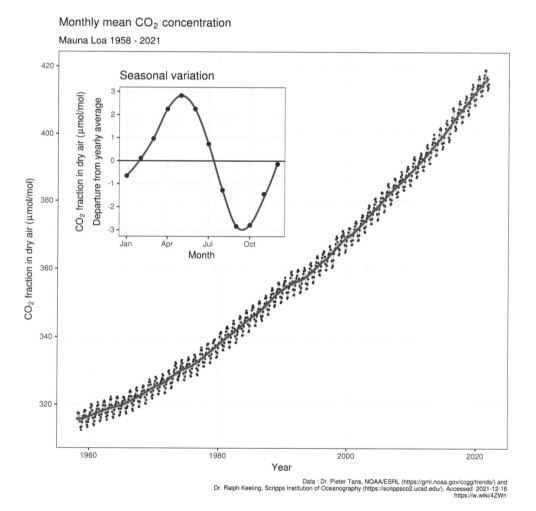

Monthly mean CO$_2$ concentration

Mauna Loa 1958 - 2021

Data : Dr. Pieter Tans, NOAA/ESRL (https://gml.noaa.gov/ccgg/trends/) and
Dr. Ralph Keeling, Scripps Institution of Oceanography (https://scrippsco2.ucsd.edu/). Accessed 2021-12-16
https://w.wiki/4ZWn

FIGURE 24.2 In 1958, Charles David Keeling used his newly invented instrument to measure atmospheric carbon dioxide at the top of Mauna Loa Observatory on the Big Island of Hawaii. That experiment has run continuously now for 66 years and clearly shows the steady and rapid increase in carbon dioxide in the atmosphere (verified by many other stations around the world). The Keeling curve has an annual cycle of decrease in the Northern Hemisphere spring (when most of the world's plants are growing after the winter and absorbing carbon dioxide) and increase in the fall (when those same plants lose their leaves and release that carbon dioxide back to the atmosphere). In 1958, the concentration of carbon dioxide was only around 315 ppmv (parts per million by volume). In 2023, the upswing of the Keeling curve passed 418 ppmv.

Source: Redrawn from several sources.

Hemisphere in the spring takes in CO_2. Carbon dioxide then increases in the fall, when the trees in the Northern Hemisphere lose their leaves and the carbon dioxide is released as the leaves decay. (Most of the world's land vegetation is in the Northern Hemisphere, so its effects are much stronger in the northern spring and fall than in the southern spring and fall).

From Keeling's initial data through every dataset that has been collected since then, the trend is clear: carbon dioxide in our atmosphere has increased at a dramatic rate in the past 150 years. It was barely 315 ppm (parts per million) when the Keeling experiment started in 1958, and as of 2023, it's over 415 ppm. Every dataset of temperature or carbon dioxide collected over a long-enough span of time confirms this. A compilation of the past 900 years' worth of temperature data from tree rings, ice cores, corals, and direct measurements of the past few centuries shows the sudden increase of temperature of the past century sticking out like a sore thumb (Figure 24.3a). This famous graph is nicknamed the "hockey stick" because it is long and relatively straight through most of its length, then bends sharply upward at the end like the blade of a hockey stick. Other graphs show that climate was very stable within a narrow range of variation through the past 1,000, 2,000, or even 10,000 years since the end of the last ice age, and even back to the end of the last glacial maximum about 18,000 years ago (Figure 24.3b). The graph clearly shows the minor warming events during the Climatic Optimum about 7,000 years ago, the Medieval Warm Period, and the slight cooling of the Little Ice Age from the 1700s and 1800s (compare it with Figure 24.2). But the magnitude and rapidity of the warming represented by the last 200 years are simply unmatched in all of geologic history. More revealing, the timing of this warming coincides with the Industrial Revolution, when humans first began massive deforestation and released carbon dioxide by burning coal, gas, and oil.

24.2.2 Melting Polar Regions

If the data from atmospheric gases were not enough, we are now seeing unprecedented changes in our planet. The polar ice caps are thinning and breaking up at an alarming rate. In 2000, for the first time ever, people flying over the North Pole in summertime saw no ice, just open water. So much for Santa's workshop! The Arctic ice cap has been frozen solid for at least the past 3 million years, and maybe longer, but now the entire ice sheet is breaking up so fast that by 2030 (and possibly sooner), less than half of the Arctic will be ice covered in the summer (Figure 24.4). As you can see from watching the news, this is an ecological disaster for everything that lives up there, from the polar bears to the seals and walruses and whales, to the animals they feed upon.

The Antarctic is thawing even faster. In February–March 2002, the Larsen B ice shelf, over 3,000 km² (the size of Rhode Island) and 220 m (700 ft) thick, broke up

in just a few months, a story typical of nearly all the ice in Antarctica. The Larsen B shelf had survived all the previous ice ages and interglacial warming episodes for the past 3 million years, and even the warmest periods of the last 10,000 years, yet it and nearly all the other thick ice sheets on the Arctic, Greenland, and Antarctic are vanishing at a rate never before seen in geologic history. And in 2017, the Larsen C ice shelf began to break apart.

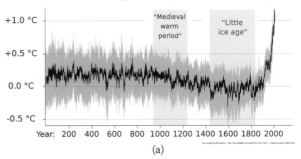

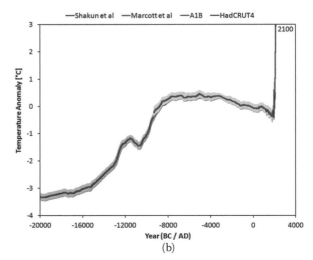

FIGURE 24.3 (a) The last thousand years of temperatures for the Northern Hemisphere, compiled from tree rings, ice cores, and historical records and directly measured by thermometers in the last century in red. The trend begins at the end of the Medieval Warm Period, around 1000 CE, and reaches its lowest point with the Little Ice Age from 1600 to 1850 CE. Then the effect of human-caused global warming kicks in, and the temperatures rise dramatically and rapidly to a level not seen since the Eocene. This curve has been nicknamed the "hockey stick" curve because the rapid inflection at the end after 900 years of stability resembles the way the blade of a hockey stick bends sharply from the handle. Note that the recent rise in global temperature is much faster and more extreme than the warmest part of the Medieval Warm Period. (b) The last 20,000 years of climate change, from the peak glacial at 20,000 to 18,000 years ago, the warming of the glacial–interglacial transition from 18,000 to 10,000 years ago to the Holocene stability of the past 10,000 years. Even in this scale, the extreme magnitude and rapidity of the heating of the past 150 years stand out as unprecedented, with more and faster warming than in any time in the geologic past.

Source: (a) Redrawn from M. Mann and IPCC. (b) Redrawn from several sources.

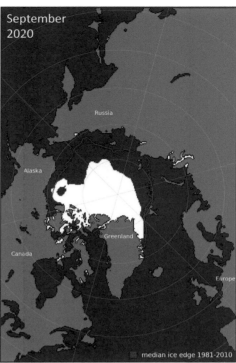

FIGURE 24.4 Disappearance of the ice caps in the Arctic over the past decades. Gigantic amounts of ice vanish every summer and fall and are not replaced by the freezing of the following winter.

Source: Courtesy NOAA.

24.3   SEA LEVEL RISE

Many people don't care about the polar ice caps, but there is a serious side effect worth considering: all that melted ice eventually ends up as more water in the ocean, causing sea level to rise, as it has many times in the geologic past. At the moment, sea level is rising about 3–4 mm per year (Figure 24.5), more than ten times the rate of 0.1–0.2 mm/ year that has occurred over the past 3,000 years. Geological data show that sea level was virtually unchanged over the past 10,000 years since the present interglacial began. A few millimeters of sea level rise here or there don't impress people, until you consider that the rate is accelerating and that most scientists predict it will rise 80–130 cm in just the next century. A sea level rise of 130 cm or 1.3 m (almost 4 ft) would drown many of the world's low-elevation cities, like Venice and New Orleans, and low-lying countries like the Netherlands or Bangladesh. A number of tiny island nations like Vanuatu and the Maldives, which barely poke out above the ocean now, are already vanishing beneath the waves. Their entire population will have to move some-place else. Already, low-lying coastal cities like Miami and Venice are now routinely being flooded during very high tides, when they never used to have the problem. If sea level rose by just 6 m (20 ft), nearly all the world's coastal plains and low-lying areas (such as the Louisiana bayous, Florida, and most of the world's river deltas) would be drowned. Most of the world's population lives in coastal cities, like New York; Boston; Philadelphia; Baltimore; Washington,

DC; Miami; Shanghai; and London. All those cities would be partially or completely underwater with such a sea level rise. If all the glacial ice caps melted completely (as they have several times before during past greenhouse worlds in the geologic past), sea level would rise by 65 m (215 ft)! The entire Mississippi Valley would flood, so you could dock your boat in Cairo, Illinois (Figure 24.6). Such a sea level rise would drown nearly every coastal region under hun-dreds of feet of water and inundate New York City, London, and Paris. All that would remain would be the tall land-marks, such as the Empire State Building, Big Ben, and the Eiffel Tower. You could tie your boats to these high spots, but most of the rest of these drowned cities would be deep under water.

24.3.1   OTHER EFFECTS OF CLIMATE CHANGE

The changes occur not only in polar ice and in rising sea level; it also has effects on all the climates around the world. About 95% of the remaining mountain glaciers left over from the Pleistocene are retreating at the highest rates ever documented—or they have already vanished. Many of those glaciers, especially in the Himalayas and Andes and Alps and Sierras, provide most of the fresh water that the populations below the mountains depend upon, yet this fresh water supply is vanishing. The permafrost that once remained solidly frozen even in the summer is now thaw-ing, damaging the Inuit villages on the Arctic coast and

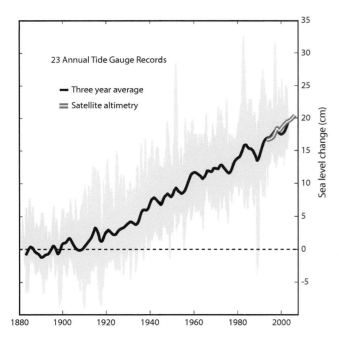

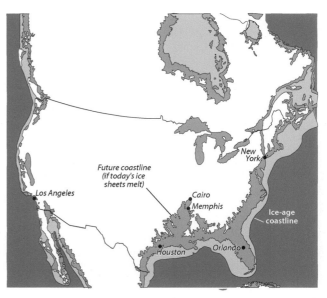

FIGURE 24.5 Data showing the rise in global sea level in the past 140 years, after 10,000 years of relative stability of the global sea level.

Source: Redrawn from several sources.

FIGURE 24.6 The effect of melting all the world's ice caps and glaciers would cause sea level to rise 65 m (215 ft). This would completely flood all the coastal cities of North America and make Florida and Louisiana virtually disappear and have a similar effect on the Low Countries of Belgium and the Netherlands, northern Germany and Poland, Denmark, and much of Great Britain.

Source: Redrawn from several sources.

threatening all our pipelines to the North Slope of Alaska. A more serious issue with the thawing permafrost is that it stores huge amounts of methane from the decay of vegetation in the soil, which is just now being released. Methane is a much more powerful greenhouse gas than carbon dioxide, so if it increases dramatically in the atmosphere, global warming would be even more out of control than it is now.

Not only is the ice vanishing, but we have also seen record heat waves over and over, killing thousands of people, as each year joins the list of the hottest years on record. The year 2020 almost tied 2016 as the hottest year on record, and 2021 was in the top 7 hottest years on record; 2022 was the fifth hottest year on record. Natural animal and plant populations are being decimated all over the globe as their environment changes. Many animals respond by moving their ranges to formerly cold climates, so now places that once did not have to worry about disease-bearing mosquitoes are infested as the climate warms and allows them to breed further north.

If you watch any of the recent movies or videos about climate change, the long litany of "things we have never seen before" and "things that have never occurred in the past 3 million years of glacial–interglacial cycles" is staggering. Still, there are many people who are not moved by the dramatic images of vanishing glaciers or by the forlorn polar bears starving to death. Many of these people have been fed misinformation by the powerful lobbies and political organizations and fossil fuel companies who want to cloud or confuse the issue.

24.4 HOW DO WE KNOW THAT HUMANS ARE THE CAUSE?

There are people who doubt climate change is happening (despite the overwhelming evidence), and others who admit it is real but do not believe humans are the cause. But the evidence is overwhelming that humans are to blame. How do we know that climate is changing in an unusual manner and not just normal "climate fluctuations"?

- **"It's just natural climatic variability."** Geologists and paleoclimatologists know a lot about past greenhouse worlds and the icehouse planet that has existed for the past 33 million years. We have a good understanding of how and why the Antarctic ice sheet first appeared at that time, and how the Arctic froze over about 3.5 million years ago, beginning the 24 glacial and interglacial episodes of the "ice ages" that have occurred since then. We know how variations in the Earth's orbit (the Milankovitch cycles) control the amount of solar radiation the Earth receives, triggering the shifts between glacial and interglacial periods. Our current warm interglacial has already lasted 10,000 years, the duration of most previous interglacials, so if it were not for global warming, we would be headed into the next glacial any time now. Instead, our pumping greenhouse gases into our atmosphere after they were long trapped in the Earth's crust

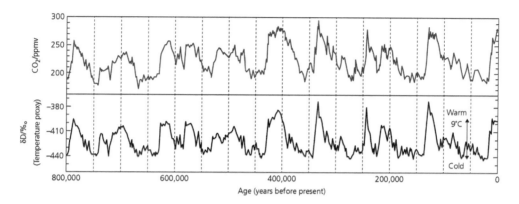

FIGURE 24.7 The record of carbon dioxide in the EPICA-1 core from Antarctica, the longest and deepest core ever taken. It spans 800,000 years into the past, through at least seven complete glacial–interglacial cycles. Natural climate variability is clearly shown by the core, which proves that the highest levels of carbon dioxide (top curve) during the warmest interglacials were only 280 ppm—and today it is over 418 ppm. The bottom curve displays deuterium (hydrogen-2), which is an indirect measure of temperature recorded in the core.

Source: Redrawn from several sources.

has pushed the planet into a "super-interglacial," already warmer than any previous warming period. We can see the "big picture" of climate variability most clearly in the EPICA-1 core from Antarctica (Figure 24.7), which shows the details of the last almost 750,000 years of glacial–interglacial cycles. *At no time during any previous interglacial did the natural carbon dioxide levels exceed 280 ppm, even at their very warmest.* Our atmospheric carbon dioxide levels are already close to 420 ppm today. The atmosphere is headed to 600 ppm within a few decades, even if we stopped releasing greenhouse gases immediately. This is decidedly *not* within the normal range of "natural climatic variability" but clearly unprecedented in earth history. Anyone who says this is "normal variability" or "climate changes all the time and humans weren't the cause" has never seen the huge amount of paleoclimatic data that show otherwise.

- **"It's just the sun, or cosmic rays, or volcanic activity, or methane."** The amount of heat that the sun provides has been cooling down since 1940, just the opposite of what some people claim (Figure 24.8). If the sun were causing global warming, it should show an increase in activity, not a decrease. Cosmic radiation causes an increase in cloud cover on the Earth, so increased cosmic rays would cool the planet, and decreased cosmic radiation would warm it. There are lots of measurements of cosmic radiation, and the result is clear: in the last 40 years, cosmic radiation has been increasing (which should cool the planet), while the temperature has been rising, the exact opposite effect expected if cosmic radiation contributed to recent warming. Nor is there any clear evidence that large-scale volcanic events (such as the 1815 eruption of Tambora in Indonesia, which

changed global climate for about a year) have any long-term effect that would explain 200 years of warming and carbon dioxide increase. Volcanoes erupt only 0.3 billion tons of carbon dioxide each year, but humans emit over 29 billion tons a year, about 100 times as much; clearly, we have a bigger effect. Methane is a more powerful greenhouse gas, but there is 200 times more carbon dioxide than methane, so carbon dioxide is still the most important agent. Every other alternative has been looked at, but the only clear-cut relationship is between human-caused carbon dioxide increase and global warming. We just can't squirm out of the blame on this one.

- **"The climate records since 1998 show cooling."** People who make this argument are cherry-picking the data. Over the short term, there was a slight cooling trend from 1998 to 2000 (Figure 24.9), because 1998 was a record-breaking El Niño year, so the next few years look cooler by comparison. This is deliberately "cherry-picking" a data point to bias the data to make it show what you want it to show. In science, "cherry-picking" data is considered dishonest and unethical—but a common tactic in the political sphere and in public relations and propaganda, where lies are the norm. As Mark Twain said, "A lie can travel around the world and back again while the truth is lacing up its boots." But since 2002, the overall long-term trend of warming (Figure 24.9B) is unequivocal. Saying that there was a "climate pause since 1998" or "cooling trend since 1998" is a clear-cut case of using out-of-context data in an attempt to distort and deny the evidence. All the 20 hottest years ever recorded on a global scale have occurred in the last 25 years, and the 9 hottest years on record are the last 9 years since 2013. The hottest years

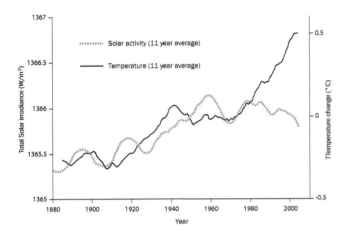

FIGURE 24.8 The trends of incoming solar radiation and temperature over the past century. If increasing solar activity explained global warming, there should be an upward trend in the solar curve—but solar input has actually decreased in the past 50 years.

Source: Redrawn from several sources.

on record are (in order of hottest first): 2023, 2016, 2020, 2019, 2015, 2022, 2017, 2018, 2021, 2014, 2010, 2013, 2005, 2009, 1998, 2012, 2007, 2006, 2003, 2002, 2004, 2001, 1997, 2008, 1995, 1999, 1990, and 2000. In other words, almost every year since 2000 has been in the top 25 hottest years list, and the rest of the list includes 1995, 1997, 1998, 1999, and 2000. Only 1996 failed to make the list (because of the short-term cooling mentioned already). More significantly, the record-breaking temperatures of the last few years (2015, 2016, and 2020 each broke or tied the record of the previous year) rocketed upward at a rate never seen in any climatic record, ancient or modern.

- **"We had heavy snow and freezing temperatures last winter."** So what? This is a classic case of how the scientifically illiterate public cannot tell the difference between *weather* (short-term seasonal changes) and *climate* (the long-term average of weather over decades and centuries and longer). *Weather* is what happens from hour to hour, or day to day; *climate* is the average of weather events over decades to centuries or longer. *Climate* is what you might look up in an almanac or on a website about the expected average temperature for a given day in the future; *weather* is what you can see when you look outdoors. Winter doesn't just vanish as the globe gets warmer; winters just become *on average* a bit warmer and shorter than usual, but large cold spells, snowstorms, and freezing events will still happen. More importantly, local weather tells us nothing about the next continent, or the global average; it is only a local effect, determined by short-term atmospheric and oceanographic conditions. (In addition, meteorologists

are *not* climate experts; they are trained only to analyze the short-term changes in the weather, and in most cases, they have no formal training or published research in climate science, which is an entirely different field. Thus, their opinions about climate change are no better-informed than anyone else's opinion unless they are also doing research in climate science.) In fact, warmer global temperatures can mean *more moisture* in the atmosphere, which increases the intensity of normal winter snowstorms. The past few years have actually had many unusually mild winters in the colder northern part of North America, which people conveniently forget when it finally does get cold. When a long cold spell from the polar vortex lingers over the northeastern United States (where all the policy and politics are dictated in cities like Washington, DC, and New York City), it is actually a result of climate change. The strong temperature difference between the pole and the equator, which powers the jet stream, gets weaker and weaker when the Arctic warms and melts. This makes the jet stream slow down and develop a lazy, looping path (called Rossby waves) over North America, which can get stalled for weeks. If the loop down from the Arctic brings a mass of cold air with it, the freezing conditions will stick around for a while—but somewhere else in North America, a warm loop of the jet stream will get stalled over the northern Midwest or Plains or the West Coast, bringing abnormally warm weather to that area while the East Coast freezes.

24.5 OUR FINGERPRINTS ON THE MURDER WEAPON

But how do we know humans are to blame? The list of lines of evidence for human causes is very long but includes a wide spectrum of different kinds of data:

- We can directly measure the amount of carbon dioxide humans are producing, and it matches exactly with the amount of increase in atmospheric carbon dioxide (Figure 24.10).
- Through carbon isotope analysis (Figure 24.11), we can show that this carbon dioxide in the atmosphere is coming directly from our burning of fossil fuels, not from natural sources. The atmosphere has many other chemical fingerprints as well which prove that the greenhouse gases were produced by humans.
- We can also measure oxygen levels that drop as we produce more carbon that then combines with oxygen to produce carbon dioxide.
- We have satellites out in space that are measuring the heat released from the planet and can actually

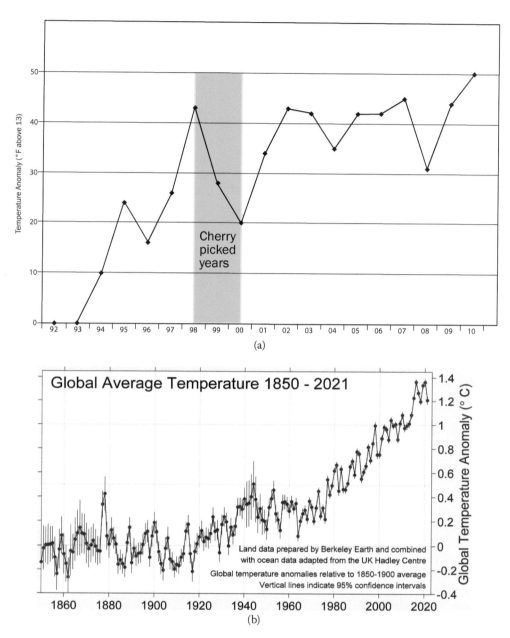

FIGURE 24.9 (a) The effect of cherry-picking the anomalously warm El Niño year of 1998 as a starting point to falsely suggest that global temperature has been cooling or steady since 1998. (b) The long-term average temperature on Earth over the past century.

Source: Redrawn from several sources.

see and measure the atmosphere get warmer in real time. They can also see the carbon dioxide emerging from power plants and cities, and it is clear it comes mostly from those sources.

- The most crucial proof emerged only the past few years: climate models of the greenhouse effect predict that there should be cooling in the stratosphere (the upper layer of the atmosphere above 10 km or 6 mi in elevation), but warming in the troposphere (the bottom layer of the atmosphere below 10 km or 6 mi), where the human-produced gases come from. In contrast, if the warming were

due to an increase in solar radiation, the stratosphere would warm first and the troposphere would be relatively cool. Indeed, our space probes have measured stratospheric cooling and upper troposphere warming (Figure 24.12), just as climate scientists had predicted, and proving it is due to greenhouse gases, not the sun.

- Finally, we can rule out any other culprits (see earlier): solar heat has been decreasing since 1940, not increasing, and there are no measurable increases in cosmic radiation, methane, volcanic gases, or any other potential cause.

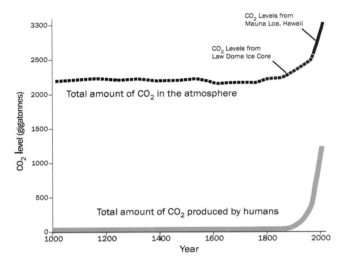

FIGURE 24.10 The measurements of carbon dioxide in the atmosphere (as determined from ice cores and then direct measurements since the Keeling experiment in 1958) exactly match the amount of carbon we have released into the atmosphere by burning fossil fuels, starting at the same time and showing the same rate of increase. This is one of many lines of evidence that our burning of fossil fuels explains the rise in carbon dioxide, since there are no other trends in nature which match the carbon dioxide increase.

Source: Redrawn from several sources.

24.6 CLIMATE CHANGE AND THE FUTURE OF CALIFORNIA

Now that we have seen the global evidence that shows climate change is real and caused by humans, let us focus on what climate change is doing (and will do) to California over the next century. The effects are already happening every year, and people are becoming more and more aware of it. Just the past decade in California has seen some extraordinary weather events, from severe drought for most of the past 23 years to record rains in 2022–2024, more numerous and hotter heat waves, smaller and smaller Sierra snowpacks in most years, longer and earlier wildfire seasons with more extreme megafires and the smoke pollution it creates. These things have already happened in the past 25 years and will continue to get worse. In addition, there are slower-paced events, like the gradual rise of sea level drowning coastal cities, ecological disasters in our ocean next to our coasts, and the spread of diseases that flourish in warmer weather, and many other events that normally don't happen in California. Let's look at these one at a time.

24.6.1 RISING TEMPERATURE AND HEAT WAVES

The first place we would notice the warming of the globe here in California are rising temperatures (Figure 25.13). As scientists at Scripps Institution of Oceanography summarized it:[1]

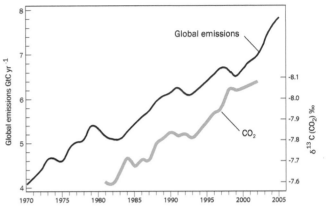

FIGURE 24.11 The greenhouse gases accumulating in our atmosphere have distinctive geochemical "fingerprints on the murder weapon" that prove than humans are the source of the carbon dioxide, not natural sources. This is demonstrated by many different geochemical indicators, such as the ratio of carbon-12 (more abundant due to burning of fossil fuels) versus carbon-13.

Source: Redrawn from several sources.

Average summer temperatures in California have risen by approximately 3 degrees F (1.8°C) since 1896, with more than half of that increase occurring since the early 1970s. If global greenhouse gas emissions continue at current rates, the state is likely to experience further warming by more than 2 degrees F more by 2040, more than 4 degrees F by 2070, and by more than 6 degrees F by 2100. Some of the most impressive impacts of warming will be felt during short period heat events (e.g. days exceeding 106.6 degrees F). For example, if emissions continue at current rates, Fresno will likely suffer 43 extreme heat days per year between 2050 and 2099; 10 times more than its yearly average between 1961 and 2005.

Another trend that has been noted is that heat waves will increase along the coast, meaning that they will become much more humid and "sticky" than previously observed in California.[2] In much of California (especially Southern California and the deserts), the arid climate means that at night, there is little moisture in the air to retain heat and the nighttime air cools rapidly and makes for comfortable sleeping. But increased humidity means that instead of cooling down at night, the air will remain warm and humid and "muggy," and the relief we expect during the nighttime will not occur as much.

But the most striking effect of the overall warming of California have been heat waves, which are getting both longer and more extreme (Figure 24.14). One study (Tamrazian and others, 2008) found that between 1996 and 2006, average annual Los Angeles maximum temperature has heated up by 5.0°F over the 100-year study period, and the average annual minimum temperature has increased by 4.2°F over the same period, and that Los Angeles is experiencing more heat waves (periods of three consecutive days above 90°F) and also more extreme heat days (days above 90°F). These numbers have increased by 3.09 and 22.8 occurrences over the study period, respectively.

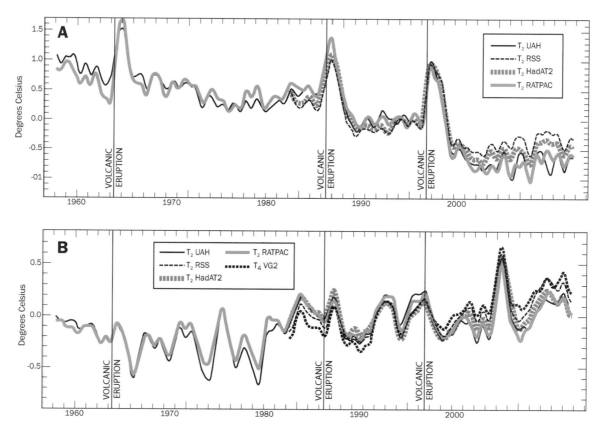

FIGURE 24.12 The crucial test that warming is coming from sources on the ground and not from solar radiation from space is to measure the relative changes in temperature in the different levels of the atmosphere. If it came from space, then the stratospheric layer high above us would warm, and the troposphere (the layer in which we live) would be relatively cool. If it came from below, the troposphere would warm and the stratosphere would remain relatively cool. Satellite measurements show definite stratospheric cooling and tropospheric warming, confirming it comes from ground sources, namely, the burning of fossil fuels. This plot also shows spikes in cooling shortly after volcanic eruptions (dashed vertical lines) due to the blocking of sunlight by volcanic ash and dust blowing in the stratosphere for several years after big eruptions.

Source: Redrawn from several sources.

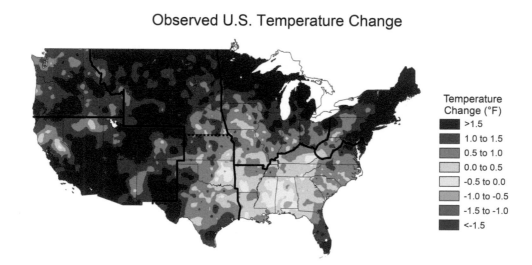

FIGURE 24.13 Trend in temperature in the United States from 1991 to 2012.

Source: Courtesy NASA (https://earthobservatory.nasa.gov/images/83624/climate-changes-in-the-united-states).

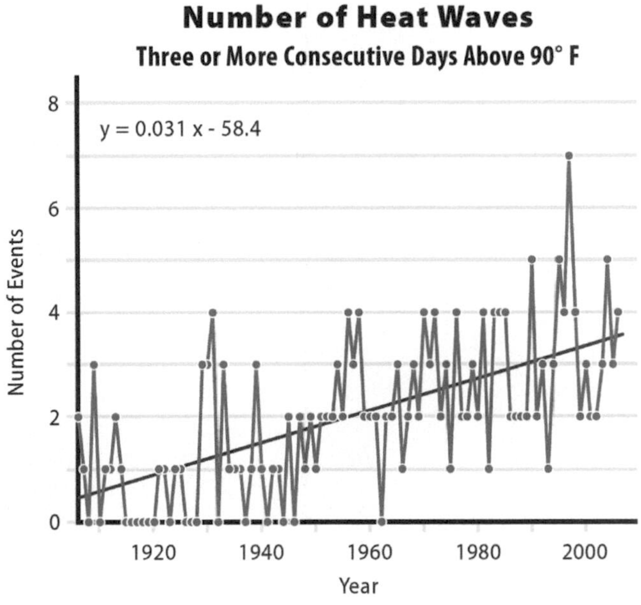

Number of Heat Waves
Three or More Consecutive Days Above 90° F

$y = 0.031x - 58.4$

FIGURE 24.14 Trend in increasing number of heat waves in Southern California.

Source: From Tamrazian et al. (2008, figure 6).

This study focused on the period from 1996 to 2006, and 2006 was a record-breaking year for heat waves that shattered the norms of that time, when 126 people died in California alone. The heat wave of 2012 then became the worst on record up to that point, killing 123 people and causing crop failures and food shortages costing over $30 billion. But the trend has only continued (Figure 24.15). July 2018 then broke records as the warmest month in California history.

More recently, in August 2020, temperature records were shattered across the western states, and severe power outages were caused by the peak demand for air-conditioning. Then over a six-day period during the middle of June 2021, a dome of hot air languished over the western United States, causing temperatures to skyrocket. From June 15 to 20, all-time maximum temperature records fell at locations in seven different states (California, Arizona, New Mexico, Texas, Utah, Colorado, Montana). In Phoenix, Arizona, the high temperature was over 115°F for a record-setting six consecutive days, topping out at 118°F on June 17.

Then a record-shattering heat wave hit in late August–September 2022, breaking not only the heat record of 134°F for Death Valley (officially now the hottest place on Earth) but also even for the relatively cool Bay Area and most areas in the state. As reported on September 7, 2022:[3]

Long-standing daily and all-time records were shattered in cities across California over the past several days. Sacramento hit 116 degrees Fahrenheit for the first time ever, surpassing a 97-year-old record of 114 degrees previously set on July 17, 1925, according to the National Weather Service. Stockton reached 115 degrees, tying the city's previous all-time record set in 2006. San Francisco was just a few degrees shy of breaking 100 degrees. In the

Bay Area, San Jose, Santa Rosa, Livermore, Redwood City, King City and Napa also experienced all-time record high temperatures. San Jose had a high of 109 degrees Tuesday, beating the old record of 108 degrees set in 2017. Santa Rosa hit 115 degrees, breaking a record set in 1913, and Napa had a high of 114 degrees. Both Livermore and King City reached 116 degrees. In the Los Angeles area, both Burbank and Long Beach set daily records Sunday. Burbank hit 110 degrees while Long Beach Airport marked a high of 109 degrees. Newport Beach broke its all-time high temperature record of 96 degrees Sunday at 97 degrees.

As I write this in 2023, we broke the summer–fall heat wave record again, and the trend has not let up in the past 20 years. It is a safe bet that heat waves will just get worse and worse. And as discussed in Chapter 18, droughts are getting worse and worse in California.

24.6.2　CALIFORNIA ON FIRE

Another consequence of the long-term increase in heat and drought in California are wildfires. Thanks to its normal pattern of a long dry season through the hot summer and fall, much of California is naturally prone to wildfires. The native plants of the chaparral, the oak-scrub woodlands, and even the pine forests of the mountains of California are mostly fire-adapted and can survive frequent small fires. Some of these plants actually require frequent small fires to clear out all the dead undergrowth and allow new plants to grow in their place. Certain types of chaparral scrub cannot germinate unless the heat of a fire breaks open their seedpods. So fire has always been a part of California's landscape.

However, as any expert on forestry and fire management will tell you, the problem is now much worse, partly due to the warming climate, and also due to misguided policies. Before 1910, fires were frequent in most the forests of the western United States, but they were small and burned out quickly. Many of them were deliberately set by Native Americans to renew the forests and to improve their hunting. Consequently, our forests were a "patchwork" of small areas of trees interspersed with meadows and other breaks in the forest cover. Anytime another fire occurred, whether by lightning or human activity, it would burn a limited area and then go out when it ran into the natural firebreaks formed by the meadows and grassy areas. These frequent small fires kept the dead undergrowth from accumulating and kept the trees spaced widely apart so fire could not jump so quickly from one tree to the next. The bigger mature trees could survive such fires, because these small fires burned through quickly and were not that hot, so the tall trees might have some scarring and burned bark, but they survived.

Then two events happened which totally transformed the forests and made fires much worse. One was the rapacious logging of the American West, especially in the late 1800s and early 1900s, which reduced beautiful old-growth forests

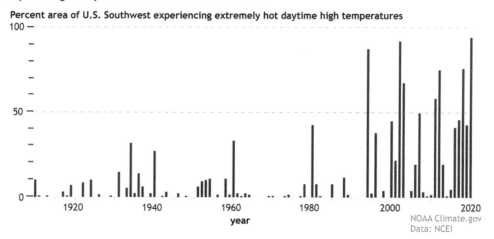

Expanding footprint of extreme summer heat

FIGURE 24.15 Red bars show the percent of the US Southwest (Colorado, Utah, Arizona, and New Mexico) having extremely warm days—daytime high temperatures in the top 10% of the historical record—each summer since 1910. The footprint of extreme heat in the Southwest has exploded in the past 30 years.

Source: NOAA Climate.gov graph, based on data from NCEI's Climate Extremes Index (from www.climate.gov/news-features/event-tracker/record-breaking-june-2021-heatwave-impacts-us-west).

to barren wastelands of stumps and slash. Loggers tended to focus on the largest, tallest, and most mature trees, because their thicker and longer trunks gave more profit per tree. These tall trees tended to prevent other trees from growing into too densely because they shaded the areas around them and had a large root network. This prevented seedlings from growing too close together and kept the density of trees low. When these larger trees were removed, younger trees could grow and ended up becoming too densely packed together.

But the biggest change happened due to a huge wildfire in Idaho and Montana in 1910, known as the "Big Burn." It devastated over 3 million acres, making it the largest forest fire in US history, killed 87 people (mostly firefighters), and burned a number of towns to the ground. It was such a big event that it influenced political opinion, and the newly formed US Forest Service (which manages most of the public forested land in the United States) adopted a policy of total fire suppression. No fires of any kind were allowed to burn out naturally, even if they might be far from any human activity and might actually help clear the dead undergrowth. For over 100 years now, this policy has allowed most forests on federal lands (and most state forests as well) to become overgrown and dense, with trees too closely spaced together and lots of dead undergrowth on the forest floor ready to act as tinder. These are the forests we have now, and they are a powder keg waiting to explode. When they do, we have megafires that burn enormous numbers of acres and burn so hot that not even the mature trees can survive.

Add to this potent mixture the effects of climate change, which make forests hotter and drier and more likely to have too many dead trees, plus the expansion of human population into what were formerly wild areas, and you have a recipe for disaster. Clearly, the policy of total fire suppression has created this nightmare and is unsustainable. Ask any expert in forest management and they will tell you that we need to have frequent small controlled burns in times and places where a fire is less likely to spread, to get rid of all the dead undergrowth and thin out the trees to their natural state. We need to restore the "power of the patchwork." But people hate the idea of *any* fires being allowed to burn and are often vocal in their resistance to smart fire policy. They seem to think that fire will magically go away if we just throw enough money and resources (and the lives of firefighters) at it.

As a result of this stupidity, we now live in an age of *megafires*, defined as any fire that burns more than 100,000 ac (40,500 ha). The forests of the world are all suffering from the same issues: bad forest management, plus the effect of global warming, so the era of increasingly catastrophic wildfires is here and will not go away soon. This was predicted as early as 1980 by some forestry experts,[4] but now it is apparent to everyone. Each year, the acreage burned by wildfires increases, and there are bigger and bigger megafires (Figure 24.16).

As one of the largest states in the union with huge areas of forest and increasing drought year after year, California sees a lion's share of these megafires, and they are getting worse year by year. All but 1 of the 10 largest wildfires in California history have occurred since 2017, and all but 2 of the 20 largest fires in California have occurred since 2003. The biggest was the deadly August Complex fire of August 2020, which burned over a million acres in six counties in northern California. The second worst was the 2021 Dixie fire, the largest single-source wildfire in California history, with 963,000 ac burned. The next four largest fires occurred in 2018, and three separate fires in 2020 are on the top 10 list. Thus, all 7 of the largest fires in California history have occurred in 2018 or later, and nearly all are in northern California, where the dense forests are now vulnerable to fires after years of heat and drought (Figure 24.17).

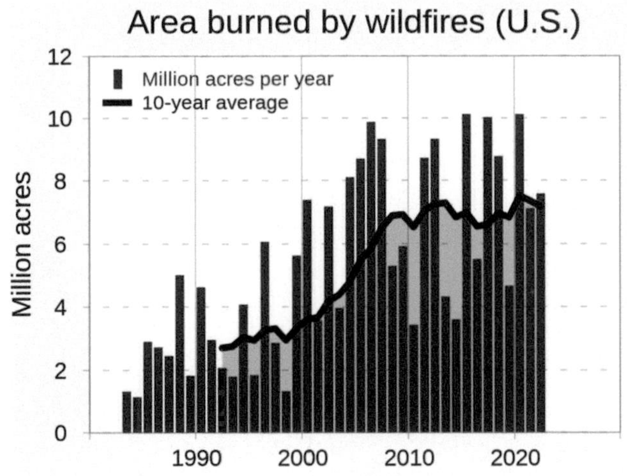

Sometimes a fire doesn't have to burn a lot of acres to be deadly or destructive. The deadliest was the infamous Camp Fire of 2018, which killed 85 people and wiped out the tiny town of Paradise, California (Figure 24.18). It was also the most destructive fire in California history, because even though it burned only 153,000 ac, it destroyed some 18,804 structures. As described in this source:[5]

AT 6:30 AM on November 8, a wildfire of astounding proportions and speed broke out in Northern California. Dubbed the Camp Fire, at one point it was burning 80 acres a minute. When it hit the town of Paradise, home to 27,000 people, those buildings became yet more fuel to power the blaze. It destroyed over 18,000 structures. For perspective, the previously most destructive wildfire in state history, the Tubbs Fire that raged through the city of Santa Rosa last year, destroyed 5,500 total structures. The death toll so far stands at 88. That makes it by far the deadliest wildfire in California history. Hundreds are still missing. "We're seeing urban conflagrations, and that's the real phase change in recent years," says Stephen Pyne, a wildfire expert at Arizona State University. It used to be that fires destroyed exurbs or scattered enclaves. "But what's remarkable is the way they're plowing over

Area burned by wildfires (U.S.)

■ Million acres per year
— 10-year average

FIGURE 24.16 Average US acreage burned annually by wildfires has almost tripled in three decades.

Source: Courtesy of Wikimedia Commons.

FIGURE 24.17 Top 20 largest California wildfires of all time (most happening since 2017).

Source: Courtesy of CalFire.

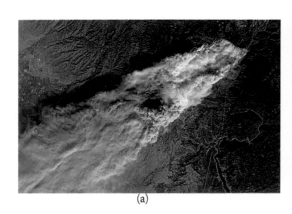
(a)

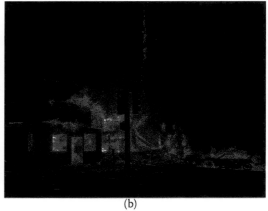
(b)

FIGURE 24.18 The 2018 Camp Fire, which wiped out the town of Paradise, California: (a) The Camp Fire as seen from the Landsat 8 satellite on November 8, 2018, with red highlighting active fire seen in infrared. (b) The fire at its peak, burning every structure in town that could be burned. (c) The Camp Fire destroyed at least 12,000 buildings.

Source: Courtesy of Wikimedia Commons.

(a)

FIGURE 24.18 (Continued)

cities, which we thought was something that had been banished a century ago." The Camp Fire horror show, which burned 70,000 acres in 24 hours, and reached over 150,000 acres, is a confluence of factors. The first is wind—lots of it, blasting in from the east. "We have a weather event, in this case a downslope windstorm, where, as opposed to the normal westerly winds, we get easterly winds that are cascading off the crest of the Sierra Nevada," says Neil Lareau, an atmospheric scientist at the University of Nevada, Reno.

Another terrible incident was the 1991 Tunnel Fire, which roared across the hilly neighborhoods east of downtown Oakland and killed 23 people. And the notorious Thomas Fire burned much of Ventura and Santa Barbara counties in 2017, killing 23 people, and then was followed by huge mudslides and debris flows the following winter which killed several people in Montecito, just east of Santa Barbara.

These stories are just a few of many. The future of California is predicted to have more and more giant fires, both due to the poor management policies of the major agencies that run our forests and especially due to climate change. More and more of California will be burning for years to come.

24.6.3 Rising Seas

Earlier in the chapter, we looked at the global rise of sea level and how it affects coastal regions outside California. We described how sea level is likely to rise between 1 and

4 ft in the next century. Even a 16 in rise could threaten coastal highways, bridges, and the San Francisco and Oakland airports. Indeed, the entire Bay Area would be very vulnerable as the expanse of San Francisco Bay rose and flooded not only the marshlands in the southern tip of the Bay but also the coastal regions of nearly every community with a shoreline on the Bay as well (Figure 24.19). The sea could also submerge wetlands in San Francisco Bay and other estuaries, which would harm local fisheries and potentially remove key intertidal feeding habitat for migratory birds.

A rise of 3 ft would increase the number of Californians living in places that are flooded by a 100-year storm from about 250,000 today to about 400,000. Along some ocean shores, homes will fall into the water as beaches, bluffs, and cliffs erode; but along shores where seawalls protect shorefront homes from erosion, beaches may erode up to the seawall and then vanish.

This is particularly striking in the low-lying areas of Southern California. If sea level rises 1.3 m (4 ft) by 2100 (as most predict it will), most of the beach communities from Long Beach to Huntington Beach to Balboa and Newport Beach would vanish beneath the waves (Figure 24.20). Finally, if all the glaciers melted and sea level rose by 70 m (230 ft), not only would almost all the coast cities of California vanish, but so would the entire Central Valley (Figure 24.21), most of which is barely 33 m (100 ft) above sea level at the low spots in the Sacramento River Delta, and the rest is below 66 m (200 ft) above sea level, except on the very edges of the Central Valley in the foothills of the mountains. Figure 24.22 shows a map if the sea level rose 80 m (260 ft), drowning almost all of the Los Angeles County and Orange

FIGURE 24.19 Flooding of central and southern San Francisco Bay with a 1 m (3.3 ft) rise in sea level.

Source: From Hanak and Moreno (2008, figure 2).

FIGURE 24.20 Map showing the flooding of the coastal communities from Long Beach to Newport if a 1.3 m (4 ft) rise of sea level occurred.

Source: Courtesy of Wikimedia Commons.

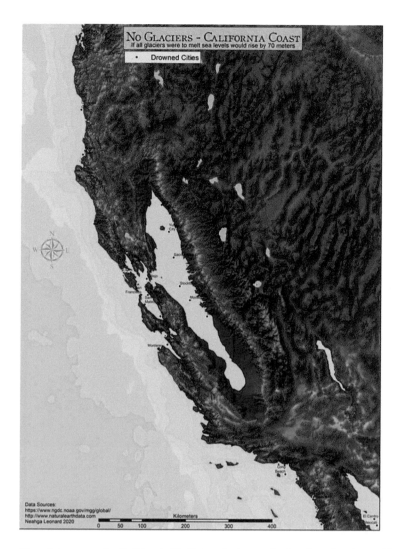

FIGURE 24.21 Drowning of low-lying areas in California if all the glaciers melted and sea level rose 70 m (215 ft). Not only would most coastal cities go under water, but so would most of the Central Valley.

Source: From several sources.

County area except for the areas that are now mountains or at least low hills taller than 260 ft above present sea level. The authors of the map have whimsically converted the place-names of the region today into "future place-names" if they are all submerged, like "Los Atlantis," "The O Sea," "Strait Outta Compton," "Drowney," and the "Miracle Mire," among others.

24.7   A PERSPECTIVE

We assume that our civilization is a permanent feature on this Earth, and that some form of our culture will persist indefinitely. But as any archeologist or historian knows, this is not true. We have many examples of extinct cultures that have vanished, leaving only a few durable artifacts, and often we know very little about how they lived—or why they failed. One only needs to look at the mysterious Etruscans, or Minoans, or Mycenaeans, or Mayans, or Anasazi. The list of failed societies goes on and on.

Jared Diamond, in his 2004 book *Collapse: How Societies Choose to Fail or Succeed*, points out that in instances where we do know why a society vanished, it is truly humbling. For example, the Easter Island culture vanished completely before the European settlers had much chance to witness their civilization, leaving only the famous huge stone heads, or *moai*, dotting the island. Diamond shows that to a large extent, the extinction of the Easter Islanders was self-inflicted: too many people, too much overexploitation of their environment when they cut down all the trees on what had been a densely forested island, and finally, starvation, disease, and warfare wiped out the survivors. As Diamond shows, such a fate could await our world civilization if we overpopulate this planet or damage our environment or overexploit our resources. After all, 99% of all species that have ever lived are now extinct. There is no good biological reason to believe that our fate will be different, especially given our accelerated pace of self-destruction.

FIGURE 24.22 Whimsical map by Jeffrey Linn showing what Southern California would look like with 215 ft of sea level rise, with whimsical names playing upon the existing geographic names of the region.

Source: From https://conspiracyofcartographers.com/mapgallery/.

NOTES

1 https://scripps.ucsd.edu/research/climate-change-resources/faq-climate-change-california

2 https://scripps.ucsd.edu/news/heat-waves-move-toward-coasts-study-finds

3 https://thehill.com/changing-america/sustainability/climate-change/3632565-five-things-to-know-about-the-brutal-california-heat-wave/

4 https://e360.yale.edu/features/the-age-of-megafires-the-world-hits-a-climate-tipping-point

5 www.wired.com/story/the-terrifying-science-behind-californias-massive-camp-fire/

RESOURCES

Diamond, J. 2004. *Collapse: How Societies Choose to Fail or Succeed.* Penguin, New York.

Flannery, T. 2006. *The Weather Makers: How Man is Changing the Climate and What it Means for Life on Earth.* Atlantic Monthly Press, New York.

Gershunov, A., and Guirguis, K. 2012. California heat waves in the present and future. *Geophysical Research Letters, 39,* L18710. https://doi.org/10.1029/2012GL052979.

Gore, A.I. 1992. *Earth in the Balance: Ecology and the Human Spirit.* Penguin, New York.

Gore, A.I. 2006. *An Inconvenient Truth.* Rodale Press, Emmaus, PA.

Hanak, E., and Moreno, G. 2008. California coastal management with a changing climate. *Preparing California for a Changing Climate,* Public Policy Institute of California, San Francisco, CA, 1–28.

Harari, Y.N. 2017. *Homo Deus: A Brief History of Tomorrow.* Harper, New York.

Heinemann, W. 2003. *Our Final Century? Will the Human Race Survive the Twenty-First Century?* Arrow, New York.

Henson, R. 2014. *The Thinking Person's Guide to Climate Change.* American Meteorological Society, Washington, DC.

Kolbert, E. 2015. *Field Notes from a Catastrophe: Man, Nature, and Climate Change.* Bloomsbury, London.

Mann, M. 2012. *The Hockey Stick and the Climate Wars: Dispatches from the Front Lines.* Columbia University Press, New York.

Mann, M, and Kump, L. 2015. *Dire Predictions: Understanding Climate Change.* DK, London.

Mann, M., and Toles, T. 2018. *The Madhouse Effect: How Climate Denial Is Threatening Our Planet, Destroying our Politics, and Driving Us Crazy.* Columbia University Press, New York.

McKibben, B. 1989. *The End of Nature.* Random House, New York.

McKibben, B. 2010. *Eaarth: Making a Life on a Tough New Planet.* Times Books, New York.

McNeill, J.R., and Engelke, P. 2016. *The Great Acceleration: An Environmental History of the Anthropocene since 1945.* Harvard Belknap Press, Cambridge, MA.

Oreskes, N., and Conway, E. 2010. *Merchants of Doubt: How a Handful of Scientists Obscured the Truth on Issues from Tobacco Smoke to Climate Change.* Bloomsbury, London.

Pearce, F. 2007. *With Speed and Violence: Why Scientists Fear Tipping Points in Climate Change.* Beacon Press, Boston, MA.

Prothero, D.R. 2013. *Reality Check: How Science Deniers Threaten our Future.* Indiana University Press, Bloomington, IN.

Purdy, J. 2015. *After Nature: Politics for the Anthropocene.* Harvard University Press, Cambridge, MA.

Rees, M. 2003. *Our Final Hour: A Scientist's Warning: How Terror, Error, and Environmental Disaster Threaten Humankind's Future in This Century—on Earth and Beyond.* Basic Books, New York.

Romm, J. 2015. *Climate Change: What Everyone Needs to Know.* Oxford University Press, Oxford, UK.

Scranton, R. 2015. *Learning to Die in the Anthropocene: Reflections on the End of Civilization.* City Lights Publishers, San Francisco, CA.

Tamrazian, A., LaDochy, S., Willis, J., and Patzert, W.C. 2008. Heat waves in southern California: Are they becoming more frequent and longer lasting? *Association of Pacific Coast Geographers Yearbook, 70*, 59–69.

Weiner, J. 1990. *The Next One Hundred Years: Shaping the Future of our Lives on Earth.* Bantam, New York.

Wilson, E.O. 2003. *The Future of Life.* Vintage, New York.

Index

Note: Page numbers in **bold** are where the term is introduced in the text.

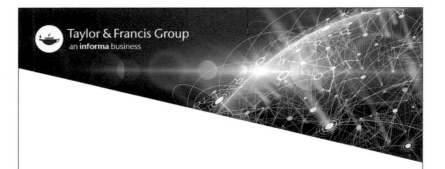